Numerical Methods for Inverse Scattering Problems

Jingzhi Li · Hongyu Liu

Numerical Methods for Inverse Scattering Problems

Science Press
Beijing

Springer

Jingzhi Li
Department of Mathematics
Southern University of Science
and Technology
Shenzhen, Guangdong, China

Hongyu Liu
Department of Mathematics
City University of Hong Kong
Kowloon, Hong Kong, SAR, China

ISBN 978-981-99-3771-4 ISBN 978-981-99-3772-1 (eBook)
https://doi.org/10.1007/978-981-99-3772-1

Jointly published with Science Press
The print edition is not for sale in China (Mainland). Customers from China (Mainland) please order the print book from: Science Press.

This Springer imprint is published by the registered company Springer Nature Singapore Pte Ltd.
The registered company address is: 152 Beach Road, #21-01/04 Gateway East, Singapore 189721, Singapore

Jingzhi Li dedicates this book to, Mother Fengying, Father Shide, Wife Ning, Children Mengya and Mengqi. They have been always very supportive of him and his mathematical interest.

Hongyu Liu dedicates this book to, Mother Kairong, Father Chunqi, for their unwavering love and support with no thought for a return.

Preface

This book is devoted to the numerical study of inverse scattering problems arising in acoustics, electromagnetism and elastodynamics. In the physical setup, one sends incident waves to interrogate an unknown or inaccessible object, called a scatterer, to generate the scattered wave fields. Then by collecting/measuring the scattered data, encoded into the so-called far-field patterns, one intends to infer geometrical/medium properties of the scatterer. This type of inverse scattering problems arises in various industrial applications of practical importance, including radar and sonar, remote sensing, geophysical prospecting, medical imaging and nondestructive testing. We are mostly concerned with the recovery and identification of the geometrical/topological information of the anomalous scatterer including its location and shape, independent of its medium property, by the associated far-field measurement. In fact, in many practical scenarios of wave imaging, one is primarily interested in visualising the anomalous scatterer, and will further explore the medium property if needed based on the imaging effect. We include acoustic, electromagnetic and elastic scattering in our study and mainly work within the time-harmonic regime.

For the inverse scattering problems described above, we present comprehensive and systematic numerical treatments in different setups. To that purpose, sophisticated numerical treatment of the direct scattering problems is an indispensable ingredient, say, for example, in generating the synthetic scattering data or in discretising a certain intermediate direct scattering problem in Newton's type methods for inverse scattering problems. Hence, we also discuss numerical methods for direct scattering problems, one based on integral equation methods and the other one based on finite element methods. Many of the numerical methods for both the direct and inverse scattering problems are the collaborative research outcome of the two authors of the book in the last one and half decades. There are many numerical results by the other researchers as well as several classical books discussing the numerical treatments of direct and inverse scattering problems. We shall discuss and comment on some of those results. Inverse scattering theory is a vital and active field of mathematical research with a colourful history as well as many ongoing studies. Hence, it is an impossible mission for us to cover all the numerical developments in the literature on this intriguing topic. Hence, our discussion is selective and concentrating. Moreover, our focus is the numerics, and we shall only briefly discuss the theoretical aspects in the book. The book can serve as a reference book for active researchers as well as an inspiring source for young researchers and graduate students who are interested in taking up research on the numerical treatments of inverse problems arising in new applications.

There are many people whose help is indispensable to the writing of this book. First, we would like to thank Dr. Chao Wang who helped us with the typesetting as well as read the first version of the book and provided many constructive comments and suggestions. Her patient and professional assistance is gratefully acknowledged. Second, we would like to thank our collaborators, postgraduate students and postdocs, for their continuing support and fruitful collaborations over the years.

Shenzhen, China Jingzhi Li
Hong Kong, China Hongyu Liu
April 2022

Acknowledgements

The work of Li was partially supported by the NSF of China No. 11971221, Guangdong NSF Major Fund No. 2021ZDZX1001, the Shenzhen Sci-Tech Fund No. RCJC20200714114556020,JCYJ20200109115422828 and JCYJ20190809150413261, National Center for Applied Mathematics Shenzhen, and SUSTech International Center for Mathematics.

The work of H. Liu was supported by the Hong Kong RGC General Research Funds (projects 11311122, 11300821 and 12301420), NSF/RGC Joint Research Fund (project N_CityU101/21) and the ANR/RGC Joint Research Fund (project A_CityU203/19).

Contents

Chapter 1
Introduction

Inverse problems are concerned with the determination of causes by known consequences. They lie at the heart of scientific inquiries and technological developments. In its abstract formulation, an inverse problem can be described by the following operator equation,

$$\mathscr{F}(\mathbf{x}) = \mathbf{y}, \tag{1.0.1}$$

where $\mathbf{x} \in \mathscr{X}$ and $\mathbf{y} \in \mathscr{Y}$, with $\mathscr{X}$ and $\mathscr{Y}$ signifying the sets of target objects and observation data, respectively. $\mathscr{X}$ usually consists of certain functions and/or geometries, which respectively signify the medium and geometrical/topological properties of the target objects; and $\mathscr{Y}$ consists of certain physical observables. The inverse problem associated with (1.0.1) can be formulated as recovering/identifying $\mathbf{x}$ or certain missing information of $\mathscr{F}$ by knowledge of $\mathbf{y}$.

In this book, we are mainly concerned with inverse problems associated with waves. In physics, mathematics and related fields, a wave is a propagating dynamic disturbance (change from equilibrium) of one or more quantities. Due to their propagating, penetrating and non-destructive nature, waves are widely used for probing the real world, including acoustic, electromagnetic, elastic and particle waves. The associated inverse problems are generally referred to as the inverse scattering problems. Inverse scattering problems lie at the core of many areas of science and technology, including radar and sonar, geophysical detection, medical imaging, non-destructive testing, remote sensing and cosmological exploration. Due to the practical importance, the study of inverse scattering problems is a central topic in applied mathematics. Moreover, new inverse scattering problems arise from various emerging technologies, which makes the corresponding study in this field challenging and intriguing. Figure 1.1 provides the schematic illustration of a typical inverse scattering scenario. The system consists of three parts: the input is a certain incident wave field; the target object which causes the perturbation/scattering of the incident wave field and is usually referred to as the *scatterer*; and the measurement of the scattered wave field away from the scatterer. In terms of the abstract formulation (1.0.1), $\mathbf{x}$

J. Li and H. Liu, *Numerical Methods for Inverse Scattering Problems*,
https://doi.org/10.1007/978-981-99-3772-1_1

Fig. 1.1 Schematic illustration of a typical inverse scattering scenario

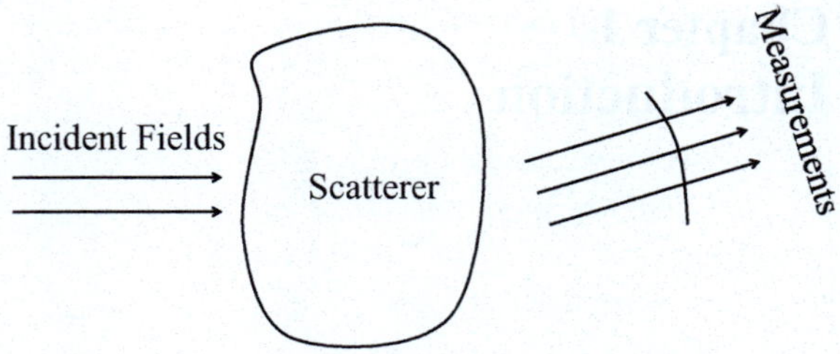

signifies the underlying scatterer, $\mathbf{y}$ is the measured wave field, and $\mathscr{F}$ is defined by the physical mechanism (namely, the PDE system) governing the wave propagation. In this book, we consider three types of waves including the acoustic, electromagnetic and mechanical waves. We are mainly concerned with the scattering of linear waves in the time-harmonic regime. The basic wave equations are, respectively, the Helmholtz equation, Maxwell's equations and Lamé equations. Depending on the specific situation, the scatterer $\mathbf{x}$ and the measurement $\mathbf{y}$ can vary and be complicated. For the inverse problems, the reconstruction can be the recovery of the geometric parameters of the target scatterer $\mathbf{x}$, e.g. its location or shape; or the recovery of the complicated medium parameters of the scatterer $\mathbf{x}$; or some other local or global features of the scatterer.

In this book, we are mainly concerned with the numerical methods for inverse scattering problems described above, including numerical methods for the forward scattering problems which are indispensable for tackling the corresponding inverse problems, and numerical reconstruction methods for the inverse problems. For the forward solvers, we mainly discuss the finite element methods, whereas for the inverse solvers, we mainly discuss various indirect approaches. In comparison, a direct approach for the inverse problem (1.0.1) usually adopts an optimization viewpoint and tackles it as follows:

$$\min_{\mathbf{x}\in\mathscr{C}} \|\mathscr{F}(\mathbf{x}) - \mathbf{y}\|_{\mathscr{Y}} + \mathscr{R}(\mathbf{x}), \tag{1.0.2}$$

where $\mathfrak{R}$ denotes a regularization functional, $\|\cdot\|_{\mathscr{Y}}$ signifies a certain norm in $\mathscr{Y}$, and $\mathscr{C}$ is a certain a-priori class in $\mathscr{X}$. The regularization term is used to tackle the ill-posedness of the inverse problem as well as to incorporate the useful a-priori knowledge into the reconstruction process. Though direct approaches are handy and generally workable, they lose the track of the speciality and particularity of the underlying specific inverse problems, which enable the development of more effective and efficient reconstruction algorithms, as shall be seen in our subsequent discussion. Here, we would like to mention that the modern approach of machine learning can also offer effective and efficient reconstructions for inverse scattering problems. We shall not cover this topic in this book and instead we refer to [12, 21, 24] for related studies in the literature.

Finally, we would like to give two related remarks. First, the inverse scattering problem depicted in Fig. 1.1 is usually referred to as an inverse problem with active measurement. In fact, the scatterer is inactive and in order to bring out the relevant information of the scatterer, one actively sends an incident wave to interact with the scatterer which then induces the scattering measurement. Inverse scattering problems with active measurements occur in a variety of applications including radar/sonar, medical imaging and geophysical exploration. In contrast, there are other situations where the scatterer is actively radiating wave fields and the measurements are passive. Inverse scattering problems with passive measurements also occur in various practical application including detection and monitoring hazardous radioactive materials, thermo- and photo-acoustic tomography and geomagnetic anomaly detection; see [4–22] and the references cited therein for related studies. In this book, we are mainly concerned with inverse scattering problems with active measurements. Second, in addition to the numerical reconstruction issue, there are more related issues on the mathematical study of inverse scattering problems. In fact, inverse problems are a unique field of research which has its own philosophy and methodologies. We refer to [10] for more discussion on the general issues of mathematical study of inverse problems including modelling, unique identifiability, stability and sensitivity, non-identifiability and invisibility, reconstruction and resolution, as well as applications.

References

1. E. Blåsten, H. Liu, Scattering by curvatures, radiationless sources, transmission eigenfunctions and inverse scattering problems. SIAM J. Math. Anal. **53**(4), 3801–3837 (2021)
2. X. Cao, H. Liu, Determining a fractional Helmholtz system with unknown source and medium parameter. Commun. Math. Sci. **17**(7), 1861–1876 (2019)
3. Y. Chang, Y. Guo, H. Liu and D. Zhang, Recovering source location, polarization, and shape of obstacle from elastic scattering data, arXiv:2301.02355
4. R. Chen, Y. Deng, Y. Gao, J. Li, H. Liu, Locating multiple magnetized anomalies by geomagnetic monitoring, arXiv:2301.12218
5. Y. Deng, J. Li, H. Liu, On identifying magnetized anomalies using geomagnetic monitoring. Arch. Ration. Mech. Anal. **231**(1), 153–187 (2019)
6. Y. Deng, J. Li, H. Liu, On identifying magnetized anomalies using geomagnetic monitoring within a magnetohydrodynamic model. Arch. Ration. Mech. Anal. **235**, 691–721 (2020)
7. Y. Deng, H. Liu, W.-Y. Tsui, Identifying variations of magnetic anomalies using geomagnetic monitoring. Discret. Continuous Dyn. Syst. Ser. A **40**(11), 6411–6440 (2020)
8. Y. Deng, H. Liu, G. Uhlmann, On an inverse boundary problem arising in brain imaging. J. Differ. Equ. **267**(4), 2471–2502 (2019)
9. Y. Deng, H. Liu, X. Wang, D. Wei, L. Zhu, Simultaneous recovery of surface heat flux and thickness of a solid structure by ultrasonic measurements. Electron. Res. Arch. **29**(5), 3081–3096 (2021)
10. H. Diao, H. Liu, *Spectral Geometry and Inverse Scattering Theory*, Book Manuscript (2022)
11. Y. Gao, H. Liu, Y. Liu, On an inverse problem for the plate equation with passive measurement. SIAM J. Appl. Math., arXiv:2301.07852
12. Y. Gao, H. Liu, X. Wang, K. Zhang, On an artificial neural network for inverse scattering problems. J. Comput. Phys. **448**, 110771 (2021)
13. Y. Guo, J. Li, H. Liu, X. Wang, Mathematical design of a novel input/instruction device using a moving emitter. Inverse Prob. **33**, 105009 (2017)

14. Y. Guo, J. Li, H. Liu, X. Wang, Two gesture-computing approaches by using electromagnetic waves. Inverse Prob. Imaging **13**, 879–901 (2019)
15. J. Li, H. Liu, S. Ma, Determining a random Schrödinger equation with unknown source and potential. SIAM J. Math. Anal. **51**(4), 3465–3491 (2019)
16. J. Li, H. Liu, S. Ma, Determining a random Schrödinger operator: both potential and source are random. Commun. Math. Phys. **381**, 527–556 (2021)
17. Y. H. Lin, H. Liu, Inverse problems for fractional equations with a minimal number of measurements. Comm. Anal. Comp. https://doi.org/10.3934/cac.2023005
18. Y.-H. Lin, H. Liu, X. Liu, Determining a nonlinear hyperbolic system with unknown sources and nonlinearity, arXiv:2107.10219
19. Y.-H. Lin, H. Liu, X. Liu, S. Zhang, Simultaneous recoveries for semilinear parabolic systems. Inverse Prob. **38**(11), 115006 (2022)
20. H. Liu, S. Ma, Determining a random source in a Schrödinger equation involving an unknown potential, arXiv:2005.04984
21. H. Liu, P. Meng, W. Shi, P. Zhang, A neural network method for time-dependent inverse source problem with limited-aperture data. J. Comput. Appl. Math. **421**, 114842 (2023)
22. H. Liu, G. Uhlmann, Determining both sound speed and internal source in thermo and photoacoustic tomography. Inverse Prob. **31**, 105005 (2015)
23. X. Wang, M. Song, Y. Guo, H. Li, H. Liu, Fourier method for identifying electromagnetic sources with multi-frequency far-field data. J. Comput. Appl. Math. **358**, 279–292 (2019)
24. W. Yin, W. Yang, H. Liu, A neural network scheme for recovering scattering obstacles with limited phaseless far-field data. J. Comput. Phys. **417**, 109594 (2020)
25. D. Zhang, Y. Guo, J. Li, H. Liu, Retrieval of acoustic sources from multi-frequency phaseless data. Inverse Prob. **34**, 094001 (2018)
26. D. Zhang, Y. Guo, J. Li, H. Liu, Locating multiple multipolar acoustic sources using the direct sampling method. Commun. Comput. Phys. **25**(5), 1328–1356 (2019)

Chapter 2
Direct Acoustic Scattering Problems

We consider the propagation of acoustic waves in a medium in $\mathbb{R}^N$, $(N = 2, 3)$. Let $x \in \mathbb{R}^N$ and $t \in \mathbb{R}_+$ be the space and time variables, respectively. Let $U(x, t)$ signify the wave field (i.e. the displacement away from the equilibrium state) and let $c(x)$, $\sigma(x)$ and $\rho(x)$ denote the wave velocity, damping coefficient and the mass density of the medium, respectively. The motion is governed by the following scalar wave equation

$$\frac{1}{c(x)}\frac{\partial^2 U(x,t)}{\partial t^2} + \sigma(x)\frac{\partial U(x,t)}{\partial t} - \nabla\cdot\left(\frac{1}{\rho(x)}\nabla U(x,t)\right) = -F(x,t), \quad (2.0.1)$$

where $F(x, t)$ signifies a source or sink. We consider the time-harmonic wave propagation, namely seeking a solution of (2.0.1) in the following form

$$U(x,t) = \mathscr{R}\{u(x)e^{-\mathrm{i}\omega t}\}, \qquad F(x,t) = \mathscr{R}\{f(x)e^{-\mathrm{i}\omega t}\},$$

where $\mathrm{i} := \sqrt{-1}$ is the imaginary unit and $\omega \in \mathbb{R}_+$ is the angular frequency. u and f stand for the pressures of the wave fields, respectively.

We consider time-harmonic wave scattering from an inhomogeneous isotropic medium supported in a bounded domain $\Omega \subset \mathbb{R}^N$, $(N = 2, 3)$ and the medium outside Ω is assumed to be uniformly homogeneous with no damping. After normalization, we can assume that $c = \tilde{c}_0$, $\rho = 1$ and $\sigma = 0$ in $\Omega^c := \mathbb{R}^N\backslash\overline{\Omega}$. We set $k := \omega/\tilde{c}_0$ to denote the normalized frequency, which is known as the wave number. We suppose that $f(x)$ is compactly supported, namely $supp(f) \subset B_{R_0}$ for some $R_0 > 0$, where and also in what follows B_r denotes a ball of radius r centred at the origin in $\mathbb{R}^N$. Factorizing out the time-dependent part, the wave equation (2.0.1) reduces to the following time-harmonic equation:

$$\nabla\cdot\left(\frac{1}{\rho}\nabla u\right)(x) + k^2\left(\frac{\widetilde{c}_0^2}{c^2} + \mathrm{i}\frac{\sigma\widetilde{c}_0}{k}\right)u(x) = f(x) \quad \text{in } \mathbb{R}^N. \qquad (2.0.2)$$

J. Li and H. Liu, *Numerical Methods for Inverse Scattering Problems*,
https://doi.org/10.1007/978-981-99-3772-1_2

There are two fundamental scattering scenarios. The first one is the scattering of time-harmonic waves by an impenetrable obstacle and the other one by a penetrable inhomogeneous medium.

2.1 Acoustic Scattering From Obstacles

Consider a bounded Lipschitz domain D in $\mathbb{R}^N$, $(N = 2, 3)$, which supports an inhomogeneous acoustic scatter, and it is assumed that $\mathbb{R}^N \backslash \overline{D}$ is connected. D represents an impenetrable obstacle located in the space, namely the wave cannot penetrate inside it. In order to generate the scattering, one sends a time-harmonic plane wave u^i, which is an entire solution to the Helmholtz equation $(-\Delta - k^2)u = 0$ in $\mathbb{R}^N$. A widely used one is the plane wave of the form

$$u^i(x; d, k) := e^{\mathrm{i}kx\cdot d}, \quad x \in \mathbb{R}^N, \tag{2.1.1}$$

where $k \in \mathbb{R}_+$ and $d \in \mathbb{S}^{N-1} := \{x \in \mathbb{R}^N; |x| = 1\}$ are the wave number and the incident direction, respectively.

The presence of the obstacle D interrupts the propagation of the incident plane wave, leading to the so-called scattered wave field u^s, which exists only in the exterior of the obstacle. The wave scattering is described by the following equation

$$\Delta u(x) + k^2 u(x) = 0 \quad \text{for} \quad x \in \mathbb{R}^N \backslash \overline{D}, \tag{2.1.2}$$

where u is the total field, formed by the linear superposition of the incident wave u^i and the scattered wave u^s, namely

$$u(x) := u^i(x) + u^s(x) \quad x \in \mathbb{R}^N \backslash \overline{D}. \tag{2.1.3}$$

The scattered wave $u^s(x)$ satisfies the *Sommerfeld radiation condition*:

$$\lim_{r\to\infty} r^{(N-1)/2} \left(\frac{\partial u^s(x)}{\partial r} - \mathrm{i}ku^s(x) \right) = 0 \quad \text{with} \quad r = |x|, \tag{2.1.4}$$

which holds uniformly in all directions $\hat{x} := x/|x| \in \mathbb{S}^{N-1}$, $x \in \mathbb{R}^N \backslash \{0\}$. The Sommerfeld radiation condition (2.1.4) characterizes the outgoing nature of the scattered wave field.

To complete the description of the direct scattering problem, we prescribe either one of the following three boundary conditions on ∂D:

$$u = 0 \quad \text{on } \partial D; \quad \frac{\partial u}{\partial \nu} = 0 \quad \text{on } \partial D; \quad \frac{\partial u}{\partial \nu} + \mathrm{i}\lambda u = 0 \quad \text{on } \partial D, \tag{2.1.5}$$

corresponding, respectively, to the case when the obstacle is *sound-soft*, *sound-hard*, and of *impedance type*. In (2.1.5), $\nu \in \mathbb{S}^{N-1}$ is the outward unit normal vector to ∂D, and $\lambda \in L^\infty(\partial D)$ ($\lambda > 0$) stands for a *surface impedance*. For a sound-soft obstacle, the pressure of the total wave vanishes on the boundary of the obstacle, resulting in a *Dirichlet boundary condition* whereas for a sound-hard obstacle, the normal velocity of the acoustic wave field vanishes on the boundary of the obstacle, which leads to a *Neumann boundary condition*. For an impedance-type obstacle, the normal velocity of the wave field is proportional to its pressure on the boundary through an impedance parameter.

2.2 Acoustic Scattering From Mediums

In this section, we consider the acoustic scattering from an inhomogeneous medium in $\mathbb{R}^N$, $N = 2, 3$. Per our discussion in deriving (2.0.2), we set

$$\gamma = \rho^{-1}, \quad q = \frac{\tilde{c}_0^2}{c^2} + \mathrm{i}\frac{\sigma\tilde{c}_0}{k} \quad \text{in } \mathbb{R}^N. \tag{2.2.1}$$

ρ and q are the density and modulus of the acoustic medium, respectively. They satisfy the physical conditions that

$$\gamma_0 \leq \gamma(x) \leq \Upsilon_0, \quad \mathscr{R}q(x) \geq \Gamma_0, \quad \mathscr{I}q(x) \geq 0 \quad \text{for } x \in \mathbb{R}^N, \tag{2.2.2}$$

where $\gamma_0, \Upsilon_0, \Gamma_0$ are positive constants. Moreover, it is assumed that $\mathrm{supp}(1-\gamma) \subset \Omega$ and $\mathrm{supp}(1-q) \subset \Omega$, where Ω is a bounded Lipschitz domain in $\mathbb{R}^N$ with a connected complement $\mathbb{R}^N \backslash \overline{\Omega}$.

Let u^i be an incident field as introduced in the previous section. Set u^s and $u = u^i + u^s$ to denote the scattered and total wave fields, respectively. Then the acoustic scattering due to the interaction of the inhomogeneous medium supported in Ω and the incident wave is described by the following system:

$$\begin{cases} \nabla(\gamma \cdot \nabla u) + k^2 q u = 0 & \text{in } \mathbb{R}^N, \\ u\big|_- = u|_+ & \text{on } \partial\Omega, \\ \gamma \dfrac{\partial u}{\partial \nu}\Big|_- = \gamma \dfrac{\partial u}{\partial \nu}\Big|_+ & \text{on } \partial\Omega, \\ \lim\limits_{r\to\infty} r^{\frac{N-1}{2}} \left(\dfrac{\partial u^s(x)}{\partial r} - \mathrm{i}k u^s(x) \right) = 0, & \end{cases} \tag{2.2.3}$$

where $\pm$ signify the traces on $\partial\Omega$ taken from outside and inside of Ω, respectively.

It is worth noting that in addition to the transmission conditions on $\partial\Omega$ in (2.2.3), more general resistive boundary conditions can be introduced and applied for acoustic mediums with specific physical properties. For their description and treatment we refer to [5].

2.3 Acoustic Scattering From Complex Scatterers

In this section we unify the treatments in Sects. 2.1 and 2.2 and introduce a more general scattering problem. Then we establish the well-posedness of the general scattering system. To that end, we let D and Ω be the Lipschitz domains introduced respectively in Sects. 2.1 and 2.2. We assume that $D \Subset \Omega$ and $\Omega\backslash\overline{D}$ is connected. Let $q \in L^\infty(\mathbb{R}^N\backslash\overline{D})$ and $\gamma \in L^\infty(\mathbb{R}^N\backslash D)$, and satisfy the physical conditions in (2.2.2) for $x \in \mathbb{R}^N\backslash\overline{D}$. Moreover, it is assumed that $\mathrm{supp}(1-\gamma) \subset \Omega\backslash\overline{D}$ and $\mathrm{supp}(1-q) \subset \Omega\backslash\overline{D}$.

With the above preparation, we consider the following scattering problem:

$$\begin{cases} \nabla\cdot(\gamma(x)\nabla u) + k^2 q(x) u = 0 & \text{in} \quad \Omega\backslash\overline{D}, \\ \Delta u^s + k^2 u^s = f & \text{in} \quad \mathbb{R}^N\backslash\overline{\Omega}, \\ u = u^i + u^s & \text{in} \quad \mathbb{R}^N\backslash\overline{\Omega}, \\ \gamma\frac{\partial u^+}{\partial\nu} = 0 & \text{on} \quad \partial D, \\ u^- = u^s + u^i, \quad \gamma\frac{\partial u^-}{\partial\nu} = \frac{\partial u^s}{\partial\nu} + \frac{\partial u^i}{\partial\nu} & \text{on} \quad \partial\Omega, \\ \lim_{r\to\infty} r^{(N-1)/2}\left\{\frac{\partial u^s}{\partial r} - iku^s\right\} = 0 & \text{for} \quad r = |x|, \end{cases} \tag{2.3.1}$$

where ν denotes the exterior unit normal to ∂D or $\partial\Omega$. We use the notations u^-, u^+ to represent the limits of u on ∂D or $\partial\Omega$, taking respectively from inside and outside D or Ω. The last limits in (2.3.1) is the Sommerfeld radiation condition.

Now we consider the well-posedness of the scattering problem (2.3.1). Here we let the parameters α and β be defined as

$$\{\alpha, \beta\} = \begin{cases} 1, 1 & \text{in} \quad \mathbb{R}^N\backslash\overline{\Omega}, \\ \gamma, q & \text{in} \quad \Omega\backslash\overline{D}. \end{cases} \tag{2.3.2}$$

Then the scattering problem (2.3.1) can be formulate as follows.

Find $u \in H^1_{\mathrm{loc}}(\mathbb{R}^N)$ such that $u = u^i + u^s$ in $\mathbb{R}^N\backslash\overline{\Omega}$ and solves the equation

$$\begin{cases} \mathscr{L}u := \nabla\cdot(\alpha\nabla u) + k^2\beta u = f & \text{in} \quad \mathbb{R}^N, \\ \lim\limits_{r\to\infty} r^{(N-1)/2}\left\{\dfrac{\partial u^s}{\partial r} - iku^s\right\} = 0 & \text{for} \quad r = |x|. \end{cases} \tag{2.3.3}$$

Now we show the well-posedness of the system (2.3.3).

Theorem 2.3.1 *Under the above assumption, the problem* (2.3.3) *is uniquely solvable.*

Proof We first show the uniqueness of the solutions for the system (2.3.3). The problem (2.3.3) is equivalent to the following transmission problem

$$\begin{cases} \nabla\cdot(\gamma(x)\nabla v)+k^2q(x)v=0 & \text{in} \quad \Omega\backslash\overline{D}, \\ \Delta u^s+k^2u^s=f & \text{in} \quad \mathbb{R}^N\backslash\overline{\Omega}, \\ \gamma\frac{\partial v}{\partial\nu}=p\in H^{-1/2}(\partial D) & \text{on} \quad \partial D, \\ v-u^s=g_1\in H^{1/2}(\partial\Omega) & \text{on} \quad \partial\Omega, \\ \gamma\frac{\partial v}{\partial\nu}-\frac{\partial u^s}{\partial\nu}=g_2\in H^{-1/2}(\partial\Omega) & \text{on} \quad \partial\Omega, \\ \lim_{|x|\to+\infty}|x|^{(N-1)/2}\left\{\frac{\partial u^s}{\partial|x|}-iku^s\right\}=0. \end{cases} \tag{2.3.4}$$

To show the uniqueness of the solution (v, u^s) to system (2.3.4), we first introduce the following auxiliary problem.

Find $(v_1, u_1)\in H^1(\Omega\backslash\overline{D})\times H^1(B_R\backslash\overline{\Omega})$ such that

$$\begin{cases} \nabla\cdot(\gamma(x)\nabla v_1)+k^2q(x)v_1=0 & \text{in} \quad \Omega\backslash\overline{D}, \\ \Delta u_1+k^2u_1=f & \text{in} \quad B_R\backslash\overline{\Omega}, \\ \gamma\frac{\partial v_1}{\partial\nu}=p & \text{on} \quad \partial D, \\ v_1-u_1=g_1 & \text{on} \quad \partial\Omega, \\ \gamma\frac{\partial v_1}{\partial\nu}-\frac{\partial u_1}{\partial\nu}=g_2 & \text{on} \quad \partial\Omega, \\ \frac{\partial u_1}{\partial\nu}=\Lambda u_1 & \text{on} \quad \partial B_R, \end{cases} \tag{2.3.5}$$

where $\Lambda: H^{1/2}(\partial B_R)\to H^{-1/2}(\partial B_R)$ is the Dirichlet-to-Neumann map defined by $\Lambda\Psi=\frac{\partial W}{\partial\nu}|_{\partial B_R}$, with $W\in H^1_{\text{loc}}(\mathbb{R}^N\backslash\overline{B}_R)$ being the unique solution to the system

$$\begin{cases} \Delta W+k^2W=0 & \text{in} \quad \mathbb{R}^N\backslash\overline{B}_R, \\ W=\Psi\in H^{1/2}(\partial B_R) & \text{on} \quad \partial B_R, \\ \lim\limits_{|x|\to+\infty}|x|^{(N-1)/2}\left\{\dfrac{\partial W}{\partial|x|}-ikW\right\}=0. \end{cases} \tag{2.3.6}$$

Then we prove the uniqueness of the problem. For the purpose we set p, g_1, g_2, f to be all zeros. Multiplying the first and second equations of (2.3.4), respectively, by $\overline{v}$ and $\overline{u}^s$, and integrating by parts in $\Omega\backslash\overline{D}$ and $B_R\backslash\overline{\Omega}$, together with use of the conditions on ∂D and $\partial\Omega$, we have

$$\begin{aligned} &-\int_{\Omega\backslash\overline{D}}\gamma|\nabla v|^2dx+\int_{\Omega\backslash\overline{D}}k^2q|v|^2dx-\int_{B_R\backslash\overline{\Omega}}|\nabla u^s|^2dx \\ &+\int_{B_R\backslash\overline{\Omega}}k^2|u^s|^2dx+\int_{\partial B_R}\frac{\partial u^s}{\partial\nu}\overline{u}^s ds=0. \end{aligned} \tag{2.3.7}$$

Taking the imaginary part of both sides of (2.3.7), we derive

$$\text{Im}\int_{\partial B_R}\frac{\partial u^s}{\partial\nu}\overline{u}^s ds=-\text{Im}\int_{\Omega\backslash\overline{D}}k^2q|v|^2dx\le 0.$$

Then by Relllich's theorem we know u^s is zero outside B_R, which with the unique continuation implies that $u^s=0$ in $\Omega\backslash\overline{D}$ and $v=0$ in D.

Finally, we show the equivalence between (2.3.4) and (2.3.5). By the definition of Λ, we see that if (v, u^s) solves the system (2.3.4), then $(v_1 = v, u_1 = u^s|_{B_R\backslash\overline{\Omega}})$ is the solution to the system (2.3.5). On the other hand, by applying the Green's representation to the solution (v_1, u_1) of (2.3.5) we obtain that

$$
\begin{aligned}
u_1(x) = &-\int_{\partial\Omega}\left(\frac{\partial u_1(y)}{\partial\nu(y)}\Phi(x,y) - u_1(y)\frac{\partial\Phi(x,y)}{\partial\nu(y)}\right)ds(y) \\
&+\int_{\partial B_R}\left(\Lambda u_1(y)\Phi(x,y) - u_1(y)\frac{\partial\Phi(x,y)}{\partial\nu(y)}\right)ds(y) - \int_{B_R\backslash\overline{\Omega}} f(y)\Phi(x,y)dy,
\end{aligned}
\tag{2.3.8}
$$

for $x \in B_R\backslash\overline{\Omega}$, where

$$
\Phi(x,y) = \frac{\mathrm{i}}{4}\left(\frac{k}{2\pi|x-y|}\right)^{(N-2)/2} H^{(1)}_{(N-2)/2}(k|x-y|) \tag{2.3.9}
$$

is the outgoing Green's function. By the definition of Λ and the radiation of $\Phi(x, y)$

$$
\int_{\partial B_R}\left(\Lambda u_1(y)\Phi(x,y) - u_1(y)\frac{\partial\Phi(x,y)}{\partial\nu(y)}\right)ds(y) = 0.
$$

Hence,

$$
u_1(x) = -\int_{\partial\Omega}\left(\frac{\partial u_1(y)}{\partial\nu(y)} - u_1(y)\frac{\partial\Phi(x,y)}{\partial\nu(y)}\right)ds(y) - \int_{B_R\backslash\overline{\Omega}} f(y)\Phi(x,y)dy.
\tag{2.3.10}
$$

It is clear that u_1 can be readily extended to an $H^1_{\mathrm{loc}}(\mathbb{R}^N\backslash\overline{\Omega})$ function, which we still denote by u_1. We can see that u_1 satisfies the Sommerfeld radiation condition, which together with the uniqueness of solution to (2.3.4) implies that $u_1 = u^s$.

Next we show the existence and stability estimate. In the following, by appropriately choosing R_0 we can assume that k^2 is not a Dirichlet eigenvalue in B_{R_0+1}. Let $\theta(x) \in C^\infty(\mathbb{R}^N)$ be a cut-off function such that $\theta(x) = 0$ for $|x| < R_0$ and $\theta(x) = 1$ for $|x| > R_0 + 1$. Setting

$$
W = u \quad \text{in } \Omega \text{ and } W = u^s + (1-\theta)u^i \quad \text{in } \mathbb{R}^N\backslash\overline{\Omega}, \tag{2.3.11}
$$

we can then verify directly that $W \in H^1_{\mathrm{loc}}(\mathbb{R}^N)$ satisfies

$$
\begin{cases}
\mathscr{L}W = g \quad \text{in } \mathbb{R}^N, \\
\lim\limits_{|x|\to\infty} |x|^{(N-1)/2}\left\{\dfrac{\partial W}{\partial |x|} - \mathrm{i}kW\right\} = 0,
\end{cases}
\tag{2.3.12}
$$

with $g = -(\Delta + k^2)(\theta u^i) + f \in L^2(B_{R_0+1}\backslash\Omega)$.

Next, we look for a solution to (2.3.12) of the following form

$$W = w - \phi(w - V), \tag{2.3.13}$$

where ϕ is C^∞ cut-off function such that $\phi = 1$ in B_{R_0} and $\phi = 0$ in $\mathbb{R}^N \backslash B_{R_0+1}$. We let $V \in H^1(B_{R_0+1})$ be the solution of the system

$$\begin{cases} \mathscr{L}V = g^* & \text{in} \quad B_{R_0+1}, \\ V = 0 & \text{on} \quad \partial B_{R_0+1} \end{cases} \tag{2.3.14}$$

and $w \in H^1_{\text{loc}}(\mathbb{R}^N)$ be the solution of the system

$$\begin{cases} (\Delta + k^2)w = g^* \quad \text{in} \ \ \mathbb{R}^N, \\ \lim\limits_{|x|\to\infty} |x|^{(N-1)/2} \left\{ \dfrac{\partial w}{\partial |x|} - ikw \right\} = 0, \end{cases} \tag{2.3.15}$$

where $g^* \in L^2(B_{R_0+1} \backslash \Omega)$ shall be determined later.

Clearly, by the classical regularity estimates we see

$$V \in H^2(B_{R_0+1} \backslash \overline{\Omega}) \quad \text{and} \quad w \in H^2_{\text{loc}}(\mathbb{R}^N).$$

By direct verification we have

$$\begin{aligned} g = (\Delta + k^2)W &= \Delta w + k^2 w + \Delta\phi(w - V) \\ &\quad + 2\nabla\phi \cdot \nabla(w - V) + \phi\big(\Delta(w - V) + k^2(w - V)\big) \\ &= g^* + Kg^*, \end{aligned} \tag{2.3.16}$$

where K is defined to be $Kg^* = \Delta\phi(w - V) + 2\nabla\phi \cdot \nabla(w - V)$.

We can show that K is compact from $L^2(B_{R_0+1} \backslash \Omega)$ to itself. We shall make use of the Fredholm theory to show the unique solvability of (2.3.16). It suffices to show the uniqueness of solution to (2.3.16). We set $g = 0$. By (2.3.12) we have $W = 0$. Hence $w = \phi(w - V)$ in $\mathbb{R}^N$ and $V = 0$ in Ω and $w = 0$ in $\mathbb{R}^N \backslash B_{R_0+1}$. It is straightforward to verify that

$$\begin{cases} (\Delta + k^2)(V - w) = 0 & \text{in} \quad B_{R_0+1}, \\ V - w = 0 & \text{on} \quad \partial B_{R_0+1}, \end{cases} \tag{2.3.17}$$

hence $V - w = 0$. Therefore $w = 0$, which then implies that $g^* = 0$. Then by the Fredholm theory we have a unique $g^* \in L^2(B_{R_0+1} \backslash \Omega)$ to (2.3.16) such that

$$\|g^*\|_{L^2(B_{R_0+1}\backslash\Omega)} \le C\|g\|_{L^2(B_{R_0+1}\backslash\Omega)} \le C\big(\|u^i\|_{H^1(B_{R_0+1}\backslash\overline{\Omega})} + \|f\|_{L^2(B_{R_0}\backslash\Omega)}\big).$$

Finally, by the classical theory on elliptic equations one can show that

$$\|u\|_{H^1(B_{R_0+1}\backslash\overline{\Omega})} \le C\left(\|f\|_{L^2(B_{R_0}\backslash\Omega)} + \|u^i\|_{H^1(B_{R_0+1}\backslash\overline{\Omega})}\right). \qquad \square$$

2.4 Green's Formula and Linear Potential Theory for Scattering Problems

Green's formula allows to represent smooth functions in a smooth, bounded domain by a superposition of certain potentials. In this section, in order to prove the uniqueness of the problem (2.2.3), we first introduce the Green's Formula as follows:

Theorem 2.4.1 *Let Ω be a C^2-smooth and bounded domain and let ν denote the unit normal vector to the boundary $\partial\Omega$ pointing towards the outside of Ω. For $u \in C^1(\overline{\Omega})$ and $v \in C^2(\overline{\Omega})$, we have the Green's first formula*

$$\int_\Omega u \cdot \Delta v dx = -\int_\Omega \nabla u \cdot \nabla v dx + \int_{\partial\Omega} u \cdot \frac{\partial v}{\partial \nu} ds, \tag{2.4.1}$$

and the Green's second formula

$$\int_\Omega u \cdot \Delta v - \Delta u \cdot v dx = \int_{\partial\Omega} u \cdot \frac{\partial v}{\partial \nu} - \frac{\partial u}{\partial \nu} \cdot v ds, \tag{2.4.2}$$

where $\frac{\partial u}{\partial \nu} = \nu \cdot \Delta u$.

For the first Green identity, $v \in C^2(\overline{\Omega})$ and $u \in C^1(\overline{\Omega})$. For the second Green's identity, $u \in C^2(\overline{\Omega})$ and $v \in C^2(\overline{\Omega})$. However, the regularity requirement can be relaxed to be $H^1(\Omega)$ and $H^2(\Omega)$ correspondingly and further relaxed for $H^1_*(\Omega) := \{u \in H^1(\Omega); Pu \in L^2(\Omega)\}$, where P is a strongly elliptic partial differential operator [6].

For $x, y \in \mathbb{R}^N$, $x \neq y$, we denote by

$$\Phi(x, y) := \begin{cases} \frac{i}{4} H_0^{(1)}(k|x-y|) & N = 2, \\ \frac{e^{ik|x-y|}}{4\pi|x-y|} & N = 3, \end{cases}$$

the fundamental solution for the Helmholtz equation $\Delta u + k^2 u = 0$, where $H_0^{(1)}$ is the Hankle functions of the first kind of order 0. If $y \in \mathbb{R}^N$ is fixed, $\Phi(\cdot, y)$ is a solution to the Helmholtz equation with respect to the variable $x \in \mathbb{R}^N \backslash \{y\}$ and similarly, if x is fixed.

The following Green's formula, which is also known as the Helmholtz representation gives representation of u.

Theorem 2.4.2 *Let Ω be a C^2-smooth and bounded domain and let ν denote the unit normal vector to the boundary $\partial\Omega$ pointing towards the exterior of Ω. Then for a function $u \in H^2(\Omega)$ we have the following Green's formula*

$$u(x) = \int_{\partial\Omega} \left\{ \frac{\partial u}{\partial \nu} \Phi(x, y) - u(y) \frac{\partial \Phi(x, y)}{\partial \nu(y)} \right\} ds(y) - \int_{\Omega} \big(\Delta u(y) + k^2 u(y)\big) \Phi(x, y) dy, \quad x \in \Omega. \tag{2.4.3}$$

Green's formula is also true if u only satisfies $u \in H^1(\Omega)$ and $\Delta u + k^2 u \in L^2(\Omega)$.

Notice that the above theorem holds true for Ω being a Lipschitz domain. For more details we refer to [6].

We now introduce the linear potential theory for scattering from a sound-soft obstacle problem

$$\begin{cases} \Delta u + k^2 u = 0 \quad \text{in } \mathbb{R}^N \backslash \overline{\Omega}, \\ u|_{\partial\Omega} = 0, \\ u = u^i + u^s, \\ \lim\limits_{r \to \infty} r^{\frac{N-1}{2}} (\partial_r - ik)\, u^s = 0. \end{cases} \tag{2.4.4}$$

We first introduce the single-layer and double-layer potentials as follows

$$u(x) = \int_{\partial\Omega} \varphi(y) \Phi(x, y) ds(y), \quad x \in \mathbb{R}^N \backslash \partial\Omega,$$

$$v(x) = \int_{\partial\Omega} \varphi(y) \frac{\partial \Phi(x, y)}{\partial \nu_y} ds(y), \quad x \in \mathbb{R}^N \backslash \partial\Omega,$$

where φ is an integrable function. By this definition, we notice that $(\Delta + k^2) u(x) = 0$ and $(\Delta + k^2) v(x) = 0$ for all $x \in \mathbb{R}^N \backslash \partial\Omega$. Meanwhile, $u(x)$ and $v(x)$ satisfy the Sommerfeld radiation condition, because $\Phi(x, y)$ satisfies the Sommerfeld radiation condition for $|y|$ bounded. We notice that for $\partial\Omega \in C^2(\Omega)$ and $\varphi(y) \in C(\partial\Omega)$, the following integral

$$\int_{\partial\Omega} \varphi(y) \Phi(x, y) ds(y) = \frac{1}{4\pi} \int_{\partial\Omega} \varphi(y) \frac{e^{ik|x-y|}}{|x - y|} ds(y), \quad x \in \partial\Omega$$

exists as an improper integral, which is also applicable for $\int_{\partial\Omega} \varphi(y) \frac{\partial \Phi(x,y)}{\partial \nu(x)} ds(y)$. Besides, we have the following identity

$$\frac{\partial u_\pm}{\partial \nu} = \int_{\partial\Omega} \varphi(y) \frac{\partial \Phi(x, y)}{\partial \nu(x)} ds(y) \mp \frac{1}{2} \varphi(x), \quad x \in \partial\Omega,$$

where $\frac{\partial u_\pm}{\partial \nu}(x) = \nu(x) \cdot \nabla u(x \pm h\nu(x))$ as $h \to +0$. Also, we have

$$v_{\pm}(x) = \int_{\partial\Omega} \varphi(y) \frac{\partial \Phi(x, y)}{\partial \nu(y)} ds(y) \pm \frac{1}{2}\varphi(x), \quad x \in \partial\Omega.$$

The jump relations holds for both u and $\frac{\partial v}{\partial \nu}$, which means that they are continuous across $\partial\Omega$. Next we define the single-layer and double-layer potential operators as follows:

$$S[\varphi](x) = \int_{\partial\Omega} \Phi(x, y)\varphi(y)ds(y), \quad x \in \partial\Omega$$

$$K[\varphi](x) = \int_{\partial\Omega} \frac{\partial \Phi(x, y)}{\partial \nu(y)} \varphi(y)ds(y), \quad x \in \partial\Omega$$

$$K^*[\varphi](x) = \int_{\partial\Omega} \frac{\partial \Phi(x, y)}{\partial \nu(x)} \varphi(y)ds(y), \quad x \in \partial\Omega.$$

We notice that K^* is the adjoint operator of K with respect to $L^2(\partial\Omega)$, i.e., $\langle K[\varphi], \psi \rangle_{L^2(\partial\Omega)} = \langle \varphi, K^*[\psi] \rangle_{L^2(\partial\Omega)}$. The operators K and K^* are also called the Neumann-Poincaré (NP) operators, which have been playing a central role in the potential theory.

Theorem 2.4.3 *Let $\partial\Omega \in C^2$, the operators S, K and K^*: $C(\partial\Omega) \to C^{0,\alpha}(\partial\Omega)$, where $\alpha \in (0, 1)$. Hence, they are compact operators from $C(\partial\Omega)$ to $C(\partial\Omega)$.*

Thus the scattered field can be expressed as

$$u^s(x) = \int_{\partial\Omega} \frac{\partial \Phi(x, y)}{\partial \nu(y)} \varphi(y)ds(y) \quad x \in \mathbb{R}^N \backslash \overline{\Omega},$$

where $\varphi = \left(\frac{I}{2} + K\right)^{-1} [-u^i|_{\partial\Omega}]$. By taking $|x| \to \infty$ and recalling

$$\Phi(x, y) = C_N \frac{e^{ik|x|}}{|x|^{\frac{N-1}{2}}} e^{-ik\hat{x}\cdot y} + \mathcal{O}\left(\frac{1}{|x|^{\frac{N+1}{2}}}\right),$$

we have

$$u^s(x) = \frac{e^{ik|x|}}{|x|^{\frac{N-1}{2}}} u^{\infty}(\hat{x}, d) + \mathcal{O}\left(\frac{1}{|x|^{\frac{N+1}{2}}}\right) \tag{2.4.5}$$

which holds uniformly in $\hat{x} := x/|x| \in \mathbb{S}^{N-1}$, where $x \in \mathbb{R}^n$ and $x \neq 0$. $u^{\infty}(\hat{x}, d)$ is known as the far-field pattern given by

$$u_{\infty}(\hat{x}) = \int_{\partial\Omega} \frac{\partial e^{-ik\hat{x}\cdot y}}{\partial \nu(y)} \varphi(y)ds(y) \tag{2.4.6}$$

$$\varphi = \left(\frac{I}{2} + K\right)^{-1} [-u^i|_{\partial\Omega}]. \tag{2.4.7}$$

The above results is subject to an "artificial" condition that k^2 is not a Neumann Laplacian eigenvalue to Ω.

We then introduce the linear potential theory for scattering from medium problem

$$\begin{cases} \Delta u + k^2 n(x) u = 0 \quad \text{in} \quad \mathbb{R}^N, \\ u = u^i + u^s, \\ \lim_{r\to\infty} r^{\frac{N-1}{2}} \left(\frac{\partial u}{\partial r} - iku \right) = 0. \end{cases} \tag{2.4.8}$$

Recall that the following problem

$$\begin{cases} (-\Delta - k^2) v = f \quad \text{in} \quad \mathbb{R}^N, \\ \lim_{r\to\infty} r^{\frac{N-1}{2}} \left(\frac{\partial v}{\partial r} - ikv \right) = 0, \end{cases} \tag{2.4.9}$$

has the unique solution which is given by

$$v = \Phi * f = \int_{\mathbb{R}^3} \Phi(x, y) f(y) dy = \int_{\mathbb{R}^3} \frac{1}{4\pi} \frac{e^{ik|x-y|}}{|x-y|} f(y) dy.$$

Thus we have that

$$u(x) = u^i(x) + k^2 \int_{\Omega} (n(x) - 1)(y) u(y) \Phi(x, y) dy. \tag{2.4.10}$$

Now we introduce the integral operator as follows:

$$K[u](x) = k^2 \int_{\Omega} (n(x) - 1)(y) u(y) \Phi(x, y) dy \quad x \in \Omega$$

with the mapping property $K : C(\overline{\Omega}) \to C^2(\overline{\Omega})$. Here, we notice that the fundamental solution $\Phi(x, y)$ is weakly singular and the integral is well-defined as an improper integral. Since the embedding $C^2(\overline{\Omega}) \hookrightarrow C(\overline{\Omega})$ is compact, the operator $K : C(\overline{\Omega}) \to C(\overline{\Omega})$ is compact. Then (2.4.10) can be written as

$$(I - K)[u](x) = u^i(x), \qquad x \in \Omega. \tag{2.4.11}$$

Here $I - K$ is a Fredholm operator. By (2.4.11), we know that

$$u = K[u] = k^2 \int_{\Omega} (n(x) - 1)(y) u(y) \Phi(x, y) dy.$$

Therefore, the solution for (2.4.8) satisfies

$$u(x) = u^i(x) + K[u](x), \qquad x \in \mathbb{R}^3, \tag{2.4.12}$$

which is known as the Lippmann-Schwinger integral, and

$$u|_\Omega = [I - K]^{-1}[u^i|_\Omega]. \tag{2.4.13}$$

Then by letting $|x| \to \infty$, we have

$$u(x) = u^i(x) + k^2 \int_\Omega (n(x)-1)(y) \cdot u(y) \cdot \frac{1}{4\pi} \frac{e^{ik|x|}}{|x|} \cdot e^{-ik\hat{x}\cdot y} dy + \mathscr{O}\left(\frac{1}{|x|^2}\right).$$

Thus it admits the following asymptotic expansion:

$$u^s(x) = \frac{e^{ik|x|}}{|x|^{(N-1)/2}} u_\infty(\hat{x}) + \mathscr{O}\left(\frac{1}{|x|^{(N+1)/2}}\right) \quad \text{as } |x| \to \infty, \tag{2.4.14}$$

where $u_\infty(\hat{x}) = \frac{k^2}{4\pi} \int_\Omega (n(x)-1)(y) \cdot u(y) \cdot e^{-ik\hat{x}\cdot y} dy$ is known as the far-field pattern of u^s and the correspondence between u^s and u_∞ is one-to-one, which is given by the following theorem.

Theorem 2.4.4 *The correspondence between u_∞ and u^s is one-to-one.*

Proof First, we let

$$u^s(x) = \frac{e^{ik|x|}}{|x|} u_\infty(\hat{x}) + \mathscr{O}\left(\frac{1}{|x|^2}\right)$$
$$\tilde{u}^s(x) = \frac{e^{ik|x|}}{|x|} \tilde{u}_\infty(\hat{x}) + \mathscr{O}\left(\frac{1}{|x|^2}\right).$$

If $u_\infty = \tilde{u}_\infty$, then $u^s - \tilde{u}^s = \mathscr{O}\left(\frac{1}{|x|^2}\right)$. By using Rellich's Theorem, we know $u^s = \tilde{u}^s$. Conversely, if $u^s = \tilde{u}^s$, it is obvious $u_\infty = \tilde{u}_\infty$. □

Remark that the far field pattern u_∞ is the "clean" data without redundacy and is independent of the coordinate system.

Let $u^i(x) = e^{ikx\cdot d}$, $d \in \mathbb{S}^{N-1}$, which is known as a time-harmonic plane wave with d signifying the incident direction. Set

$$u^k_\infty(\hat{x}, d) = u_\infty(\hat{x}; e^{ikx\cdot d}), \tag{2.4.15}$$

which denotes the far-field pattern associated with the plane wave. $u^k_\infty(\hat{x}, d)$, $(\hat{x}, d) \in \mathbb{S}^{N-1} \times \mathbb{S}^{N-1}$, is known as the scattering amplitude at $k \in \mathbb{R}_+$. We already know that

$$u_\infty^k(\hat{x}, d) = \frac{k^2}{4\pi}\int_\Omega (n(x)-1)(y)\cdot u(y)\cdot e^{-ik\hat{x}\cdot y}dy$$
$$u|_\Omega = (I-K)^{-1}[u^i|_\Omega].$$

It can be proved that $u_\infty^k(\hat{x}, d)$ is analytic in $\hat{x}$, d and k. Hence, if u_∞^k is known for $k \in (a, b)$, it is known for all $k \in \mathbb{R}_+$ by analytic continuation. Similarly, if $u_\infty(\hat{x})$ is known on any given open subset $\Gamma \subset \mathbb{S}^{N-1}$, it is known on the whole unit shpere $\mathbb{S}^{N-1}$, which is the same for d.

Define the Herglotz operator $\mathscr{H} : L^2(\mathbb{S}^{N-1}) \mapsto \mathscr{A}(\mathbb{R}^N)$ with $\mathscr{A}(\mathbb{R}^N)$ denoting the space of analytic functions over $\mathbb{R}^N$:

$$\mathscr{H}(g)(x) := \int_{\mathbb{S}^{N-1}} e^{ikx\cdot d} g(d)\, ds(d), \quad g \in L^2(\mathbb{S}^{N-1}). \tag{2.4.16}$$

$v_g(x) = \mathscr{H}(g)(x)$ is known as a Herglotz wave, which is the linear superposition of the plane waves with a density function $g(d)$. We have the following denseness property of the Hergoltz waves.

Lemma 2.4.1 ([8]) *Let $\Omega \Subset \mathbb{R}^n$ be a bounded Lipschitz domain and $\mathbf{H}_k$ be the space of all the Herglotz wave functions of the form* (2.4.16). *Define*

$$\mathbf{S}_k(\Omega) = \{u \in C^\infty(\Omega) \mid \Delta u + k^2 u = 0\}$$

and

$$\mathbf{H}_k(\Omega) = \{u|_\Omega \mid u \in \mathbf{H}_k\}.$$

Then $\mathbf{H}_k(\Omega)$ is dense in $\mathbf{S}_k(\Omega) \cap L^2(\Omega)$ with respect to the topology induced by the $H^1(\Omega)$-norm.

Using a Herglotz wave v_g as an incident field, by the linearity of the Helmholtz system (2.2.3) with respect to the incident wave, one readily has

$$u_\infty(\hat{x}; v_g) = u_\infty(\hat{x}; \mathscr{H}(g(d))) = \mathscr{H}(u_\infty^k(\hat{x}, d)) = \int_{\mathbb{S}^{N-1}} u_\infty^k(\hat{x}, d) g(d)\, ds(d). \tag{2.4.17}$$

Define the far-field operator $\mathscr{F} : L^2(\mathbb{S}^{N-1}) \mapsto L^2(\mathbb{S}^{N-1})$ as follows:

$$\mathscr{F}(g)(\hat{x}) = \int_{\mathbb{S}^{N-1}} u_\infty^k(\hat{x}, d) g(d)\, ds(d). \tag{2.4.18}$$

By Lemma 2.4.1, we readily see that the far-field operator $\mathscr{F}$ actually contains all the possible scattering information from the scatterer $(\Omega, n(x)^2 - 1)$. Hence, the direct scattering problem is to determine the far-field operator $\mathscr{F}$ for a given scatterer $(\Omega, n(x)^2 - 1)$. Reversely, the inverse scattering problem is to determine the scatterer $(\Omega, n(x)^2 - 1)$ by knowledge of the far-field operator:

$$\mathscr{F} \to (\Omega, n(x)^2 - 1), \tag{2.4.19}$$

which is equivalent to determining $(\Omega, n(x)^2 - 1)$ by knowledge of the associated scattering amplitude $u_\infty^k(\hat{x}, d)$. By introducing an operator $\mathscr{S}$ which sends to the scatterer $(\Omega, n(x)^2 - 1)$ to the associated scattering amplitude $u_\infty^k(\hat{x}, d)$, the inverse scattering problem can be recast as the following operator equation

$$\mathscr{S}(\Omega, n(x)^2 - 1) = u_\infty^k(\hat{x}, d). \tag{2.4.20}$$

It is directly verified that the inverse scattering problem (2.4.19)/(2.4.20) is nonlinear.

2.5 Numerical Methods for Acoustic Scattering Problems

2.5.1 Nyström Method

In this subsection, we present the most efficient numerical method for acoustic scattering problems. We assume that the boundary curve ∂D possesses a regular analytic and 2π-periodic parametric representation of the form

$$x(t) = \big(x_1(t), x_2(t)\big), \qquad 0 \le t \le 2\pi, \tag{2.5.1}$$

in counterclockwise orientation satisfying $|x'(t)|^2 > 0$ for all t.

By the linear potential theory, we know that the solution of the exterior Dirichlet problem in the form of acoustic surface potential can be expressed as

$$u(x) = \int_{\partial D} \left\{ \frac{\partial \Phi(x, y)}{\partial \nu(y)} - i\eta\Phi(x, y) \right\} \varphi(y) ds(y), \quad x \in \mathbb{R}^N \backslash \partial D, \tag{2.5.2}$$

with a density $\varphi \in C(\partial D)$ and a real coupling parameter $y \neq 0$. Then from the jump relations we see that the potential u given above solves the exterior Direchlet problem provided the density is a solution of the integral equation

$$\varphi + K\varphi - i\eta S\varphi = 2f. \tag{2.5.3}$$

Then, by straightforward calculations using the $H_1^{(1)} = -H_0^{(1)'}$ of the Hankel functions, we transform (2.5.3) into the parametirc form

$$\Psi(t) - \int_0^{2\pi} \big\{L(t, \tau) + i\eta M(t, \tau)\big\}\Psi(\tau) d\tau = g(t), \quad 0 \le t \le 2\pi, \tag{2.5.4}$$

where $\Psi(t) := \varphi(x(t))$, $g(t) := 2f(x(t))$, and the kernels are given by

$$L(t,\tau) := \frac{ik}{2}\{x_2'(\tau)[x_1(\tau)-x_1(t)] - x_1'(\tau)[x_2(\tau)-x_2(t)]\}\frac{H_1^{(1)}(k|x(t)-x(\tau)|)}{|x(t)-x(\tau)|},$$

$$M(t,\tau) := \frac{i}{2}H_0^{(1)}(k|x(t)-x(\tau)|)|x'(\tau)|$$

for $t \neq \tau$. We notice that the kernels L and M have logarithmic singularities at $t = \tau$ by an expansion for the Neumann functions. Therefore, we split the kernels into

$$L(t,\tau) = L_1(t,\tau)\ln\left(4\sin^2\frac{t-\tau}{2}\right) + L_2(t,\tau),$$

$$M(t,\tau) = M_1(t,\tau)\ln\left(4\sin^2\frac{t-\tau}{2}\right) + M_2(t,\tau),$$

where

$$L_1(t,\tau) := \frac{k}{2\pi}\{x_2'(\tau)[x_1(t)-x_1(\tau)] - x_1'(\tau)[x_2(t)-x_2(\tau)]\}\frac{J_1(k|x(t)-x(\tau)|)}{|x(t)-x(\tau)|},$$

$$L_2(t,\tau) := L(t,\tau) - L_1(t,\tau)\ln\left(4\sin^2\frac{t-\tau}{2}\right),$$

$$M_1(t,\tau) := -\frac{1}{2\pi}J_0(k|x(t)-x(\tau)|)|x'(\tau)|,$$

$$M_2(t,\tau) := M(t,\tau) - M_1(t,\tau)\ln\left(4\sin^2\frac{t-\tau}{2}\right)$$

The kernels L_1, L_2, M_1 and M_2 are analytic and we can derive the diagonal terms by expansions

$$L_2(t,\tau) = L(t,t) = \frac{1}{2\pi}\frac{x_1'(t)x_2''(t) - x_2'(t)x_1''(t)}{|x'(t)|^2},$$

$$M_2(t,t) = \left\{\frac{i}{2} - \frac{C}{\pi} - \frac{1}{\pi}\ln\left(\frac{k}{2}|x'(t)|\right)\right\}|x'(t)|$$

for $0 \le t \le 2\pi$. We note that despite the continuity of the kernel L, for numerical accuracy it is advantageous to separate the logarighmic part of L since the derivatives of L fail to be continuous at $t = \tau$.

Hence, we have to numerically solve an integral equation of the form

$$\Psi(t) - \int_0^{2\pi} K(t,\tau)\Psi(\tau)d\tau = g(t), \quad 0 \le t \le 2\pi, \tag{2.5.5}$$

where the kernel can be written in the form

$$K(t,\tau) = K_1(t,\tau)\ln\left(4\sin^2\frac{t-\tau}{2}\right) + K_2(t,\tau) \tag{2.5.6}$$

with analytic functions K_1 and K_2 with an analytic right-hand side g.

The Nyström method consists in the straightforward approximation of integrals by quadrature formulas. For the 2π-periodic integrands, we choose an equidistant set of knots $t_j := \pi j/n$, $j = 0, \dots, 2n-1$, and use the quadrature rule

$$\int_0^{2\pi} \ln\left(4\sin^2\frac{t-\tau}{2}\right) f(\tau)d\tau \approx \sum_{j=0}^{2n-1} R_j^{(n)}(t) f(t_j), \quad 0 \le t 2\pi, \tag{2.5.7}$$

with the quadrature weights given by

$$R_j^{(n)}(t) := -\frac{2\pi}{n}\sum_{m=1}^{n-1}\frac{1}{m}\cos m(t-t_j) - \frac{\pi}{n^2}\cos n(t-t_j), \quad j = 0, \dots, 2n-1,$$

and the trapezoidal rule

$$\int_0^{2\pi} f(\tau)d\tau \approx \frac{\pi}{n}\sum_{j=0}^{2n-1} f(t_j). \tag{2.5.8}$$

Both these numerical integration formulas are obtained by replacing the integrand f by its trigonometric interpolation polynomial and then integrating exactly.

In the Nyström method, by applying the quadrature rule (2.5.7) to $f = K_1(t,\cdot)\Psi$ and (2.5.8) to $f = K_2(t,\cdot)$, the integral equation (2.5.4) is replaced by the approximating equation

$$\Psi^{(n)}(t) - \sum_{j=0}^{2n-1}\left\{R_j^{(n)}(t)K_1(t,t_j) + \frac{\pi}{n}K_2(t,t_h)\right\}\Psi^{(n)}(t_j) = g(t) \tag{2.5.9}$$

for $0 \le t \le 2\pi$. Then by solving a finite dimensional linear system we can obtain the solution of (2.5.9). In particular, for any solution of (2.5.9) the values $\Psi_i^{(n)} = \Psi^{(n)}(t_i), i = 0, \dots, 2n-1$, at the quadrature points trivially satisfy the linear system

$$\Psi_i^{(n)} - \sum_{j=0}^{2n-1}\left\{R_{|i-j|}^{(n)}K_1(t_i,t_j) + \frac{\pi}{n}K_2(t_i,t_j)\right\}\Psi_j^{(n)} = g(t_i), \quad i = 0, \dots, 2n-1, \tag{2.5.10}$$

where

$$R_j^{(n)} := R_j^{(n)}(0) = -\frac{2\pi}{n}\sum_{m=1}^{n-1}\frac{1}{m}\cos\frac{mj\pi}{n} - \frac{(-1)^j\pi}{n^2}, \quad j = 0, \ldots, 2n-1.$$

Conversely, given a solution $\Psi_i^{(n)}$, $i = 0, \ldots, 2n-1$ of the system (2.5.10), the function $\Psi^{(n)}$ defined by

$$\Psi^{(n)}(t) := \sum_{j=0}^{2n-1}\left\{R_j^{(n)}(t)K_1(t, t_j) + \frac{\pi}{n}K_2(t, t_j)\right\}\Psi_j^{(n)} + g(t), \quad 0 \le t \le 2\pi, \tag{2.5.11}$$

satisfies the approximating equation (2.5.9). The formula (2.5.11) may be viewed as a natural interpolation of the values $\Psi_i^{(n)}$, $i = 0, \ldots, 2n-1$, at the quadrature points to obtain the approximating function $\Psi_i^{(n)}$ and goes back to Nyström.

For the solution of the large linear system (2.5.10), we recommend the use of the fast iterative two-grid or multi-grid methods.

Provided the integral equation (2.5.5) is uniquely solvable and the kernels K_1 and K_2 and the right-hand side g are continuous, a rather involved error analysis shows that the approximating linear system (2.5.10), i.e., the approximating equation (2.5.9), is uniquely solvable for all sufficiently large n. Meanwhile, the approximate solution s $\Psi^{(n)}$ converge uniformly to the solution Ψ of the integral equation as $n \to \infty$. Also, the convergence order of the quadrature errors for (2.5.7) and (2.5.8) carries over to the error $\Psi^{(n)} - \Psi$, which, in particular, means that the case of analytic kernels K_1 and K_2 and analytic right-hand sides g the approximation error decreases exponentially, i.e., there exist positive constants C and σ such that

$$|\Psi^{(n)}(t) - \Psi(t)| \le Ce^{-n\sigma}, \quad 0 \le t \le 2\pi, \tag{2.5.12}$$

for all n. In particular, the constants in (2.5.12) are computable but usually they are difficult to evaluate. In most practical cases, it is sufficient to judge the accuracy of the computed solution by doubling the number $2n$ of knots and then comparing the results for the coarse and the fine grid with the aid of the exponential convergence order, i.e., by the fact that doubling the number $2n$ of knots will double the number of correct digits in the approximate solution.

Numerical Example 1

For a numerical example 1, we consider the scattering of a plane wave by a cylinder with a non-convex kite-shaped cross section with boundary ∂D shown in Fig. 2.1 and described by the parametric representation

$$x(t) = (\cos t + 0.65\cos 2t - 0.65, 1.5\sin t), \quad 0 \le t \le 2\pi.$$

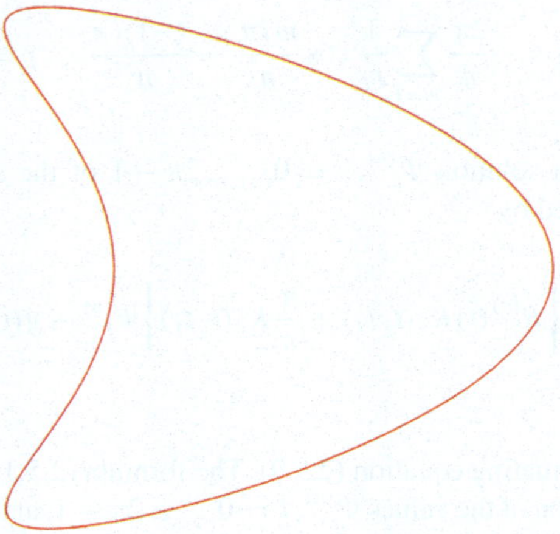

Fig. 2.1 Kite-shaped domain for Numerical Example 1

Table 2.1 Numerical results for Nyström's method

	n	$\mathrm{Re}\, u_\infty(d)$	$\mathrm{Im}\, u_\infty(d)$	$\mathrm{Re}\, u_\infty(-d)$	$\mathrm{Im}\, u_\infty(-d)$
k = 1	8	–1.6259	0.6017	1.3944	0.0940
	16	–1.6273	0.6021	1.3969	0.0951
	32	–1.6274	0.6022	1.3969	0.0950
	64	–1.6275	0.6022	1.3969	0.0950
k = 3	8	–2.0486	1.8264	0.4202	1.6358
	16	–2.1137	1.2441	0.2507	1.5911
	32	–2.1127	1.2430	0.2508	1.5910
	64	–2.1128	1.2430	0.2508	1.5910
k = 5	8	–3.1116	2.8098	0.2976	–0.7549
	16	–2.4165	1.9705	–0.2031	0.0674
	32	–2.4755	1.6874	–0.1994	0.0602
	64	–2.4755	1.6875	–0.1995	0.0602

From the asymptotics for Hankel functions, it can be deduced that the far field pattern of the combined potential in two dimensions is given by

$$u_\infty(\hat{x}) = \frac{e^{-\mathrm{i}\frac{\pi}{4}}}{\sqrt{8\pi k}} \int_{\partial D} \{k\nu(y)\cdot\hat{x} + \eta\} e^{-\mathrm{i}k\hat{x}\cdot y} \varphi(y)ds(y), \quad |\hat{x}| = 1, \tag{2.5.13}$$

which can be evaluated again by the trapezoidal rule after solving the integral equation for φ. Table 2.1 gives some approximate values for the far field pattern $u_\infty(d)$ and $u_\infty(-d)$ and the backward direction $-d$. The direction d of the incident wave is $d = (1, 0)$ and the coupling parameter is $\eta = k$. Note that the exponential convergence is clearly exhibited.

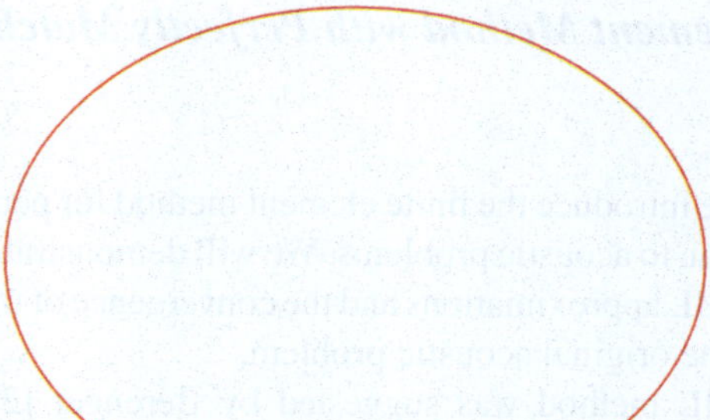

Fig. 2.2 Circle domain for numerical example 2

Table 2.2 Numerical results for Nyström's method

	n	$\mathrm{Re}\, u_\infty(d)$	$\mathrm{Im}\, u_\infty(d)$	$\mathrm{Re}\, u_\infty(-d)$	$\mathrm{Im}\, u_\infty(-d)$
k = 1	8	–1.3342	0.3337	0.1818	0.7627
	16	–1.3343	0.3337	0.1818	0.7627
	32	–1.3344	0.3337	0.1818	0.7627
	64	–1.3344	0.3337	0.1818	0.7627
k = 3	8	–1.6192	0.7977	–0.6731	–0.2646
	16	–1.6174	0.7973	–0.6731	–0.2646
	32	–1.6174	0.7974	–0.6731	–0.2646
	64	–1.6174	0.7974	–0.6731	–0.2646
k = 5	8	–1.8852	1.3045	0.6278	–0.3534
	16	–1.8493	1.0989	0.6210	–0.3524
	32	–1.8494	1.0990	0.6210	–0.3524
	64	–1.8494	1.0990	0.6210	–0.3524

Numerical Example 2

For a numerical Example 2, we consider the scattering of a plane wave by a cylinder with a circle cross section with boundary ∂D shown in Fig. 2.2 and described by the parametric representation

$$x(t) = (\cos t, \sin t), \quad 0 \leq t \leq 2\pi.$$

The numerical for this example is provided in (Table 2.2). The matlab code for this numerical example can be found in supplementary material.

2.5.2 Finite Element Method with Perfectly Matched Layer (PML)

In this subsection we introduce the finite element method for perfectly matched layer (PML) approximation to acoustic problems. We will demonstrate both the solvability of the continuous PML approximations and the convergence of the resulting solutions to the solutions of the original acoustic problem.

The original PML method was suggested by Bérenger [2] which proposed a PML technique for solving with the time dependent Maxwell equations. Under the assumption that the exterior solution is composed of outgoing waves only, the basic idea of the PML technique is to surround the computation domain by a layer of finite thickness with specially designed model medium that would either slow down or attenuate all the waves that propagate from inside the computational domain.

In this section we consider a finite element method for PML technique for the following acoustic scattering problem with perfectly conducting boundary:

$$\begin{cases} \Delta u + k^2 u = 0 \quad \text{in } \mathbb{R}^2 \backslash \overline{D}, \\ \dfrac{\partial u}{\partial \nu} = -g \quad \text{on } \Gamma_D, \\ \lim\limits_{r\to\infty} \sqrt{r}\left(\dfrac{\partial u}{\partial r} - iku\right) = 0, \quad \text{for } r = |x|. \end{cases} \tag{2.5.14}$$

Here $D \subset \mathbb{R}^2$ is a bounded domain with Lipschitz boundary Γ_D, $g \in H^{-1/2}(\Gamma_D)$ is determined by the incoming wave, and ν is the unit outer normal to Γ_D. We assume the wave number $k \in \mathbb{R}$ is a constant. The results in this section can be easily extended to solve the scattering problem with other boundary conditions such as Dirichlet or the impedance boundary condition on Γ_D, or to solve the acoustic wave propagation through inhomogeneous media with a variable wave number $k^2(x)$ inside some bounded domain.

Let D be contained in the interior of the circle $B_R = \{x \in \mathbb{R}^2 : |x| < R\}$. We start by introducing an equivalent variational formulation of (2.5.14) in the bounded domain $\Omega_R = B_R \backslash \overline{D}$. In the domain $\mathbb{R}^2 \backslash \overline{B}_R$, the solution u of (2.5.14) can be written under the polar coordinates as follows:

$$u(r,\theta) = \sum_{n\in\mathbb{Z}} \frac{H_n^{(1)}(kr)}{H_n^{(1)}(kR)} \hat{u}_n e^{in\theta}, \quad \hat{u}_n = \frac{1}{2\pi}\int_0^{2\pi} u(R,\theta)e^{-in\theta}d\theta, \tag{2.5.15}$$

where $H_n^{(1)}$ is the Hankel function of the first kind and order n. The series in (2.5.15) converges uniformly for $r > R$. Let $T : H^{1/2}(\Gamma_R) \to H^{-1/2}(\Gamma_R)$, where $\Gamma_R = \partial B_R$, be the Dirichlet-to-Neumann operator defined as follows: $\forall f \in H^{1/2}(\Gamma_R)$,

$$Tf = \sum_{n\in\mathbb{Z}} k \frac{H_n^{(1)'}(kR)}{H_n^{(n)}(kR)} \hat{f}_h e^{in\theta}, \quad \hat{f}_h = \frac{1}{2\pi}\int_0^{2\pi} f e^{-in\theta} d\theta. \tag{2.5.16}$$

It is known that T is well-defined and the solution u written in (2.5.15) satisfies

$$\left.\frac{\partial u}{\partial \nu}\right|_{\Gamma_R} = Tu.$$

Let $a : H^1(\Omega_R) \times H^1(\Omega_R) \to \mathbb{C}$ be the sesquilinear form:

$$a(\varphi, \psi) = \int_{\Omega_R} (\nabla\varphi \cdot \nabla\overline{\psi} - k^2\varphi\overline{\psi})dx - \langle T\varphi, \psi\rangle_{\Gamma_R}, \tag{2.5.17}$$

where $\langle\cdot,\cdot\rangle_{\Gamma_R}$ stands for the inner product on $L^2(\Gamma_R)$ or the duality pairing between $H^{-1/2}(\Gamma_R)$ and $H^{1/2}(\Gamma_R)$. For a given $g \in H^{-1/2}(\Gamma_D)$, the scattering problem (2.5.14) is equivalent to the following weak formulation: Find $u \in H^1(\Omega_R)$ such that

$$a(u, \psi) = \langle g, \psi\rangle_{\Gamma_D} \quad \forall\psi \in H^1(\Omega_R). \tag{2.5.18}$$

The existence of a unique solution of the variational problem (2.5.18) is known.

Now we introduce the absorbing PML layer. We surround the domain Ω_R with a PML layer $\Omega^{PML} = \{x \in \mathbb{R}^2 : R < |x| < \rho\}$. The specially designed model medium in the PML layer should basically be so chosen that either the wave never reaches its external boundary or the amplitude of the reflected wave is so small that it does not essentially contaminate the solution in Ω_R. Here we assume $\rho \le CR$ for some generic fixed constant $C > 0$.

Let $\alpha(r) = 1 + i\sigma(r)$ be the model medium property which satisfies

$$\sigma \in C(\mathbb{R}), \quad \sigma \ge 0, \quad \text{and } \sigma = 0 \quad \text{for} \quad r \le R.$$

Denote by $\overline{r}$ the complex radium defined by

$$\tilde{r} = \tilde{r}(r) = \begin{cases} r & \text{if} \quad r \le R \\ \int_0^r \alpha(t)dt = r\beta(r) & \text{if} \quad r \ge R \end{cases} \tag{2.5.19}$$

Following [4], we introduce the PML equation

$$\nabla\cdot(A\nabla w) + \alpha\beta k^2 w = 0 \quad \text{in} \quad \Omega^{PML}, \tag{2.5.20}$$

where $A = A(x)$ is a matrix which satisfies, in polar coordinates,

$$\nabla\cdot(A\nabla) = \frac{1}{2}\frac{\partial}{\partial r}\left(\frac{\beta r}{\alpha}\frac{\partial}{\partial r}\right) + \frac{\alpha}{\beta r^2}\frac{\partial^2}{\partial\theta^2}. \tag{2.5.21}$$

The PML solution $\hat{u}$ in $\Omega_\rho = B_\rho \backslash \overline{D}$ is defined as the solution of the following system

$$\begin{cases} \nabla \cdot (A\nabla \hat{u}) + \alpha\beta k^2 \hat{u} = 0 & \text{in} \quad \Omega_\rho, \\ \frac{\partial \hat{u}}{\partial \nu} = -g & \text{on} \quad \Gamma_D, \\ \hat{u} = 0 & \text{on} \quad \Gamma_\rho. \end{cases} \tag{2.5.22}$$

This problem can be reformulated in the bounded domain Ω_R by imposing the boundary condition

$$\frac{\partial \hat{u}}{\partial \nu} = \hat{T}\hat{u},$$

where the operator $\hat{T} : H^{1/2}(\Gamma_R) \to H^{-1/2}(\Gamma_R)$ is defined as follows: Given $f \in H^{1/2}(\Gamma_R)$,

$$\hat{T} f = \left.\frac{\partial \xi}{\partial \nu}\right|_{\Gamma_R},$$

where $\xi \in H^1(\Omega^{PML})$ satisfies

$$\begin{cases} \nabla \cdot (A\nabla \xi) + \alpha\beta k^2 \xi = 0 & \text{in} \quad \Omega^{PML}, \\ \xi = f & \text{on} \quad \Gamma_R, \\ \xi = 0 & \text{on} \quad \Gamma_\rho. \end{cases} \tag{2.5.23}$$

The existence and uniqueness of the solutions of the PML problem (2.5.23) can be found in [3].

Based on the operator $\hat{T}$, we introduce the sesquilinear form $\hat{a} : H^1(\Omega_R) \times H^1(\Omega_R) \to \mathbb{C}$ by

$$\hat{a}(\varphi, \psi) = \int_{\Omega_R} (A\nabla\varphi \cdot \nabla\overline{\psi} - k^2\alpha\beta\varphi\overline{\psi})dx - \left\langle \hat{T}\varphi, \psi \right\rangle_{\Gamma_R}. \tag{2.5.24}$$

Then for any given $g \in H^{-1/2}(\Gamma_D)$ the weak formulation for (2.5.22) is to find $\hat{u} \in H^1(\Omega_R)$ such that

$$\hat{a}(\hat{u}, \psi) = \langle g, \psi \rangle_{\Gamma_D} \qquad \forall \psi \in H^1(\Omega_R). \tag{2.5.25}$$

The well-posedness of the PML problem (2.5.25) and the convergence of its solution to the solution of the original scattering problem can be found in [3].

Now we introduce the finite element approximations of the PML problems (2.5.22). From now on we assume $g \in L^2(\Gamma_D)$. Let $b : H^1(\Omega_\rho) \times H^1(\Omega_\rho) \to \mathbb{C}$ be the sesquilinear form given by

$$b(\varphi, \psi) = \int_{\Omega_\rho} (A\nabla\varphi \cdot \nabla\overline{\psi} - \alpha\beta k^2 \varphi\overline{\psi})dx. \tag{2.5.26}$$

Let Γ_ρ^h, which consists of piecewise segments whose vertics lie on Γ_ρ, be an approximation of Γ_ρ. Let Ω_ρ^h be the subdomain of Ω_ρ bounded by Γ_D and Γ_ρ^h. Let T_h be a regular triangulation of the domain Ω_ρ^h. We assume the elements $T \in \mathcal{T}_h$ may have one curved edge align with Γ_D so that $\Omega_\rho^h = \cup_{T\in\mathcal{T}_h} T$.

Let $V_h \subset H^1(\Omega_\rho^h)$ be the conforming linear finite element space over Ω_ρ^h, and $\widetilde{V}_h = \{v_h \in V_h : v_h = 0 \text{ on } \Gamma_\rho\}$. In the following we will always assume that the functions in $\widetilde{V}_h$ are extended to the domain Ω_ρ by zero so that any function $v_h \in \widetilde{V}_h$ is also a function in $H_0^1(\Omega_\rho)$. The finite element approximation to the PML problem (2.5.22) reads as follows: Find $u_h \in \widetilde{V}_h$ such that

$$b(u_h, \psi_h) = \int_{\Gamma_D} g\overline{\psi}_h ds \qquad \forall \psi_h \in \widetilde{V}_h. \tag{2.5.27}$$

Following the general theory in [1], the existence of unique solution of discrete problem (2.5.27) and the finite element convergence analysis depend on the following discrete inf-sup condition

$$\sup_{0\neq\psi_h\in\widetilde{V}_h} \frac{|b(\varphi_h, \psi_h)|}{\|\psi_h\|_{H^1(\Omega_\rho)}} \geq \hat{\mu}\|\varphi_h\|_{H^1(\Omega_\rho)} \qquad \forall \varphi_h \in \widetilde{V}_h, \tag{2.5.28}$$

where the constant $\hat{\mu} > 0$ is independent of the finite element mesh size. Since the continuous problem (2.5.26) has a unique solution, the sesquilinear form $b : H_0^1(\Omega_\rho) \times H_0^1(\Omega_\rho) \to \mathbb{C}$ satisfies the continuous inf-sup condition. Then a general argument of Schatz [7] implies (2.5.28) is valid for sufficiently small mesh size $h < h^*$. Based on (2.5.28), appropriate a priori error estimate can also be derived which depends on the regularity of the PML solution $\hat{u}$.

Now we present some numerical experiments to demonstrate the theoretical results established in above.

First, the numerical experiment are conducted in $\mathbb{R}^2$. We consider a kite-shaped scatterer D which is given by

$$x(t) = (\cos t + 0.65\cos 2t - 0.65, 1.5\sin t), \qquad 0 \leq t \leq 2\pi. \tag{2.5.29}$$

Experimental settings are shown in Fig. 2.3a. For the setting with respect to the scattering measurement, a PML layer of width 1 is attached to the square $(-4, 4)^2$ to truncate the whole space into a finite domain with scattering boundary conditions enforced on the outer boundary. We fix the wave number $k = 3$ and the incident direction $d = (1, 0)^T$. The incident field, total field and the scatter field is given in Fig. 2.3b, c and d, respectively.

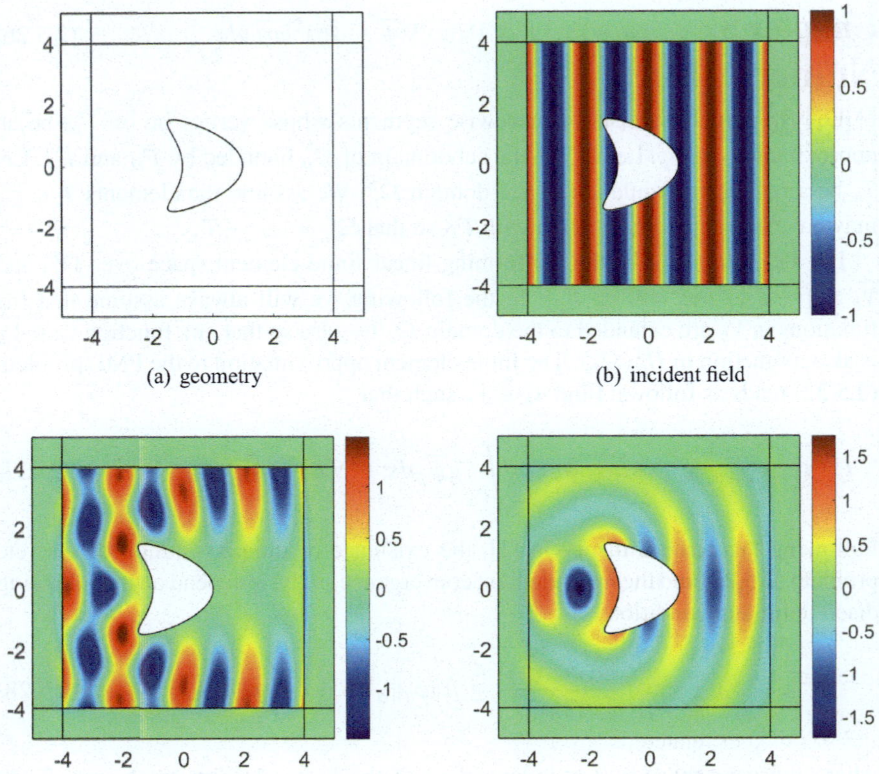

(a) geometry (b) incident field

(c) total field (d) scatter field

Fig. 2.3 The incident direction is $d = (1, 0)$ and the wavenumber is $k = 3$

References

1. I. Babuska, A. Aziz, Survey Lectures on Mathematical Foundations of the Finite Element Method, in *The Mathematical Foundations of the Finite Element Method with Application to Partial Differential Equations*. ed. by A. Aziz (Academic, New York, 1973), pp.5–359
2. J.-P. Berenger, A perfectly matched layer for the absorption of electromagnetic waves. J. Comput. Phys. **114**, 185–200 (1994)
3. Z. Chen, X. Liu, An adaptive perfectly matched layer technique for time-harmonic scattering problems. SIAM J. Numer. Anal. **43**, 645–671 (2005)
4. F. Collino, P.B. Monk, The perfectly matched layer in curvilinear coordinates. SIAM J. Sci. Comput. **92**, 2061–2090 (1998)
5. D. Colton, R. Kress, *Inverse Acoustic and Electromagnetic Scattering Theory*, 2nd edn. (Springer, New York, 1998)
6. W. McLean, *Strongly Elliptic Systems and Boundary Integral Equations* (Cambridge University Press, Cambridge, 2000)
7. A.H. Schatz, An observation concerning Ritz-Galerkin methods with indefinite bilinear forms. Math. Comput. **28**, 959–962 (1974)
8. N. Weck, Approximation by Herglotz wave functions, mathematical methods in the applied. Science **27**(2), 155–162 (2004)

Chapter 3
Numerical Inverse Acoustic Scattering Problems

3.1 Overview

We would like to point out that many numerical reconstruction methods have been developed for inverse scattering problems in various scenarios, such as the linear sampling method, factorization method, MUSIC-type methods, time reversal, topological-optimization-type methods, interior-resonance-based methods, and machine learning methods etc.; we refer the readers to [2–9, 11, 15–17, 23–25, 35–37, 41, 42, 59, 74–76, 79, 81–83, 85] and the references therein for these methods and some other related developments.

The inverse problems are to recover the scatterer, namely (Ω, n) if it is a medium, or (Ω, λ) if it is an obstacle of impedance type, or Ω if it is a sound-soft or sound-hard obstacle, by the knowledge of the far-field pattern $a_k(\hat{x}, d)$. If one introduces an operator $\mathscr{F}$ which maps the scatterer to the corresponding far-field pattern, the inverse scattering problem can be formulated as the following operator equation

$$\mathscr{F}(\mathscr{O}) = a_k(\hat{x}, d), \quad k \in \mathbb{R}_+,\ \hat{x}, d \in \mathbb{S}^{N-1}, \tag{3.1.1}$$

where $\mathscr{O}$ represents (Ω, n), (Ω, λ) or Ω, depending on the type of the concerned inverse problem. It is widely known that the operator equation (3.1.1) is nonlinear and ill-posed (cf. [26]). In (3.1.1), $a_k(\hat{x}, d)$ is given by the measurement data which are usually recorded by some physical apparatus. The data shall be called *a single measurement* if $a_k(\hat{x}, d)$ is given for a fixed $k \in \mathbb{R}_+$ and a fixed $d \in \mathbb{S}^{N-1}$, but for all $\hat{x} \in \mathbb{S}^{N-1}$. That is, for a single far-field measurement, one collects the far-field data in every possible observation direction by sending a single detecting plane wave. If multiple detecting plane waves are used, e.g., many different d's or k's are used, then the corresponding scattering data shall be referred to as *multiple measurements*. We note that as $a_k(\cdot, d)$ is an analytic function on the unit sphere, it is known on the whole sphere as long as it is known on any open subset of the unit sphere by the analytic continuation.

J. Li and H. Liu, *Numerical Methods for Inverse Scattering Problems*,
https://doi.org/10.1007/978-981-99-3772-1_3

We define the operator $\mathscr{F}$ which maps the boundary of the obstacle to the corresponding far-field pattern, then the inverse problem can be expressed as the following operator equation

$$\mathscr{F}(\partial\Omega) = u_\infty(\hat{x}; d). \tag{3.1.2}$$

It is easily seen that $\mathscr{F}$ is nonlinear. Moreover, the operator equation (3.1.2) is widely known to be severely ill-posed in the sense of Hadamard (see, e.g., [26]). For the inversion, one may naturally solve (3.1.2) by Newton-type's methods or formulate it in terms of a non-linear optimization problem. All these direct methods are formulated in the framework of Tikhonov regularization and conducted in an iterative manner due to the ill-posedness and nonlinearity of (3.1.2). Hence, in order to achieve satisfactory approximate solutions, they typically require a good initial guess for the iteration and the computational accounts are formidable. Furthermore, those direct approaches require a priori information on the physical properties of the underlying scatterer, e.g., it is sound-soft or sound-hard, or of impedance type. For many practical applications, such information is generally not available, e.g., the detection of hostile objects or buried obstacles. We refer to [26] for a general discussion and related literatures on the aforementioned direct methods. In order to avoid those disadvantages, Colton and Kirsch developed in [24] a "simple" method for shape reconstruction in inverse scattering problems which is nowadays known as the *linear sampling method* (LSM). The idea is to characterize the boundary $\partial\Omega$ of the scatterer by the behavior of an indicator function, which is implied in a first kind linear Fredholm integral equation with the far-field pattern (namely, the measurement data) as its integral kernel. Then finding the profile of the scatterer reduces to capturing the behavior of the indicator function. The LSM is computationally very fast (see [51]), and is easy for implementation, more importantly, it is independent of the physical properties of the underlying scatterer. There has been a considerable study during the last ten years on the LSM and several qualitative methods have been proposed following the spirits of the LSM, e.g., the factorization method (see [42]) and another variety of methods (see [75]). We refer to the review papers [27, 75] and the monographs [17, 74] for detailed expositions of the LSM and related literatures.

The LSM provides the boundary information of an unknown scatterer through the magnitudes of the L^2-norm of the indicator function at a sampling grid. When the magnitude is relatively large at a point, we count the point lying outside the scatterer; otherwise a point is lying inside the scatterer. So it is extremely critical to define a cut-off value in an LSM to count if the magnitude of the indicator function is large enough at a sampling point. Unfortunately, to our knowledge, there are no deterministic strategies available in the literature to tell how to choose a cut-off value, and all the existing methods are heuristic. This brings a great difficulty and inconvenience in practically realizing the LSM, in particular considering the fact that the numerical reconstructions are very sensitive to the choices of the cut-off value. Another theoretical barrier in implementing an LSM is how to prevent the interior eigenvalue problem, which may cause the failure of an LSM. In the discussion of strengthened linear sampling methods in Sect. 3.2, we shall show a practically easy implementable technique on how to choose the cut-off value, while overcoming the interior eigenvalue problem in the LSM at the same time. These problems are of much practical interests to both researchers and practitioners on linear sampling methods. The proposed technique is to artificially introduce a reference ball as an extra obstacle component to the underlying scattering system. Due to the scattering coupling between the reference ball and the original unknown scatterers, one can expect the closely related cut-off values for both the reference ball and the other unknown scatterers. The crux of the technique is as follows: since we know the reference ball in advance, the cut-off value around the boundary of the reference ball can be determined and then it can be employed as the cut-off value to identify the profiles of other unknown scatterers. Moreover, the reference ball can be suitably chosen such that the square of the wavenumber k is not a Laplacian eigenvalue for the ball. In doing so, we can show that the interior eigenvalue problem for the LSM is automatically eliminated. As an interesting and important theoretical by-product, by letting the radius of the reference ball tend to zero, we have further shown that the interior eigenvalue problem for the LSM only exists theoretically but not in practice as long as noise presents. Finally, we present extensive numerical experiments to show the feasibility and effectiveness of the discussed method.

In Sect. 3.3, we discuss three inverse scattering schemes for locating multiple multiscale acoustic scatterers in a very general and practical setting. For all of the three locating schemes, only one single far-field measurement is used. The number of the multiple scatterer components may be unknown, and each scatterer component is allowed to be an inhomogeneous medium with an unknown content or an impenetrable obstacle of sound-soft, sound-hard or impedance type. Moreover, the scatterers could be multiscale, i.e., some scatterers may be of regular size, and some others may be of small size in terms of the wavelength of the detecting acoustic wave. If a scatterer component is of regular size, it is required that its shape (not necessarily its orientation, size and location) should be from an admissible class which is known in advance. The locating schemes are based on some indicator functions, and are computationally cheap and robust against the measurement noise. Rigorous mathematical justifications are provided for each scheme, and numerical experiments are presented to demonstrate its robustness and effectiveness.

It is noted that the inverse problem is formally posed with a fixed $\xi \in \mathbb{R}^n$ and all $\hat{x} \in \mathbb{S}^{n-1}$. Hence, there is a widespread belief that one can recover Ω by using the far-field pattern corresponding to a single incident plane wave $e^{ix\cdot\xi}$, which is referred to as a single far-field measurement. However, this still remains to be a longstanding problem with very limited progress in the literature. If the obstacle is of small size; roughly speaking, smaller than half of the detecting wavelength, the unique recovery result was established in [14, 28]. If the obstacle is extremely "rough" in the sense that its boundary is nowhere analytic, the unique recovery result was established in [39]. If the obstacle is of general polyhedral type, the corresponding uniqueness study by at most a few far-field measurements is almost complete [1, 18–21, 34, 54, 56–58, 60, 65–67, 69]. We also refer to [3, 12–14, 31, 32, 61–64] for related unique identifiability results on recovering the shape of medium scatterers. Recently, some qualitative numerical schemes of recovering the obstacles by a single far-field measurement were proposed in [8, 49, 53], where certain restrictive a priori assumptions have to be imposed on the obstacles. Another challenging issue in the study of inverse scattering problems is the recovery by phaseless data, say the modulus of the far-field pattern, $|u^\infty(\hat{x})|$. There is also very limited progress in the literature in this aspect (cf. [43, 46, 58, 84]). Here, we would like to note that due to the following translation relation (cf. [46, 53])

$$u^\infty(\hat{x}; k, d, z + D) = e^{ik(d-\hat{x})\cdot z} u^\infty(\hat{x}; k, d, D), \quad z \in \mathbb{R}^n,$$

where $z + D := \{z + x; x \in D\}$, one has $|u^\infty(\hat{x}; k, d, z + D)| = |u^\infty(\hat{x}; k, d, D)|$. Hence, by using the phaseless data, there is such an obstruction that the location of the obstacle can not be uniquely recovered.

This chapter follows the work [52, 53, 58].

3.2 Strengthened Linear Sampling Methods

By adding a reference ball as an extra artificial obstacle component to the underlying scattering system, we discuss a very simple but robust and effective technique to choose the crucial cut-off value required in the linear sampling method (LSM) for inverse scattering problems. The reference ball technique causes little extra computational costs to the LSM, but brings in a practically very important by-product, i.e. it eliminates the interior eigenvalue problem automatically, a well-known barrier when applying the LSM. Some mathematical justifications of the technique are provided, and numerical experiments are also presented to demonstrate its feasibility and effectiveness.

This section is concerned with the inverse scattering problems of imaging obstacles by acoustic or electromagnetic far-field measurements. We take as our model problem the inverse acoustic obstacle scattering by time-harmonic plane waves. We will illustrate how to extend all the results for the acoustic case to the electromagnetic system at the end of the section. This section concerns the inverse scattering problem of recovering an impenetrable obstacle by the corresponding acoustic wave detection. The problem has its physical origin in radar/sonar, geophysical exploration, non-destructive testing and medical imaging (cf. [26, 40]).

In the next subsection, we will present some theoretical analysis to reveal the rationale behind the reference ball technique to choose the cut-off value and rigorously prove the elimination of the interior eigenvalue problems of the LSM when noise exists. In Sect. 3.2.2, some numerical experiments are performed to illustrate the feasibility and effectiveness of the discussed method.

3.2.1 Strengthened Linear Sampling Method with a Reference Ball

We start with the basics of the LSM. First, we introduce the far-field operator $F : L^2(\mathbb{S}^{n-1}) \mapsto L^2(\mathbb{S}^{n-1})$:

$$(Fg)(\hat{x}) := \int_{\mathbb{S}^{n-1}} u_\infty(\hat{x}, d; \mathbf{D}) g(d) ds(d), \qquad \hat{x} \in \mathbb{S}^{n-1}. \tag{3.2.1}$$

and a function Φ_∞ induced by the incident field $u^i(x) = \exp\{ikx \cdot d\}$:

$$\Phi_\infty(\hat{x}, z) = \gamma \exp\{-ik\hat{x} \cdot z\} \tag{3.2.2}$$

with $\gamma = 1/4\pi$ in $\mathbb{R}^3$ and $\gamma = e^{i\pi/4}/\sqrt{8\pi k}$ in $\mathbb{R}^2$. The linear sampling method is to use the solution g to the following *far-filed equation*

$$(Fg)(\hat{x}) = \Phi_\infty(\hat{x}, z), \qquad \hat{x} \in \mathbb{S}^{n-1},\ z \in \mathbb{R}^n \tag{3.2.3}$$

as an indicator function. As the far-field operator F has smooth kernel, it is compact in $L^2(\mathbb{S}^{n-1})$, and the Eq. (3.2.3) does not have a solution in general. But under appropriate assumptions one can use density argument to solve the equation approximately to find an approximate function $\tilde{g}$ whose behavior can be used to characterize the boundary of the underlying scatterer. The assumption is that k^2 is not a Laplacian eigenvalue for $\mathbf{D}$, namely k^2 is not a Dirichlet eigenvalue for $-\Delta$ in $H_0^1(\mathbf{D})$ when $\mathbf{D}$ is sound-soft, and is not a Neumann eigenvalue for $-\Delta$ in $H^1(\mathbf{D})$ when $\mathbf{D}$ is sound-hard. The following theorem forms the basis of the linear sampling method (see, e.g., Theorem 4.1 in [27]).

Theorem 3.2.1 *Assume that k^2 is not a Dirichlet eigenvalue for $-\Delta$ in $H_0^1(\mathbf{D})$ when $\mathbf{D}$ is sound-soft and is not a Neumann eigenvalue for $-\Delta$ in $H^1(\mathbf{D})$ when $\mathbf{D}$ is sound-hard. Then*

1. *For $z \in \mathbf{D}$ and a fixed $\varepsilon > 0$ there exists a $g_\varepsilon^z \in L^2(\mathbb{S}^{N-1})$ such that*

$$\|Fg_\varepsilon^z - \Phi_\infty(\cdot, z)\|_{L^2(\mathbb{S}^{N-1})} < \varepsilon$$

and

$$\lim_{z \to \partial\mathbf{D}} \|g_\varepsilon^z\|_{L^2(\mathbb{S}^{N-1})} = \infty.$$

2. *For $z \in \mathbb{R}^N \backslash \bar{\mathbf{D}}$ and any given $\varepsilon > 0$, every $g_\varepsilon^z \in L^2(\mathbb{S}^{N-1})$ that satisfies*

$$\|Fg_\varepsilon^z - \Phi_\infty(\cdot, z)\|_{L^2(\mathbb{S}^{N-1})} < \varepsilon$$

ensures

$$\lim_{\varepsilon \to 0} \|g_\varepsilon^z\|_{L^2(\mathbb{S}^{N-1})} = \infty.$$

We would like to remark that the blow-up behavior of g^z will occur disregarding the physical property of the underlying obstacle, namely, the obstacle in Theorem 3.2.1 could be associated with the general boundary condition. The LSM elegantly turns the reconstruction of the shape of obstacle **D** into the process of numerically determining the indicator function g^z in Theorem 3.2.1. The general procedure can be stated as follows (see also Chap. 4, [17]):

Algorithm LSM

1. Select a mesh $\mathscr{T}_h$ of sampling points in a region Ω which contains **D**.
2. Use the Tikhonov regularization and the Morozov discrepancy principle to compute an approximate solution g^z to the far-field equation (3.2.3) for each mesh point z in $\mathscr{T}_h$.
3. Select a cut-off value c; then count $z \in \mathbf{D}$ if $\log \|g^z\|_{L^2(\mathbb{S}^{N-1})} \leq c$; and $z \notin \mathbf{D}$ if $\log \|g^z\|_{L^2(\mathbb{S}^{N-1})} > c$.

The LSM will reconstruct the shape of the scatterer provided one could solve (at least approximately) Eq. (3.2.3) for a "valid" indicator function g^z; we refer to [17, 24, 51] for numerous numerical illustrations. It is emphasized that the LSM remains valid when there is noise attached to the far-field data. In view of practical applications, the far-field data is provided by measurement, the noise is unavoidable. Hence it is of crucial importance that the LSM should be robust to noise, and this is guaranteed by intensive theoretical and numerical analyses in existing work mentioned earlier.

From the above exposition, one may notice two important issues involved in the LSM, which deserve our special attention, i.e., the interior eigenvalue problem (see Theorem 3.2.1) and the selection of the cut-off value. In the rest of this section, we shall investigate these two issues in detail and discuss some techniques to effectively deal with the issues.

To guarantee the feasibility of the LSM, theoretical analysis must require that the interior eigenvalue problem cannot occur (see, e.g., [10]). To gain a better understanding of the interior eigenvalue problem, we introduce the Herglotz wave function

$$v_g(x) := \mathscr{H} g(x) = \int_{\mathbb{S}^{n-1}} e^{ikx\cdot d} g(d)\, ds(d), \quad g(d) \in L^2(\mathbb{S}^{n-1}), \; x \in \mathbb{R}^n. \tag{3.2.4}$$

$v_g(x)$ is a solution to the Helmholtz differential operator $(\Delta + k^2)$ in $\mathbb{R}^n$, thus is an entire function. Define the Herglotz operator $\mathscr{H} : L^2(\mathbb{S}^{n-1}) \to H^\alpha(\partial\mathbf{D})$ by

$$\mathscr{H} g(x) := \mathscr{B}\mathscr{H} g(x), \quad x \in \partial\mathbf{D}, \tag{3.2.5}$$

where $\alpha = 1/2$ if the scatterer **D** is sound-soft and $\alpha = -1/2$ if Neumann or impedance type boundary condition is presented on part of the boundary $\partial\mathbf{D}$. By further introducing the boundary operator $\mathscr{G} : H^\alpha(\partial\mathbf{D}) \to L^2(\mathbb{S}^{n-1})$ which maps the boundary data of the scattered field, namely $\mathscr{B}u^s|_{\partial\mathbf{D}}$ for the direct scattering problem (2.3.3), to the corresponding far-field pattern. We have the factorization

$$F = -\mathscr{G}\mathscr{H}.$$

The following theorem summarizes the properties of the operators $\mathscr{G}$ and $\mathscr{H}$ (see [10]).

Theorem 3.2.2 *$\mathscr{H}$ has dense range and $\mathscr{G}$ is injective and both F and $\mathscr{G}$ are compact. Suppose that $\mathscr{H}$ is also injective, then F is injective and both F and $\mathscr{G}$ has dense range.*

It is shown in [10] that provided the injectivity and denseness properties of the operators $F, \mathscr{G}$ and $\mathscr{H}$ in Theorem 3.2.2 are satisfied, then the LSM works to provide a valid reconstruction. It can be easily seen from the theorem that the key point relies on the injectivity of the Herglotz operator $\mathscr{H}$. Obviously, the non-injectivity of $\mathscr{H}$ amounts to the existence of a nontrivial Herglotz wave function $v_g(x) = \mathscr{H}g(x)$ which satisfies

$$(\Delta + k^2)v_g = 0 \quad \text{in } \mathbf{D}, \quad \mathscr{B}v_g = 0 \quad \text{on } \partial\mathbf{D},$$

that is, $v_g(x)$ is a Laplacian eigenfunction in $\mathbf{D}$. We would like to remark that the same interior eigenvalue problem occurs in using the LSM to solve inverse electromagnetic obstacle scattering problems. In order to overcome this problem, a linear combination of far-field operators corresponding to different polarizations of the incident field is proposed. However, it is then necessary to develop new approximation properties of Herglotz function and electromagnetic Herglotz pairs. Furthermore, the approach used there relies heavily on the polarization of the incident and scattered fields and hence is only applicable to the case of electromagnetic scattering. Next, we are going to present a simple technique to get rid of the interior eigenvalue problem. We present the technique for the inverse acoustic scattering problem. But as we will see in following chapters, the technique works also for the inverse electromagnetic scattering problems.

Suppose that $\mathbf{D}$ is the unknown obstacle to be recovered by the scattering measurement. We artificially introduce a scatterer $\mathbf{B}$ and fix its shape, position and physical property, such that the underlying scatterer to be recovered becomes $\mathbf{D} \cup \mathbf{B}$. For practical consideration, we choose $\mathbf{B}$ to be a sound-soft ball with radius r and call it *a reference ball*. Furthermore, we let $\mathbf{B}$ be suitably chosen such that k^2 is not a Dirichlet eigenvalue for $-\Delta$ in $\mathbf{B}$. In fact, by the Funk-Heck formula (see [26]), we know that if k^2 is a Dirichlet eigenvalue for $\mathbf{B}$ then $f(kr) = 0$, where f represents the Bessel functions J_n in $\mathbb{R}^2$ and spherical Bessel functions j_n in $\mathbb{R}^3$, with the order n being a non-negative integer. Hence, the requirement that k^2 is not a Dirichlet eigenvalue for $\mathbf{B}$ can be easily fulfilled.

To reconstruct the unknown scatterer $\mathbf{D}$ by the LSM, with the artificially added known obstacle $\mathbf{B}$, one needs to solve the following far-field operator equation in place of (3.2.3):

$$\int_{\mathbb{S}^{n-1}} u_\infty(\hat{x}, d; \mathbf{B} \cup \mathbf{D}) g^z(d)\, ds(d) = \Phi_\infty(\hat{x}, z), \quad \hat{x} \in \mathbb{S}^{n-1}, z \in \Omega, \tag{3.2.6}$$

where Ω is the sampling region that does not contain $\mathbf{B}$. Since we are only interested in recovering the unknown scatterer $\mathbf{D}$ and know all the information about $\mathbf{B}$ in advance, the reference ball is excluded in the sampling region Ω to reduce the computational costs. Hence no extra computational accounts are introduced for our modified LSM, which will be referred in the sequel to as the *strengthened LSM*. We note that the integral kernel $u_\infty(\hat{x}, d; \mathbf{D})$, the far-field pattern produced by $\mathbf{D}$ in the original LSM in (3.2.3) is replaced by $u_\infty(\hat{x}, d; \mathbf{D} \cup \mathbf{B})$, the far-field pattern produced by $\mathbf{D}$ and $\mathbf{B}$. In doing this, the interior eigenvalue problem for the strengthened LSM based on (3.2.6) is automatically eliminated, disregarding the unknown scatterer $\mathbf{D}$, as concluded from the following lemma.

Theorem 3.2.3 *The strengthened LSM does not have the interior eigenvalue problem.*

Proof By Theorem 3.2.2, we only need to show the injectivity of the operator $\mathscr{H}$. It is verified directly that if

$$\mathscr{H} g(x) = 0 \,, \quad x \in \partial \mathbf{D} \cup \partial \mathbf{B},$$

then the Herglotz wave function $v_g(x) = \mathscr{H} g(x)$ is a Dirichlet eigenfunction in $\mathbf{B}$, which implies $v_g(x) = 0$ for $x \in \mathbf{B}$ since k^2 is not a Dirichlet eigenvalue. So we have $v_g(x) = 0$ for $x \in \mathbb{R}^n$ by analytic continuation, which further implies $g(x)$ is identically zero by Theorem 3.15 in [26]. That is, the Herglotz operator $\mathscr{H}$ is injective. □

We would like to remark that the injectivity of $\mathscr{H}$ is obviously irrelevant to $\mathbf{D}$. This is very important since one can eliminate the interior eigenvalue problem by using the strengthened LSM without knowing any a priori information on the unknown scatterer $\mathbf{D}$.

Concerning the interior eigenvalue problem, we further have an important and interesting observation. If one let the reference ball be infinitesimal such that the scattering effect due to $\mathbf{B}$ is minuscular, that is, for a fixed incident wave $u^i = \exp\{ikx \cdot d\}$,

$$\|u_\infty(\hat{x}; \mathbf{B} \cup \mathbf{D}) - u_\infty(\hat{x}; \mathbf{D})\|_{L^2(\mathbb{S}^{n-1})} < \delta(r), \tag{3.2.7}$$

where r is the radius of $\mathbf{B}$ and $\delta(r) \to 0$ as $r \to 0$. We will give a rigorous verification to the above estimate (3.2.7) later (see Theorem 3.2.4), but we now apply it to show an interesting observation that interior eigenvalue problem for the original LSM (3.2.3) only exists theoretically but not practically. Suppose that k is given such that the LSM (3.2.3) has the interior eigenvalue problem. Since the integral kernel $u_\infty(\hat{x}; \mathbf{D})$ is provided by the measurement data and hence the noise is inevitable. That is, in practice, the integral kernel should take the form $u_\infty(\hat{x}; \mathbf{D}) + \varepsilon$ with small $\|\varepsilon\|_{L^2(\mathbb{S}^{n-1})} > 0$ representing the noise level. By (3.2.7), we know that there exists a ball $\mathbf{B}$ with sufficiently small radius $r > 0$ such that

$$u_\infty(\hat{x}; \mathbf{D}) + \varepsilon = u_\infty(\hat{x}; \mathbf{D} \cup \mathbf{B}) + \mathscr{O}(\varepsilon). \tag{3.2.8}$$

Hence, using the original LSM with (3.2.3) to recover $\mathbf{D}$ when noise is present amounts to applying the strengthened LSM (3.2.6) with a fictitious ball $\mathbf{B}$ and the same level noise to recover $\mathbf{D}$. It is well-known that for the fixed k, k^2 is not a Dirichlet eigenvalue for $-\Delta$ in $\mathbf{B}$ when $\mathbf{B}$ is sufficiently small (see, e.g. [66]). From our earlier discussion, the strengthened LSM would not encounter the interior eigenvalue problem, neither would the original LSM do when noise presents. In conclusion, we have shown by a heuristic argument that

Proposition 3.2.1 *The interior eigenvalue problem for the LSM exists only theoretically but not practically when noise presents.*

As can be seen from Sect. 3.2.2, our numerical experiments are in full consistency with the assertion in Proposition 3.2.1. In fact, our numerical results have shown that when there is interior eigenvalue problem and no noise is attached to the far-field data, the LSM does fail to work; whereas if noise is presented to the far-field data, the LSM would work normally and still gives a valid reconstruction. Clearly, a key ingredient to our above study is the estimate given in (3.2.7). We next gives a rigorous theoretical justification to (3.2.7), where to ease our study we only consider the three dimensional problem and assume that $\mathbf{D}$ is sound-soft with C^2-continuous boundary $\partial\mathbf{D}$. We would make essential use of the single-and double-layer potential operators S and K, defined respectively by

$$(S\varphi)(x) := 2\int_{\partial\mathbf{D}\cup\partial\mathbf{B}} \Phi(x,y)\varphi(y)\,ds(y), \quad x\in\partial\mathbf{D}\cup\partial\mathbf{B}, \tag{3.2.9}$$

$$(K\varphi)(x) := 2\int_{\partial\mathbf{D}\cup\partial\mathbf{B}} \frac{\partial\Phi(x,y)}{\partial\nu(y)}\varphi(y)\,ds(y), \quad x\in\partial\mathbf{D}\cup\partial\mathbf{B}, \tag{3.2.10}$$

where $\varphi\in C(\partial\mathbf{D}\cup\partial\mathbf{B})$ and $\Phi(x,y)=e^{ik|x-y|}/|x-y|$ is the fundamental solution to the Helmholtz equation. We refer to [26] for related mapping properties of these two operators.

Theorem 3.2.4 *Let $\mathbf{B}$ be a reference ball with sufficiently small radius $r>0$ such that $dist(\mathbf{B},\mathbf{D})>c_0>0$. Then, we have*

$$\mathscr{F}(\partial\mathbf{D}\cup\partial\mathbf{B}) = \mathscr{F}(\partial\mathbf{D}) + \mathscr{O}(r). \tag{3.2.11}$$

Proof Clearly, we only need to show that for a fixed $u^i=\exp\{ikx\cdot d\}$,

$$u_\infty(\mathbf{D}\cup\mathbf{B}) = u_\infty(\mathbf{D}) + \mathscr{O}(r). \tag{3.2.12}$$

We know that $u^s\in C^2(\mathbb{R}^3\backslash\overline{\mathbf{D}\cup\mathbf{B}})\cap C(\mathbb{R}^3\backslash(\mathbf{D}\cup\mathbf{B}))$ and can be represented in the form (see [26])

$$u^s(x;\mathbf{D}\cup\mathbf{B}) = \int_{\partial\mathbf{D}\cup\partial\mathbf{B}} \left\{\frac{\partial\Phi(x,y)}{\partial\nu(y)} - \mathrm{i}\Phi(x,y)\right\}\varphi(y)\,ds(y) \tag{3.2.13}$$

with a density $\varphi \in C(\partial\mathbf{D} \cup \partial\mathbf{B})$ satisfying

$$\varphi + K\varphi - \mathrm{i}S\varphi = f, \quad f(x) = -2u^i(x), \ x \in \partial\mathbf{D} \cup \partial\mathbf{B}. \tag{3.2.14}$$

In the following, we set

$$\varphi_1(x) := \varphi(x), \ x \in \partial\mathbf{D}; \qquad \varphi_2(x) := \varphi(x), \ x \in \partial\mathbf{B},$$

and

$$(S_1\varphi_1)(x) = \int_{\partial\mathbf{D}} \Phi(x,y)\varphi_1(y)ds(y), \quad (K_1\varphi_1)(x) = \int_{\partial\mathbf{D}} \frac{\partial\Phi(x,y)}{\partial\nu(y)}\varphi_1(y)ds(y)$$
$$(S_2\varphi_2)(x) = \int_{\partial\mathbf{B}} \Phi(x,y)\varphi_2(y)ds(y), \quad (K_2\varphi_2)(x) = \int_{\partial\mathbf{B}} \frac{\partial\Phi(x,y)}{\partial\nu(y)}\varphi_2(y)ds(y).$$

Then, Eq. (3.2.14) can be reformulated as

$$[\varphi_1 + K_1\varphi_1 - \mathrm{i}S_1\varphi_1 + K_2\varphi_2 - \mathrm{i}S_2\varphi_2](x) = f(x), \quad x \in \partial\mathbf{D} \tag{3.2.15}$$
$$[\varphi_2 + K_2\varphi_2 - \mathrm{i}S_2\varphi_2 + K_1\varphi_1 - \mathrm{i}S_1\varphi_1](x) = f(x), \quad x \in \partial\mathbf{B} \tag{3.2.16}$$

Since $dist(\partial\mathbf{D}, \partial\mathbf{B}) > c_0 > 0$, one can show directly that

$$\|K_2 - \mathrm{i}S_2\|_{C(\partial\mathbf{B})\to C(\partial\mathbf{D})} = \mathscr{O}(r), \qquad \|K_1 - \mathrm{i}S_1\|_{C(\partial\mathbf{D})\to C(\partial\mathbf{B})} = \mathscr{O}(1). \tag{3.2.17}$$

Next, for $x \in \mathbb{S}^2$, we define

$$(K_0\varphi)(x) = \int_{\mathbb{S}^2} \frac{\partial\Phi_0(x,y)}{\partial\nu(y)}\varphi(y)\,ds(y), \quad (S_0\varphi)(x) = \int_{\mathbb{S}^2} \Phi_0(x,y)\varphi(y)\,ds(y), \tag{3.2.18}$$

where $\Phi_0(x,y) = 1/|x-y|$ and $\varphi \in C(\mathbb{S}^2)$. By coordinate and using the fact the r is infinitesimal, we can prove by straightforward calculations that

$$\|\frac{1}{r^2}S_1 - S_0\|_{C(r\mathbb{S}^2)\to C(\mathbb{S}^2)} = \mathscr{O}(r), \quad \|K_2 - K_0\|_{C(r\mathbb{S}^2)\to C(\mathbb{S}^2)} = \mathscr{O}(r). \tag{3.2.19}$$

Next, we set

$$f_1(x) := f(x), \ x \in \partial\mathbf{D}; \qquad f_2(x) := f(x), \ x \in \partial\mathbf{B}.$$

By changing to polar coordinate and using the results in (3.2.17) and (3.2.19), we have from (3.2.16) that

$$\varphi_2(r\hat{x}) = [I + K_0 + \mathscr{O}(r)]^{-1}[f_2(r\hat{x}) - (K_1 - \mathrm{i}S_1)\varphi_1(r\hat{x})], \quad \hat{x} \in \mathbb{S}^2, \tag{3.2.20}$$

where we note that $I + K_0 : C(\mathbb{S}^2) \to C(\mathbb{S}^2)$ is continuous and has a bounded inverse Then, by plugging (3.2.20) into (3.2.15) and using the relations in (3.2.17), we further have

$$\varphi_1 = [I + K_1 - \mathrm{i}S_1 + \mathscr{O}(r)]^{-1}(f_1 + \mathscr{O}(r)), \tag{3.2.21}$$

which, by noting $I + K_1 - \mathrm{i}S_1$ is invertible (see [26]), gives

$$\varphi_1 = \tilde{\varphi}_1 + \mathscr{O}(r), \tag{3.2.22}$$

where

$$\tilde{\varphi}_1 = [I + K_1 - \mathrm{i}S_1]^{-1} f_1.$$

Equation (3.2.22) in turn implies by (3.2.20) that

$$\varphi_2 = \mathscr{O}(1). \tag{3.2.23}$$

Finally, by (3.2.13) we have

$$\begin{aligned} u_\infty(\hat{x}; \mathbf{D} \cup \mathbf{B}) &= \frac{1}{4\pi} \int_{\partial\mathbf{D}\cup\partial\mathbf{B}} \varphi(y) \left\{ \frac{\partial}{\partial\nu(y)} e^{-iky\cdot\hat{x}} - \mathrm{i}e^{-iky\cdot\hat{x}} \right\} ds(y) \\ &= u_\infty(\hat{x}; \mathbf{D}) + \mathscr{O}(r), \qquad \hat{x} \in \mathbb{S}^2, \end{aligned}$$

where we have made use of the estimates in (3.2.22) and (3.2.23) and the fact that

$$u_\infty(\hat{x}; \mathbf{D}) = \frac{1}{4\pi} \int_{\partial\mathbf{D}} \tilde{\varphi}_1(y) \left\{ \frac{\partial}{\partial\nu(y)} e^{-iky\cdot\hat{x}} - \mathrm{i}e^{-iky\cdot\hat{x}} \right\} ds(y).$$

The proof is completed. □

The estimate (3.2.11) can be straightforwardly extended to hold in two dimensions. Moreover, by using the mapping properties of single- and double-layer potential operators in [71], one can show similar results to Theorem 3.2.4 when the scatterer has only Lipschitz continuous boundary $\partial\mathbf{D}$, as well as the scatterer is associated with more general boundary condition.

Next, we are going to present a more essential feature of the reference ball technique, i.e. it provides an effective deterministic approach to select the desired cut-off value for the LSM. In fact, by introducing the reference ball, the scattering due to the unknown obstacle **D** and the reference ball **B** are interrelated and interacted on each other, it is natural to expect that the scattering information on **D** is partially revealed by the scattering information of **B**. Therefore, when implementing the strengthened LSM with (3.2.6), one can first calculate the indicator function near the known boundary $\partial\mathbf{B}$, whose behavior can be used as a reference for the selection of the cut-off value in the strengthened LSM to identify the unknown boundary $\partial\mathbf{D}$. This motivates the following strengthened LSM.

Algorithm of strengthened LSM

1. Introduce a sound-soft reference ball **B** to the scattering system, and collect the far-field data.
2. Select a mesh $\mathcal{T}_h$ of sampling points in a region Ω which contains **D** but excludes **B**; also select a small sampling mesh $\mathcal{T}_h'$ around $\partial\mathbf{B}$.
3. Use the Tikhonov regularization and the Morozov discrepancy principle to compute an approximate solution g^z to the far-field equation (3.2.6) for each mesh point z of $\mathcal{T}_h \cup \mathcal{T}_h'$.
4. Select an appropriate cut-off value $c := \log \|g^z\|_{L^2(\mathbb{S}^{n-1})}$ from the mesh points $z \in \mathcal{T}_h'$.
5. Count $z \in \mathbf{D}$ if $\log \|g^z\|_{L^2(\mathbb{S}^{n-1})} \le c$; and $z \notin \mathbf{D}$ if $\log \|g^z\|_{L^2(\mathbb{S}^{n-1})} > c$.

We observe that in order for the reference ball to give a reasonable cut-off value, the scattering interaction between the unknown obstacle **D** and the reference ball **B** should not be too weak. Hence by Theorem 3.2.4, the reference ball should not be very small in size compared to that of the unknown obstacle. Moreover, we have the following results which indicate that the reference ball should not be too far away from the unknown obstacle in order for the ball **B** to yield some notable effect on the scattering by **D**. To ease our study, we still consider the scatterer **D** to be sound-soft with a C^2-continuous boundary $\partial\mathbf{D}$.

Theorem 3.2.5 *Let* **B** *be a fixed reference ball such that* $dist(\mathbf{B}, \mathbf{D}) > \rho > 0$ *with* ρ *sufficiently large, then we have*

$$\mathcal{F}(\partial\mathbf{D} \cup \partial\mathbf{B}) = \mathcal{F}(\partial\mathbf{D}) + \mathcal{F}(\partial\mathbf{B}) + \mathcal{O}(\frac{1}{\rho}). \tag{3.2.24}$$

Proof It suffices to prove that for a fixed $u^i = \exp\{ikx \cdot d\}$,

$$u_\infty(\mathbf{D} \cup \mathbf{B}) = u_\infty(\mathbf{D}) + u_\infty(\mathbf{B}) + \mathcal{O}(\frac{1}{\rho}). \tag{3.2.25}$$

Below we shall adopt the same notation as that in the proof of Theorem 3.2.4. Still, we express $u^s(\mathbf{D} \cup \mathbf{B})$ in the form (3.2.13) with $\varphi \in C(\partial\mathbf{D} \cup \partial\mathbf{B})$ satisfying (3.2.14). Since $dist(\partial\mathbf{D}, \partial\mathbf{B}) > \rho$ with ρ sufficiently large, it is straightforward to verify that

$$\|K_2 - \mathrm{i}S_2\|_{C(\partial\mathbf{B}) \to C(\partial\mathbf{D})} = \mathcal{O}(\frac{1}{\rho}), \qquad \|K_1 - \mathrm{i}S_1\|_{C(\partial\mathbf{D}) \to C(\partial\mathbf{B})} = \mathcal{O}(\frac{1}{\rho}). \tag{3.2.26}$$

Then by a similar argument to that in the proof of Theorem 3.2.4, we derive

$$\varphi_1 = \tilde{\varphi}_1 + \mathcal{O}(\frac{1}{\rho}), \qquad \varphi_2 = \tilde{\varphi}_2 + \mathcal{O}(\frac{1}{\rho}), \tag{3.2.27}$$

where $\tilde{\varphi}_1$ and $\tilde{\varphi}_2$ are given by

$$\tilde{\varphi}_1 = [I + K_1 - \mathrm{i}S_1]^{-1} f_1, \qquad \tilde{\varphi}_2 = [I + K_2 - \mathrm{i}S_2]^{-1} f_2.$$

Therefore we have

$$\begin{aligned} u_\infty(\hat{x}; \mathbf{D} \cup \mathbf{B}) &= \frac{1}{4\pi}\Bigg[\int_{\partial\mathbf{D}} \varphi_1(y)\Big\{\frac{\partial}{\partial\nu(y)} e^{-iky\cdot\hat{x}} - \mathrm{i}e^{-iky\cdot\hat{x}}\Big\}\, ds(y) \\ &\quad + \int_{\partial\mathbf{B}} \varphi_2(y)\Big\{\frac{\partial}{\partial\nu(y)} e^{-iky\cdot\hat{x}} - \mathrm{i}e^{-iky\cdot\hat{x}}\Big\}\, ds(y)\Bigg] \\ &= u_\infty(\hat{x}; \mathbf{D}) + u_\infty(\hat{x}; \mathbf{B}) + \mathscr{O}(\frac{1}{\rho}), \qquad \forall \hat{x} \in \mathbb{S}^2, \end{aligned}$$

where we have made use of the estimates in (3.2.27) and the fact that

$$\begin{aligned} u_\infty(\hat{x}; \mathbf{D}) &= \frac{1}{4\pi}\int_{\partial\mathbf{D}} \tilde{\varphi}_1(y)\Big\{\frac{\partial}{\partial\nu(y)} e^{-iky\cdot\hat{x}} - \mathrm{i}e^{-iky\cdot\hat{x}}\Big\}\, ds(y), \\ u_\infty(\hat{x}; \mathbf{B}) &= \frac{1}{4\pi}\int_{\partial\mathbf{B}} \tilde{\varphi}_2(y)\Big\{\frac{\partial}{\partial\nu(y)} e^{-iky\cdot\hat{x}} - \mathrm{i}e^{-iky\cdot\hat{x}}\Big\}\, ds(y), \end{aligned}$$

This completes the proof of Theorem 3.2.5. □

Clearly, Theorem 3.2.5 asserts that as the distance between **B** and **D** increases, the scattering coupling between **B** and **D** becomes weaker. Therefore, the reference ball should not be in very far distance from the unknown obstacle in order to yield a good cut-off value. However, we have observed from lots of numerical experiments (see next section) that the cut-off value of the reference ball is still reasonably good for the unknown scatterers even for moderately long distances. We would also like to point out that in some extreme circumstances, although the correlation of the reference ball and the unknown scatterer decreases as the scattering coupling turns weaker, possibly due to the large size ratio and/or far distance between the scatterers, the mismatch between their cut-off values digresses slowly. In this sense, the reference ball technique can still provide an approximate reference range for the selection of cut-off value.

It is also worth noting that one may take other objects with known geometry and physical property to be the reference obstacle in our technique, not necessarily to be the ball as we are considering in this section.

3.2.2 Numerical Experiments and Discussion

In this subsection, we present some numerical tests to illustrate the applicability and effectiveness of the technique on the choice of the cut-off value aided with the reference ball, namely the strengthened LSM discussed in the previous section. All the programs in our experiments are written in Matlab and run on a Pentium 3GHz PC.

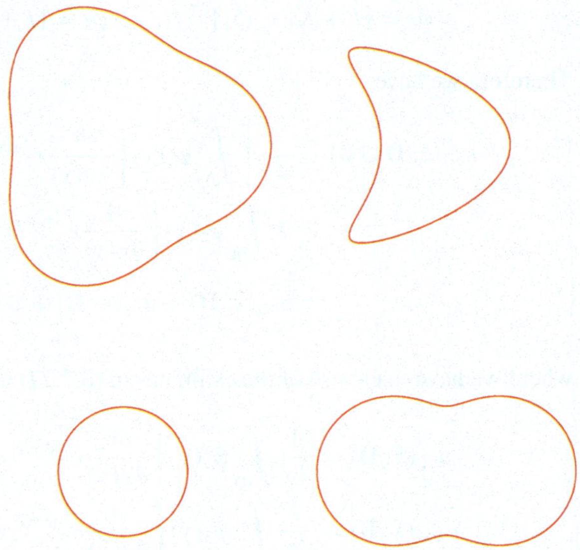

Fig. 3.1 Scatterers under consideration. Top left: pear-shaped scatterer $\mathbf{D_p}$; Top right: kite-shaped scatterer $\mathbf{D_k}$; Bottom left: reference ball scatterer $\mathbf{B}$; Bottom right: peanut-shaped scatterer $\mathbf{D_t}$

The scatterers under concern in system (2.3.3) include a unit reference ball of radius 1 centered at the origin, a pear-shaped, a kite-shaped, and a peanut-shaped one, which are denoted by $\mathbf{B}$, $\mathbf{D_p}$, $\mathbf{D_k}$, $\mathbf{D_t}$, respectively, or a finite combination of the previous four shapes possibly at different locations; see Fig. 3.1 for the exact shapes of the obstacles. These scatterers can be parameterized as follows:

$$\texttt{Ball} \qquad x(t) = (\cos t, \sin t), \quad 0 \le t \le 2\pi, \tag{3.2.28}$$

$$\texttt{Pear} \qquad x(t) = (2 + 0.3\cos 3t)(\cos t, \sin t), \quad 0 \le t \le 2\pi, \tag{3.2.29}$$

$$\texttt{Kite} \qquad x(t) = (\cos t + 0.65\cos 2t - 0.65, 1.5\sin t), \quad 0 \le t \le 2\pi, \tag{3.2.30}$$

$$\texttt{Peanut} \qquad x(t) = \sqrt{3\cos^2 t + 1}(\cos t, \sin t), \quad 0 \le t \le 2\pi. \tag{3.2.31}$$

There are totally four groups of numerical tests to be considered, and they are respectively referred to as **BK**, **BPT**, **Neu** and **Eig**. The synthetic far-field data of the direct problems are generated by solving the combined-layer potential operator equation with the Nyström's method (see Sect. 3.5, Chap. 3 in [26, 45]), which is exponentially convergent for analytic boundaries. We compute the far-field patterns at 64 equidistantly distributed observation points $(\cos t_j, \sin t_j)$, $t_j = 2j\pi/64$, $j = 0, 1, \ldots, 63$, corresponding to 64 equidistantly distributed incident directions $(\cos\tau_j, \sin\tau_j)$, $\tau_j = 2j\pi/64$, $j = 0, 1, \ldots, 63$, around the unit circle. The far-field patterns are generated by solving the linear Fredholm integral equation of first kind over a family of increasingly finer meshes along the boundary of the scatterer until the relative error is very small, e.g. less than 10^{-3}, compared with the noise level we added. It is noted that we can achieve at least six significant digits by using 64 equi-distant nodal points of the parametric form of each scatterer component in our

Table 3.1 Experimental parameters for the experiments

	BK	BPT	Neu	Eig
Noise level δ	0.05	0.05	0.02	0–0.02
Incident wave number k	3	3	1	10.173
No. of incident directions	64			
No. of observation directions	64			

numerical tests. Except for the Eig test, these far-field patterns are accurate enough to be viewed as noise-free. Such far-field patterns generated are then subjected pointwise to certain uniform random noise. The uniform random noise in magnitude as well as in direction is added according to the following formula,

$$u_\infty = u_\infty + \delta r_1 |u_\infty| \exp(\mathrm{i}\pi r_2)$$

where r_1 and r_2 are two uniform random numbers, both ranging from –1 to 1, and δ represents the noise level. For each mesh point z, the corresponding far-field equation (3.2.3) or (3.2.6) is solved by using the Tikhonov regularization method, with the corresponding regularization parameters determined by the Morozov discrepancy principle [77].

All the parameters chosen for the experiments are listed in Table 3.1. And for the purpose of comparisons, we always plot the exact boundaries of the scatterers using red smooth curves of double thinkness in all the figures of this section.

It is noted that for some experiments, we have taken two cut-off values $V_{cut,1}$ and $V_{cut,2}$, instead of only one cut-off value V_{cut}. For practical purposes, it is better to take a range of cut-off values, i.e. $[V_{cut,1}, V_{cut,2}]$, which enables us to get a better buffer region to locate the boundary of the underlying object instead of one single geometric boundary.

3.2.2.1 Example BK

In Example **BK**, the unknown scatterer in system (2.3.3) are chosen to be a kite-shaped object $\mathbf{D_k}$ after positive 10-unit displacement in both longitude and latitude directions. Both the kite and the reference ball are of sound-soft kind.

First, at the readers' disposal, we solve the far-field equation (3.2.3) on a rather coarse sampling mesh over the whole region $(-3, 13)^2$ containing both $\mathbf{B}$ and $\mathbf{D_k}$ (see Fig. 3.2) to find g^z with z being a sampling grid point in order to have a global view of the behavior of this g^z.

Hereafter, we remind that the value at the sampling point indicates the logarithm of the L^2-norm over the unit circle $\mathbb{S}^1$ of the indicator function g^z, i.e. $\log \|g^z\|_{L^2(\mathbb{S}^1)}$, and that the sampling mesh over the whole region is only for the auxiliary purpose for visualization, which can be avoided in practice. As we mentioned in the remark

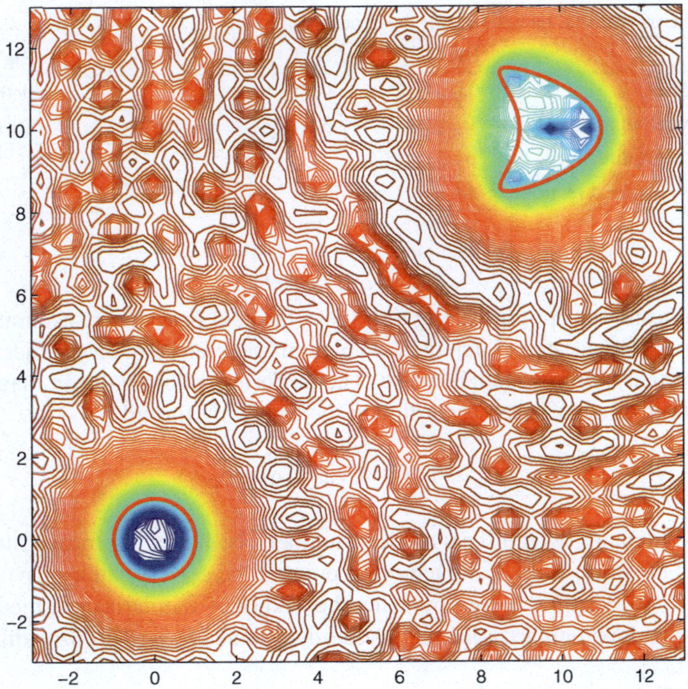

Fig. 3.2 Auxiliary global view of the indicator data $\log \|g^z\|_{L^2(\mathbb{S}^1)}$ in Example BK with a reference ball and a kite displaced at (10, 10)

of Sect. 3.2, there is always no need to sample the overall region. We only have to focus on the small subregions where the unknown scatterers may lie in.

For this example, we merely sample a small region $(-1.5, 1.5)^2$ around the reference ball, label the characteristic contour curves with sufficient resolution as in Fig. 3.3a, and read out the cut-off values V_{cut}'s, i.e. -0.51328 and -0.20004, of the isolines which best approximate the reference ball from Fig. 3.3b. In the sequel, the small sampling region are always covered by a 65×65 uniform mesh if not claimed explicitly. Then, the rest of the section is to sample another small region $(7.5, 12.5)^2$ which potentially contains the unknown scatterer, and then to display the isolines of the indicator data set $\log \|g^z\|_{L^2(\mathbb{S}^1)}$ with the same isoline values as that of the reference ball as shown in Fig. 3.3c, where the shape of the unknown obstacle can be quickly identified as the isoline with the cut-off value $V_{cut} = -0.20004$ from Fig. 3.3d, which is a reasonable reconstruction with the observed far-field data with five percent noise except some deterioration in the non-convex part of the kite. It is worth remarking that the other isoline with $V_{cut} = -0.51328$ can be dropped safely due to its irregular nature of no geometric meaning. This rule of thumb will apply to all the rests of our experiments, i.e. we identify a more regular curve as our boundary profile of the unknown obstacle in the buffer region.

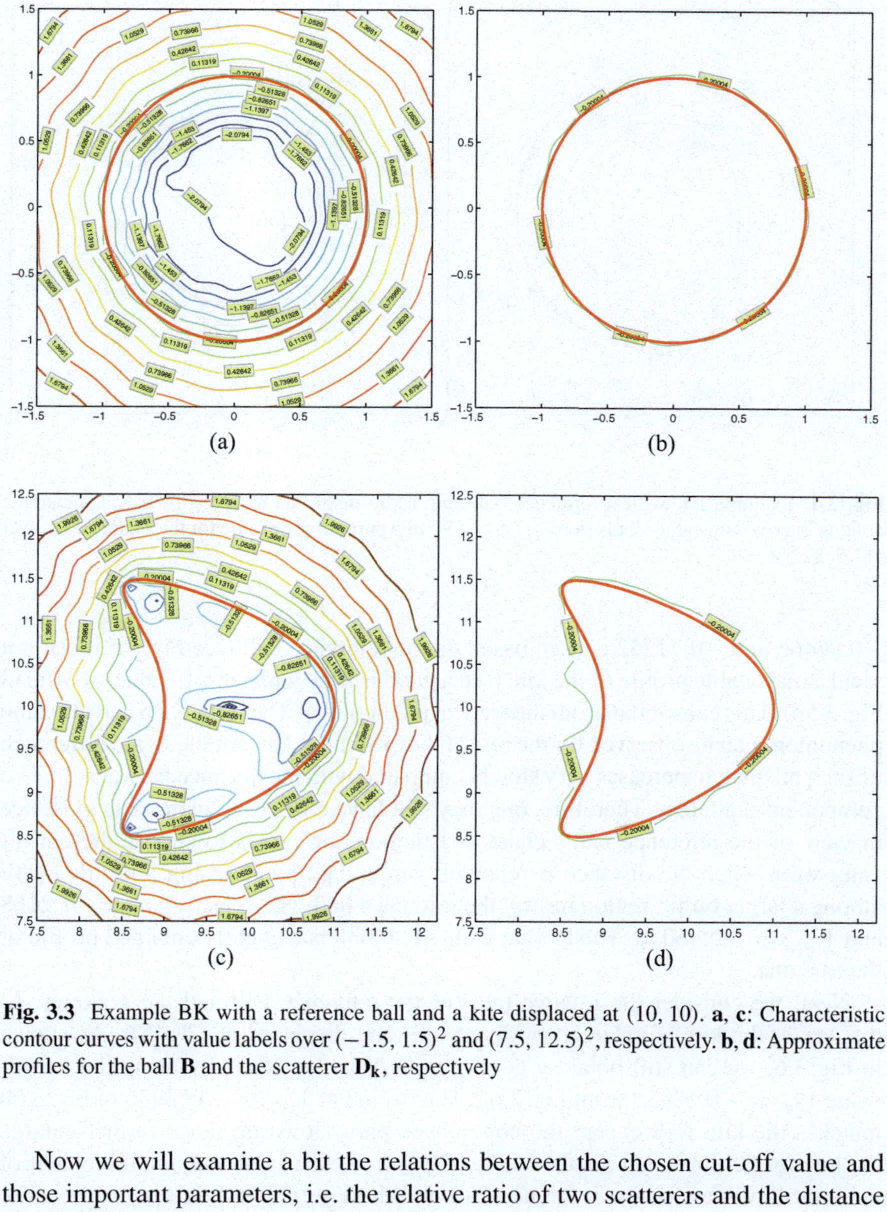

Fig. 3.3 Example BK with a reference ball and a kite displaced at (10, 10). **a**, **c**: Characteristic contour curves with value labels over $(-1.5, 1.5)^2$ and $(7.5, 12.5)^2$, respectively. **b**, **d**: Approximate profiles for the ball **B** and the scatterer $\mathbf{D_k}$, respectively

Now we will examine a bit the relations between the chosen cut-off value and those important parameters, i.e. the relative ratio of two scatterers and the distance between the reference ball and the unknown scatterer.

Let us move the kite away from the reference ball gradually. As shown in in Fig. 3.4, one can still identify a good profile of the scatterer in Fig. 3.4b by choosing the cut-off value $V_{cut} = -0.084982$ from Fig. 3.4a. However, as the distance increases up to 100, which is about fifty times larger than the wavelength $2\pi/3$, we see from Fig. 3.5 that the cut-off values of the ball and the kite do not match each other any more along their boundaries. More precisely, the cut-off value between

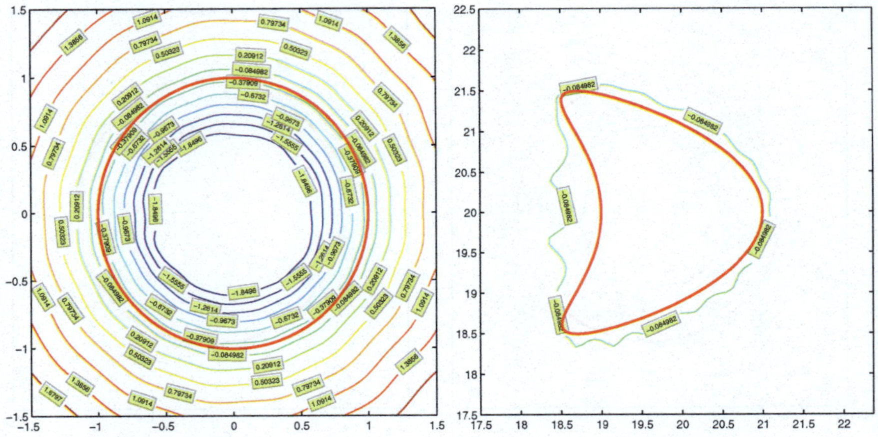

(a) (b)

Fig. 3.4 Example BK with a reference ball and a kite displaced at (20, 20). **a**: Characteristic contour curves with value labels over $(-1.5, 1.5)^2$; **b**: Approximate profile for the scatterer $\mathbf{D_k}$ over $(17.5, 22.5)^2$

−0.96466 and −0.71252 chosen based on the reference ball (see Fig. 3.5a) can not yield a reasonable profile of the kite (see a good isoline with cut-off value −0.46038 Fig. 3.5b). This consolidates our theoretical prediction of Theorem 3.2.5 and the same phenomena can be observed for the rest of the examples. Nevertheless, the difference of such mismatch increases very slowly compared with the distance between the two component scatterers. Therefore, one may still have chances to gain some guidance in view of the reference ball's characteristic contour curves to choose the cut-off value even when the distance is relatively large, e.g., by choosing a profile curve among a larger buffer region nearby the reference ball, viz, select $V_{cut,1} = -1.2168$ and $V_{cut,2} = -0.46038$. The isoline with −0.46038 is a slightly enlarged profile of the true one.

Next, we consider the relative ratio of the reference ball and the scatterer. At this stage, the kite is enlarged by five times and displaced at (20, 20). As shown in Fig. 3.6, we can still obtain a good profile of the scatterer by using the cut-off value $V_{cut} = -0.50936$ from Fig. 3.6a. The isoline of $V_{cut} = -0.50936$ in Fig. 3.6b matches the kite well except the non-convex part. However, if we further enlarge the kite by 10 times and displace it at (30, 30), we see again the non-alignment of the cut-off values for the reference ball and the kite along their boundaries. The cut-off value between −0.34582 and −0.11521 chosen based on the reference ball (see Fig. 3.7a) can not yield a reasonable profile of the kite (see a good isoline with cut-off value 0.34603 Fig. 3.7b). The reason for these phenomena can be explained by the weak coupling between the two scatterers due to the large size-ratio of them. This observation confirms our assertion of Theorem 3.2.4. Once again we see that the mismatch between the two cut-off values for the reference ball and the kite is insignificant even though their size ratio reaches nearly 15 times. This signifies a

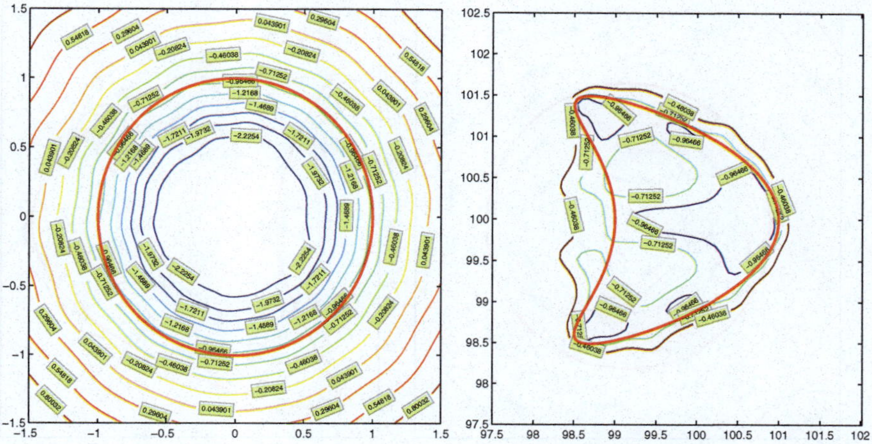

(a) (b)

Fig. 3.5 Example BK with a reference ball and a kite displaced at (100, 100). **a**: Characteristic contour curves with value labels over $(-1.5, 1.5)^2$; **b**: Approximate profiles for the scatterer $\mathbf{D_k}$ over $(17.5, 22.5)^2$

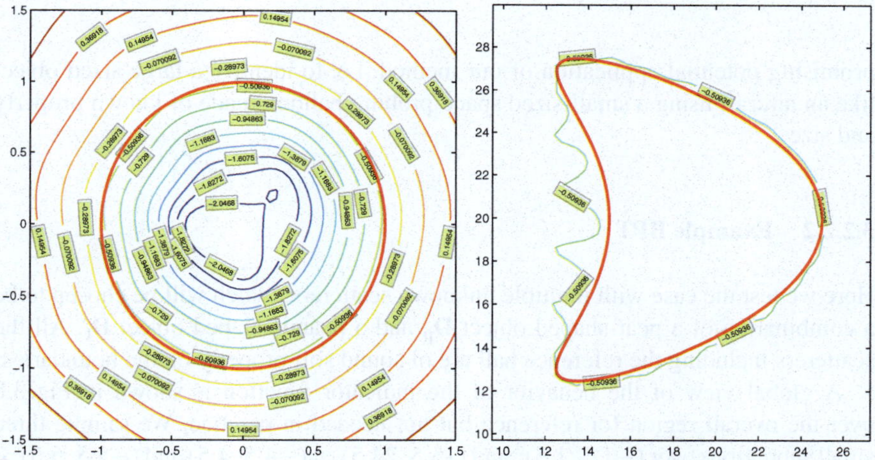

(a) (b)

Fig. 3.6 Example BK with a reference ball and a kite displaced at (20, 20) and enlarged by 5 times. **a**: Characteristic contour curves with value labels over $(-1.5, 1.5)^2$; **b**: Approximate profile for the scatterer $\mathbf{D_k}$ over $(17.5, 22.5)^2$

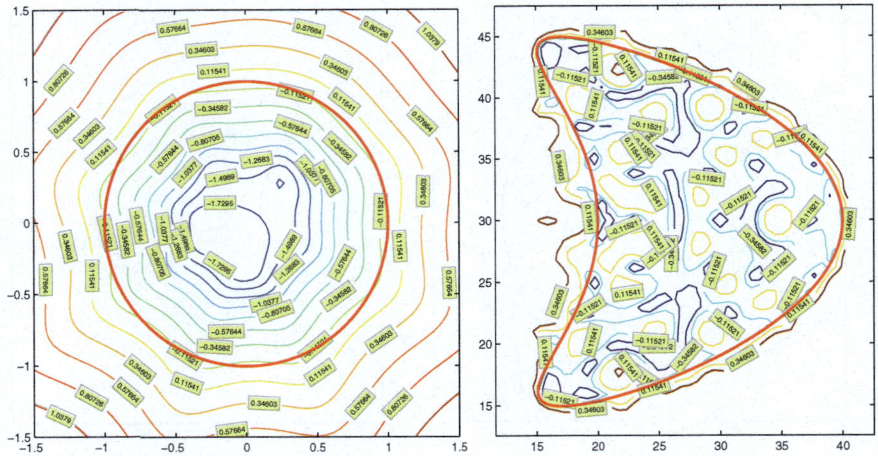

(a) (b)

Fig. 3.7 Example BK with a reference ball and a kite displaced at (30, 30) and enlarged by 10 times. **a**: Characteristic contour curves with value labels over $(-1.5, 1.5)^2$; **b**: Approximate profile for the scatterer $\mathbf{D_k}$ over $(17.5, 22.5)^2$

promising potential application of our method, i.e. to identify a large-sized object like an aircraft using a small-sized space-probing balloon made of known property and size.

3.2.2.2 Example BPT

Here we test the case with multiple unknown scatterers, which will be chosen to be a combination of a pear-shaped object $\mathbf{D_p}$ and a peanut-shaped object $\mathbf{D_t}$. All the scatterers including the reference ball are of sound soft property on the boundaries.

A global view of the behavior of the indicator function is shown in Fig. 3.8 over the overall region for reference but not needed in practice. We sample three small subregions over $(-1.5, 1.5)^2$ and $(15.5, 24.5) \times (-4.5, 4.5)$ and $(-5.5, 5.5) \times (-14.5, 25.5)$, respectively, as plotted in characteristic contour curves in Fig. 3.9a, c and e. As one can tell from Fig. 3.9b, the isoline with the value $V_{cut} = -0.94582$ matches the reference ball quite well and thus $V_{cut} = -0.94582$ is chosen as the promising cut-off value. After checking the isolines with the same groups of contour values, one can be assured that the profiles of the pear-shape and peanut-shape obstacles can be well captured by the isoline of the chosen cut-off value $V_{cut} = -0.94582$ as shown in Fig. 3.9d and f. This example shows that reference ball technique can be carried over from single scatterer detection to that of multiple component scatterers.

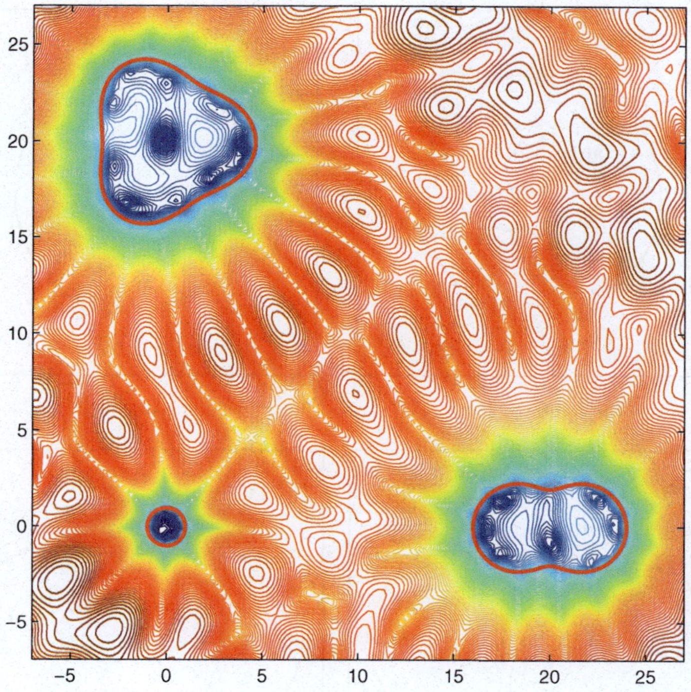

Fig. 3.8 Auxiliary global view of the indicator data $\log \|g^z\|_{L^2(\mathbb{R}^{d-1})}$ in Example BPT with a reference ball, a pear displaced at (0, 20), and and a peanut displaced at (20, 0)

3.2.2.3 Example Neu

We examine the effect of the different physical property on the obstacle boundaries in this example. First, the component scatterers are chosen to be a sound-hard pear-shaped object $\mathbf{D_p}$ and the reference ball will be kept sound-soft. Again we present an auxiliary coarse view of the indicator data over the whole region over $(-3, 13)^2$ in Fig. 3.10 and two fine views over two sampling regions $(-1.5, 1.5)^2$ and $(7.5, 12.5)^2$ as in Fig. 3.11a and c, respectively.

As shown in Fig. 3.11b the isoline with value $V_{cut} = -0.81606$ is a good approximation of the reference ball. With this $V_{cut} = -0.81606$ in hand, an isoline with the same value can be found in Fig. 3.11d, which is quite consistent to the original boundary of the pear.

Next, we reverse the physical properties of the two scatterers. In other words, we employ a sound-hard reference ball to detect the unknown sound-soft scatterer at this time (see Fig. 3.12 for a global view of the indicator data). As shown in Fig. 3.13, a cut-off value $V_{cut} = -0.86808$ chosen from Fig. 3.13b gives a nice profile of the scatterer in Fig. 3.13d.

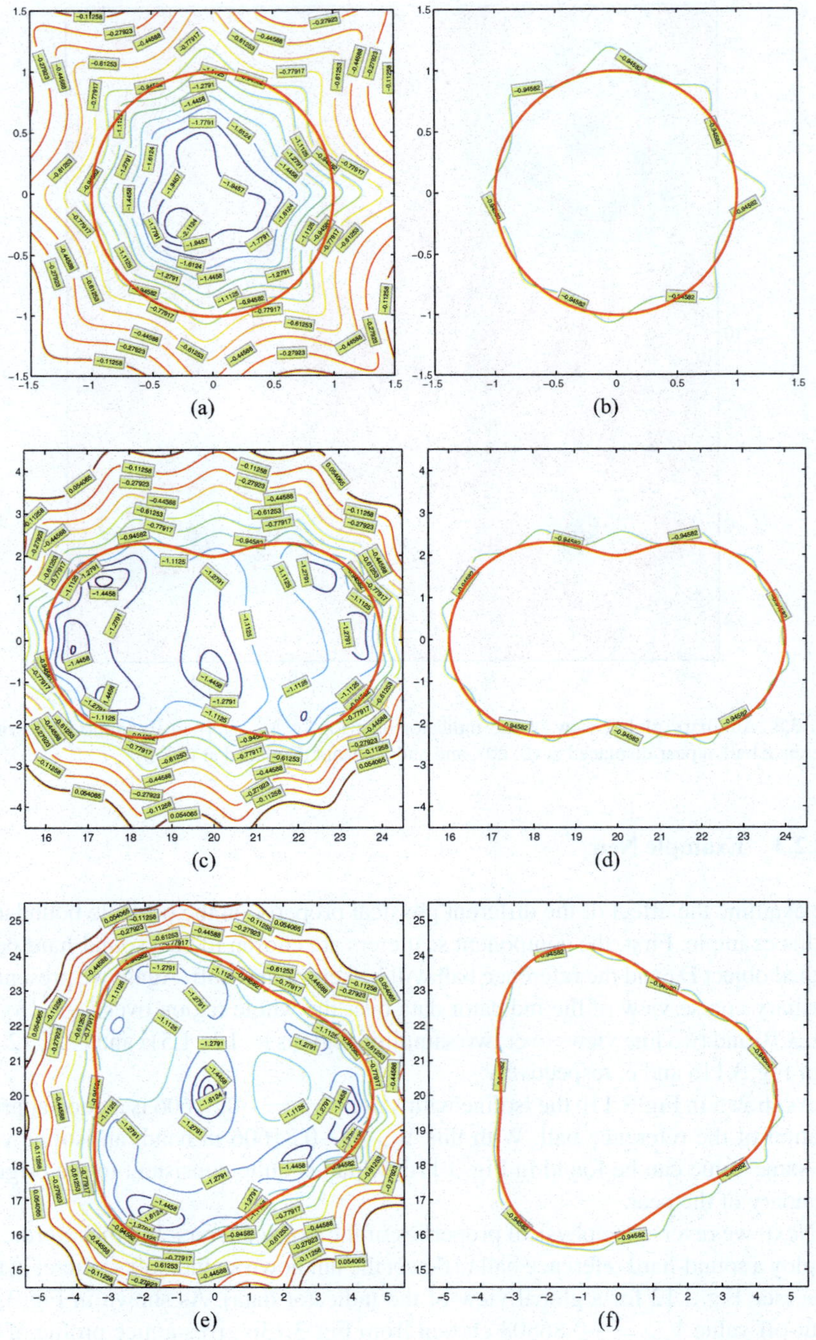

Fig. 3.9 Example BPT with a reference ball, a pear displaced at (0, 20), and and a peanut displaced at (20, 0). **a**, **c**, **e**: Characteristic contour curves with with value labels over $(-1.5, 1.5)^2$ and $(15.5, 24.5) \times (-4.5, 4.5)$ and $(-5.5, 5.5) \times (-14.5, 25.5)$, respectively. **b**, **d**, **f**: Approximate profiles for the ball $\mathbf{B}$ and the scatterer $\mathbf{D_t}$ and $\mathbf{D_p}$, respectively

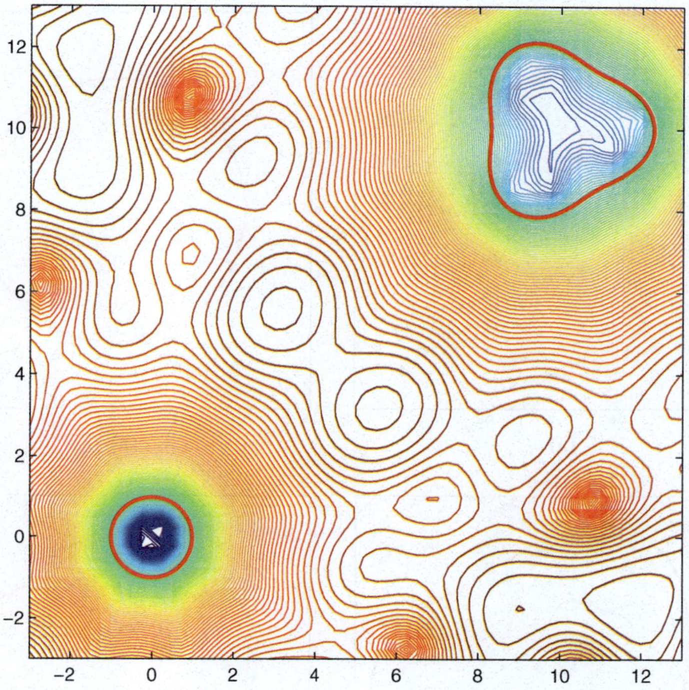

Fig. 3.10 Auxiliary global view of the indicator data $\log \|g^z\|_{L^2(\mathbb{S}^1)}$ in Example Neu with a sound-soft reference ball and a sound-hard pear-shaped scatterer displaced at (10, 10)

3.2.2.4 Example Eig

The aim of this example is to give illustrations to Proposition 3.2.1, namely when there is noise present in the measurement data, the interior eigenvalue problem associated the LSM can be bypassed. Here we deliberately choose the scatterer to be a sound soft unit circle **B** with radius 1 centered at the origin.

By the Funk-Heck formula (see [26]), we know that if k satisfies $J_n(k) = 0$, where $J_n(x)$ is the bessel function of order n, then there is a non-trivial Herglotz function wave function which is a Dirichlet eigenvalue for **B**. Hence the conditions in Theorem 3.2.2 are not satisfied, namely, we would encounter the interior eigenvalue problem when using the LSM to solve the inverse scattering problem. It is well known for the exact zeros of the bessel functions of order n, i.e. $J_n(x)$. We list in Table 3.2 the first three roots $\lambda_{n,m}$ $(m = 1, 2, 3)$ for Bessel functions of the first kind of order n from 0 up to 3 with reference to [73, 78, 80]. Thus we know the exact Dirichlet eigenvalues for **B** are $\lambda_{n,m}^2$ for $n = 0, 1, 2, \ldots$ and $m = 1, 2, \ldots$.

To observe the failure of the interior eigenvalue problem in the LSM, we need the Dirichlet eigenvalue and the far-field pattern to be extremely accurate. Let us take the wavenumber $k = 10.173468135062720$, thus k^2 is a typical Dirichlet eigenvalue for the the ball **B**. First, we solve (3.2.1) for the noise-free case, viz, $\delta = 0$. The

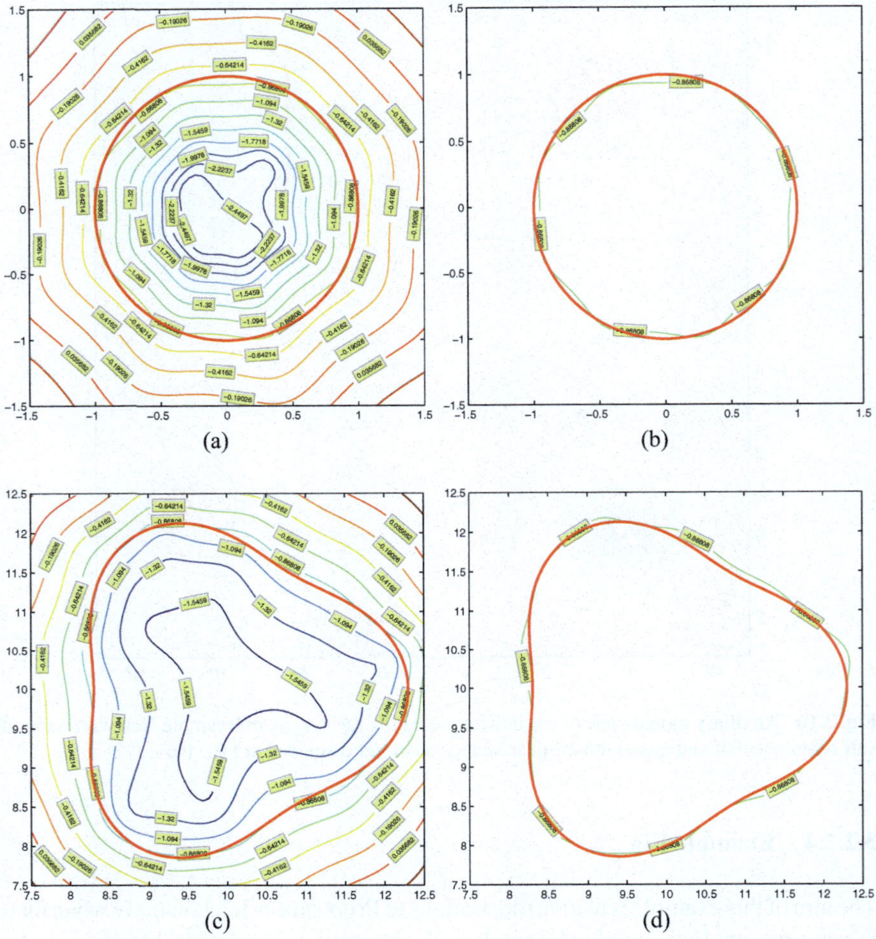

Fig. 3.11 Example Neu with a sound-soft reference ball and a sound-hard pear-shaped scatterer displaced at (10, 10). **a, c**: Characteristic contour curves with with value labels over $(-1.5, 1.5)^2$ and $(7.5, 12.5)^2$, respectively. **b, d**: Approximate profiles for the ball $\mathbf{B}$ and the scatterer $\mathbf{D_p}$, respectively

far-field pattern is computed using 256 nodal points for the discretization of the circle. The computed pattern achieves up to ten significant digits compared with the exact formula given in [66]. In particular when $\delta = 0$, the regularization parameter is taken to be very small, e.g., 10^{-10} since the Morozov principle does not apply for this case. It can be observed that the indicator data $\log \|g^z\|_{L^2(\mathbb{S}^1)}$ exhibits an oscillatory behavior as plotted in Fig. 3.14a. Through further scanning its characteristic contour curves in Fig. 3.14b, one may easily find that those indicator values are extremely small and show ups and downs in numerics. The difference between those indicator function values is so small that one can hardly identify a correct boundary profile. Even if one can choose a correct cut-off value, e.g., $V_{cut} = -5.8523$ or -5.5091, it is

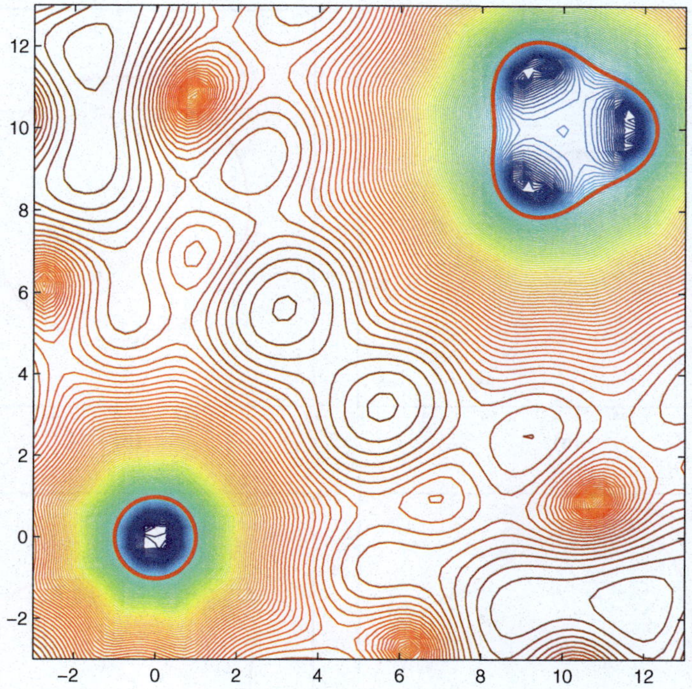

Fig. 3.12 Auxiliary global view of the indicator data $\log \|g^z\|_{L^2(\mathbb{S}^1)}$ in Example Neu with a sound-hard reference ball and a sound-soft pear-shaped scatterer displaced at (10, 10)

still impossible for one to guess the correct profile among several concentric contour curves. That is, the LSM indeed fails when the interior eigenvalue problem occurs.

Next with the same wave number k, we add 0.1% noise or $\delta = 0.001$. As predicted by our perturbation Proposition 3.2.1, the LSM is able to work even for this tiny little noise case. The blow-up behavior happens again from the inside to the outside. The boundary of the scatterer can be well resolved by an isoline with cut-off value 0.053259, which can be easily determined by our technique of introducing another reference ball around.

Moreover, when we increase the noise level $\delta = 0.02$, the LSM works again and yields a nice profile of the scatterer with an isoline of cut-off value -0.39772.

3.2.3 Conclusion

By adding a ball with known geometrical and physical properties as an extra artificial obstacle component to the underlying scattering system, we have discussed a very simple but robust and effective technique to choose the crucial cut-off value required in the LSM for inverse acoustic and electromagnetic scattering problems.

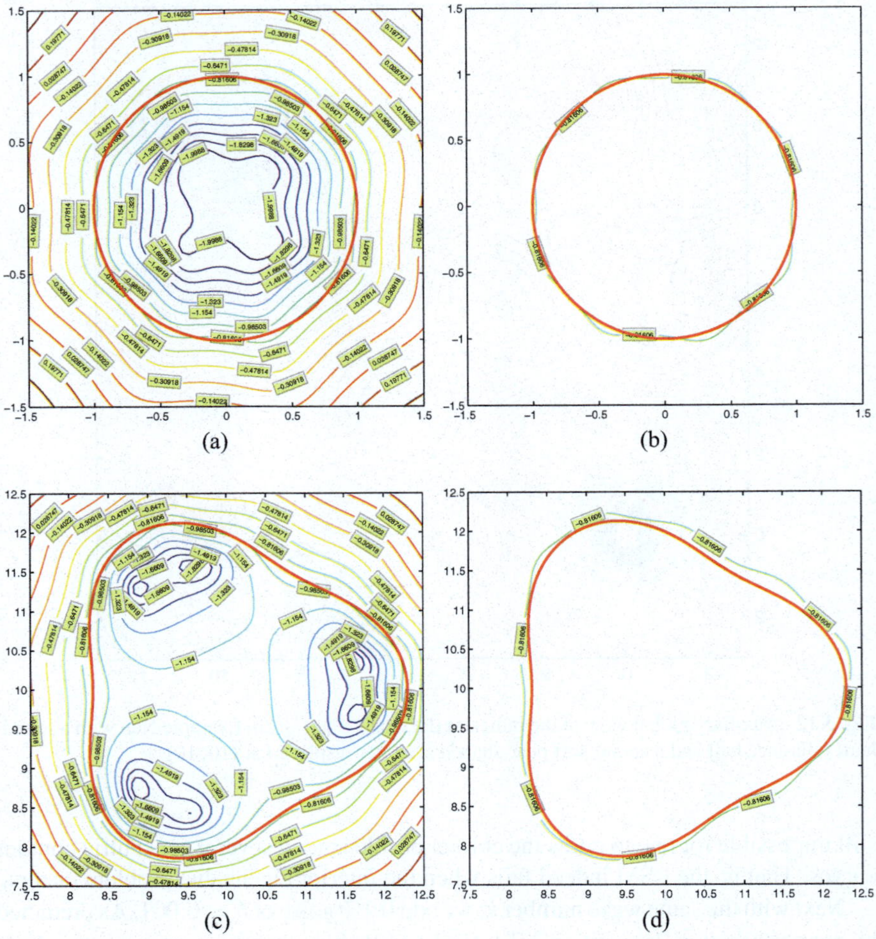

Fig. 3.13 Example Neu with a sound-hard reference ball and a sound-soft pear-shaped scatterer displaced at (10, 10). **a**, **c**: Characteristic contour curves with with value labels over $(-1.5, 1.5)^2$ and $(7.5, 12.5)^2$, respectively. **b**, **d**: Approximate profiles for the ball **B** and the scatterer $\mathbf{D_p}$, respectively

Table 3.2 First three roots $\lambda_{n,m}$, $m = 1, 2, 3$ for Bessel functions of the first kind of order n from 0 up to 3

m	1	2	3
$J_0(x)$	2.404825557695772	5.520078110286311	8.653727912911011
$J_1(x)$	3.831705970207512	7.015586669815619	10.173468135062720
$J_2(x)$	5.135622301840684	8.417244140399864	11.619841172149059
$J_3(x)$	6.380161895923983	9.761023129981668	13.015200721698434

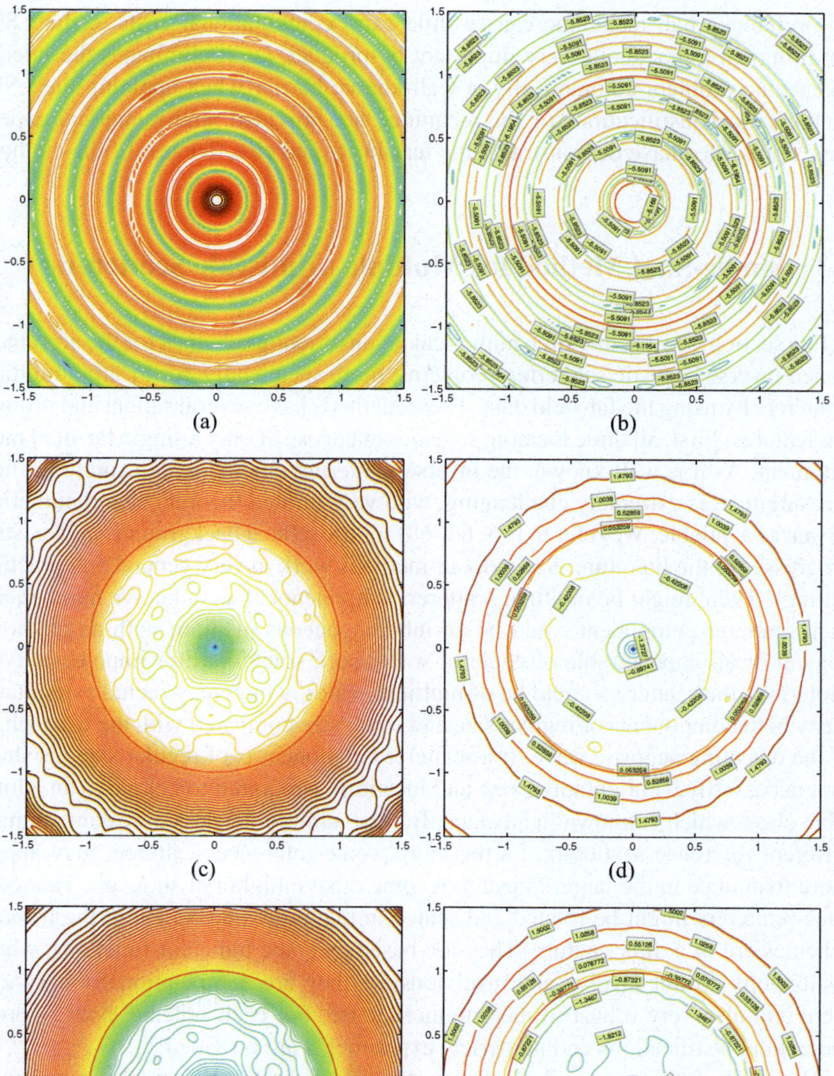

Fig. 3.14 Example BK with a reference ball and a kite displaced at (30, 30) and enlarged by 10 times. **a**, **c**, **e**: Contour plot over the sampling region $(-3, 3)^2$ and $(7.5, 12.5)^2$, respectively. **b**, **d**, **f**: Characteristic contour curves with with value labels over $(-1.5, 1.5)^2$ and $(7.5, 12.5)^2$, respectively

The reference ball technique causes little extra computational costs to the LSM, but brings in a practically very important by-product, i.e. it eliminates the interior eigenvalue problem automatically, a well-known barrier when applying the LSM. Mathematical justifications of the technique are provided, and numerical experiments presented have demonstrated the feasibility and effectiveness of the method.

3.3 Single-Shot Method for Multiple Multiscale Scatterers

In this section, we discuss three numerical reconstruction schemes for the aforementioned inverse acoustic scattering problems, more specifically, for locating multiple scatterers by using the far-field data. These methods have several salient and promising features. First, all three locating schemes make use of only a single far-field measurement. As it is well-known, the inverse scattering method with a single far-field measurement is extremely challenging, with very limited theoretical and numerical advances available. We refer to [56, 65, 68] for the related backgrounds and existing progresses in the literature. Second, our methods work in very general and practical settings. There might be multiple scatterer components of an unknown number, and each scatterer component could be an inhomogeneous medium with an unknown content or an impenetrable obstacle of sound-soft, sound-hard or impedance type. Moreover, the scatterers could be of multiple scales, which may include simultaneously both components of regular size and small size compared with the wavelength of the detecting acoustic wave. If a scatterer component is of regular size, its shape (not necessarily its orientation, size and location) is required to belong to an admissible class, which is known in advance. But the admissible class may contain many different reference scatterers. Furthermore, some reference scatterers may appear more than once in the target object, but some others might not show up. The reference scatterers might be rotated and scaled in the target object. Third, the locating schemes are of a direct nature. They are based on some indicator functions, whose evaluations do not involve any inversions, so they are computationally very efficient and also very robust to measurement noise. For each scheme, both rigorous theoretical justifications and numerical experiments are provided.

Our study follows a similar spirit to the one of the locating methods that were proposed in [49, 50] for electromagnetic (EM) scattering problems governed by the time-harmonic Maxwell equations. The methods in [49, 50] are based on two imaging functionals, respectively, for locating small-size and regular-size EM media or perfectly conducting (PEC) obstacles. A local re-sampling technique was developed in [50] to concatenate the two imaging functionals for locating multiscale EM scatterers. In this section we discuss three schemes, Schemes I, II and III, respectively for locating small-size, regular-size and multiscale acoustic scatterers. Due to the distinct physical nature of the acoustic scattering problems, some new ingredients and techniques are needed. In defining the imaging functional of locating small EM scatterers in [49], only the EM monopoles are involved. However, the acoustic scattering from small scatterers exhibit more complicated behaviors. In order to obtain the imaging functional for Scheme I that works independently of the physical properties of the underlying acoustic objects, both the acoustic monopole and dipoles should be incorporated (see definition (3.3.9) in next section). For locating regular-size scatterers, we present our acoustic study in a much more comprehensive manner than that in [49, 50] for the EM case. Indeed, the regular-size EM scatterers were all assumed in [49, 50] to be PEC obstacles, while the regular-size scatterers can be both inhomogeneous media and impenetrable obstacles of different kinds in the current acoustic case. Finally, in order to concatenate Schemes I and II to obtain Scheme III for locating multiscale acoustic scatterers, a local tuning technique is proposed. The local tuning technique generalizes the local re-sampling technique proposed in [50]. In fact, the local re-sampling technique was only used for tuning the locations of scatterers, whereas the local tuning technique here can be used for adjusting the orientations, scales as well as locations; see more discussions in Sect. 3.3.3. The local tuning technique concatenates Schemes I and II in a nice manner to produce Scheme III, which can be used for locating multiscale acoustic scatterers in a very general and practical setting. It is remarked that the local tuning technique can be directly extended to strengthen the method proposed in [50] for locating multiscale EM scatterers to enable it to work in a more general setting as considered here.

Compared with most of the existing methods, which rely on multiple scattered field measurements, the methods discussed in this section are more general in the sense that they combine all of the following features together: only one single far-field measurement is used; the scatterers are allowed to be a multiscale mixed set of inhomogeneous media and impenetrable obstacles; rigorous mathematical justifications are established under general settings; some iterative-type refining and local tuning strategies are introduced for quantitatively improving the reconstructions. More relevant discussions on the comparisons of our method with others are provided in Sects. 3.3.1 and 3.3.2.

In Sects. 3.3.1, 3.3.2 and 3.3.3, we shall discuss *Schemes I, II* and *III* respectively for locating multiple small-size scatterers, multiple regular-size scatterers and multiple multiscale scatterers. For each of the three schemes, rigorous mathematical justifications and numerical results are also provided.

3.3.1 Locating Small Scatterers

Throughout the rest of the subsection, we assume the incident acoustic wavenumber $k = \mathcal{O}(1)$. That is, the wavelength of the incident plane wave is given by $\lambda = 2\pi/k = \mathcal{O}(1)$, hence the size of a scatterer can be expressed in terms of its Euclidean diameter. In this section, we discuss an imaging scheme, referred to as *Scheme I*, to locate multiple small scatterers in terms of the incident wavelength.

We first introduce the class of small acoustic scatterers for our current study. Let $l \in \mathbb{N}$, and D_j, $1 \leq j \leq l$, be bounded simply-connected C^2 domains in $\mathbb{R}^N$ containing the origin. For $\rho \in \mathbb{R}_+$, we define $\rho D_j := \{\rho x \,|\, x \in D_j\}$ and set

$$\Omega_j^s = z_j + \rho D_j, \quad z_j \in \mathbb{R}^N, \quad 1 \le j \le l.$$

Each Ω_j^s is referred to as a *small* scatterer component. If it is sound-soft or sound-hard, we further write it, respectively, as

$$\Omega_j^{s,s} := z_j + \rho D_j^s \text{ and } \Omega_j^{s,h} := z_j + \rho D_j^h. \tag{3.3.1}$$

If Ω_j^s is of impedance type, we let λ_j be the surface impedance on $\partial\Omega_j^s$, and denote by

$$(\Omega_j^s; \lambda_j) := \Omega_j^{s,i} = z_j + \rho D_j^i, \tag{3.3.2}$$

where

$$D_j^i := (D_j; \lambda_j(\cdot + z_j)) \text{ and } \rho D_j^i = (\rho D_j; \lambda_j(\frac{\cdot}{\rho} + z_j)). \tag{3.3.3}$$

If Ω_j^s is an inhomogeneous medium, we let n_j be its refractive index, and denote by

$$(\Omega_j^s; n_j) := \Omega_j^{s,m} = z_j + \rho D_j^m, \tag{3.3.4}$$

where

$$D_j^m := (D_j; n_j(\cdot + z_j)) \text{ and } \rho D_j^m := (\rho D_j; n_j(\frac{\cdot}{\rho} + z_j)). \tag{3.3.5}$$

In the sequel, we set

$$\Omega^{s,t} := \bigcup_{j=1}^{l_t} \Omega_j^{s,t}, \quad t = s, h, i \text{ or } m, \tag{3.3.6}$$

where l_t, ρ and t denote respectively the number of components in the scatterer $\Omega^{s,t}$, the relative size of each component in $\Omega^{s,t}$, and the type of the scatterer, which can be sound-soft, sound-hard, of impedance type, or a medium. For $\Omega_j^{s,t}$ introduced in (3.3.6), we shall impose the following qualitative assumptions

$$\rho \ll 1 \quad \text{and} \quad L = \min_{1 \le j, j' \le l_t, j \ne j'} \operatorname{dist}(z_j, z_{j'}) \gg 1. \tag{3.3.7}$$

These conditions mean that the relative size of each scatterer component is small compared with the wavelength of the detecting/incident wave, and all the components must be well separated in the case of multiple components. It is remarked that ρ and L in (3.3.7) should be different with different type of scatterer components; see also Remark 3.3.1 at the end of this section. But we will always use the same ρ and L for the ease of notations.

3.3.1.1 Scheme I

In the sequel, we present *Scheme I* to locate the multiple components of $\Omega^{s,t}$ introduced in (3.3.6). The imaging scheme works in a very general and practical setting. First, we assume very little *a priori* knowledge of the scatterer. Both its type and the number of the components, i.e., t and l_t, are not required to be known in advance. Second, if the scatterer is a medium or of impedance type, the refractive indices or the surface impedances of its components are not required to be known a priori. Third, in a certain generic situation, the underlying scatterer $\Omega^{s,t}$ could be composed of mixed-type components, namely some of its components could be media while the others are obstacles of different type. We shall give some more remarks about this point at the end of this subsection. Finally, we would like to point out that our numerical experiments could speak a bit more about the qualitative assumptions (3.3.7): Scheme I can produce satisfactory reconstructions, as long as the size of the scatterer is smaller than half of the detecting wavelength while the distance between any two different components is bigger than half of the detecting wavelength. Nevertheless, in the extreme situation where the distance between two scatterer components is smaller than half of the detecting wavelength, Scheme I can still produce some qualitative reconstruction of the profile of the two scatterers but it may not be able to clearly separate them; we refer to Fig. 4.4 in [49] for reconstructing two close-by electromagnetic scatterers, and Scheme I produces similar reconstructions for the current acoustic case.

We are now ready to present our first locating scheme. To begin with, we let

$$a(\hat{x}; \Omega^{s,t}) := a_k(\hat{x}, d; \Omega^{s,t}), \quad \hat{x} \in \mathbb{S}^{N-1}, \tag{3.3.8}$$

denote the scattering amplitude of $\Omega^{s,t}$ in (3.3.6) due to a single incident plane wave $e^{ikx \cdot d}$ with fixed $k \in \mathbb{R}_+$ and $d \in \mathbb{S}^{N-1}$. Then we introduce the following real-valued index function $I_1(z)$ for $z \in \mathbb{R}^N$:

$$I_1(z) := \frac{1}{\|a(\hat{x}; \Omega^{s,t})\|^2_{L^2(\mathbb{S}^{N-1})}} \sum_{n=0}^{1} \sum_{p=-n}^{n} \left| \left\langle a(\hat{x}; \Omega^{s,t}), e^{ik(d-\hat{x})\cdot z} Y_n^p(\hat{x}) \right\rangle_{L^2(\mathbb{S}^{N-1})} \right|^2, \tag{3.3.9}$$

where $\langle u, v\rangle_{L^2(\mathbb{S}^{N-1})} = \int_{\mathbb{S}^{N-1}} u \cdot \overline{v} \, ds(\hat{x})$. In (3.3.9), $Y_n^p(\hat{x})$ for $n \in \mathbb{N} \cup \{0\}$ and $p = -n, \ldots, n$ are the spherical harmonics which form a complete orthonormal system in $L^2(\mathbb{S}^{N-1})$ (cf. [26]). It is emphasized that there is no harmonic function $Y_1^0(\hat{x})$ in the two-dimensional case, so it should be removed from the summation in (3.3.9) in defining $I_1(z)$. The next theorem about the indicating behavior of $I_1(z)$ is the crux of developing our Scheme I.

Theorem 3.3.1 *Let $\Omega^{s,t}$ and $I_1(z)$ be described as in* (3.3.6) *and* (3.3.9) *respectively. Set*

$$M_j := \frac{\|a(\hat{x}; \Omega_j^{s,t})\|^2_{L^2(\mathbb{S}^{N-1})}}{\|a(\hat{x}; \Omega^{s,t})\|^2_{L^2(\mathbb{S}^{N-1})}}, \quad j = 1, \ldots, l_t.$$

Then the following asymptotic expansion holds:

$$M_j = M_j^0 + \mathscr{O}\left(\frac{1}{L} + \rho^{N-2}(\ln\rho)^{N-3}\right), \quad j = 1, \ldots, l_t, \tag{3.3.10}$$

where M_j^0 is a positive number independent of L and ρ. Moreover, there exists an open neighborhood of z_j, $neigh(z_j)$, $1 \le j \le l_t$, such that

$$I_1(z) \le M_j^0 + \mathscr{O}\left(\frac{1}{L} + \rho^{N-2}(\ln\rho)^{N-3}\right) \quad \textit{for } z \in neigh(z_j), \tag{3.3.11}$$

where the equality holds only at $z = z_j$. That is, z_j is a local maximizer of $I_1(z)$ in $neigh(z_j)$.

In order to prove Theorem 3.3.1, we first present two crucial lemmas.

Lemma 3.3.1 *Let $\Omega^{s,t}$ and $a(\hat{x}; \Omega^{s,t})$ be given in* (3.3.6) *and* (3.3.8) *respectively. Then it holds that*

$$a(\hat{x}; \Omega^{s,t}) = a(\hat{x}; \bigcup_{j=1}^{l_t} \Omega_j^{s,t}) = a(\hat{x}; \bigcup_{j=1}^{l_t}(z_j + \rho D_j^t)) = \sum_{j=1}^{l_t} e^{ik(d-\hat{x})\cdot z} a(\hat{x}; \rho D_j^t) + \mathscr{O}\left(\frac{1}{L}\right). \tag{3.3.12}$$

Proof First, one has

$$a(\hat{x}; \Omega^{s,t}) = a(\hat{x}; \bigcup_{j=1}^{l_t}(z_j + \rho D_j^t)) = \sum_{j=1}^{l_t} a(\hat{x}; z_j + \rho D_j^t) + \mathscr{O}\left(\frac{1}{L}\right), \tag{3.3.13}$$

which was proved in [52] when $\Omega^{s,t}$ is a sound-soft obstacle. Following a similar argument, one can demonstrate (3.3.13) when $\Omega^{s,t}$ is a sound-hard or an impedance obstacle, or an inhomogeneous medium. On the other hand, it is straightforward to verify that

$$a_k(\hat{x}, d; z_j + \rho D_j^t) = e^{ik(d-\hat{x})\cdot z_j} a_k(\hat{x}, d; \rho D_j^t),$$

which together with (3.3.13) readily gives (3.3.12). □

The results in the following lemma can be found in [8, 30, 36, 44, 55].

Lemma 3.3.2 *Let D be a bounded simply-connected C^2 domain containing the origin and ρD^t be a scatterer of type $t = s, h, i$ or m, as described in* (3.3.1)–(3.3.5). *Then there exists $\rho_0 \in \mathbb{R}_+$ such that for $\rho < \rho_0$,*

$$a_k(\hat{x}, d; \rho D^t) = c_0 E(\rho) Y_0^0(\hat{x}) + \mathscr{O}(\rho^{N-2}(\ln\rho)^{N-3} E(\rho)), \quad t = s, i \textit{ or } m, \tag{3.3.14}$$

where c_0 is constant depending only on D, k, d and t, but independent of ρ. In (3.3.14),

$$E(\rho) := \rho^{N-2}(\ln \rho)^{N-3} \quad \textit{when } t = s; \quad \rho^{N-1} \quad \textit{when } t = i; \quad \rho^{N} \quad \textit{when } t = m.$$

In the case when $t = h$, we have

$$a_k(\hat{x}, d; \rho D^h) = \rho^N \sum_{n=0}^{1} \sum_{p=-n}^{n} c_n^p Y_n^p(\hat{x}) + \mathscr{O}(\rho^{N+1}), \tag{3.3.15}$$

where the coefficients c_n^p are constants depending only on D, k, d, but independent of ρ, and $Y_1^0(\hat{x})$ should be removed from the summation in (3.3.15) in the two-dimensional case.

Proof of Theorem 3.3.1 We first consider the three-dimensional sound-hard case. By Lemmas 3.3.1 and 3.3.2 we can easily see that

$$\begin{aligned} a(\hat{x}; \Omega^{s,h}) &= \sum_{j=1}^{l} a(\hat{x}; \Omega_j^{s,h}) + \mathscr{O}\left(\frac{1}{L}\right) \\ &= \sum_{j=1}^{l_h} e^{\mathrm{i}k(d-\hat{x})\cdot z_j} \left[\rho^3 \sum_{n=0}^{1} \sum_{p=-n}^{n} c_{n,j}^p Y_n^p(\hat{x}) + \mathscr{O}(\rho^4) \right] + \mathscr{O}\left(\frac{1}{L}\right). \end{aligned} \tag{3.3.16}$$

Next, without loss of generality, we only consider the indicating behavior of $I_1(z)$ in $B_\rho(z_1)$, a ball of radius ρ centered at z_1. Clearly, we have

$$|z_j - z| \ge L \gg 1 \quad \text{for } z \in B_\rho(z_1) \text{ and } j = 2, 3, \ldots, l_h\,. \tag{3.3.17}$$

Hence, by using (3.3.16) and (3.3.17) one can show by direct calculations that

$$\begin{aligned} &\left| \left\langle a(\hat{x}; \Omega^{s,t}), e^{\mathrm{i}k(d-\hat{x})\cdot z} Y_{n'}^{p'}(\hat{x}) \right\rangle_{L^2(\mathbb{S}^2)} \right| \\ &= \rho^3 \left| \left\langle e^{\mathrm{i}k(d-\hat{x})\cdot z_1} \sum_{n=0}^{1} \sum_{p=-n}^{n} c_{n,1}^p Y_n^p(\hat{x}), e^{\mathrm{i}k(d-\hat{x})\cdot z} Y_{n'}^{p'}(\hat{x}) \right\rangle_{L^2(\mathbb{S}^2)} + \mathscr{O}\left(\frac{1}{L} + \rho\right) \right| \\ &\le \rho^3 \left(|c_{n'}^{p'}| + \mathscr{O}\left(\frac{1}{L} + \rho\right) \right) \quad \text{for } z \in B_\rho(z_1),\ n' = 0, 1,\ q' = -n', ..., n'\,, \end{aligned} \tag{3.3.18}$$

where we have used (3.3.17) and the decaying property of oscillatory integrals for the equality relation, and the orthogonality of spherical harmonics and the Cauchy-Schwartz inequality for the inequality relation. Furthermore, due to the Cauchy-Schwartz inequality, one can verify that the equality in the last estimate of (3.3.18) holds only at $z = z_1$. On the other hand, we have

$$\|a(\hat{x}; \Omega^{s,t})\|_{L^2(\mathbb{S}^2)}^2 = \rho^6 \sum_{j=1}^{l} \left(\sum_{n=0}^{1} \sum_{p=-n}^{n} |c_{n,j}^p|^2 + \mathscr{O}\left(\frac{1}{L} + \rho\right) \right). \tag{3.3.19}$$

By (3.3.18)–(3.3.19), it is straightforward to verify the statement of the theorem by taking

$$M_1^0 = \frac{\sum_{n=0}^{1}\sum_{p=-n}^{n}|c_{n,1}^p|^2}{\sum_{j=1}^{l}\sum_{n=0}^{1}\sum_{p=-n}^{n}|c_{n,j}^p|^2}.$$

The other cases with $t = s$, i and m can be proved by following a similar argument to the above case with $t = h$, and using Lemmas 3.3.1 and 3.3.2. □

Using Theorem 3.3.1, we are now ready to formulate our first imaging scheme of locating multiple small scatterer components.

Scheme I

---

1. For an unknown scatterer $\Omega^{s,t}$ in (3.3.6), collect the far-field data by sending a single incident plane wave $e^{ikx\cdot d}$ with fixed k and d.
2. Select a sampling region with a mesh $\mathscr{T}_h$ containing $\Omega^{s,t}$.
3. For each sampling point $z \in \mathscr{T}_h$, compute the index value $I_1(z)$.
4. Locate all the significant local maxima of $I_1(z)$ on $\mathscr{T}_h$, which represent the locations of the scatterer components.

Remark 3.3.1 As it can be seen from Theorem 3.3.1, the indicating behavior of $I_1(z)$ is independent of the type of the underlying scatterer, which can be obstacle or inhomogeneous medium. Indeed, Scheme I can be extended to a much more general situation where the underlying scatterer Ω^s might be composed of mixed-type scatterers from different $\Omega^{s,t}$ for $t \in \{s, h, i, m\}$. We illustrate the situation by taking a special example, say, Ω^s consists of two components, a sound-soft Ω_1^s and a sound-hard Ω_2^s in three dimensions. Suppose that the relative sizes of Ω_1^s and Ω_2^s are, respectively, ρ_1 and ρ_2. According to Lemma 3.3.2, the scattering strength due to Ω_1^s is of order ρ_1 whereas that due to Ω_2^s is of order ρ_2^3. If $\rho_1 \approx \rho_2$, then $\rho_2^3 \ll \rho_1$, hence the scattering information from the sound-hard component is annihilated in the scattering data due to the sound-soft component. In this case, one cannot expect a reasonable locating by using Scheme I. However, if $\rho_2^3 \sim \rho_1$, then it is straightforward to verify that the local maximum behavior in Theorem 3.3.1 holds for the locations of the two mixed-type scatterer components, hence Scheme I works to locate both of them. This observation holds for the general case with multiple scatterer components of different type. As long as the scattering strengths from different components are comparable, Scheme I is effective to locate all of them.

Remark 3.3.2 In [76], an orthogonality sampling method was proposed where the indicator function is related to our indicator function $I_1(z)$ in (3.3.9). Indeed, if only a single far-field measurement is used, the indicator function in [76] actually corresponds to the first term in the sum defining $I_1(z)$ in (3.3.9). It can be easily verified from Lemma 3.3.2 and the proof of Theorem 3.3.1, if the underlying scatterer is sound-soft, then the two indicator functions would produce the same local maximum behavior. However, if one wants to produce an indicator function for locating a small-size scatterer independent of the physical properties of its components as discussed in the previous remark, then both the acoustic monopole and dipoles should be involved. As the orthogonality sampling method and topological derivative-based approaches (cf. [8, 37]) are closely relate, similar observations can be made for the latter approaches.

3.3.1.2 Numerical Experiments

In this subsection, we present some numerical tests to verify the applicability of Scheme I in both two and three dimensions. In all the tests, the exact far-field data are obtained by solving the forward equation (2.1.2) or (2.2.3) using the quadratic finite elements on a truncated circular (2D) or spherical (3D) domain enclosed by a PML layer. The forward equation is solved on a sequence of successively refined meshes till the relative error of two successive finite element solutions between the two adjacent meshes is below 0.1%. Then the scattered data are transformed into the far-field data by employing the Kirchhoff integral formula on a closed circle (2D) or surface (3D) enclosing the scatterer. For scatterers of small size, we always add to the exact far-field data a uniform noise of 5% and use it as the measurement data in our numerical tests.

Example 1 The true scatterer consists of three components, a sound-soft square component (in red) with side length 0.2 located at $(-1,\ 2)$, a sound-hard circular component (in white) with radius 0.5 located at $(1.5,\ 0)$, and a medium square component (in yellow) with side length 0.2 located at $(-2,\ -1.5)$. They are respectively shown in Fig. 3.15a.

We set the wave number k to be 3 and choose the incident direction $d = (1,\ 0)$, namely from left to right. The detecting wave length is larger than the sizes of all the components. Figure 3.16a shows the indicating behavior using the indicator function (3.3.9) of Scheme I, and the three components of the unknown scatterer are located very well using a single detecting plane wave field. By further increasing the wave number k to be 6 and adopting a different incident direction $d = (0,\ 1)$, namely from bottom to top, we find that each component of the scatterer is highlighted as a local maximum as shown in Fig. 3.16b. It is pointed out that Scheme I applies to such a complex scenario with scatterer components of mixed types when the scattering of each component are comparable, which implies that the size of the sound-hard component should be relatively larger than those of its sound-soft and medium ones in light of Remark 3.3.1.

Example 2 The true scatterer consists of two sound-hard circular disks of radius 0.1, located at $(-1,\ 0)$, $(1,\ 0)$ as shown in Fig. 3.15b.

Through this example, we show that Scheme I is totally independent of incident directions. It is found that we can always locate this pair of scatterer components with only one measurement data from an arbitrary incident direction. For instance, we show the cases when $k = 9$ but $d = (1,\ 0)$, $(1,\ 1)$ and $(0,\ 1)$ in Fig. 3.17.

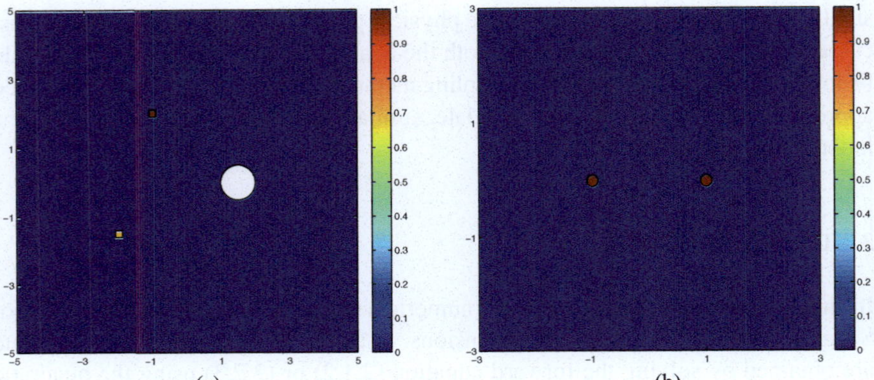

(a) (b)

Fig. 3.15 True scatterer components in **a** Example 1 and **b** Example 2

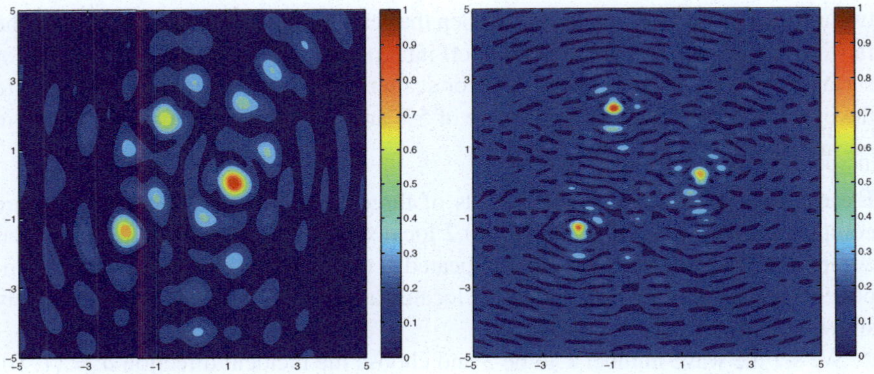

(a) $k = 3, d = (1, 0)$ (b) $k = 6, d = (0, 1)$

Fig. 3.16 Imaging of the scatterer components in Example 1 by Scheme I

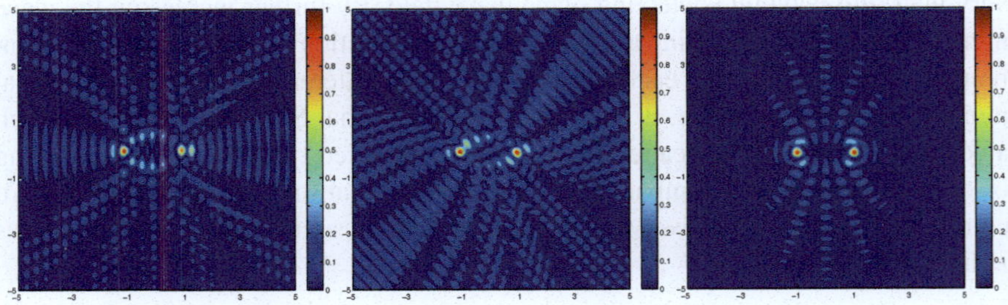

$k = 9, d = (1, 0)$ $k = 9, d = (1, 1)$ $k = 9, d = (0, 1)$

Fig. 3.17 Imaging of scatterer components in Example 2 by Scheme I

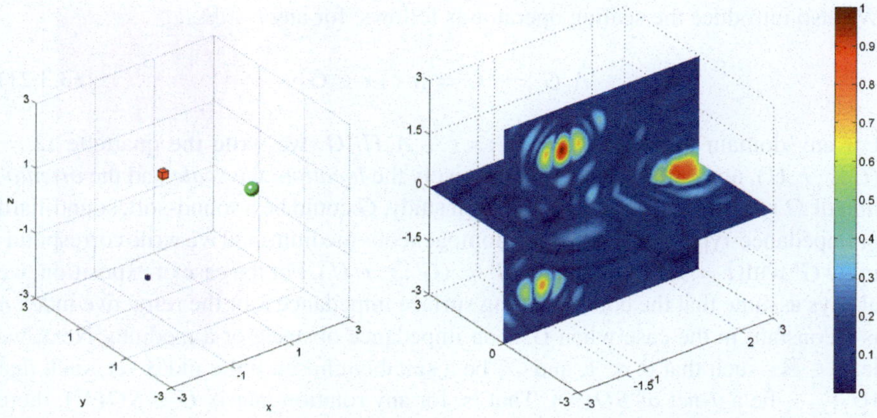

(a) (b)

Fig. 3.18 **a** True scatterer components in Example 3. **b** Imaging of the scatterer components in Example 3 by Scheme I

Example 3 We try further a complex scatterer in 3D. The wave number of the incident wave field is set to be $k = 5$. The true scatterer (see Fig. 3.18a) consists of three components, namely a sound-soft cube (in red) with side length 0.2 centered at $(-1,\ 0,\ 1.5)$, a sound-hard sphere (in green) with radius 0.2 centered at $(2,\ 0,\ 0)$, and a medium cube (in blue) with side length 0.1 centered at $(-1.5,\ 0,\ -1.5)$.

The resulting indicator function value distribution is plotted on a pair of orthogonal slice planes $x = 0$ and $y = 0$ in Fig. 3.18b. As one can see, three scatterer components are well located, and their positions are visualized in the highlighted part (local maxima). Clearly, the positions of the respective detected components match quite well with the ones of the exact components.

In summary, we have observed from Examples 1–3 that Scheme I is able to locate multiple small scatterer components of an unknown number robustly and efficiently.

3.3.2 *Locating Scatterers of Regular Size*

3.3.2.1 Scheme II

In this section, we consider the locating of multiple scatterers of regular size. We first fix some notations that shall be used throughout the rest of the section. Let G be a bounded simply-connected Lipschitz domain in $\mathbb{R}^n$ containing the origin, and $U \in SO(N)$ a rotation matrix in $\mathbb{R}^N$, and we define

$$\Pi_U G := UG = \{Ux \mid x \in G\}. \tag{3.3.20}$$

We also introduce the scaling operator as follows: for any $r \in \mathbb{R}_+$,

$$\Lambda_r G := rG = \{rx \mid x \in G\}. \tag{3.3.21}$$

For any domain Ω of the form $\Omega := z + \Lambda_r \Pi_U G$, we write the quartette $\Omega := (G; z, r, U)$, and call z, r and U respectively the *location*, the *scale* and the *orientation* of Ω relative to G. In our subsequent study, G could be a sound-soft, sound-hard or impedance-type obstacle, or an inhomogeneous medium, and we write correspondingly G^t with $t = s, h, i$ or m, and $\Omega^t = (G^t; z, r, U)$. For the ease of exposition, we always assume that the corresponding surface impedance λ or the refractive index n is a constant in the case when G is an impedance obstacle or a medium. Next, we let $\theta \in \mathbb{R}_+$ such that $\theta \ll 1$, and $\mathcal{T}_1$ be a suitably chosen finite index set, such that $\{U_j\}_{j\in\mathcal{T}_1}$ be a θ-net of $SO(N)$. That is, for any rotation matrix $U \in SO(N)$, there exists $j \in \mathcal{T}_1$ such that $\|U_j - U\| \le \theta$. We define

$$\Pi_\theta G := \{\Pi_{U_j} G\}_{j\in\mathcal{T}_1}. \tag{3.3.22}$$

In a similar manner, for Λ_r with $r \in [R_0, R_1]$, we let $\tau \ll 1$ and $\mathcal{T}_2$ be a suitably chosen finite index set such that $\{r_j\}_{j\in\mathcal{T}_2}$ be an τ-net of $[R_0, R_1]$. Define,

$$\Lambda_\tau G = \{\Lambda_j G\}_{j\in\mathcal{T}_2}. \tag{3.3.23}$$

With the above preparations, we are now ready to introduce the multiple scatterers of regular size for our subsequent study. Let $l_t \in \mathbb{N} \cup \{0\}$, $t = s, h, i$ or m, and G_j, $j \in \mathbb{N}$, be a bounded simply-connected Lipschitz domain in $\mathbb{R}^n$ containing the origin. We write

$$\mathcal{A}_t = \{G_j^t\}_{j=1}^{l_t},\ t = s, h;\quad \mathcal{A}_i = \{G_j^i\}_{j=1}^{l_i} = \{(G_j^i, \lambda_j)\}_{j=1}^{l_i}; \tag{3.3.24}$$

$$\mathcal{A}_m = \{G_j^m\}_{j=1}^{l_m} = \{(G_j^m, n_j)\}_{j=1}^{l_m};\ \mathcal{A} = \bigcup_{t=s,h,i,m} \mathcal{A}_t := \{G_j\}_{j=1}^{l':=l_s+l_h+l_i+l_m}. \tag{3.3.25}$$

Let $l \in \mathbb{N}$ and set

$$\Omega^r = \bigcup_{j=1}^{l} \Omega_j^r,\quad \Omega_j^r := (\Sigma_j; z_j, r_j, U_j) \text{ with } \Sigma_j \in \mathcal{A},\ \ j = 1, \ldots, l. \tag{3.3.26}$$

For Ω^r introduced in (3.3.26), we assume that

$$r_j \in [R_0, R_1],\ \ R_0 \approx 1,\ \ R_1 \approx 1,\ \ \text{and}\ \ L = \min_{1\le j,j'\le l, j\ne j'} \operatorname{dist}(z_j, z_{j'}) \gg 1. \tag{3.3.27}$$

Ω^r represents the multiple scatterers of regular size in our current study, and we shall present Scheme II to locate all the multiple components. We assume that the admissible class $\mathcal{A}$ is known in advance. In the physical situation, this means that

Ω^r might be composed of multiple regular-size components of an unknown number, and each component could be from a different type: obstacles of different types or a medium. If the scatterer component is an obstacle (sound-soft, sound-hard or impedance type), then its shape must be from a known admissible class. Moreover, the surface impedance of an impedance-type component must also be known a priori. If the scatterer is a medium, then both its shape and content should be known from the admissible class. It is emphasized that the number of admissible class, namely l' in $\mathscr{A}$, and the number of the unknown scatterer components, namely l in Ω^r, are not necessarily the same. This means that a certain shaped admissible scatterer might appear more than once or does not appear in Ω^r. For the regular-size multiple scatterers Ω^r described above, we shall discuss *Scheme II* to determine the location, the size/scale and the orientation of each of its components, by using a single far-field measurement, i.e. $a(\hat{x}; \Omega^r) := a_k(\hat{x}, d; \Omega^r)$ with fixed $k \in \mathbb{R}_+$ and $d \in \mathbb{S}^{N-1}$.

In order to present our *Scheme II*, we first augment the admissible class $\mathscr{A}$ as

$$\widetilde{\mathscr{A}} = \Pi_\theta \Lambda_\tau \mathscr{A} = \bigcup_{j=1}^{l'} \{\Pi_\theta \Lambda_\tau G_j\} := \{\widetilde{G}_j\}_{j=1}^{l''}. \tag{3.3.28}$$

We make the following two assumptions about the augmented admissible class $\widetilde{\mathscr{A}}$:

1. $a_k(\hat{x}, d; \widetilde{G}_j) \neq a_k(\hat{x}, d; \widetilde{G}_{j'})$ for $j \neq j'$ and $1 \leq j, j' \leq l''$, $\hat{x} \in \mathbb{S}^{N-1}$;
2. $\|a_k(\hat{x}, d; \widetilde{G}_j)\|_{L^2(\mathbb{S}^{N-1})} \geq \|a_k(\hat{x}, d; \widetilde{G}_{j'})\|_{L^2(\mathbb{S}^{N-1})}$ for $j < j'$ and $1 \leq j, j' \leq l''$.

Assumption (ii) can be fulfilled by reordering if necessary. For assumption (i), we recall the following well-known conjecture in the theory of the inverse acoustic scattering problem:

$$a_k(\hat{x}, d; G) = a_k(\hat{x}, d; \widetilde{G}) \quad \text{if and only if} \quad G = \widetilde{G}, \tag{3.3.29}$$

where G and $\widetilde{G}$ are two obstacles, with k and d fixed. Equation (3.3.29) states that a single far-field measurement can uniquely determine an acoustic obstacle. There is a widespread belief that (3.3.29) holds true, but there is only limited progress in the literature (cf. [56, 65, 68]). Clearly, if (3.3.29) holds true, and if $\widetilde{\mathscr{A}}$ contains only obstacle scatterers, assumption (ii) is always fulfilled. On the other hand, (3.3.29) does not holds true in general for inhomogeneous medium scatterers, hence if there are medium scatterers presented in $\widetilde{\mathscr{A}}$, assumption (ii) can not be always fulfilled. Nevertheless, since $\widetilde{\mathscr{A}}$ is known, assumption (ii) can always be verified in advance.

Now we introduce the following l'' indicator functions for identifying the multiple scatterers of Ω^r in (3.3.26):

$$I_2^j(z) = \frac{\left|\langle a(\hat{x}; \Omega^r), e^{\mathrm{i}k(d-\hat{x})\cdot z} a(\hat{x}; \widetilde{G}_j)\rangle_{L^2(\mathbb{S}^{N-1})}\right|}{\|a(\hat{x}; \widetilde{G}_j)\|^2_{L^2(\mathbb{S}^{N-1})}}, \quad \widetilde{G}_j \in \widetilde{\mathscr{A}}, \quad j = 1, 2, \ldots, l''. \tag{3.3.30}$$

Next, we present a key theorem on the indicating behavior of these indicator functions, which forms the basis of our Scheme II.

Theorem 3.3.2 *Suppose that $\widetilde{G}_1 \in \widetilde{\mathscr{A}}$ is of the following form*

$$\widetilde{G}_1 = (G_{j_0}; r_{p_0}, U_{q_0}) = \Pi_{U_{q_0}} \Lambda_{r_{q_0}} G_{j_0}, \quad G_{j_0} \in \mathscr{A}, \; U_{q_0} \in \mathscr{T}_1, \; r_{p_0} \in \mathscr{T}_2.$$

Suppose that in Ω^r, there exists $J_0 \subset \{1, 2, \ldots, l\}$ such that for $j \in J_0$, the component $\Omega^r_j = (\Sigma_j; z_j, r_j, U_j)$ satisfies

$$(i)\ \Sigma_j = G_{j_0}; \quad (ii)\ \|U_j - U_{q_0}\| \leq \theta; \quad (iii)\ \|r_j - r_{p_0}\| \leq \tau; \tag{3.3.31}$$

whereas for $j \in \{1, 2, \ldots, l\} \backslash J_0$, at least one of the conditions in (3.3.31) is not fulfilled by the scatterer component Ω^r_j. Then for each z_j, $j = 1, 2, \ldots, l$, there exists an open neighborhood of z_j, $neigh(z_j)$, such that

(i) if $j \in J_0$, then

$$I_2^1(z) \leq 1 + \mathscr{O}\left(\frac{1}{L} + \theta + \tau\right) \quad \forall z \in neigh(z_j). \tag{3.3.32}$$

Moreover, the equality holds in the above relation only when $z = z_j$. That is, z_j is a local maximum point for $I_2^1(z)$.

(ii) if $j \in \{1, 2, \ldots, l\} \backslash J_0$, then there exists $\varepsilon_0 > 0$ such that

$$I_2^1(z) \leq 1 - \varepsilon_0 + \mathscr{O}\left(\frac{1}{L} + \theta + \tau\right) \quad \forall z \in neigh(z_j). \tag{3.3.33}$$

In order to prove Theorem 3.3.2, we first derive a key lemma as follows.

Lemma 3.3.3 *Let G be a bounded simply-connected domain in $\mathbb{R}^N$ containing the origin, which supports an acoustic scatterer G^t, $t = s, h, i$ or m. Then we have*

$$a_k(\hat{x}, d; \Pi_U G^t) = a_k(U^T \hat{x}, U^T d; G) \quad for \quad U \in SO(n) \tag{3.3.34}$$

and

$$a_k(\hat{x}, d; \Lambda_r G^t) = r a_{kr}(\hat{x}, d; G) \quad for \quad r \in \mathbb{R}_+ . \tag{3.3.35}$$

Proof Equations (3.3.34) and (3.3.35) can be readily verified by a change of variables in the corresponding scattering systems. □

Proof of Theorem 3.3.2 First it follows by Lemma 3.3.1 that

$$
\begin{aligned}
a_k(\hat{x}, d; \Omega^r) &= a_k(\hat{x}, d; \bigcup_{j=1}^{l} \Omega_j^r) = \sum_{j=1}^{l} a_k(\hat{x}, d; \Omega_j^r) + \mathscr{O}\left(\frac{1}{L}\right) \\
&= \sum_{j=1}^{l} a_k(\hat{x}, d; (\Sigma_j; z_j, r_j, U_j)) + \mathscr{O}\left(\frac{1}{L}\right) \\
&= \sum_{j=1}^{l} a_{kr_j}(U_j^T \hat{x}, U_j^T d; \Sigma_j) e^{ik(d-\hat{x})\cdot z_j} + \mathscr{O}\left(\frac{1}{L}\right).
\end{aligned}
\tag{3.3.36}
$$

Then by (3.3.31) and Lemma 3.3.3, we have that for $j_0 \in J_0$,

$$
\begin{aligned}
a_{kr_j}(U_j^T \hat{x}, U_j^T d; \Sigma_j) &= a_{kr_{p_0}}((U_{q_0})^T \hat{x}, (U_{q_0})^T d; G_{j_0}) e^{ik(d-\hat{x})\cdot z} + \mathscr{O}(\theta + \tau) \\
&= a_k(\hat{x}, d; \widetilde{G}_1) + \mathscr{O}(\theta + \tau).
\end{aligned}
\tag{3.3.37}
$$

Hence we obtain using (3.3.36) and (3.3.37) that

$$
a(\hat{x}; \Omega^r) = \sum_{j \in J_0} a(\hat{x}; \widetilde{G}_1) e^{ik(d-\hat{x})\cdot z_j} + \sum_{j \in \{1,\ldots,l\} \backslash J_0} a(\hat{x}; \Omega_j^r) + \mathscr{O}\left(\frac{1}{L} + \theta + \tau\right).
\tag{3.3.38}
$$

For $j_0 \in J_0$, by (3.3.38) we can show that for $z \in neigh(z_{j_0})$,

$$
\begin{aligned}
&\left| \langle a(\hat{x}; \Omega^r), e^{ik(d-\hat{x})\cdot z} a(\hat{x}; \widetilde{G}_1) \rangle_{L^2(\mathbb{S}^{N-1})} \right| \\
&= \left| \langle a(\hat{x}; \widetilde{G}_1) e^{ik(d-\hat{x})\cdot z_{j_0}}, e^{ik(d-\hat{x})\cdot z} a(\hat{x}; \widetilde{G}_1) \rangle_{L^2(\mathbb{S}^{N-1})} \right| + \mathscr{O}\left(\frac{1}{L} + \theta + \tau\right) \\
&\leq \| a(\hat{x}; \widetilde{G}_1) \|_{L^2(\mathbb{S}^{N-1})}^2 + \mathscr{O}\left(\frac{1}{L} + \theta + \tau\right).
\end{aligned}
\tag{3.3.39}
$$

For the equality relation in (3.3.39), we have made use of the Riemann-Lebesgue Lemma about oscillatory integrals by noting $|z_j - z| \geq L \gg 1$ for $1 \leq j \leq l, j \neq j_0$ and $z \in neigh(z_{j_0})$ by means of (3.3.27). For the last relation in (3.3.39), we have applied the Cauchy-Schwartz inequality, and it is easily seen that the equality holds only at $z = z_{j_0}$. These observations clearly imply (3.3.32) for $z \in neigh(z_{j_0})$ and $j_0 \in J_0$. On the other hand, by a similar argument, together with assumption (i) on $\widetilde{\mathscr{A}}$, we can directly verify that

$$
I_2^1(z_j) < 1 + \mathscr{O}\left(\frac{1}{L} + \theta + \tau\right), \quad j \in \{1, 2, \ldots, l\} \backslash J_0,
\tag{3.3.40}
$$

which readily implies (3.3.33). □

Based on Theorem 3.3.2, we are now ready to formulate Scheme II for locating the multiple scatterer components of regular size in Ω^r successively.

Scheme II

1. For the admissible scatterer class $\mathscr{A}$ in (3.3.25), formulate the augmented admissible class $\widetilde{\mathscr{A}}$ as that given in (3.3.28).
2. Collect in advance the far-field patterns associated with the admissible reference scatterer space $\widetilde{\mathscr{A}}$ corresponding to a single incident plane wave $e^{ikx\cdot d}$ with fixed k and d, and reorder $\widetilde{\mathscr{A}}$ if necessary so that assumptions (i) and (ii) are satisfied.
3. For an unknown scatterer Ω^r in (3.3.26), collect the far-field data corresponding to the single incident plane wave as specified in (2).
4. Select a sampling region with a mesh $\mathscr{T}_h$ containing Ω^r.
5. Set $j = 1$.
6. For each sampling point $z \in \mathscr{T}_h$, compute the index value $I_2^j(z)$.
7. Locate all those significant local maxima of $I_2^j(z)$ satisfying $I_2^j(z) \approx 1$ for the scatterer components of the form $z + \widetilde{G}_j$.
8. Remove all the sampling points inside those N_{id} identified components $z + \widetilde{G}_j$ found in (7) from $\mathscr{T}_h$. Subtract the individual far-field patterns associated with those already reconstructed components in (7) and their respective identified locations z_j from the far field as follows:

$$a(\hat{x}; \Omega^r) := a(\hat{x}; \Omega^r) - \sum_{j=1}^{N_{id}} e^{ik(d-\hat{x})\cdot z_j} a(\hat{x}; \widetilde{G}_j) \tag{3.3.41}$$

   Note that the updated far-field pattern in (3.3.41) is still denoted by $a(\hat{x}; \Omega^r)$, which will be further checked by subsequent reference components in (3.3.30).
9. If $j = l''$, i.e., the maximum number of the unknown component reaches, then stop the reconstruction; otherwise, set $j := j + 1$, and go to (6).

Remark 3.3.3 In (3.3.26), it is assumed that the admissible class $\mathscr{A}$ contains exactly the base scatterer Σ_j of the unknown scatterer component Ω_j^r in Ω^r. However, our Scheme II relies on the augmented admissible class $\widetilde{\mathscr{A}}$, which may contain only an approximate scatterer configuration to the target scatterer component Ω_j^r in Ω^r. Hence, if the admissible class $\mathscr{A}$ contains only an approximate base scatterer to Σ_j of the unknown scatterer component Ω_j^r in Ω^r, Scheme II would still work, and in fact, it can be easily justified from the proof of Theorem 3.3.2. This point will be also illustrated by our numerical experiments in Example 5 in Sect. 3.3.2.2.

Remark 3.3.4 The introduction of a known admissible class $\mathscr{A}$ in our algorithm is related to the dictionary matching algorithms ,where some a priori known base shapes form a dictionary for the reconstruction. We also remark that comparable indicator functions are used in [11] for reconstructing the acoustic scatterers at small scale and regular scale, respectively.

3.3.2.2 Numerical Experiments

We proceed now with some numerical tests using Scheme II to detect multiple scatter components of regular size. The synthetic far-field data is generated in the same manner as stated in Sect. 3.3.1.2, then a uniform noise of 3% is added to the synthetic data.

Two geometries will be considered for the scatterer components in our numerical tests. They are characterized by the following 2D parametric curves

$$\textbf{Peanut}: \ \{(x, y) : x = \sqrt{3\cos^2(s) + 1}\cos(s), \ y = \sqrt{3\cos^2(s) + 1}\sin(s), \ 0 \leq s \leq 2\pi\},$$
$$\textbf{Kite}: \ \{(x, y) : x = \cos(s) + 0.65\cos(2s) - 0.65, \ y = 1.5\sin(s), \ 0 \leq s \leq 2\pi\}.$$

These will be denoted by **P** and **K**, respectively, for short. The candidate data set $\widetilde{\mathscr{A}}$ includes the far-field data of both reference components **P** and **K**, and is further lexicographically augmented by a collection of a priori known orientations and sizes. More precisely, the augmented data set is obtained by rotating **P** and **K** in the x-y plane every 45° as shown in Figs. 3.19 and 3.20, respectively, and by scaling **P** and **K** by one half, one and twice.

For imaging of regular-size scatterers, we adopt a technique from image contrast enhancement by increasing the order of power of the indicator function $I_2^j(z)$, namely $(I_2^j(z))^\alpha$ is employed as the indicator, where α is taken in our experiments to be 2, 3

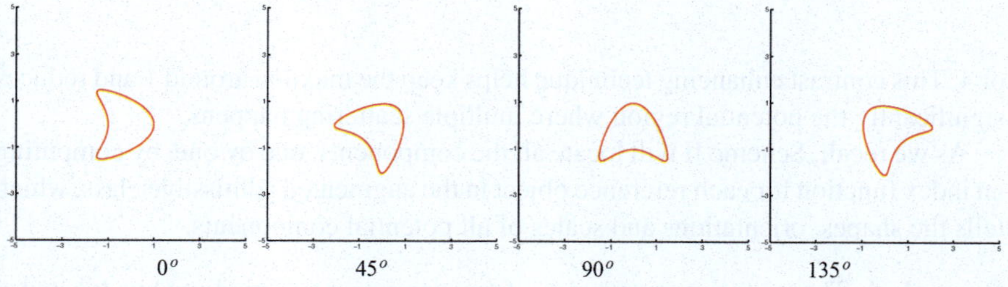

Fig. 3.19 Basic scatterer components : a reference kite with first four orientations

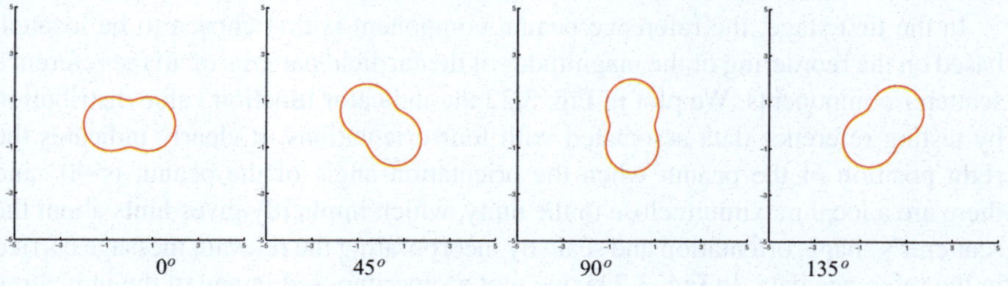

Fig. 3.20 Basic scatterer components : a reference peanut with four orientations

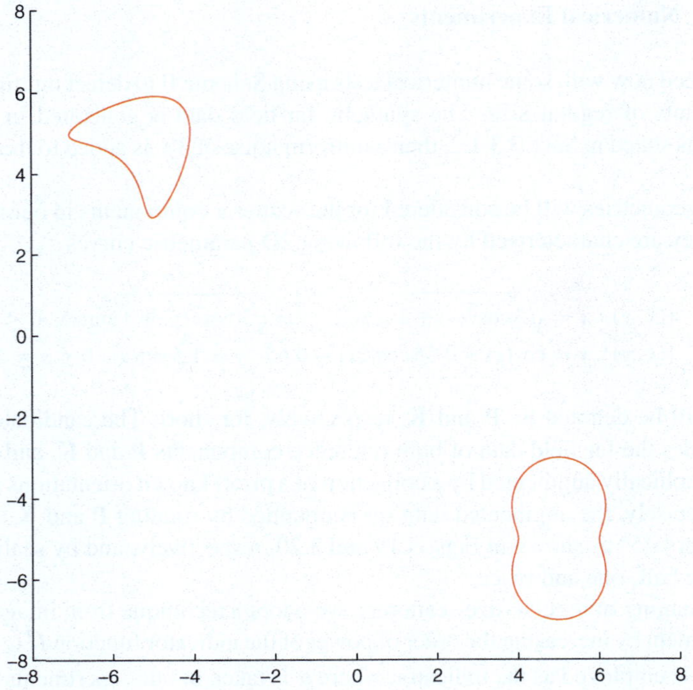

Fig. 3.21 True scatterer components in Example 4

or 4. This contrast enhancing technique helps keep the maxima around 1 and reduces significantly the potential region where multiple scattering happens.

As we recall, Scheme II will locate all the components, one by one, by computing an index function for each reference object in the augmented admissible class, which tells the shapes, orientations and scales of all potential components.

Example 4 The true scatterer consists of two components, a medium kite located at $(-5,\ 5)$ and a medium peanut located at $(5,\ -5)$, see Fig. 3.21. The wavenumber of the incident field and the incident direction are set to be $k = 5$ and $d = (1,\ 1)/\sqrt{2}$.

In the first stage, the reference peanut component is first chosen to be located, based on the reordering of the magnitudes of the far-field patterns of all the reference scatterer components. We plot in Fig. 3.22 the indicator function value distribution by testing reference data associated with four orientations. It clearly indicates the right position of the peanut when the orientation angle of the peanut is 90° and there are a local maximum close to the unity, which implicitly gives hints about the scatterer's shape, orientation and scale by incorporating the relevant message carried in the reference data. In Fig. 3.22a, we plot a superimposed image of the indicating value distribution of the four aforementioned images by taking the maximum of four indicator function values pointwise. After obtaining the position of the peanut

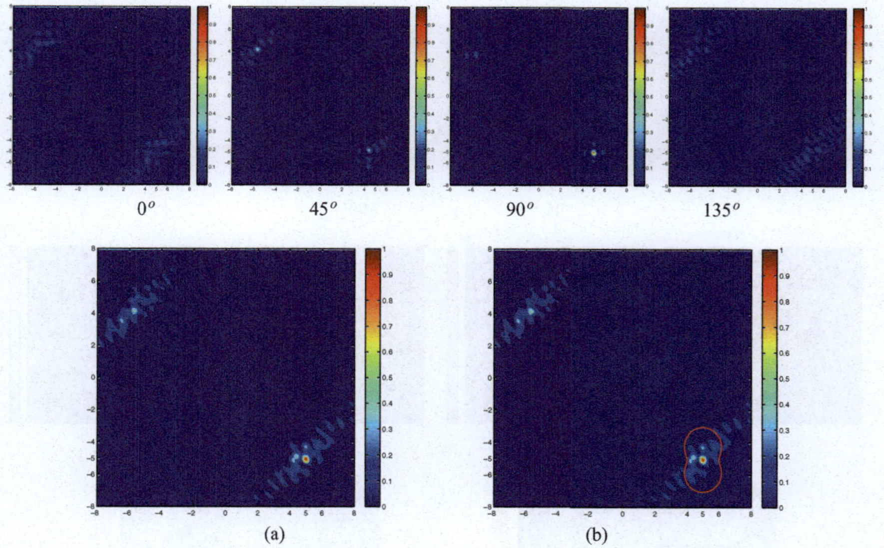

Fig. 3.22 Example 4. From left to right in the first row: Imaging of the indicator function value distribution by testing the reference far-field data of the basis peanut ($k = 5$) at four orientations of 0, 45, 90 and 135°, respectively; **a** a superimposed image of the four indicating images aforementioned by taking the maximum of four indicator functions pointwise; **b** Reconstruction of the peanut component

component, it is now possible to identify the first unknown scatterer component in Fig. 3.22b.

Once the peanut component is found, then we proceed by subtracting the far-field contribution of the detected peanut component from the total far-field pattern. We can then find the kite's position reasonably; see Fig. 3.23. We see that only the configuration with 45° maximizes the indicator function to achieve nearly the unity and indicates the position of the detected kite component very clearly. In Fig. 3.23a, we plot a superimposed image of maximizing the eight aforementioned image values. After obtaining the position of the kite component, it is now possible to identify the second unknown scatterer component in Fig. 3.22b by combining the information carried implicitly in the reference data, which tells us not only the shape of the scatterer but also its size and orientation.

Example 5 To further study the reliability of our locating scheme from a generic admissible reference class, we keep the admissible class unchanged but perturb pointwisely the parametric forms of the true scatterers **P** and **K** by 5% uniform noise with respect their respective centers as shown in Fig. 3.24. Now the admissible set is only an approximate class to describe the scatterer components.

Scheme II is repeated for such an interesting setting. The measured far-field pattern is first compared with the reference peanut data set according to (3.3.30). The highlighted red dot as shown in Fig. 3.25 when the orientation is 90° tells us the

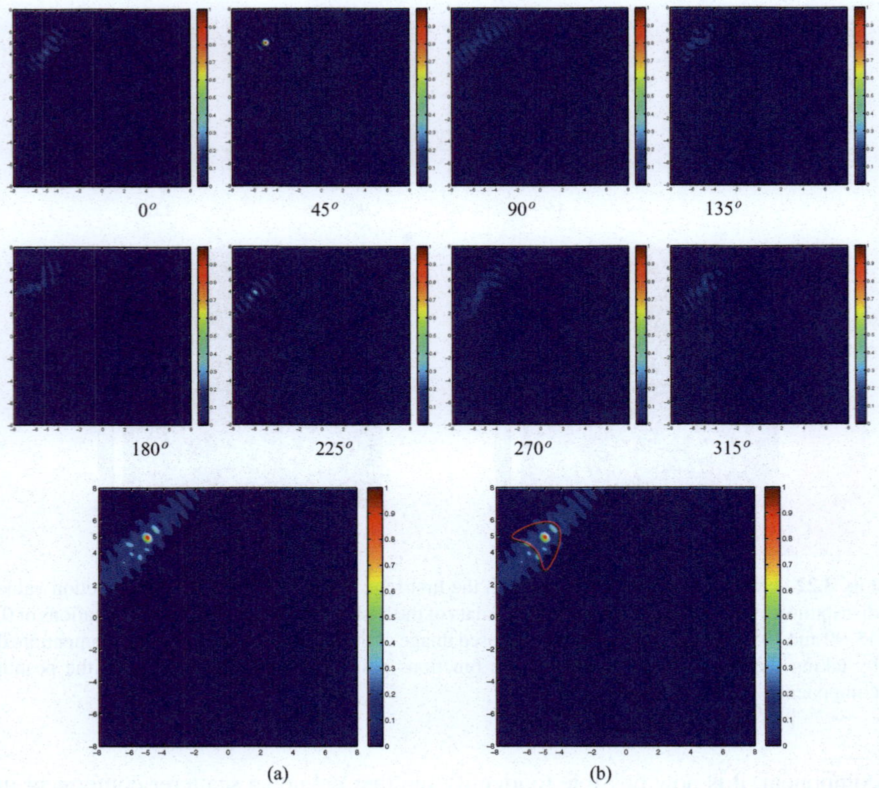

Fig. 3.23 Example 4. From left to right in the first two rows: Imaging of the indicator function value distribution by testing the reference far-field data of the basis kite ($k = 5$) at eight orientations of 0, 45, 90, 135, 180, 225, 270 and 315°, respectively; **a** a superimposed image of the eight indicating images aforementioned by taking the maximum of eight indicator functions pointwise; **b** Reconstruction of the kite component

rough position. Compared with Fig. 3.22, there exist some small ripples in the contour plots in Fig. 3.25 which are rather weak and do not affect the locating accuracy. Those weak ripples can be understood as some additional noise caused by the small geometric difference between the unknown components and their approximate class.

After subtracting the corresponding far-field pattern detected in the first step, we proceed with the locating of the perturbed kite components. The indicating contour plots associated with the first four orientations are shown in Fig. 3.26. Except for some oscillating ripples, the correct position of the perturbed kite component can be well located by the red dot when the orientation is 45°.

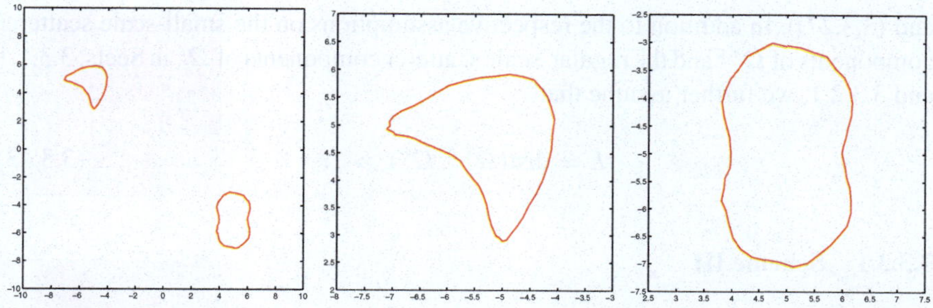

Fig. 3.24 Perturbed scatterer components (left) and their respective zoomed-in image (middle, right) in Example 5

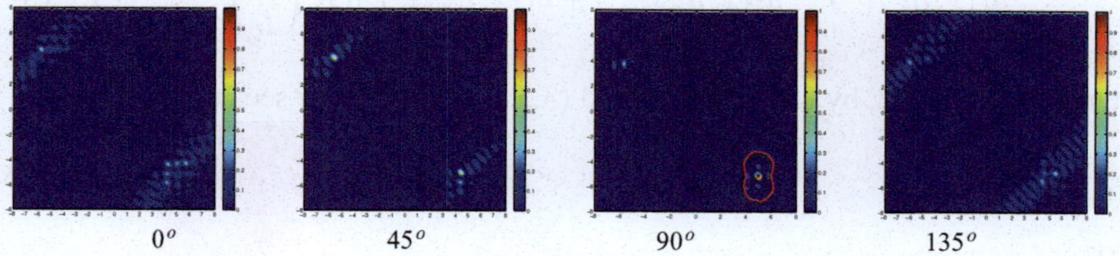

Fig. 3.25 Example 5. From left to right: Imaging of the indicator function value distribution by testing the reference far-field data of the basis kite ($k = 5$) at first four orientations of 0, 45, 90 and 135°, respectively; The highlighted red dot reveals the reconstruction of the perturbed peanut component at the orientation of 90°

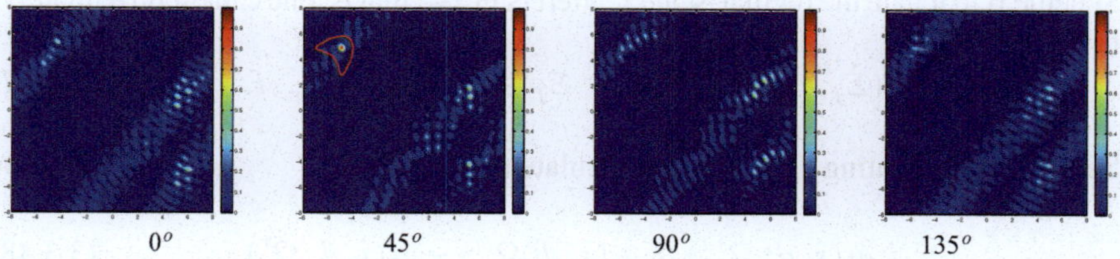

Fig. 3.26 Example 5. From left to right: Imaging of the indicator function value distribution by testing the reference far-field data of the basis peanut ($k = 5$) at four orientations of 0, 45, 90 and 135°, respectively. The highlighted red dot reveals the reconstruction of the perturbed kite component at the orientation of 45°

3.3.3 Locating Scatterers of Multiple Scales

In this section, we consider locating multiple scatterers of multiple scales of the form

$$\Omega^m = \Omega^{s,t} \cup \Omega^r, \tag{3.3.42}$$

where $\Omega^{s,t}$ is the scatterer of small size described in Sect. 3.3.1.1 (cf. (3.3.6) and (3.3.7)) and Ω^r is the scatterer of regular size described in Sect. 3.3.2.1 (cf. (3.3.26) and (3.3.27)). In addition to the respective assumptions on the small-scale scatterer components of $\Omega^{s,t}$ and the regular-scale scatterer components of Ω^r in Sects. 3.3.1.1 and 3.3.2.1, we further assume that

$$L = \text{dist}(\Omega^{s,t}, \Omega^r) \gg 1. \tag{3.3.43}$$

3.3.3.1 Scheme III

By Lemma 3.3.1, one has

$$a_k(\hat{x}, d; \Omega^m) = a_k(\hat{x}, d; \Omega^{s,t}) + a_k(\hat{x}, d; \Omega^r) + \mathcal{O}\left(\frac{1}{L}\right). \tag{3.3.44}$$

On the other hand, by Lemma 3.3.2 and (3.3.16), one further sees that

$$|a_k(\hat{x}, d; \Omega^{s,t})| \ll |a_k(\hat{x}, d; \Omega^r)|. \tag{3.3.45}$$

Hence we know

$$a_k(\hat{x}, d; \Omega^r) \approx a_k(\hat{x}, d; \Omega^m). \tag{3.3.46}$$

Therefore, it is natural to locate the scatterers Ω^m of multiple scales in a two-stage manner as follows. First, by (3.3.46), one uses $a_k(\hat{x}, d; \Omega^m)$ as the far-field data for Scheme II to locate the regular-scale scatterers in Ω^r; that is, one can (approximately) find

$$\Omega_j^r = (\Sigma_j; z_j, r_j, U_j) \text{ with } \Sigma_j \in \mathscr{A}, \quad j = 1, \ldots, l. \tag{3.3.47}$$

Second, after locating Ω_j^r, one can calculate that

$$a_k(\hat{x}, d; \Omega^{s,t}) \approx a_k(\hat{x}, d; \Omega^m) - a_k(\hat{x}, d; \Omega^r). \tag{3.3.48}$$

Then, using the far-field data obtained above to Scheme I, one can locate the multiple small-scale scatterers in $\Omega^{s,t}$. However, we would like to emphasize the following two facts: first, the size contrast between Ω^r and $\Omega^{s,t}$ can not be too sharp, since otherwise by (3.3.45) and (3.3.46), the scattering information of $\Omega^{s,t}$ might be overwhelmed by that of Ω^r; second, one should have fine reconstructions of Ω_j^r's in the first stage. Indeed, in the first-stage reconstruction, instead of (3.3.47), one has

$$\widehat{\Omega}_j^r := (\Sigma_j; \widehat{z}_j, \widehat{r}_j, \widehat{U}_j) \approx (\Sigma_j; z_j, r_j, U_j), \quad \Sigma_j \in \mathscr{A}, \quad j = 1, \ldots, l, \tag{3.3.49}$$

where $\widehat{z}_j, \widehat{r}_j$ and $\widehat{U}_j$ are, respectively, approximations to z_j, r_j and U_j. Hence, by (3.3.48), the far-field data used in the second stage of the reconstruction is

$$a_k(\hat{x}, d; \Omega^{s,t}) \approx a_k(\hat{x}, d; \Omega^m) - a_k(\hat{x}, d; \Omega^r) + (a_k(\hat{x}, d; \Omega^r_j) - a_k(\hat{x}, d; \widehat{\Omega}^r_j)). \tag{3.3.50}$$

If the reconstructed scatterer $(\Sigma_j; \widehat{z}_j, \widehat{r}_j, \widehat{U}_j)$ is not close enough to the true scatterer $(\Sigma_j; z_j, r_j, U_j)$, the error produced by $a_k(\hat{x}, d; \Omega^r_j) - a_k(\hat{x}, d; \widehat{\Omega}^r_j)$ would dominate over the weak scattering from $\Omega^{s,t}$. In order to overcome this error-sensitivity problem, we discuss the following *local tuning technique* to be incorporated into the above two-stage reconstruction of Ω^m.

Local tuning technique. Let $\{U_j\}_{j\in\mathscr{T}_1}$ and $\{r_j\}_{j\in\mathscr{T}_2}$ be the two given sets of rotations and scalings and $\mathscr{T}_h$ be the sampling mesh introduced in Sect. 3.3.2.1, and $(\Sigma_j; \widehat{z}_j, \widehat{U}_j, \widehat{r}_j)$, $j = 1, \ldots, l$ be the reconstructed scatterers described above. For a properly chosen $\delta \in \mathbb{R}_+$, let $\mathscr{N}_1^j$, $\mathscr{N}_2^j$, and $\mathscr{N}_3^j$ be, respectively, δ-neighborhoods of $\widehat{z}_j$, $\widehat{U}_j$ and $\widehat{r}_j$, $j = 1, \ldots, l$. Then let $\{\mathscr{T}_{h'}, \{U_j\}_{j\in\mathscr{T}_1'}, \{r_j\}_{j\in\mathscr{T}_2'}\}$ be an arbitrary refined mesh of $\{\mathscr{T}_h \cap \mathscr{N}_1^j, \{U_j\}_{j\in\mathscr{T}_1} \cap \mathscr{N}_2^j, \{r_j\}_{j\in\mathscr{T}_3} \cap \mathscr{N}_3^j\}$, then we call

$$\widehat{\widehat{\Omega}}^r_j := (\Sigma_j; \widehat{\widehat{z}}_j, \widehat{\widehat{r}}_j, \widehat{\widehat{U}}_j) \quad \text{for } \widehat{\widehat{z}}_j \in \mathscr{T}_{h'}, \quad \widehat{\widehat{r}}_j \in \{r_j\}_{j\in\mathscr{T}_2'}, \quad \widehat{\widehat{U}}_j \in \{U_j\}_{j\in\mathscr{T}_1'} \tag{3.3.51}$$

a *local tuneup* of $\widehat{\Omega}^r_j = (\Sigma_j; \widehat{z}_j, \widehat{r}_j, \widehat{U}_j)$ relative to $\{\mathscr{T}_{h'}, \{r_j\}_{j\in\mathscr{T}_2'}, \{U_j\}_{j\in\mathscr{T}_1'}\}$ for $j = 1, 2, \ldots, l$.

Now we define

$$\widehat{\widehat{\Omega}}^r := \bigcup_{j=1}^{l} \widehat{\widehat{\Omega}}^r_j, \tag{3.3.52}$$

where each $\widehat{\widehat{\Omega}}^r_j$ is a local tuneup relative to $\left\{\mathscr{T}_{h'}, \{r_j\}_{j\in\mathscr{T}_2'}, \{U_j\}_{j\in\mathscr{T}_1'}\right\}$ for $j = 1, 2, \ldots, l$. We call $\widehat{\widehat{\Omega}}^r$ a local tuneup of $\widehat{\Omega}^r$ relative to $\{\mathscr{T}_{h'}, \{r_j\}_{j\in\mathscr{T}_2'}, \{U_j\}_{j\in\mathscr{T}_1'}\}$ for $j = 1, 2, \ldots, l$.

With the above preparations, the local tuning can be proceeded as follows:

For each local tuneup $\widehat{\widehat{\Omega}}^r$, we compute

$$\widehat{a}_k(\hat{x}, d) := a_k(\hat{x}, d; \Omega^m) - a_k(\hat{x}, d; \widehat{\widehat{\Omega}}^r), \tag{3.3.53}$$

then apply the resulting far-field data to Scheme I. By running through all the local tuneups relative to $\left\{\mathscr{T}_{h'}, \{r_j\}_{j\in\mathscr{T}_2'}, \{U_j\}_{j\in\mathscr{T}_1'}\right\}$ according to the above procedure, one can locate all the clustered local maximum points on $\mathscr{T}_h$, which represents the locations of the small scatterer components of $\Omega^{s,t}$.

By using the local tuning technique, one can not only locate the small-size scatterers components of $\Omega^{s,t}$, but also improve the reconstruction of the regular-size scatterers. Indeed, it can be easily seen that the local tuneup, $\widehat{\widehat{\Omega}}^r$, which is used in the local tuning that can produce the clustered local maximum points for Scheme I, is a more accurate updating of the reconstruction $\widehat{\Omega}^r$.

In summary, we are now ready to formulate Scheme III for locating the multiple multiscale scatterers of Ω^m in (3.3.42).

Scheme III

---

1. Collect a single far-field measurement $a_k(\hat{x}, d; \Omega^m)$ corresponding to $\Omega^m = \Omega^{s,t} \cup \Omega^r$ in (3.3.42).
2. Select a sampling region with a mesh $\mathscr{T}_h$ containing Ω^m.
3. Let Ω^r be given as in (3.3.26). Apply Scheme II with $a_k(\hat{x}, d; \Omega^m)$ as the far-field data, and locate the rough scatterer components of Ω^m,

$$\widehat{\Omega}_j^r = (\Sigma_j; \widehat{z}_j, \widehat{r}_j, \widehat{U}_j), \quad \Sigma_j \in \mathscr{A}, \quad j = 1, \ldots, l, \tag{3.3.54}$$

   where $\widehat{z}_j \in \mathscr{T}_h, \widehat{r}_j \in \{r_j\}_{j\in\mathscr{T}_2}$ and $\widehat{U}_j \in \{U_j\}_{j\in\mathscr{T}_1}$ (cf. Sect. 3.3.2.1).
4. Apply the *local tuning technique* as stated below to update $\widehat{\Omega}_j^r$, $j = 1, \ldots, l$, and locate the small-scale components of $\Omega^{s,t}$.
   a. For each reconstructed $\widehat{\Omega}_j^r$ in (3.3.54), formulate the refined local sampling meshes $\mathscr{T}_{h'}$, $\{r_j\}_{j\in\mathscr{T}_2'}$ and $\{U_j\}_{j\in\mathscr{T}_1'}$, $j = 1, 2, \ldots, l$.
   b. For a local tuneup given in (3.3.51)–(3.3.52), calculate the far-field pattern $\widehat{a}_k(\hat{x}, d)$ according to (3.3.53).
   c. Using $\widehat{a}_k(\hat{x}, d)$ obtained in b) as the far-field data, apply Scheme I to locate the significant local maximum points on $\mathscr{T}_h \backslash \cup_{j=1}^{l} \mathscr{N}_1^j$.
   d. Repeat b) and c) for all possible local tuneups relative to $\{\mathscr{T}_{h'}, \{r_j\}_{j\in\mathscr{T}_2'}, \{U_j\}_{j\in\mathscr{T}_1'}\}$. The clustered local maximum points on $\mathscr{T}_h \backslash \cup_{j=1}^{l} \mathscr{N}_1^j$ are the positions corresponding to the scatterer components of $\Omega^{s,t}$.
   e. Update $\widehat{\Omega}^r$ to the local tuneup $\widehat{\widehat{\Omega}}{}^r$ which generates the clustered local maximum points in d).

Remark 3.3.5 By concatenating Schemes I and II with the local tuning technique, Scheme III can effectively locate multiple multiscale scatterers. For the practical consideration, one can easily see from our discussions that as long as the scattering strengths from the small scatterer components of $\Omega^{s,t}$ are more significant than the measurement noise involved, Scheme III can produce reasonable reconstructions for multiple multiscale scatterers.

3.3.3.2 Numerical Experiments

In this subsection, we test some 3D multiscale imaging problem using Scheme III. The wave number k is chosen to be 5 and the incident direction is $d = (0, 0, -1)$. The synthetic data are obtained for the revolving solids of the 2D shapes **K** and **P** along the x-axis, which are for short still denoted by **K** and **P** without ambiguities. As for each reference component, we rotate it every 90° in the x-y, y-z and z-x planes; see, e.g., four different orientations of the peanut in the z-x plane are shown in Fig. 3.27. Three different sizes of the reference components are tested, namely scaled by a factor of 0.3, 1 and 1.5 respectively.

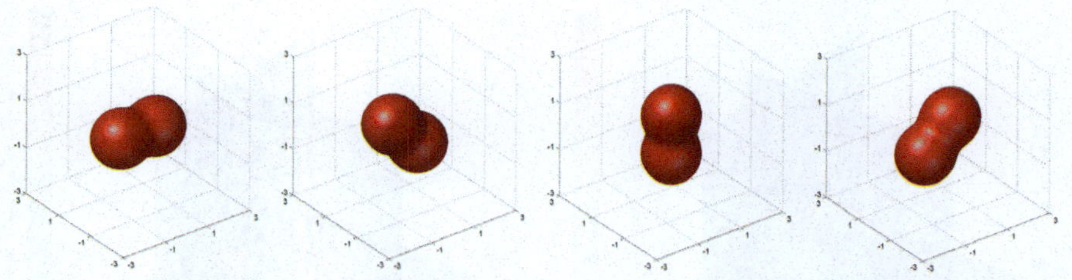

Fig. 3.27 Basic Scatterers: Peanut (multiscale)

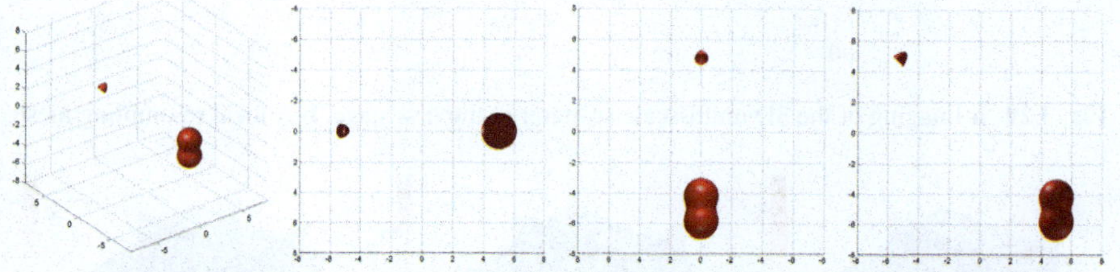

Fig. 3.28 True scatterer components in Example 6

Example 6 We consider a 3D multiscale scatterer consisting of two components, a small sound-hard kite scaled by a factor 0.3 located at $(-5,\ 0,\ 5)$ and a large sound-hard peanut with no scaling and located at $(5,\ 0,\ -5)$, see Fig. 3.28.

In the first stage, we extract the information of the regular-size component using the indicator function of Scheme II by computing the inner product with a priori known far-field patterns associated with those reference scatterer components with different orientations and sizes. We can find the approximate position of the larger peanut component of regular size when the reference scatterer is at its upright position, see Fig. 3.29a.

Next, we adopt the local tuning (resampling) technique discussed in Sect. 3.3.3.1 to search a small cubic mesh around the rough position of the peanut determined by the local maximum, which is shown in Fig. 3.29b.

In the final stage, the location of the smaller component can be obtained by performing a local searching of each grid point in the cubic mesh. In Fig. 3.30, as the search grid points approach gradually from $(4.8\,,0\,,-5)$ to $(5\,,0\,,-5)$ (from left to right), the value distribution of the indicator function in Scheme I displays an interesting change of the highlighted position. In the right plot of Fig. 3.30, the red dot indicates an approximate position of the smaller kite component, which agrees with the exact one very well. In such a way, the smaller sound-hard kite could be positioned, and it helps us finely tune the position of the peanut and update it to be around $(5\,,0\,,-5)$. They could now be well approximated in Fig. 3.31 by combining the relevant shape, scale and size information hidden in the reference data.

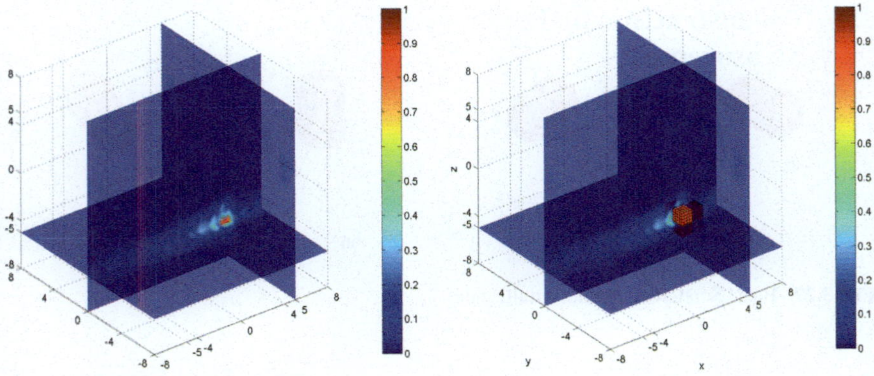

(a) (b)

Fig. 3.29 **a** Imaging of the 3D multi-scale scatterer in the first stage. **b** A local resampling mesh

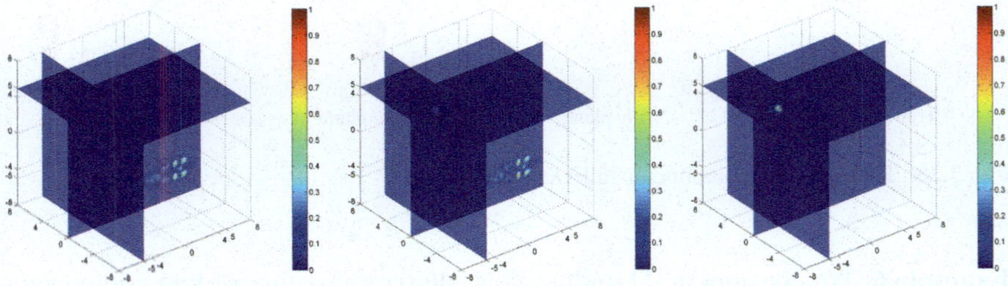

Fig. 3.30 Example 6. Indicating behavior of local re-sampling technique. From left to right: Imaging of indicator function value distribution at the sampling points (4.8, 0, −5), (4.9, 0, −5) and (5, 0, −5), respectively

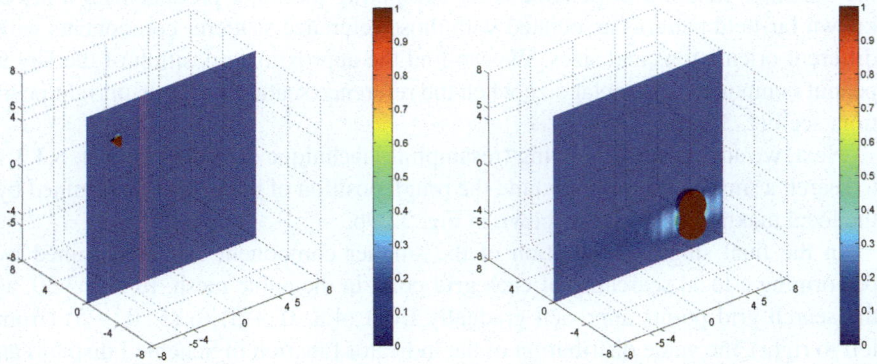

Fig. 3.31 Example 6. From left to right: local sampling procedure at the sampling points (−4.8, 0, 5), (−4.9, 0, 5) and (5, −5), respectively. Imaging of the position of the reconstructed kite component

3.4 Reconstruction by Phaseless Backscattering Measurements

In this section, we discuss an inverse scattering scheme of recovering a polyhedral obstacle in $\mathbb{R}^n$, $n = 2, 3$, by only a few high-frequency acoustic backscattering measurements. The obstacle could be sound-soft or sound-hard. It is shown that the modulus of the far-field pattern in the backscattering aperture possesses a certain local maximum behavior, from which one can determine the exterior normal directions of the front sides/faces. Then by using the phaseless backscattering data corresponding to a few incident plane waves with suitably chosen incident directions, one can determine the exterior unit normal vector of each side/face of the obstacle. After the determination of the exterior unit normals, the recovery is reduced to a finite-dimensional problem of determining a location point of the obstacle and the distance of each side/face away from the location point. For the latter reconstruction, we need make use of the far-field data with phases. Numerical experiments are also presented to illustrate the effectiveness of the discussed scheme.

We follow the treatments in [58]. The obstacle could be sound-soft or sound-hard. The crux is the observation that the modulus of the far-field pattern in the backscattering aperture possesses a certain local maximum behavior, from which one can determine the exterior normal directions of the front sides/faces. Then by using the modulus of the backscattering data corresponding to a few incident plane waves with a high wavenumber and a few suitably chosen incident directions, one can determine the exterior unit normal vector of each side/face of the obstacle. After the determination of the exterior unit normals, the recovery is reduced to a finite-dimensional problem of determining a location point of the obstacle and the distance of each side/face away from the location point. For the latter reconstruction, the far-field data with phases would be used. Our study is based on the high-frequency asymptotics, namely the Kirchhoff or the physical optics approximation. However, our numerical experiments show that the high-frequency requirement could be relaxed to a certain extent. Moreover, in order to simplify the discussion, we focus on convex polyhedral obstacles in the present study. However, through our theoretical arguments, it can be expected that the method developed would work for non-convex obstacles, but under certain geometrical constraints. We focus on showing the inverse scattering scheme for convex obstacles in this section.

In Sect. 3.4.1, we consider the physical optics approximation on the high-frequency scattering from an admissible polyhedral obstacle, and derive the local maximum behavior of the modulus of the corresponding far-field pattern. In Sect. 3.4.2, we present the recovery scheme in detail. Section 3.4.3 is devoted to the numerical experiments to validate the applicability and effectiveness of the discussed method. Some discussion is presented in Sect. 3.4.3.

3.4.1 Physical Optics Approximation

Throughout the present subsection, we let $k \in \mathbb{R}_+$ and $d \in \mathbb{S}^{n-1}$ be fixed. Let D be a convex polygon in $\mathbb{R}^2$ or a convex polyhedron in $\mathbb{R}^3$, such that

$$\partial D = \bigcup_{j=1}^{m} C_j, \tag{3.4.1}$$

where each C_j represents an open side/face of ∂D, and is referred to as a *cell*. In what follows, D is called a polyhedral obstacle. Let $\nu(x) \in \mathbb{S}^{n-1}$, $x \in \partial D$ denote the exterior unit normal vector to ∂D, and we set

$$\nu_j := \nu(x) \ \text{ when } x \in C_j, \quad j = 1, 2, \ldots, m. \tag{3.4.2}$$

Clearly, ν_j is a constant unit vector.

Define

$$\partial D^+ := \{x \in \partial D;\ \nu(x) \cdot d \geq 0\} \ \text{ and } \ \partial D^- := \{x \in \partial D;\ \nu(x) \cdot d < 0\}$$

to be, respectively, the back-face and front-face of ∂D with respect to the incident direction d.

Let h_0, h_1 and h_2 be fixed a priori positive constants. It is assumed that

$$(i) \ \ k \cdot \operatorname{diam}(D) \gg 1; \tag{3.4.3}$$

$$(ii) \ \ \min_{1 \leq j \leq m} \operatorname{diam}_{\mathbb{R}^{n-1}}(C_j) \geq h_0; \tag{3.4.4}$$

$$(iii) \ \ h_1 \leq \min_{1 \leq j, j' \leq m,\ j \neq j'} \angle(\nu_j(y), \nu_{j'}(y)) \leq h_2 \ \text{ for } \ y \in \partial D. \tag{3.4.5}$$

Roughly speaking, (3.4.5) implies that the obstacle should not be very "round" or "sharp", and a generic condition which can guarantee this assumption is that the angle between any two adjacent cells is bounded below and above by certain constants (depending on the obstacle). Assumption (i) means that we are considering the acoustic scattering in the high-frequency regime. If a polyhedral obstacle D satisfies the above three assumptions, then it is called an *admissible* obstacle with respect to the incident plane wave $e^{ikx\cdot d}$.

Denote

$$\mathbb{S}^{n-1}_+ := \{\hat{x} \in \mathbb{S}^{n-1};\ \hat{x} \cdot d \geq 0\} \ \text{ and } \ \mathbb{S}^{n-1}_- := \{\hat{x} \in \mathbb{S}^{n-1};\ \hat{x} \cdot d < 0\}$$

the forward-scattering and backscattering apertures, respectively.

Let $C_j \subset \partial D^-$ be a front-cell of ∂D, and $\nu_j \in \mathbb{S}^{n-1}_-$ denote its unit normal vector pointing to the exterior of D. Define $\hat{x}_j \in \mathbb{S}^{n-1}$ satisfying

$$(d - \hat{x}_j) \parallel \nu_j$$

to be the critical observation direction with respect to d and ν_j; see Fig.3.32 for a 2D illustration. We note that one clearly has $\hat{x}_j \in \mathbb{S}_-^{n-1}$. It is directly calculated that the critical direction is given by

$$\hat{x}_j = d - 2(d \cdot \nu_j)\nu_j. \tag{3.4.6}$$

On the other hand, for the subsequent use, we note that by using (3.4.6) and the fact that $d \cdot \nu_j < 0$, one has by innerly producting both sides of (3.4.6) with d

$$d \cdot \nu_j = -\sqrt{\frac{1 - \hat{x}_j \cdot d}{2}}. \tag{3.4.7}$$

Hence, by combining (3.4.6) and (3.4.7), one further has

$$\nu_j = \frac{\hat{x}_j - d}{\sqrt{2(1 - \hat{x}_j \cdot d)}}. \tag{3.4.8}$$

Next we present the major result motivating the recovery scheme that we are going to present in the next section. Before that, we give a numerical example by plotting the real part of the scattered wave field and the square power of the modulus of the associated far-field pattern corresponding to a sound-soft triangle due to an incident plane wave impinging from southeast to northwest; see Fig. 3.33 for the illustration.

Clearly, one can see a certain local maximum behavior from Fig. 3.33b. The observation direction (black arrow) associated with the largest magnitude of phaseless data points to the incident/forward-scattering direction, which gives no information of the

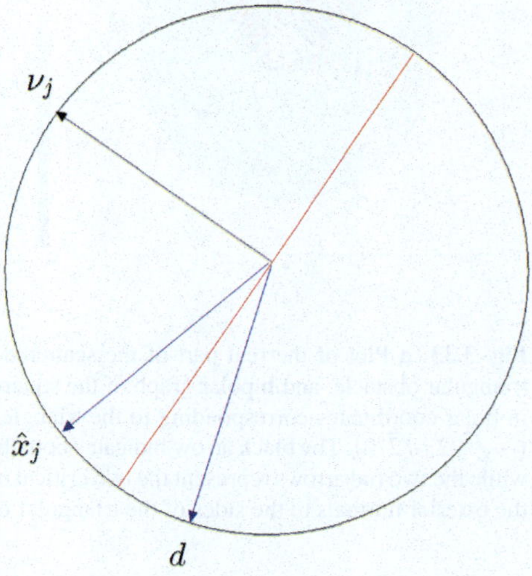

Fig. 3.32 2D illustration of the relation between the incident direction d, the exterior unit normal vector ν_j and the critical observation direction $\hat{x}_j$

obstacle. While the other two critical observation directions (red arrows) associated with the second and third largest magnitude of phaseless data provide profound information of the unknown obstacle, which signify the major reflection angles due to the physical optics approximation in the high frequency scattering. The exterior unit normal direction can thus be determined by connecting the arrow heads of incident and critical observation directions by yellow lines. Those unit vectors (green arrows) parallel to the yellow lines are the desired normal directions of those sides of the triangular obstacle in the backscattering aperture as shown in Fig. 3.33b. We shall present a mathematical justification for this local maximum behavior, which also forms the basis of our subsequent recovery scheme.

We first give some discussion on the Kirchhoff or the physical optics approximation (cf. [22, 26, 38, 47, 70, 72]), which shall play a key role in our mathematical justification of the local maximum behavior observed in Fig. 3.33. It states that for the scattering from a convex polyhedral obstacle due to a high-frequency plane wave, the wave field near the boundary of the obstacle is composed of two parts: the contribution of the incident and reflected waves where they are present, and the contribution of the diffraction due to the corners and/or the edges of the obstacle. The Kirchhoff or the physical optics approximation takes the first contribution as the total wave field near the boundary of the obstacle. However, we would like to remark the study on rigorously justifying such approximation is still not fully understood and we refer to [22] for an excellent account on the existing progress in the literature. For the current study, we assume that the physical optics approximation holds true.

Throughout the rest of the section, we let a cell C be parameterized as

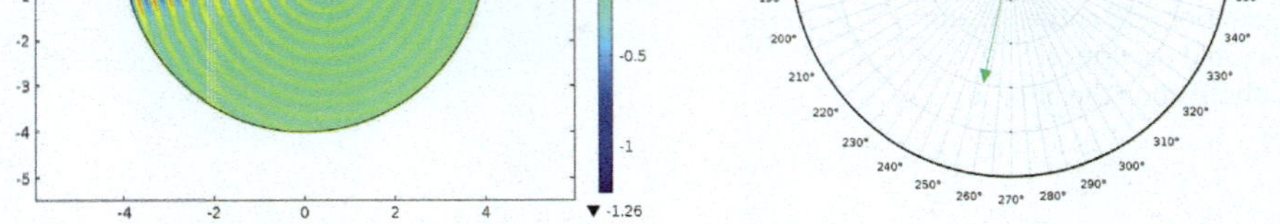

(a) (b)

Fig. 3.33 **a** Plot of the real part of the scattered wave $\Re(u^s(x))$ in the vicinity of a sound-soft triangular obstacle, and **b** polar graph of the square power of the phaseless far-field data $|u^\infty(\hat{x})|^2$ in polar coordinates corresponding to the triangle due to an incident plane wave $e^{ikx\cdot d}$ with $d = (-\sqrt{2}/2, \sqrt{2}/2)$. The black arrow indicates both the incident and the forward-scattering directions, while the two red arrows represent the two critical observation directions. The green arrows indicate the exterior normals to the sides of the triangular obstacle

$$\langle x, \nu \rangle = l, \quad x \in C, \tag{3.4.9}$$

where $\nu \in \mathbb{S}^{n-1}$ is the unit normal vector to C pointing to the exterior of the obstacle, and $l \geq 0$ denotes the distance between the origin and the line/plane containing C. Let C^0 denote the affine cell of C defined by

$$\langle x, \nu \rangle = 0, \quad x \in C^0. \tag{3.4.10}$$

In what follows, we let Π_{C^0} denote the Euclidean reflection with respect to C^0. Now we consider the scattering locally near a boundary cell, say C_j, due to a plane wave $e^{ikx\cdot d}$. Let

$$v_j(x) := e^{ik(x-x_0)\cdot(\Pi_{C_j^0} d)} \cdot e^{ikx_0\cdot d}, \quad 1 \leq j \leq m, \tag{3.4.11}$$

where $x_0 \in C_j$ is a fixed point, and $x \in \mathbb{R}^n \backslash \overline{D}$. It is easily verified that $v_j(x)$ satisfies $(\Delta + k^2)v_j = 0$. Moreover, one can further verify that

$$u^i(x) - v_j(x) = 0 \quad \text{and} \quad \frac{\partial(u^i + v_j)}{\partial \nu}(x) = 0 \quad \text{on } C_j. \tag{3.4.12}$$

$v_j(x)$ is the reflected wave field of $u^i(x)$ with respect to the cell C_j. Using the physical optics approximation, we have

Lemma 3.4.1 *Let D be an admissible polyhedral obstacle with respect to the plane wave field $u^i = e^{ikx\cdot d}$, as described in* (3.4.1)–(3.4.5). *Using the physical optics approximation, one has*

$$\frac{\partial u}{\partial \nu}(x) \approx \begin{cases} 2\frac{\partial u^i}{\partial \nu}(x), & x \in C_j \subset \partial D^-, \ 1 \leq j \leq m, \\ 0, & x \in C_{j'} \subset \partial D^+, \ 1 \leq j' \leq m, \end{cases} \tag{3.4.13}$$

if D is sound-soft; and

$$u(x) \approx \begin{cases} 2u^i(x), & x \in C_j \subset \partial D^-, \ 1 \leq j \leq m, \\ 0, & x \in C_{j'} \subset \partial D^+, \ 1 \leq j' \leq m, \end{cases} \tag{3.4.14}$$

if D is sound-hard.

Proof According to our earlier discussion on the physical optics approximation, one takes

$$\frac{\partial u}{\partial \nu_j}(x) \approx \frac{\partial(u - v^j)}{\partial \nu_j}(x), \quad x \in C_j \subset \partial D^-, \tag{3.4.15}$$

if D is sound-soft. Then by using (3.4.11), one directly verifies (3.4.13) for $x \in \partial D^-$, whereas if $C_{j'} \subset \partial D^+$, it is not illuminated, where one then takes $\partial u/\partial \nu_{j'} \approx 0$. If D is sound-hard, then one takes

$$u(x) \approx u(x) + v^j(x), \quad x \in C_j \subset \partial D^-, \tag{3.4.16}$$

which readily verifies (3.4.14). □

In the sequel, we let

$$\Phi(x, y) = \frac{e^{\mathrm{i}k|x-y|}}{4\pi|x-y|}, \quad n = 3; \quad \frac{\mathrm{i}}{4} H_0^{(1)}(k|x-y|), \quad n = 2; \quad x \neq y,$$

where $H_0^{(1)}$ denotes the zeroth-order Hankel function of the first kind. Φ is the fundamental solution to $-\Delta - k^2$. The following lemma is needed and its proof can be found in [26].

Lemma 3.4.2 *For the scattering of a plane wave field u^i in (2.1.1) from an obstacle D, we have*

$$u(x; D, u^i) = u^i(x) + \int_{\partial D} \left\{ \frac{\partial \Phi(x, y)}{\partial \nu(y)} u(y) - \Phi(x, y) \frac{\partial u}{\partial \nu}(y) \right\} ds(y), \quad x \in \mathbb{R}^n \backslash \overline{D}, \tag{3.4.17}$$

and

$$u^\infty(\hat{x}; D, u^i) = \gamma(n, k) \left[\int_{\partial D} \left\{ \frac{\partial e^{-\mathrm{i}k\hat{x}\cdot y}}{\partial \nu(y)} u(y) - e^{-\mathrm{i}k\hat{x}\cdot y} \frac{\partial u}{\partial \nu}(y) \right\} ds(y) \right], \tag{3.4.18}$$

where the dimensional parameter γ is given by

$$\gamma(n, k) = \frac{1}{4\pi} \quad \text{when } n = 3; \quad \frac{e^{\mathrm{i}\frac{\pi}{4}}}{\sqrt{8\pi k}} \quad \text{when } n = 2. \tag{3.4.19}$$

Lemma 3.4.3 *Using the physical optics approximation in Lemma 3.4.1, for the scattering of a plane wave u^i in (2.1.1) from an admissible polyhedral obstacle D, one has that if D is sound-soft*

$$u^\infty(\hat{x}) \approx -2\gamma(n, k) \int_{\partial D^-} \frac{\partial e^{\mathrm{i}ky\cdot d}}{\partial \nu(y)} e^{-\mathrm{i}k\hat{x}\cdot y} \, ds(y), \tag{3.4.20}$$

whereas if D is sound-hard

$$u^\infty(\hat{x}) \approx 2\gamma(n, k) \int_{\partial D^-} \frac{\partial e^{-\mathrm{i}k\hat{x}\cdot y}}{\partial \nu(y)} e^{\mathrm{i}ky\cdot d} \, ds(y). \tag{3.4.21}$$

Proof This is a straightforward consequence of (3.4.18) in Lemma 3.4.2 and, (3.4.13) and (3.4.14) in Lemma 3.4.1. □

By using Lemma 3.4.3, we are now in a position to present the local maximum behavior of $|u_\infty(\hat{x}; D, k, d)|$.

Theorem 3.4.1 *Let D be an admissible sound-soft or sound-hard polyhedral obstacle with respect to the incident plane wave $e^{ikx\cdot d}$ as described earlier. Suppose that $C_j \subset \partial D^-$ is a front cell of the obstacle, and ν_j is the unit normal vector to C_j pointing to the exterior of D, $1 \le j \le m$. Let $\hat{x}_j \in \mathbb{S}^{n-1}$ be the critical observation direction with respect to d and ν_j. Under the physical optics approximation, $\hat{x}_j$ is a local maximum point of $|u^\infty(\hat{x}; D, e^{ikx\cdot d})|$.*

Proof We first consider the case that D is an admissible sound-soft polyhedral obstacle. By (3.4.20) in Lemma 3.4.3, we have

$$
\begin{aligned}
u^\infty(\hat{x}; D, k, d) &\approx -2\gamma(n,k)\int_{\partial D^-}\frac{\partial e^{iky\cdot d}}{\partial\nu(y)}e^{-ik\hat{x}\cdot y}\,ds(y)\\
&= -2\gamma(n,k)\int_{\partial D^-} ik\nu(y)\cdot d e^{iky\cdot(d-\hat{x})}\,ds(y)\\
&= -2\gamma(n,k)\cdot ik\left[\int_{C_j}\nu_j\cdot d\, e^{iky\cdot(d-\hat{x})}\,ds(y) + \int_{\partial D^-\backslash C_j}\nu(y)\cdot d\, e^{iky\cdot(d-\hat{x})}\,ds(y)\right]\\
&= \widetilde{\gamma}(n,k)\cdot[I_1(\hat{x}) + I_2(\hat{x})],
\end{aligned}
\tag{3.4.22}
$$

where $\widetilde{\gamma}(n,k) := -2\gamma(n,k)\cdot ik$ and

$$
I_1(\hat{x}) := \int_{C_j}\nu_j\cdot d\, e^{iky\cdot(d-\hat{x})}\,ds(y), \quad I_2(\hat{x}) := \int_{\partial D^-\backslash C_j}\nu(y)\cdot d\, e^{iky\cdot(d-\hat{x})}\,ds(y).
$$

We next analyze the behavior of $I_\alpha(\hat{x})$, $\alpha = 1, 2$, in a small neighborhood of $\hat{x}_j$ on $\mathbb{S}^{n-1}$, say Γ_j, $1 \le j \le m$. Since

$$
(d - \hat{x}_j) \parallel \nu_j, \tag{3.4.23}
$$

by assumption (3.4.5), we see that there exists a non-asymptotic constant $\varepsilon_0 \in \mathbb{R}_+$ such that

$$
|\tau(y)\cdot(d - \hat{x}_j)| \ge \varepsilon_0, \quad y \in \partial D^-\backslash C_j, \tag{3.4.24}
$$

where $\tau(y)$ represents a unit tangent vector on ∂D. Hence, by (3.4.24) and (3.4.4), one has by direct calculations that

$$
|I_2(\hat{x})| \sim \frac{1}{k^{n-1}h_0} \ll 1, \quad \hat{x} \in \Gamma_j. \tag{3.4.25}
$$

On the other hand, for $\hat{x} \in \Gamma_j$, we have

$$
\begin{aligned}
I_1(\hat{x}) &= \int_{C_j}\nu_j\cdot d\, e^{iky\cdot(d-\hat{x})}\,ds(y)\\
&= e^{iky_0\cdot(d-\hat{x})}\int_{C_j^0}\nu_j\cdot d\, e^{iky\cdot(d-\hat{x})}\,ds(y),
\end{aligned}
\tag{3.4.26}
$$

where y_0 is any fixed point on C_j. Since

$$y \cdot (d - \hat{x}_j) = 0 \ \text{ for } \ y \in C_j^0,$$

one clearly sees that $|I_1(\hat{x})|$ achieves its local maximum value at $\hat{x}_j$, which in combination with (3.4.25) completes the proof of the theorem for the sound-soft case.

The sound-hard case can be shown by following a similar argument. By using (3.4.21), one has

$$\begin{aligned} u^\infty(\hat{x}; D, k, d) &\approx -2\gamma(n,k) \cdot ik \int_{\partial D^-} \nu(y) \cdot \hat{x}\, e^{iky\cdot(d-\hat{x})}\, ds(y) \\ &= \widetilde{\gamma}(n,k) \cdot [J_1(\hat{x}) + J_2(\hat{x})], \end{aligned} \tag{3.4.27}$$

where

$$J_1(\hat{x}) = \int_{C_j} \nu_j \cdot \hat{x}\, e^{iky\cdot(d-\hat{x})}\, ds(y), \quad J_2(\hat{x}) = \int_{\partial D^- \setminus C_j} \nu(y) \cdot \hat{x}\, e^{iky\cdot(d-\hat{x})}\, ds(y). \tag{3.4.28}$$

We consider the behavior of $J_\alpha(\hat{x})$, $\alpha = 1, 2$, in a small neighborhood of $\hat{x}_j$ on $\mathbb{S}^{n-1}$, say Σ_j, $1 \le j \le m$. Following a similar argument to (3.4.26), together with the fact in (3.4.23), one can see that $|J_1(\hat{x})|$ achieves its local maximum value at $\hat{x}_j$. On the other hand, by a similar argument in deriving (3.4.25), one has that

$$|J_2(\hat{x})| \ll 1 \ \text{ for } \ \hat{x} \in \Sigma_j, \ \ 1 \le j \le m. \tag{3.4.29}$$

Hence, $\hat{x}_j$ is local maximum point of $|u^\infty(\hat{x}; D, e^{ikx\cdot d})|$.

The proof is complete. □

Remark 3.4.1 In Theorem 3.4.1, we only consider the local maximum behavior of $|u^\infty(\hat{x})|$ in the backscattering aperture. On the other hand, by (3.4.22), one clearly sees that $|u^\infty(\hat{x})|$ achieves its (global) maximum value at $\hat{x} = d$. However, this maximum behavior in the forward-scattering aperture gives us no useful information on the obstacle, which is demonstrated and verified in Fig. 3.33.

Remark 3.4.2 It is noted from (3.4.27) that for a sound-hard obstacle if $\hat{x} \perp \nu_j$, then both J_1 and J_2 vanish, and hence such an $\hat{x}$ should be a local minimum point of $|u^\infty(\hat{x}; D, k, d)|$. However, it cannot be guaranteed that such an $\hat{x}$ belongs to the backscattering aperture. Nevertheless, it could be used in combination with the critical observation direction as a double indicator.

3.4.2 Recovery Scheme

Based on our study in the previous subsection, we shall present a recovery scheme of identifying an admissible polyhedral obstacle D. Let $x_0 \in D$ be a fixed location point, and ∂D be described in (3.4.1). Let each boundary cell C_j of ∂D, $1 \le j \le m$, be parameterized as follows

$$\langle x - x_0, \nu_j \rangle = l_j, \quad x \in C_j, \quad 1 \le j \le m, \tag{3.4.30}$$

where $\nu_j \in \mathbb{S}^{n-1}$ denotes the exterior unit normal to C_j, and l_j denotes the distance between the origin and the line/plane containing the cell $C_j - x_0 := \{x - x_0; x \in C_j\}$. Clearly, x_0, ν_j and l_j, $j = 1, 2, \ldots, m$, uniquely determine the obstacle D. Hence, our recovery scheme consists of steps in recovering those ingredients.

We first consider the reconstruction of the location point x_0. This is based the single-shot locating scheme developed in the work [53]. For completeness and self-containedness, we briefly present the scheme in what follows. The next theorem forms the basis of the method.

Theorem 3.4.2 ([53]) *Let $\widetilde{k} \in \mathbb{R}_+$ be a wavenumber such that $\widetilde{k} \cdot \mathrm{diam}(D) \ll 1$. For any fixed $d \in \mathbb{S}^{n-1}$, let $\widetilde{u}^\infty(\hat{x}) = \widetilde{u}^\infty(\hat{x}; \widetilde{k}, d, D)$ be the far-field pattern of D due to the incident plane wave $e^{i\widetilde{k}x\cdot d}$. Consider the following functional,*

$$I(z) := \frac{1}{\|\widetilde{u}^\infty(\hat{x})\|^2_{L^2(\mathbb{S}^{n-1})}} \sum_{\alpha=0}^{1} \sum_{\beta=-\alpha}^{\alpha} \left| \left\langle \widetilde{u}^\infty(\hat{x}), e^{i\widetilde{k}(d-\hat{x})\cdot z} Y_\alpha^\beta(\hat{x}) \right\rangle_{L^2(\mathbb{S}^{n-1})} \right|, \tag{3.4.31}$$

where $z \in \mathbb{R}^n$, and Y_α^β denotes the spherical harmonics of order $\alpha \in \mathbb{N}$. Then $I(z)$ achieves its maximum value at the point $z = x_0$.

By Theorem 3.4.2, the reconstruction of x_0 can be proceeded as follows. First, one chooses a single incident plane wave $e^{i\widetilde{k}x\cdot d}$ with a sufficiently small wavenumber $\widetilde{k} \in \mathbb{R}_+$, and then collects the corresponding far-field data. With the far-field data, one can calculate the indicator functional $I(z)$ in (3.4.31) for any z from a sampling mesh containing D. Finally, the location point x_0 can be determined by locating the maximum point of $I(z)$.

We next consider the recovery of the exterior unit normal vectors, ν_j, $j = 1, 2, \ldots, m$, and this is based on our results in Sect. 3.4.1, particularly Theorem 3.4.1 and the discussion following its proof. To that end, we let $k \in \mathbb{R}_+$ be a sufficiently large wavenumber such that (3.4.3) is fulfilled. Then, let d_α, $\alpha = 1, 2, \ldots, p$, be a few properly chosen incident directions. The basic requirement is that the union of the front-faces with respect to d_α, $\alpha = 1, 2, \ldots, p$, should cover the whole boundary ∂D. That is, the number of the far-field measurements needed for the reconstruction depends on the complexity of the scattering obstacle, and as a generic principle, the more measurements are used, the more accurate and stable reconstructions can be expected. Furthermore, we assume that D is admissible with respect to each d_α, $1 \le \alpha \le p$. As remarked earlier, this assumption is generically satisfied by an overall not very "round" nor very "sharp" obstacle.

For each α, $1 \le \alpha \le q$, one calculates $|u^\infty(\hat{x}; k, d_\alpha, D)|$ for $\hat{x} \in \mathbb{S}_-^{n-1}$. Here, it is remarked that in our subsequent numerical experiments, we actually calculate $|u^\infty(\hat{x}; k, d_\alpha, D)|^2$ for $\hat{x} \in \mathbb{S}_-^{n-1}$ in order to achieve sharp contrasts at the local maximum points. Clearly, according to Theorem 3.4.1, by locating the local maximum points of $|u^\infty(\hat{x}; k, d_\alpha, D)|$ for $\hat{x} \in \mathbb{S}_-^{n-1}$, one can then determine the corresponding critical observation directions, say,

$$\hat{x}_j^\alpha \in \mathbb{S}_-^{n-1}, \quad j = 1, 2, \ldots, m_\alpha. \tag{3.4.32}$$

Then by the formula (3.4.8), one can determine the unit vectors using d_α and $\hat{x}_j^\alpha$ in (3.4.32). We denote them by ν_j^α, $j = 1, 2, \ldots, m_\alpha$, and they clearly correspond to the exterior normal directions to some boundary cells of ∂D. Since the incident directions d_α, $\alpha = 1, 2, \ldots, p$ were such chosen that the union of the front-faces with respect to d_α, $\alpha = 1, 2, \ldots, p$ covers the whole boundary ∂D. By performing the recovery described above, one can clearly recover all of the exterior unit normal vectors ν_j to the boundary cells C_j of ∂D, $j = 1, 2, \ldots, m$. There is an obvious byproduct that the number of cells of the obstacle is also determined. Here, since multiple incident directions d_α's are used, as an implementation issue, we would like to remark that some of the critical observation directions, say ν_j^α and $\nu_{j'}^{\alpha'}$ with $\alpha \neq \alpha'$, may actually yield the same exterior unit normal vector to ∂D. In the practical reconstruction, due to the measurement error, this is actually the case that ν_j^α and $\nu_{j'}^{\alpha'}$ only slightly differ from each other. In such a case, we let $\hat{x}_j^\alpha$ and $\hat{x}_{j'}^{\alpha'}$ be the critical observation directions, respectively, corresponding to ν_j^α and $\nu_{j'}^{\alpha'}$ with $\alpha \neq \alpha'$. We then compare the amplitudes of $|u^\infty(\hat{x}_j^\alpha; k, d_\alpha)|$ and $|u^\infty(\hat{x}_{j'}^{\alpha'}; k, d_{\alpha'})|$, and say that ν_j^α is more effective than $\nu_{j'}^{\alpha'}$ if there holds $|u^\infty(\hat{x}_j^\alpha; k, d_\alpha)| > |u^\infty(\hat{x}_{j'}^{\alpha'}; k, d_{\alpha'})|$. If there are more than two reconstructed unit normal vectors that differ slightly among each other, we shall choose the most effective one as the *effective* normal direction. This strategy proves to be very effective in our subsequent numerical experiments.

Finally, after the determination of x_0 and ν_j, $j = 1, 2, \ldots, m$, the corresponding inverse problem reduces to a finite dimensional problem of finding l_j, $j = 1, 2, \ldots, m$ and we have richer data set $u^\infty(\hat{x}; k, d_\alpha)$, $\hat{x} \in \mathbb{S}_-^{n-1}$ and $\alpha = 1, 2, \ldots, p$, to that purpose. In the following, we take the sound-soft obstacle as an illustration to present the rest of the recovery scheme. Suppose $\hat{x}_j^\alpha$ is a critical observation direction corresponding to C_j with respect to a certain d_α, $1 \leq \alpha \leq p$. Then by (3.4.22) and (3.4.25), we have

$$u^\infty(\hat{x}_j^\alpha; k, d_\alpha) \approx \widetilde{\gamma}(n, k) \int_{C_j} \nu_j \cdot d_\alpha \, e^{iky\cdot(d_\alpha - \hat{x}_j^\alpha)} \, ds(y). \tag{3.4.33}$$

Equation (3.4.33) clearly gives a nonlinear equation with l_j as its unknown; see (3.4.30). By supplementing m effective normal directions reconstructed from the previous step and their respective critical observation angles, one can arrive at a nonlinear system of m equations with m unknowns l_j, $j = 1, 2, \ldots, m$. The cells can be determined all at once by solving the finite dimensional problem. It is emphasized that the chosen m nonlinear equation (3.4.33) hold only approximately in the practical construction due to errors, thus we convert it into a nonlinear least-squares minimization problem to yield a more stable and accurate identification; see (3.4.34).

Summarizing the above discussion, we present the recovery scheme of reconstructing an admissible polyhedral obstacle D in the following; see Fig. 3.34 for a schematic illustration.

Recovery Scheme.

Step 1. By using the far-field pattern $\widetilde{u}^{\infty}$ corresponding to a single incident plane wave $e^{i\widetilde{k}x\cdot d}$ with a sufficiently small wavenumber $\widetilde{k} \in \mathbb{R}_+$ and the indicator functional $I(z)$ in (3.4.31) to locate the position x_0 of the obstacle D.

Step 2. Choose the incident plane waves $e^{ikx\cdot d_\alpha}$ with a sufficiently large wavenumber $k \in \mathbb{R}_+$ and p incident directions $d_\alpha, \alpha = 1, 2, \ldots, p$, and collect the corresponding backscattering far-field patterns. Determine the m effective exterior normal vectors $\nu_j,\ j = 1, 2, \ldots, m$ and their associated critical observation angles using the local maximum behavior of the modulus the backscattering far-field patterns.

Step 3. Given the distances l_j from x_0 along the direction $\nu_j,\ j = 1, 2, \ldots, m$, one can determine the perpendicular points P_j in terms of l_j, x_0 and ν_j; see Fig. 3.34.

Step 4. Extend the line/plane at P_j perpendicular to the respective $\nu_j,\ j = 1, 2, \ldots, m$. Determine the crossing points/lines among those extended lines/planes. The cells of the polyhedral obstacle can thus be determined to be $C_j, j = 1, 2, \ldots, m$.

Step 5. Select the first m critical observation angles $\hat{z}_\beta,\ \beta := \beta(\alpha, j) = 1, 2, \ldots, m$ from all of the $\hat{x}_j^\alpha, \alpha = 1, 2, \ldots, p,\ j = 1, 2, \ldots, m$, according to the decreasing order of magnitudes of local backscattering maxima and convert the corresponding m nonlinear equations associated with (3.4.33) with m unknowns l_j's into a least-squares minimization problem in (3.4.34). It is noted that the integral to the right hand side of (3.4.33) can be approximated by the trapezoidal quadrature rule with a sufficiently fine step size.

Fig. 3.34 Schematic illustration of the recovery scheme

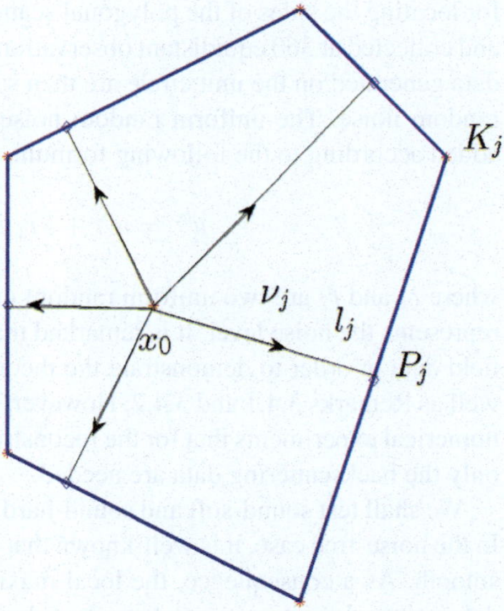

$$\min F(t) = \sum_{\beta=1}^{m} \left| u^{\infty}(\hat{z}_\beta; k, d_\alpha) - \widetilde{\gamma}(n,k) \int_{C_j} \nu_j \cdot d_\alpha \, e^{iky\cdot(d_\alpha - \hat{z}_\beta)} \, ds(y) \right|^2 ,$$
$$t = (l_1, l_2, \ldots, l_m)^T \in \mathbb{R}_+^m. \tag{3.4.34}$$

Step 6. The minimum of the multivariable nonlinear least-squares minimization problem (3.4.34) is obtained by employing a derivative-free trust region method via a local quadratic surrogate model-based search algorithm. Interested readers may refer to [29] and the references therein.

3.4.3 Numerical Experiments and Discussions

In this subsection, we present some numerical tests to verify the applicability and effectiveness of the discussed recovery scheme in two dimensions. In the sequel, the forward equation (2.1.2)–(2.1.5) is first solved by using the quadratic finite element discretization on a truncated circular domain enclosed by a perfectly matched layer (PML). The forward solver is iterated on a sequence of successively refined meshes till the relative error of two successive finite element solutions between the two adjacent meshes is below 0.1%. Then the scattered data are transformed into the far-field data by employing the Kirchhoff integral formula on a closed circle (2D) enclosing the scatterer (cf. [26]).

First of all, let us fix the parameter settings. For the positioning purpose in Step 1 of the discussed recovery scheme, we take $\widetilde{k} = 1$ to detect the location x_0 of the unknown polygonal obstacle. A few detecting waves with the incident directions chosen among the set $\{d_j = (\cos(j\pi/4), \sin(j\pi/4))\}$, $j = 1, 2, \ldots, 8$, are sent off for locating the sides of the polygonal scatterer. Then the far-field data are measured and collected at 360 equidistant observation angles along the unit circle. The far-field data generated on the unit circle are then subjected pointwisely to a certain uniform random noise. The uniform random noise in magnitude as well as in direction is added according to the following formula,

$$u^{\infty} = u^{\infty} + \delta\, r_1 |u^{\infty}| \exp(i\pi\, r_2)\,, \tag{3.4.35}$$

where r_1 and r_2 are two uniform random numbers, both ranging from -1 to 1, and δ represents the noise level. It is remarked that we have collected the full-aperture far-field data in order to demonstrate the theoretical observations in Theorem 3.4.1, as well as Remarks 3.4.1 and 3.4.2. However, it can be easily seen from our subsequent numerical experiments that for the reconstruction of the exterior unit normal vectors, only the backscattering data are needed.

We shall test sound-soft and sound-hard, noise-free and noisy cases, respectively. In the noise-free case, it is well-known that the far-field data is analytic and thus very smooth. As a consequence, the local maximum behavior of the phaseless far-field data is very clear from its polar graph as shown in Fig. 3.33 when there exists no noise.

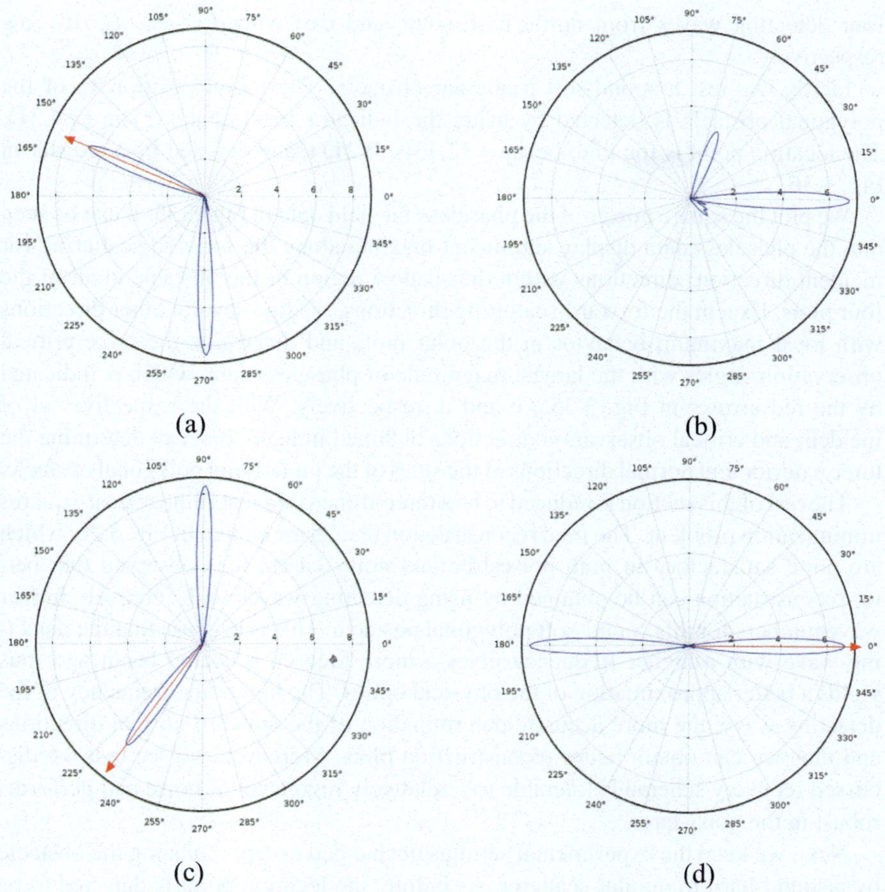

Fig. 3.35 Plot of $|u^\infty(\hat{x})|^2$ in polar coordinates corresponding to a sound-soft triangle due to incident plane waves $e^{ikx\cdot d}$ with $k = 6\pi$ and $d = d_2, d_4, d_6, d_8$ from **a–d**. The selected critical observation angles $\hat{z}_1$, $\hat{z}_2$ and $\hat{z}_3$ are highlighted by red arrows

While in the noisy case, we always add to the exact far-field data a uniform noise of 5% and use it as the noisy measurement data, which is inevitable from a practical viewpoint. In the noisy case, the local maximum behavior of the phaseless far-field data might be corrupted by the ups and downs of the random noise, which cause fictitious and/or more local maxima than the exact ones. To deal with the fictitious reconstruction, we add a preprocessing step to filter the raw noisy data. In our tests, a Fourier filtering stage is applied to the noisy far-field data in advance. The filtered measurement data are then used in the discussed recovery scheme, which yield much better reconstructions than simply using the raw data.

Example 7 (*A triangle*) The obstacle is chosen to be a triangle with three vertices displaced at $(1, 0)$, $(2.5, -0.5)$ and $(2.5, 1)$, respectively. In this test, we send off four detecting waves from north, east, south and east with $d = d_2,\ d_4,\ d_6,\ d_8$, respectively.

Firstly, we test a sound-soft triangular obstacle. The initial position x_0 of the polygonal obstacle is detected by using the indicator functional $I(z)$ in (3.4.31). The location point is found to be $x_0 = (2.136,\ 0.217)$, and denoted by a red star in Fig. 3.36.

We plot the square power of the phaseless far-field data in Fig. 3.35. It can be seen that the phaseless data display significant maxima along the forward-scattering (or incident direction) directions within the shadow region of the obstacle in all of the four plots. Except the forward-scattering directions, we find several other directions with local maximum behavior in the polar plots and determine the three critical observation angles with the largest magnitude of phaseless data, which is indicated by the red arrows in Fig. 3.35a, c and d, respectively. With the respective set of incident and critical observation directions obtained in hand, one can determine the three exterior unit normal directions of the sides of the underlying polygonal obstacle.

The rest of this section is reduced to be a three-dimensional nonlinear least-squares minimization problem. The final reconstruction results are shown in Fig. 3.36, which are quite satisfactory in both noise-free and noisy cases. It is observed that better reconstruction can be obtained by using detecting waves with relatively higher wavenumbers for this sound-soft polygonal obstacle. This is because that the detecting wave with a higher frequency gives a more focused reflected beam and thus yields a better approximation of the physical optics. The higher the frequency of the detecting wave, the more accurate determination of the outward normal directions and thus we can obtain better reconstruction plots. Moreover, we see that the discussed recovery scheme is tolerable to a relatively high level of noise and performs robust in the noisy case.

Next, we keep the experimental settings unchanged except replacing the obstacle by a sound-hard triangular scatterer. As before, the location point is detected to be $x_0 = (1.9307,\ 0.1412)$ using the indicator functional $I(z)$ in (3.4.31), denoted by a red star in Fig. 3.38. In Fig. 3.37, we plot the square power of the phaseless far-field data and indicate the first three critical observation angles within the backscattering aperture. The final reconstruction results are given in Fig. 3.38, which show that the discussed recovery scheme performs as good as in the sound-soft case.

Example 8 (*A convex hexagon*) In the second example, the obstacle is chosen to be a sound-soft hexagon with six vertices displaced at (4, 2.5), (3, 3) (1, 2), (0.5, 0) (2, −1) and (4.5, −0.5), respectively. The location point is detected to be $x_0 = (2.582,\ 0.759)$ using the indicator functional $I(z)$ in (3.4.31). This example is much more challenging since there are multiple facets to be determined.

It is pointed out that the hexagonal obstacle has six sides and we only send off four detecting waves along the incident directions $d = d_1,\ d_3,\ d_5,\ d_7$. In this case, the unknown number of sides is larger than that of incident directions. But from Fig. 3.39, one can determine sufficient critical observation angles from those polar plots. We identify six critical observation angles as indicated in red arrows in Fig. 3.39. The final

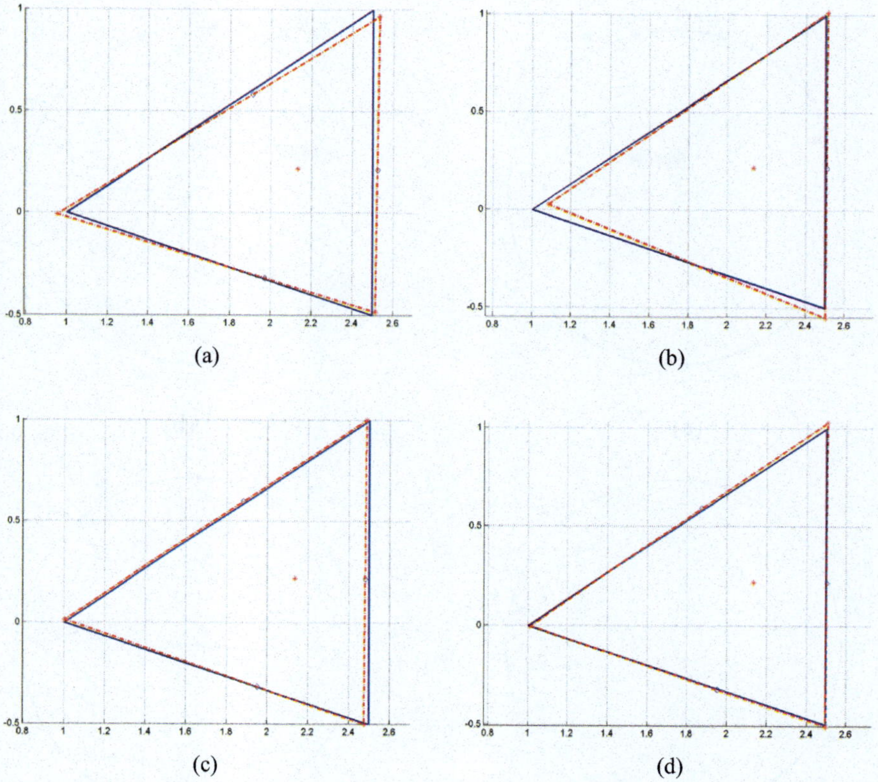

Fig. 3.36 Reconstruction of the triangular sound-soft obstacle with **a** $k = 6\pi$ with 5% noise, **b** $k = 6\pi$ without noise, **c** $k = 10\pi$ with 5% noise and **d** $k = 10\pi$ without noise

reconstruction results are obtained by solving a six-dimensional nonlinear problem and are shown in Fig. 3.40. It is again observed that the reconstruction performs better with higher frequencies.

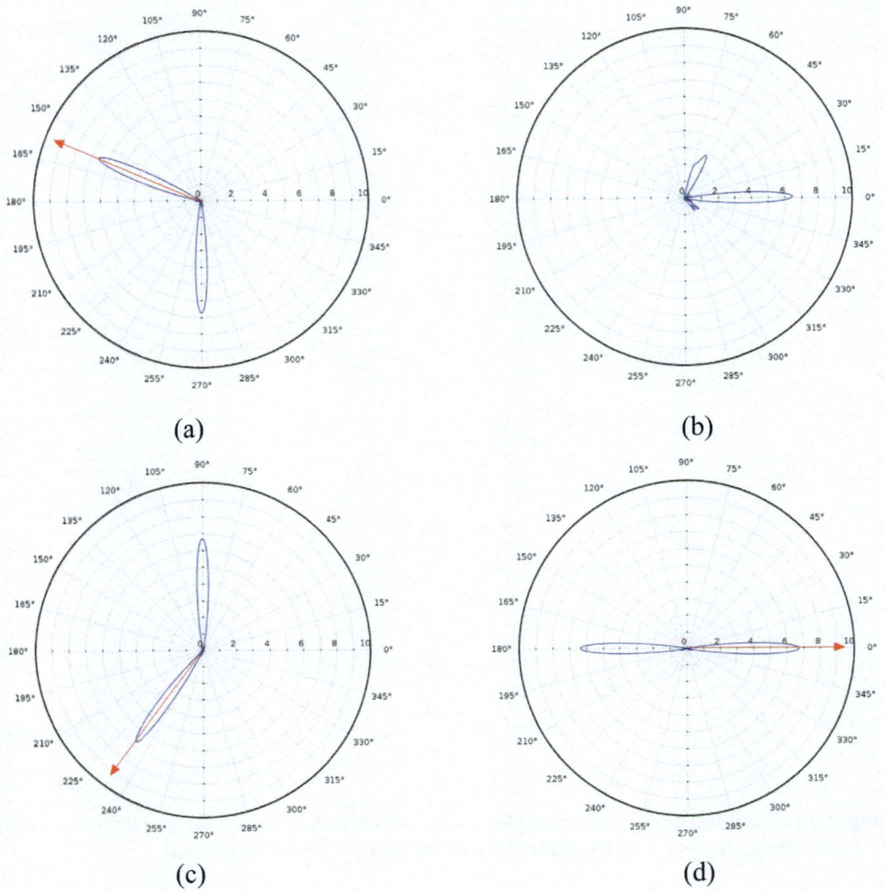

Fig. 3.37 Plot of $|u^\infty(\hat{x})|^2$ in polar coordinates corresponding to a sound-hard triangle due to incident plane waves $e^{ikx\cdot d}$ with $k = 6\pi$ and $d = d_2, d_4, d_6, d_8$ from **a–d**. The selected critical observation directions $\hat{z}_1$, $\hat{z}_2$ and $\hat{z}_3$ are highlighted by red arrows

3.4.4 Concluding Remarks

In this section, we present an inverse scattering scheme of recovering a sound-hard or sound-soft polyhedral obstacle by only a few far-field measurements. It has been a very challenging issue in the literature on recovering an obstacle by minimum measurement data. We believe that the results in this section make some important contribution to this challenging issue. The discussed scheme proceeds with two steps. First, one uses the local maximum behavior of the modulus of the backscattering far-field data to determine the exterior normal direction of each of the side/face of the obstacle. Then one can solve a small-scale finite dimensional algebraic system to completely recover the obstacle. In order to justify the local maximum behavior

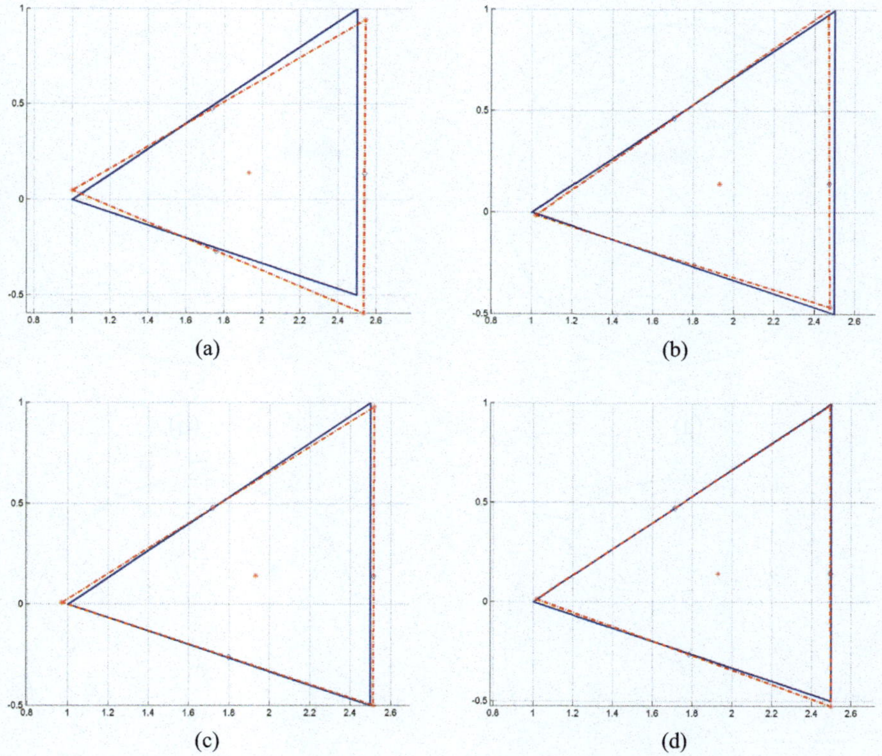

Fig. 3.38 Reconstruction of the triangular sound-hard obstacle with **a** $k = 6\pi$ with 5% noise, **b** $k = 6\pi$ without noise, **c** $k = 10\pi$ with 5% noise, and **d** $k = 10\pi$ without noise

of the modulus of the far-field data, we made essential use of the high-frequency asymptotics of the acoustic scattering. Finally, we mention two possible extensions for future investigations. First, throughout the section, we have been mainly concerned with the recovery of sound-soft and sound-hard obstacles. That is, for the Helmholtz system (2.1.2)–(2.1.5), one imposes either the homogeneous Dirichlet boundary condition or the homogeneous Neumann boundary condition. Physically, for a sound-soft obstacle, the pressure of the total wave vanishes on the boundary, and for a sound-hard obstacle, the normal velocity of the acoustic wave vanishes on the boundary. More generally, allowing an obstacles for which the normal velocity on the boundary is proportional to the excess pressure on the boundary leads to an impedance boundary condition of the form

$$\frac{\partial u}{\partial \nu} + \mathrm{i}\lambda u = 0 \quad \text{on } \partial D,$$

where $\lambda \in \mathbb{R}_+$ denotes the boundary impedance parameter. For an impedance obstacle, the reflection near the boundary of the obstacle is generally weakened, compared

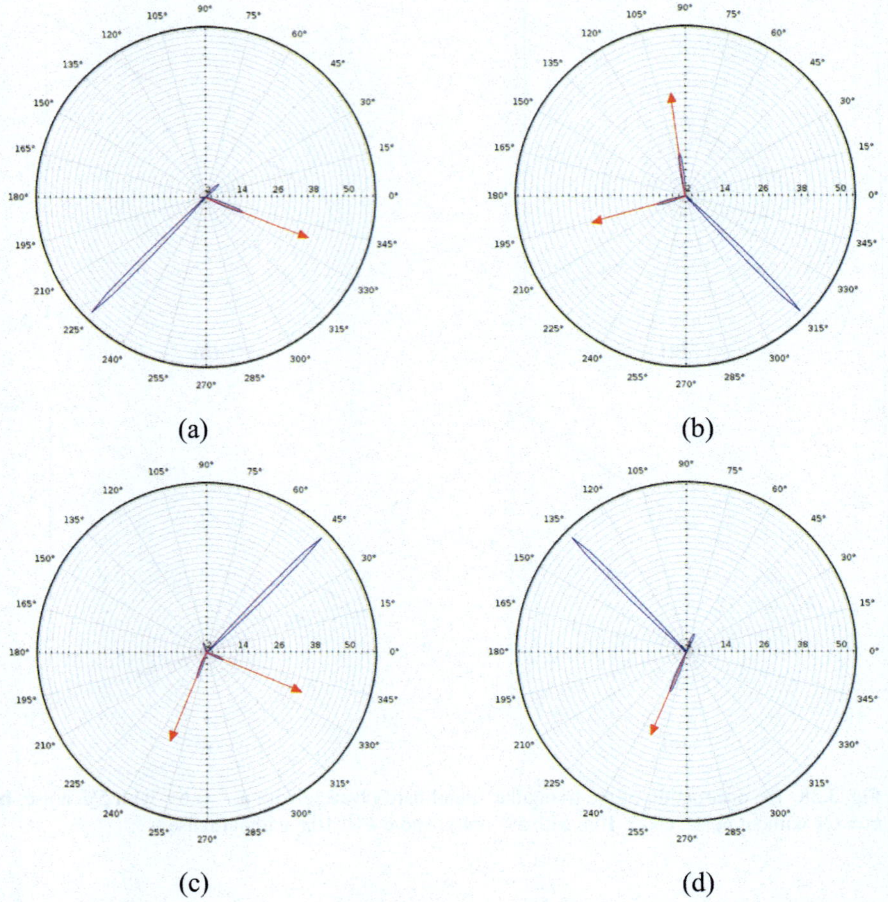

Fig. 3.39 Plot of $|u^\infty(\hat{x})|^2$ in polar coordinates corresponding to a convex sound-soft hexagon due to incident plane waves $e^{ikx\cdot d}$ with $k = 6\pi$ and $d = d_1, d_3, d_5, d_7$ from **a–d**. The selected critical observation directions $\hat{z}_1$, $\hat{z}_2$, $\hat{z}_3$, $\hat{z}_4$, $\hat{z}_5$ and $\hat{z}_6$ are highlighted by red arrows

to the sound-soft and sound-hard cases. Hence, the local maximum behavior in Theorem 3.4.1 will be deteriorated for an impedance obstacle. Second, our method can be extended to the inverse electromagnetic scattering problem of reconstructing conducting obstacles.

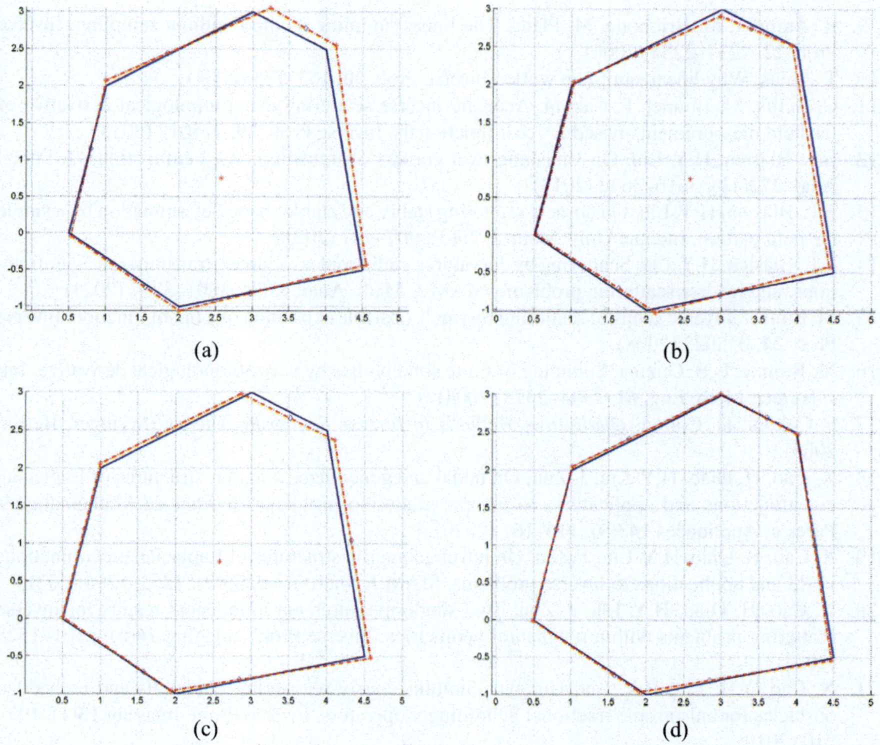

Fig. 3.40 Reconstruction of the sound-soft hexagonal obstacle with **a** $k = 6\pi$ with 5% noise, **b** $k = 6\pi$ without noise, **c** $k = 10\pi$ with 5% noise, and **d** $k = 10\pi$ without noise

References

1. G. Alessandrini, L. Rondi, Determining a sound-soft polyhedral scatterer by a single far-field measurement. Proc. Am. Math. Soc. **133**(6), 1685–1691 (2005)
2. H. Ammari, T. Boulier, J. Garnier, Modeling active electrolocation in weakly electric fish. SIAM J. Imag. Sci. **5**, 285–321 (2013)
3. H. Ammari, Y.T. Chow, H.Y. Liu, Localized sensitivity analysis at high-curvature boundary points of reconstructing inclusions in transmission problems. SIAM J. Math. Anal. **54**(2), 1543–1592 (2022)
4. H. Ammari, J. Garnier, V. Jugnon, H. Kang, Stability and resolution analysis for a topological derivative based imaging functional. SIAM J. Control. Optim. **50**(1), 48–76 (2012)
5. H. Ammari, P. Garapon, F. Jouve, H. Kang, M. Lim, S. Yu, A new optimal control approach for the reconstruction of extended inclusions. SIAM J. Control. Optim. **51**, 1372–1394 (2013)
6. H. Ammari, J. Garnier, H. Kang, M. Lim, K. Solna, Multistatic imaging of extended targets. SIAM J. Imag. Sci. **5**(2), 564–600 (2012)
7. H. Ammari, E. Iakovleva, D. Lesselier, G. Perrusson, MUSIC-type electromagnetic imaging of a collection of small three-dimensional inclusions. SIAM J. Sci. Comput. **29**, 674–709 (2007)
8. H. Ammari, H. Kang, *Reconstruction of Small Inhomogeneities from Boundary Measurements*, Lecture Notes in Mathematics, vol. 1846. (Springer, Berlin, 2004)

9. R. Aramini, M. Brignone, M. Piana, The linear sampling method without sampling. Inverse Prob. **22**, 2237–2254 (2006)
10. T. Arens, Why linear sampling works. Inverse Prob. **20**, 163–173 (2004)
11. C. Bellis, M. Bonnet, F. Cakoni, Acoustic inverse scattering using topological derivative of far-field measurements-based L^2 cost functionals. Inverse Prob. **29**, 075012 (2013)
12. E.L. Blåsten, H.Y. Liu, On vanishing near corners of transmission eigenfunctions. J. Funct. Anal. **273**(11), 3616–3632 (2017)
13. E.L. Blåsten, H.Y. Liu, On corners scattering stably and stable shape determination by a single far-field pattern. Indiana Univ. Math. J. **70**(3), 907–947 (2021)
14. E.L. Blåsten, H.Y. Liu, Scattering by curvatures, radiationless sources, transmission eigenfunctions, and inverse scattering problems. SIAM J. Math. Anal. **53**(4), 3801–3837 (2021)
15. M. Bonnet, Inverse acoustic scattering by small-obstacle expansion of a misfit function. Inverse Prob. **24**, 035022 (2008)
16. M. Bonnet, B.B. Guzina, Sounding of finite solid bodies by way of topological derivative. Int. J. Numer. Meth. Eng. **61**, 2344–2373 (2004)
17. F. Cakoni, D. Colton, *Qualitative Methods in Inverse Scattering Theory* (Springer, Berlin, 2006)
18. X. Cao, H. Diao, H.Y. Liu, J. Zou, On nodal and generalized singular structures of Laplacian eigenfunctions and applications to inverse scattering problems. Journal de Mathématiques Pures et Appliquées **143**(9), 116–161 (2020)
19. X. Cao, H. Diao, H.Y. Liu, J. Zou, On novel geometric structures of Laplacian eigenfunctions in R3 and applications to inverse problems. SIAM J. Math. Anal. **53**(2), 1263–1294 (2021)
20. X. Cao, H. Diao, H.Y. Liu, J. Zou, Two single-measurement uniqueness results for inverse scattering problems within polyhedral geometries. Inverse Probl. Imaging **16**(6), 1501–1528 (2022)
21. X. Cao, Y.H. Lin, H.Y. Liu, Hongyu, Simultaneously recovering potentials and embedded obstacles for anisotropic fractional Schrödinger operators. Inverse Probl. Imaging **13**(1), 197–210 (2019)
22. S.N. Chandler-Wilde, S. Langdon, Acoustic scattering: high frequency boundary element methods and unified transform methods, arxiv:1410.6137
23. Y.T. Chow, Y. Deng, Y. He, H.Y. Liu, X. Wang, Surface-localized transmission eigenstates, super-resolution imaging, and pseudo surface plasmon modes. SIAM J. Imag. Sci. **14**(3), 946–975 (2021)
24. D. Colton, J. Coyle, P. Monk, Recent developments in inverse acoustic scattering theory. SIAM Rev. **42**, 369–414 (2000)
25. D. Colton, A. Kirsch, A simple method for solving inverse scattering problems in the resonance region. Inverse Prob. **12**, 383–393 (1996)
26. D. Colton, R. Kress, *Inverse Acoustic and Electromagnetic Scattering Theory*, 2nd edn. (Springer, New York, 1998)
27. D. Colton, R. Kress, Using fundamental solutions in inverse scattering. Inverse Prob. **22**, R49–R66 (2006)
28. D. Colton, B.D. Sleeman, Uniqueness theorems for the inverse problem of acoustic scattering. MA J. Appl. Math. **31**, 253–259 (1983)
29. A.R. Conn, K. Scheinberg, L.N. Vicente, *Introduction to Derivative-Free Optimization* (SIAM, Philadelphia, 2009)
30. G. Dassios, R. Kleinman, *Low Frequency Scattering* (Clarendon Press, Oxford, 2000)
31. Y. Deng, C. Duan, H.Y. Liu, On vanishing near corners of conductive transmission eigenfunctions. Res. Math. Sci. **9**(1), 2 (2022)
32. H. Diao, X. Cao, H. Liu, On the geometric structures of transmission eigenfunctions with a conductive boundary condition and applications. Comm. Partial Differential Equ. **46**(4), 630–679 (2021)
33. M.H. Ding, H. Liu, G.H. Zheng, Shape reconstructions by using plasmon resonances. ESAIM: Math. Model. Numer. Anal. **56**(2), 705–726 (2022)

34. J. Elschner, M. Yamamoto, Uniqueness in determining polygonal sound-hard obstacles with a single incoming wave. Inverse Prob. **22**, 355 (2006)
35. Y. Gao, H.Y. Liu, X. Wang, K. Zhang, On an artificial neural network for inverse scattering problems. J. Comput. Phys. **448**, 110771 (2022)
36. R. Griesmaier, Multi-frequency orthogonality sampling for inverse obstacle scattering problems. Inverse Prob. **27**, 085005 (2011)
37. B.B. Guzina, M. Bonnet, Small-inclusion asymptotic of misfit functionals for inverse problems in acoustics. Inverse Prob. **22**, 1761–1785 (2006)
38. D.P. Hewett, S. Langdon, J.M. Melenk, A high frequency hp boundary element method for scattering by convex polygons. SIAM J. Numer. Anal. **51**, 629–653 (2013)
39. N. Honda, G. Nakamura, M. Sini, Analytic extension and reconstruction of obstacles from few measurements for elliptic second order operators. Math. Ann. **355**, 401–427 (2013)
40. V. Isakov, *Inverse Problems for Partial Differential Equations. Applied Mathematical Sciences*, vol. 127, 2nd edn. (Springer, New York, 2006)
41. K. Ito, B. Jin, J. Zou, A direct sampling method to an inverse medium scattering problem. Inverse Prob. **28**, 025003 (2012)
42. A. Kirsch, N. Grinberg, *The Factorization Method for Inverse Problems* (Oxford University Press, Oxford, 2008)
43. M. Klibanov, Phaseless inverse scattering problems in three dimensions. SIAM J. Appl. Math. **74**, 392–410 (2014)
44. I. Kocyigit, H.Y. Liu, H. Sun, Regular scattering patterns from near-cloaking devices and their implications for invisibility cloaking. Inverse Prob. **29**, 045005 (2013)
45. R. Kress, On the numerical solution of a hypersingular integral equation in scattering theory. J. Comput. Appl. Math. **61**, 345–360 (1995)
46. R. Kress, W. Rundell, *Inverse Obstacle Scattering with Modulus of the Far Field Pattern as Data, Inverse Problems in Medical Imaging and Nondestructive Testing (Oberwolfach, 1996), 75–92* (Springer, Vienna, 1997)
47. P.D. Lax, R.S. Phillips, *Scattering Theory* (Academic, 1967)
48. J. Li, H.Y. Liu, Recovering a polyhedral obstacle by a few backscattering measurements. J. Differential Equ. **259**(5), 2101–2120 (2015)
49. J. Li, H.Y. Liu, Z. Shang, H. Sun, Two single-shot methods for locating multiple electromagnetic scatterers. SIAM J. Appl. Math. **73**, 1721–1746 (2013)
50. J. Li, H.Y. Liu, Q. Wang, Locating multiple multiscale electromagnetic scatterers by a single far-field measurement. SIAM J. Imag. Sci. **6**, 2285–2309 (2013)
51. J. Li, H.Y. Liu, J. Zou, Multilevel linear sampling method for inverse scattering problems. SIAM J. Sci. Comput. **30**, 1228–1250 (2008)
52. J. Li, H.Y. Liu, J. Zou, Strengthened linear sampling method with a reference ball. SIAM J. Sci. Comput. **31**, 4013–4040 (2009)
53. J. Li, H.Y. Liu, J. Zou, Locating multiple multiscale acoustic scatterers. SIAM Multiscale Model. Simul. **12**, 927–952 (2014)
54. H.Y. Liu, Inverse obstacle scattering: Uniqueness and reconstruction algorithms, Thesis (Ph.D.)–The Chinese University of Hong Kong (Hong Kong) (2007), p. 168. ISBN: 978-0549-40176-6
55. H.Y. Liu, Virtual reshaping and invisibility in obstacle scattering. Inverse Prob. **25**, 045006 (2009)
56. H.Y. Liu, On recovering polyhedral scatterers with acoustic far-field measurements. IMA J. Appl. Math. **74**, 264–272 (2009)
57. H.Y. Liu, On local and global structures of transmission eigenfunctions and beyond. J. Inverse Ill-Posed Prob. **30**(2), 287–305 (2022)

58. H.Y. Liu, X. Liu, Recovery of an embedded obstacle and its surrounding medium from formally determined scattering data. Inverse Prob. **33**(6), 065001 (2017)
59. H.Y. Liu, X. Liu, X. Wang, Y. Wang, On a novel inverse scattering scheme using resonant modes with enhanced imaging resolution. Inverse Prob. **35**(12), 125012 (2019)
60. H.Y. Liu, M. Petrini, L. Rondi, J. Xiao, Stable determination of sound-hard polyhedral scatterers by a minimal number of scattering measurements. J. Differential Equ. **262**(3), 1631–1670 (2017)
61. H.Y. Liu, C.-H. Tsou, Stable determination of polygonal inclusions in Calderón's problem by a single partial boundary measurement. Inverse Prob. **36**(8), 085010 (2020)
62. H.Y. Liu, C.-H. Tsou, Stable determination by a single measurement, scattering bound and regularity of transmission eigenfunctions. Calc. Variat. Partial Differ. Equ. **61**(3), 91 (2022)
63. H.Y. Liu, C.-H. Tsou, W. Yang, On Calderón's inverse inclusion problem with smooth shapes by a single partial boundary measurement. Inverse Prob. **37**(5), 055005 (2021)
64. H.Y. Liu, H. Zhao, C. Zou, Determining scattering support of anisotropic acoustic mediums and obstacles. Commun. Math. Sci. **13**(4), 987–1000 (2015)
65. H.Y. Liu, J. Zou, Uniqueness in an inverse acoustic obstacle scattering problem for both sound-hard and sound-soft polyhedral scatterers. Inverse Prob. **22**, 515–524 (2006)
66. H.Y. Liu, J. Zou, Zeros of the Bessel and spherical Bessel functions and their applications for uniqueness in inverse acoustic obstacle scattering. IMA J. Appl. Math. **72**, 817–831 (2007)
67. H.Y. Liu, J. Zou, On unique determination of partially coated polyhedral scatterers with far field measurements. Inverse Prob. **23**(1), 297–308 (2007)
68. H.Y. Liu, J. Zou, On uniqueness in inverse acoustic and electromagnetic obstacle scattering problems. J. Phys: Conf. Ser. **124**, 012006 (2008)
69. H.Y. Liu, J. Zou, Uniqueness in determining multiple polygonal scatterers of mixed type. Discret. Continuous Dyn. Syst. Ser. B **9**(2), 375–396 (2008)
70. A. Majda, High frequency asymptotics for the scattering matrix and the inverse problem of acoustical scattering. Commun. Pure Appl. Math. **29**, 261–291 (1976)
71. W. McLean, *Strongly Elliptic Systems and Boundary Integral Equations* (Cambridge University Press, Cambridge, 2000)
72. R.B. Melrose, M.E. Taylor, Near peak scattering and the corrected Kirchhoff approximation for a convex obstacle. Adv. Math. **55**, 242–315 (1985)
73. F.W.J. Olver, *Bessel Functions Part III, Zeros and Associated Values* (Cambridge University Press, Cambridge, 1960)
74. R. Potthast, *Point Sources and Multipoles in Inverse Scattering Theory* (CRC Press, 2001)
75. R. Potthast, A survey on sampling and probe methods for inverse problems. Inverse Prob. **22**(2), R1–R47 (2006)
76. R. Potthast, A study on orthogonality sampling. Inverse Prob. **26**, 074015 (2010)
77. A. Tacchino, J. Coyle, M. Piana, Numerical validation of the linear sampling method. Inverse Prob. **18**(3), 511–527 (2002)
78. C.J. Tranter, *Bessel Functions with Some Physical Applications* (English Universities Press, London, 1968)
79. G. Uhlmann, (ed.), *Inside Out: Inverse Problems and Applications*, vol. 47 (MSRI Publications, Cambridge University Press, 2003)
80. G.N. Watson, *A Treatise on the Theory of Bessel Functions*, 2nd edn. (Cambridge Press, Cambridge, 1944)
81. W. Yin, W. Yang, H.Y. Liu, A neural network scheme for recovering scattering obstacles with limited phaseless far-field data. J. Comput. Phys. **417**, 109594 (2020)
82. Y. Yin, W. Yin, P. Meng, H.Y. Liu, On a hybrid approach for recovering multiple obstacles. Commun. Comput. Phys. **31**(3), 869–892 (2022)
83. Y. Yin, W. Yin, P. Meng, H.Y. Liu, The interior inverse scattering problem for a two-layered cavity using the Bayesian method. Inverse Probl. Imaging **16**(4), 673–690 (2022)

84. D. Zhang, Y. Guo, F. Sun, H.Y. Liu, Unique determinations in inverse scattering problems with phaseless near-field measurements. Inverse Probl. Imaging **14**(3), 569–582 (2020)
85. P. Zhang, P. Meng, W. Yin, H.Y. Liu, A neural network method for time-dependent inverse source problem with limited-aperture data. J. Comput. Appl. Math. **421**, 114842 (2023)

Chapter 4
Direct Electromagnetic Scattering Problems

Up until now, we have considered only the direct and inverse scattering problem for time-harmonic acoustic waves. In the following two chapters, we want to extend these results to scattering for time-harmonic electromagnetic (EM) waves. As in our analysis on acoustic scattering, we begin with an outline of the solution of the direct problem.

The propagation of EM waves in an isotropic medium in $\mathbb{R}^N$, $(N = 2, 3)$ is governed by the Maxwell equations

$$\begin{cases} \nabla \wedge \mathscr{E} + \mu \frac{\partial \mathscr{H}}{\partial t} = 0, \\ \nabla \wedge \mathscr{H} - \varepsilon \frac{\partial \mathscr{E}}{\partial t} = \sigma \mathscr{E}. \end{cases} \tag{4.0.1}$$

Here, the electric field $\mathscr{E}$ and the magnetic field $\mathscr{H}$ are mappings from space and time, $\mathbb{R}^N \times \mathbb{R}$, into $\mathbb{R}^N$. The parameters ε, μ and σ are, respectively, the electric permittivity, the magnetic permeability and the electric conductivity. We should note that $\varepsilon(x)$ and $\sigma(x)$ are functions of the space variables. If an EM inhomogeneity is presented in the homogeneous space, the propagation of the incident waves (E^i, H^i) will be perturbed, leading to the so-called *scattering*.

We consider with the time-harmonic EM waves from an inhomogeneous isotropic medium supported in a bounded domain $\Omega \subset \mathbb{R}^N$, $(N = 2, 3)$. We consider the medium outside Ω to be homogeneous, so we may assume the vanishing conductivity $\sigma = 0$ and that $\varepsilon = \varepsilon_0 > 0$ in $\Omega^c := \mathbb{R}^N \backslash \overline{\Omega}$. The time-harmonic EM waves have the following form:

$$\begin{cases} \mathscr{E}(x, t) = \mathscr{R}\{\varepsilon_0^{-1/2} E(x) e^{-i\omega t}\}, \\ \mathscr{H}(x, t) = \mathscr{R}\{\mu_0^{-1/2} H(x) e^{-i\omega t}\}, \end{cases}$$

J. Li and H. Liu, *Numerical Methods for Inverse Scattering Problems*,
https://doi.org/10.1007/978-981-99-3772-1_4

with frequency $\omega > 0$ and $\varepsilon_0 > 0$ and $\mu_0 > 0$ are constant values in vacuum and dimensionless. By introducing relative values $\mu_r(x), \varepsilon_r(x) \in \mathbb{R}$ and the refractive index of the medium $n(x) \in \mathbb{C}$ defined by

$$\mu_r = \frac{\mu}{\mu_0}, \quad \varepsilon_r = \frac{\varepsilon}{\varepsilon_0}, \quad n(x) = \varepsilon_r + \mathrm{i}\frac{\sigma}{\omega\varepsilon_0},$$

we derive that the complex valued fields E and H must satisfy the the reduced Maxwell equations as follows

$$\begin{cases} \nabla \wedge E - \mathrm{i}k\mu_r H = 0, \\ \nabla \wedge H + \mathrm{i}kn(x)E = 0, \end{cases} \tag{4.0.2}$$

where $k = \omega\sqrt{\varepsilon_0\mu_0}$ is the wave number.

There two fundamental scattering scenarios. In the first one, we will consider the scattering of time-harmonic waves by obstacles and its outside is surrounded by a homogeneous medium. In the second one the scattering from an inhomogeneous medium with sufficiently large homogeneous sphere.

4.1 Electromagnetic Scattering From Obstacles

Consider a bounded C^2 domain Ω in $\mathbb{R}^N$, $(N = 2, 3)$, which supports an inhomogeneous EM scatter, and it is assumed that $\mathbb{R}^N \backslash \overline{\Omega}$ is connected. Ω represents an impenetrable obstacle located in the space and it is assumed to be unknown/inaccessible. In order to identify Ω, we sends the time-harmonic EM plane waves of the form

$$E^i(x) = pe^{ikx\cdot d}, \quad H^i(x) = \frac{1}{ik}\nabla \wedge E^i(x), \quad x \in \mathbb{R}^3, \tag{4.1.1}$$

where E^i and H^i are, respectively, the electric and magnetic fields, and $k \in \mathbb{R}_+$, $d \in \mathbb{S}^2$, $p \in \mathbb{R}^3 \setminus \{0\}$ with $p \perp d$ are, respectively, the wave number, incident direction and polarization vector.

Since Ω is an obstacle, the presence of the obstacle Ω interrupts the propagation of the plane wave, leading to the so-called scattered wave fields $E^s(x)$ and $H^s(x)$, which exists only in the exterior of the obstacle. The wave fields can not penetrate and the direct scattering problem is described by

$$\nabla \wedge E(x) - \mathrm{i}kH(x) = 0, \quad \nabla \wedge H(x) + \mathrm{i}kE(x) = 0, \quad x \in \mathbb{R}^3 \backslash \overline{\Omega}. \tag{4.1.2}$$

The total EM wave fields outside the inhomogeneity, namely in $\mathbb{R}^3 \backslash \overline{\Omega}$, are composed of two parts: the incident wave fields E^i, H^i and the scattered wave fields E^s, H^s. That is, we have

$$E(x) = E^i(x) + E^s(x), \quad H(x) = H^i(x) + H^s(x), \quad x \in \mathbb{R}^3 \backslash \overline{\Omega}. \tag{4.1.3}$$

The scattered EM fields are radiating, characterized by the Silver-Müller radiation condition

$$\lim_{|x| \to +\infty} |x| \left| (\nabla \wedge E^s)(x) \wedge \frac{x}{|x|} - \mathrm{i}kE^s(x) \right| = 0,$$

which holds uniformly in all directions $\hat{x} := x/|x| \in \mathbb{S}^2$, $x \in \mathbb{R}^3 \backslash \{0\}$.

For a perfectly conducting obstacle, both the interior fields E^i and $H^i := H|_\Omega$ would vanish inside Ω, i.e. the EM fields cannot penetrate inside the object and are governed by

$$\begin{cases} \nabla \wedge E - \mathrm{i}kH = 0 & \text{in } \mathbb{R}^3 \backslash \overline{\Omega}, \\ \nabla \wedge H + \mathrm{i}kE = 0 & \text{in } \mathbb{R}^3 \backslash \overline{\Omega}, \\ \nu \wedge E = 0 & \text{on } \partial\Omega, \\ E = E^i + E^s & \text{in } \mathbb{R}^3 \backslash \overline{\Omega}, \\ H = H^i + H^s & \text{in } \mathbb{R}^3 \backslash \overline{\Omega}, \\ \lim_{|x| \to +\infty} |x| \left| (\nabla \wedge E^s)(x) \wedge \frac{x}{|x|} - \mathrm{i}kE^s(x) \right| = 0, \end{cases} \tag{4.1.4}$$

where ν denotes the outward unit normal vector to $\partial\Omega$. The scattered field E^s has the same asymptotic development.

4.2 Electromagnetic Scattering From Mediums

In this section we consider the EM scattering from an inhomogeneous medium in $\mathbb{R}^N$, $(N = 2, 3)$. We assume that the inhomogeneity is compactly supported. Let Ω be a bounded C^2 domain in $\mathbb{R}^N$ which supports an inhomogeneous isotropic EM medium characterized by the electric permittivity $\varepsilon(x)$, magnetic permeability $\mu(x)$, and conductivity $\sigma(x)$. Both $\varepsilon(x)$ and $\mu(x)$ are positive scalar functions and $\sigma(x)$ is a non-negative scalar function. It is assumed that $\Omega_e := \mathbb{R}^N \backslash \overline{\Omega}$ is connected and Ω_e represents the uniformly homogeneous background space. We take the detecting/incident wave field to be a single pair of time-harmonic EM plane waves,

$$E^i(x) = p e^{\mathrm{i}\omega x \cdot d}, \quad H^i(x) = \frac{1}{\mathrm{i}\omega} \nabla \wedge E^i(x), \quad x \in \mathbb{R}^N, \tag{4.2.1}$$

where $\omega \in \mathbb{R}_+$ denotes the frequency, $\wedge$ denotes the exterior product, $d \in \mathbb{S}^{N-1}$ denotes the impinging direction, and $p \in \mathbb{R}^N$ denotes the polarization with $p \cdot d = 0$.

The propagation of time harmonic EM waves in the medium is governed by the following equations,

$$\nabla \wedge E - \mathrm{i}kH = 0, \quad \nabla \wedge H + \mathrm{i}kn(x)E = 0 \quad \text{ in } \mathbb{R}^N, \tag{4.2.2}$$

where $k = \omega\sqrt{\varepsilon_0\mu_0}$ is the wave number and $n(x) = \varepsilon_r + \mathrm{i}\frac{\sigma}{\omega\varepsilon_0}$ is the refractive index function of the medium.

The impingement of the wave fields $E^i(x)$ and $H^i(x)$ on the EM scatterer Ω generates the EM scattering. Let E, H and $E^s = E - E^i, H^s = H - H^i$ respectively denote the total and scattered wave fields. Then the direct electromagnetic scattering from mediums can be described by the following system

$$\begin{cases} \nabla \wedge E - \mathrm{i}kH = 0 & \text{in } \mathbb{R}^3, \\ \nabla \wedge H + \mathrm{i}knE = 0 & \text{in } \mathbb{R}^3, \\ E = E^i + E^s, \quad H = H^i + H^s, \\ \lim_{|x| \to \infty} (H^s(x) \wedge x - |x| E^s(x)) = 0, \end{cases} \tag{4.2.3}$$

where the last equation is called the Silver-Müller radiation condition which holds uniformly in the angular variable $|x|$ and characterizes the outgoing nature of the scattered field.

4.3 Electromagnetic Scattering From Complex Scatterers

In this section we unify the treatments introduced in Sects. 4.1 and 4.2 as a consistent problem and give a proof of the well-posedness of this general system.

Let us consider the scattering due to an inhomogeneous EM medium compactly supported in a bounded domain $(\Omega; \varepsilon, \mu, \sigma)$ in $\mathbb{R}^N$, $(N = 2, 3)$ and an impenetrable PEC obstacle $D \Subset \Omega$. We consider that the medium outside Ω to be homogeneous and no damping present, so we may assume that $\varepsilon(x) = \varepsilon_0, \mu(x) = \mu_0$ and $n(x) = 1$ in $\Omega^c := \mathbb{R}^N \backslash \overline{\Omega}$. For notational convenience, we set

$$n(x) = \varepsilon_r + \mathrm{i}\frac{\sigma}{\omega\varepsilon_0},$$

and $E^s(x) = E(x) - E^i(x), H^s(x) = H(x) - H^i(x)$ are the scattered field outside the medium region Ω. Throughout the rest of the section, we assume that Ω and D are both bounded C^2 domains such that $\mathbb{R}^N \backslash \overline{\Omega}$ and $\Omega \backslash \overline{D}$ are connected. The EM scattering is governed by the following Maxwell system

$$\begin{cases} \nabla \wedge E - ik\mu H = 0 & \text{in } \Omega\backslash\overline{D}, \\ \nabla \wedge H + ikn(x)E = 0 & \text{in } \Omega\backslash\overline{D}, \\ \nabla \wedge E - ikH = 0 & \text{in } \mathbb{R}^N\backslash\overline{\Omega}, \\ \nabla \wedge H + ikE = 0 & \text{in } \mathbb{R}^N\backslash\overline{\Omega}, \\ \nu \wedge E^+ = 0 & \text{on } \partial D, \\ E^- = E^s + E^i, \quad H^- = H^s + H^i & \text{on } \partial\Omega, \\ \lim_{|x|\to+\infty} |x| \left| (\nabla \wedge E^s)(x) \wedge \frac{x}{|x|} - ikE^s(x) \right| = 0. \end{cases} \tag{4.3.1}$$

We use the notations E^-, E^+ and H^-, H^+ to represent the limits of u on ∂D or $\partial\Omega$, taking respectively from inside and outside D or Ω. The last limits in (2.3.1) is the Silver-Müller radiation condition.

Now we consider the well-posedness of the scattering problem (4.3.1). Here we let the parameters α and β be defined as

$$\{\alpha, \beta\} = \begin{cases} 1, 1 & \text{in } \mathbb{R}^N\backslash\overline{\Omega}, \\ \mu, n(x) & \text{in } \Omega\backslash\overline{D}. \end{cases} \tag{4.3.2}$$

Then the scattering problem (4.3.1) can be formulate as follows.

Find $E \in H^1_{\text{loc}}(\mathbb{R}^N)$ such that $E = E^i + E^s$ in $\mathbb{R}^N\backslash\overline{\Omega}$ and solves the equation

$$\begin{cases} \mathscr{L}E := \nabla \wedge (\alpha\nabla \wedge E) + k^2\beta E = 0 \quad \text{in } \mathbb{R}^N, \\ \lim_{r\to\infty} r^{(N-1)/2} \left\{ \frac{\partial E^s}{\partial r} - ikE^s \right\} = 0 \qquad \text{for } r = |x|. \end{cases} \tag{4.3.3}$$

We first show the uniqueness of the solutions for the system (4.3.3). To this end, we first introduce the unique continuation principle which is stated as follows.

Theorem 4.3.1 *If $u \in C_0^2(\mathbb{R}^3)$ satisfies $|\Delta u(x)| \leq M|u(x)|$ for all $x \in \mathbb{R}^3$ with a constant M, then u vanishes in all of $\mathbb{R}^3$. This is also true, if $u = (u_1, \ldots, u_l)$ is a vector function, $\Delta u := (\Delta u_1, \ldots, \Delta u_l)$ and $|\cdot|$ denotes the euclidean norm of a vector in $\mathbb{C}^l$.*

Theorem 4.3.2 *Under the above assumption, the problem* (4.3.3) *is uniquely solvable.*

Proof We first show the uniqueness of the solutions for the system (4.3.3). The problem (4.3.3) is equivalent to the following transmission problem

$$\begin{cases} \nabla \wedge (\mu(x)\nabla \wedge F) + k^2 n(x) F = 0 & \text{in} \quad \Omega\backslash\overline{D}, \\ \nabla \wedge \nabla \wedge E^s + k^2 E^s = 0 & \text{in} \quad \mathbb{R}^N\backslash\overline{\Omega}, \\ \mu\nu \wedge F = p \in H^{-1/2}(\partial D) & \text{on} \quad \partial D, \\ F - E^s = g_1 \in H^{1/2}(\partial\Omega) & \text{on} \quad \partial\Omega, \\ \mu\nu \wedge F - \nu \wedge E^s = g_2 \in H^{-1/2}(\partial\Omega) & \text{on} \quad \partial\Omega, \\ \lim_{|x|\to+\infty} |x|^{(N-1)/2}\left\{\frac{\partial E^s}{\partial|x|} - \mathrm{i}kE^s\right\} = 0. \end{cases} \tag{4.3.4}$$

To show the uniqueness of the solution (F, E^s) to system (4.3.3), we first introduce the following auxiliary problem.

Find $(F_1, E_1) \in H(\mathrm{curl})(\Omega\backslash\overline{D}) \times H(\mathrm{curl})(B_R\backslash\overline{\Omega})$ such that

$$\begin{cases} \nabla \wedge (\gamma(x)\nabla \wedge F_1) + k^2 q(x) F_1 = 0 & \text{in} \quad \Omega\backslash\overline{D}, \\ \nabla \wedge \nabla \wedge E_1 + k^2 E_1 = f & \text{in} \quad B_R\backslash\overline{\Omega}, \\ \mu\nu \wedge F_1 = p & \text{on} \quad \partial D, \\ F_1 - E_1 = g_1 & \text{on} \quad \partial\Omega, \\ \mu\nu \wedge F_1 - \nu \wedge E_1 = g_2 & \text{on} \quad \partial\Omega, \\ \nu \wedge E_1 = \Lambda E_1 & \text{on} \quad \partial B_R, \end{cases} \tag{4.3.5}$$

where $\Lambda : H^{1/2}(\mathrm{curl})(\partial B_R) \to H^{-1/2}(\mathrm{curl})(\partial B_R)$ is the Dirichlet-to-Neumann map defined by $\Lambda\Psi = \nu \wedge W|_{\partial B_R}$, with $W \in H(\mathrm{curl})_{\mathrm{loc}}(\mathbb{R}^N\backslash\overline{B}_R)$ being the unique solution to the system

$$\begin{cases} \nabla \wedge \nabla \wedge W + k^2 W = 0 & \text{in} \quad \mathbb{R}^N\backslash\overline{B}_R, \\ W = \Psi \in H^{1/2}(\mathrm{curl})(\partial B_R) & \text{on} \quad \partial B_R, \\ \lim\limits_{|x|\to+\infty} |x|^{(N-1)/2}\left\{\dfrac{\partial W}{\partial|x|} - \mathrm{i}kW\right\} = 0. \end{cases} \tag{4.3.6}$$

Then we show the uniqueness of the solution (F, E^s) to system (4.3.5). For the purpose we set p, g_1, g_2 to be all zeros. Multiplying the first and second equations of (4.3.4), respectively, by $\overline{F}$ and $\overline{E}^s$, and integrating by parts in $\Omega\backslash\overline{D}$ and $B_R\backslash\overline{\Omega}$, together with use of the conditions on ∂D and $\partial\Omega$, we have

$$\int_{\Omega\backslash\overline{D}} \mu|\nabla \wedge F|^2 dx + \int_{\Omega\backslash\overline{D}} k^2 n|F|^2 dx - \int_{B_R\backslash\overline{\Omega}} |\nabla \wedge E^s|^2 dx$$
$$+ \int_{B_R\backslash\overline{\Omega}} k^2|E^s|^2 dx + \int_{\partial B_R} \nu \wedge E^s \cdot \overline{E}^s ds = 0. \tag{4.3.7}$$

Taking the imaginary part of both sides of (4.3.7), we derive

$$\mathrm{Im}\int_{\partial B_R} \nu \wedge E^s \cdot \overline{E}^s ds = -\mathrm{Im}\int_{\Omega\backslash\overline{D}} k^2 n|F|^2 dx \leq 0.$$

Then by Rellich's theorem we know E^s is zero outside B_R, which with the unique continuation implies that $E^s = 0$ in $\Omega\backslash\overline{D}$ and $F = 0$ in D.

Then, we show the equivalence between (4.3.4) and (4.3.5). By the definition of Λ, we see that is (F, E^s) solves (4.3.4), then $(F_1 = F, E_1 = E^s|_{B_R\backslash\overline{\Omega}})$ is the solution to the system (4.3.5). On the other hand, by applying the Green's representation to the solution (F_1, E_1) of (4.3.5) and using a similar way as the acoustic problem, we can obtain that $F_1 = F$ and $E_1 = E^s$.

Next we show the existence and stability estimate. In the following, by appropriately choosing R_0 we can assume that k^2 is not an eigenvalue in B_{R_0+1}. Let $\theta(x) \in C^\infty(\mathbb{R}^N)$ be a cut-off function such that $\theta(x) = 0$ for $|x| < R_0$ and $\theta(x) = 1$ for $|x| > R_0 + 1$. Setting

$$W = E \quad \text{in } \Omega \text{ and } W = E^s + (1-\theta)E^i \quad \text{in } \mathbb{R}^N\backslash\overline{\Omega}, \tag{4.3.8}$$

we can then verify directly that $W \in H_{\mathrm{loc}}(\mathrm{curl}; \mathbb{R}^N)$ satisfies

$$\begin{cases} \mathscr{L}W = g & \text{in } \mathbb{R}^N, \\ \lim\limits_{|x|\to\infty} |x|^{(N-1)/2}\left\{\dfrac{\partial W}{\partial|x|} - ikW\right\} = 0, \end{cases} \tag{4.3.9}$$

with $g = (\nabla\wedge\nabla\wedge + k^2)(\theta E^i) \in [L^2(B_{R_0+1}\backslash\Omega)]^3$.

Next, we look for a solution to (4.3.9) of the following form

$$W = w - \phi(w - V), \tag{4.3.10}$$

where is C^∞ cut-off function such that $\phi = 1$ in B_{R_0} and $\phi = 0$ in $\mathbb{R}^N\backslash B_{R_0+1}$. We let $V \in H(\mathrm{curl}; B_{R_0+1})$ be the solution of the system

$$\begin{cases} \mathscr{L}V = g^* & \text{in } B_{R_0+1}, \\ V = 0 & \text{on } \partial B_{R_0+1} \end{cases} \tag{4.3.11}$$

and $w \in H_{\mathrm{loc}}(\mathrm{curl}; \mathbb{R}^N)$ be the solution of the system

$$\begin{cases} (\nabla\wedge\nabla\wedge + k^2)w = g^* & \text{in } \mathbb{R}^N, \\ \lim\limits_{|x|\to\infty} |x|^{(N-1)/2}\left\{\dfrac{\partial w}{\partial|x|} - ikw\right\} = 0, \end{cases} \tag{4.3.12}$$

where $g^* \in [L^2(B_{R_0+1}\backslash\Omega)]^3$ shall be determined later.

Clearly, by the classical regularity estimates we see

$$V \in H^2(\text{curl}; B_{R_0+1}\backslash\overline{\Omega}) \quad \text{and} \quad w \in H^2_{\text{loc}}(\text{curl}; \mathbb{R}^N).$$

By direct verification we have

$$\begin{aligned} g = (\nabla \wedge \nabla \wedge +k^2)W &= \nabla \wedge \nabla \wedge w + k^2 w + \nabla \wedge \nabla \wedge \phi(w - V) \\ &\quad + 2\nabla \wedge \phi \cdot \nabla \wedge (w - V) + \phi\big(\nabla \wedge \nabla \wedge (w - V) + k^2(w - V)\big) \\ &= g^* + Kg^*, \end{aligned} \tag{4.3.13}$$

where K is defined to be $Kg^* = \nabla \wedge \nabla \wedge \phi(w - V) + 2\nabla \wedge \phi \cdot \nabla \wedge (w - V)$.

We can show that K is compact from $[L^2(B_{R_0+1}\backslash\Omega)]^3$ to itself. We make use of the Fredholm theory to show the unique solvability of (4.3.13). It suffices to show the uniqueness of solution to (4.3.13). We set $g = 0$. By (4.3.9) we have $W = 0$. Hence $w = \phi(w - V)$ in $\mathbb{R}^N$ and $V = 0$ in Ω and $w = 0$ in $\mathbb{R}^N\backslash B_{R_0+1}$. It is straightforward to verify that

$$\begin{cases} (\nabla \wedge \nabla \wedge +k^2)(V - w) = 0 & \text{in} \quad B_{R_0+1}, \\ V - w = 0 & \text{on} \quad \partial B_{R_0+1}, \end{cases} \tag{4.3.14}$$

hence $V - w = 0$. Therefore $w = 0$, which then implies that $g^* = 0$. Then by the Fredholm theory we have a unique $g^* \in [L^2(B_{R_0+1}\backslash\Omega)]^3$ to (4.3.13) such that

$$\|g^*\|_{L^2(B_{R_0+1}\backslash\Omega)} \le C\|g\|_{L^2(B_{R_0+1}\backslash\Omega)} \le C\|E^i\|_{H^1(B_{R_0+1}\backslash\overline{\Omega})}.$$

Finally, by the classical theory on elliptic equations one can show that

$$\|E\|_{H(\text{curl}; B_{R_0+1}\backslash\overline{\Omega})} \le C\|E^i\|_{H(\text{curl}; B_{R_0+1}\backslash\overline{\Omega})}.$$

□

4.4 Green's Theorem and Representation Formulas

As in the acoustic case we first introduce the Green's theorem for vector valued functions.

Theorem 4.4.1 *Let D be a bounded domain of class C^1 and let ν denote teh unit normal vector to the boundary ∂D directed into the exterior of D. Then, for $E \in C^1(\overline{D})$ and $F \in C^2(\overline{D})$, we have the following Green's first vector theorem*

$$\int_D E \cdot \Delta F + \nabla \wedge E \cdot \nabla \wedge F + \nabla \cdot E\nabla \cdot F dx = \int_{\partial D} \nu \times E \cdot \nabla \wedge F + \nu \cdot E\nabla \cdot F ds, \tag{4.4.1}$$

and for $E, F \in C^2(\overline{D})$ we have the Green's second vector theorem

$$\int_D E \cdot \Delta F - F \cdot \Delta E dx = \int_{\partial D} \nu \times E \cdot \nabla \wedge F$$
$$+ \nu \cdot E \nabla \cdot F - \nu \times F \cdot \nabla \wedge E - \nu \cdot F \nabla \cdot E ds. \tag{4.4.2}$$

We now provide a basic representation theorem for vector fields which is known as the Stratton-Chu formula.

Theorem 4.4.2 *Let D be a bounded domain of class C^2 and let ν denote the unit normal vector to the boundary ∂D directed into the exterior of D. For vector fields $E, H \in C^1(D) \cap C(\overline{D})$, we have the Stratton-Chu formula*

$$E(x) = -\nabla \wedge \int_{\partial D} \nu(y) \wedge E(y)\Phi(x, y)ds(y) + \nabla \int_{\partial D} \nu(y) \cdot E(y)\Phi(x, y)ds(y)$$
$$- ik \int_{\partial D} \nu(y) \wedge H(y)\Phi(x, y)ds(y) + \nabla \wedge \int_D \{\nabla \wedge E(y) - ikH(y)\}\Phi(x, y)dy$$
$$- \nabla \int_D \nabla \cdot E(y)\Phi(x, y)dy + ik \int_D \{\nabla \wedge H(y) + ikE(y)\}\Phi(x, y)dy \tag{4.4.3}$$

with $\Phi(x, y) = \frac{e^{ik|x-y|}}{4\pi|x-y|}$ being the fundamental solution to the Helmholtz equation $\Delta u + k^2 u = 0$. Here the volume integrals exist as improper integrals. A similar formula holds with the roles of E and H interchanged.

If E and H are a solution for the Maxwell equations, the Stratton-Chu formula can be reformulated as follows.

Theorem 4.4.3 *Let D be the same as in Theorem 4.4.2 and let $E, H \in C^1(D) \cap C(\overline{D})$ be a solution to the Maxwell equations*

$$\nabla \wedge E - ikH = 0, \qquad \nabla \wedge H + ikE = 0, \qquad \text{in } D.$$

Then we have the following Stratton-Chu formulas

$$E(x) = -\nabla \wedge \int_{\partial D} \nu(y) \wedge E(y)\Phi(x, y)ds(y)$$
$$+ \frac{1}{ik} \nabla \wedge \nabla \wedge \int_{\partial D} \nu(y) \wedge H(y)\Phi(x, y)ds(y), \tag{4.4.4}$$

and

$$H(x) = -\nabla \wedge \int_{\partial D} \nu(y) \wedge H(y)\Phi(x, y)ds(y)$$
$$- \frac{1}{ik} \nabla \wedge \nabla \wedge \int_{\partial D} \nu(y) \wedge E(y)\Phi(x, y)ds(y), \tag{4.4.5}$$

for $x \in D$.

The representation in the Theorem 4.4.3 indicates that the solutions to Maxwell equations is analytic. Therefore, by using the identity $\nabla \wedge \nabla \wedge = -\Delta + \nabla\nabla \cdot$, we can obtain that $\nabla \cdot E = \nabla \cdot H = 0$ and $\Delta E + k^2 E = 0$, $\Delta H + k^2 H = 0$ in D, where E and H are solutions to the Maxwell equations in D. Conversely, if E (or H) is a solution to the vector Helmholtz equation satisfying $\nabla \cdot E = 0$ (or $\nabla \cdot H = 0$). Then E and $H = \frac{1}{ik}\nabla \wedge E$ (or H and $E = -\frac{1}{ik}\nabla \wedge H$) satisfy the Maxwell equations.

Since we already know the representation for the solutions to Maxwell equations, now we move our attention to that satisfying the Silver-Müller radiation condition as follows.

$$\lim_{|x|\to\infty} (H \wedge x - |x|E) = 0, \tag{4.4.6}$$

where the limit holds uniformly in all directions $x/|x|$. This will provide us a representation for solutions to the Maxwell equations in exterior domains.

For a radiating solution to the Maxwell equations in the exterior of a ball, we have the following theorem.

Theorem 4.4.4 *Let $E, H \in C^1(\mathbb{R}^3\backslash\overline{D}) \cap C(\mathbb{R}^3\backslash D)$ be a radiating solution to the Maxwell equations*

$$\nabla \wedge E - ikH = 0, \qquad \nabla \wedge H + ikE = 0 \quad \text{in } \mathbb{R}^3\backslash\overline{D}.$$

Then we have the Stratton-Chu formulas

$$\begin{aligned} E(x) = {}& \nabla \wedge \int_{\partial D} \nu(y) \wedge E(y)\Phi(x, y)ds(y) \\ & - \frac{1}{ik}\nabla \wedge \nabla \wedge \int_{\partial D} \nu(y) \wedge H(y)\Phi(x, y)ds(y), \end{aligned} \tag{4.4.7}$$

and

$$\begin{aligned} H(x) = {}& \nabla \wedge \int_{\partial D} \nu(y) \wedge H(y)\Phi(x, y)ds(y) \\ & + \frac{1}{ik}\nabla \wedge \nabla \wedge \int_{\partial D} \nu(y) \wedge E(y)\Phi(x, y)ds(y), \end{aligned} \tag{4.4.8}$$

for all $x \in \mathbb{R}^3\backslash\overline{D}$. For $x \in D$ the right hand side of the above tow equations vanish.

Let a be a constant vector. Then

$$E_m(x) := \nabla \wedge_x a\Phi(x, y), \quad H_m(x) := \frac{1}{ik}\nabla \wedge E_m(x) \tag{4.4.9}$$

represent the EM field generated by a magnetic dipole located at the point y and solve the Maxwell equations for $x \neq y$. Similarly,

$$H_e(x) := \nabla \wedge_x a\Phi(x, y), \quad E_e(x) := -\frac{1}{ik}\nabla \wedge H_e(x) \tag{4.4.10}$$

represent the EM field generated by an electric dipole. The fields (4.4.9) and (4.4.10) may be considered as fundamental solutions to the Maxwell equations.

Then straightforward calculations show that each Cartesian component of E and H satisfies the Sommerfeld radiation condition. It is also possible to see that solutions to the Maxwell equations for which each Cartesian component satisfies the Sommerfeld radiation condition also satisfy the Silver-Müller radiation condition. We then conclude that for solutions to the Maxwell equations, the Silver-Müller radiation condition is equivalent to the Sommerfeld radiation condition for the Cartesian.

Similar to the acoustic case, for a constant vector a, we have

$$\nabla \wedge_x a\frac{e^{ik|x-y|}}{|x-y|} = ik\frac{e^{ik|x|}}{|x|}\left\{e^{-ik\hat{x}\cdot y}\hat{x} \times a + O\left(\frac{|a|}{|x|}\right)\right\}, \tag{4.4.11}$$

$$\nabla \wedge_x \nabla \wedge_x a\frac{e^{ik|x-y|}}{|x-y|} = k^2\frac{e^{ik|x|}}{|x|}\left\{e^{-ik\hat{x}\cdot y}\hat{x} \times (a \times \hat{x}) + O\left(\frac{|a|}{|x|}\right)\right\}, \tag{4.4.12}$$

as $|x| \to \infty$ uniformly for all $y \in \partial D$. Substituting this into (4.4.7) and (4.4.8) we obtain the electric far field pattern and magnetic far field pattern, respectively.

$$E_\infty(\hat{x}) = \frac{ik}{4\pi}\hat{x} \times \int_{\partial D} \{\nu(y) \times E(y) + [\nu(y) \times H(y)] \times \hat{x}\}e^{-ik\hat{x}\cdot y}ds(y), \tag{4.4.13}$$

$$H_\infty(\hat{x}) = \frac{ik}{4\pi}\hat{x} \times \int_{\partial D} \{\nu(y) \times H(y) + [\nu(y) \times E(y)] \times \hat{x}\}e^{-ik\hat{x}\cdot y}ds(y). \tag{4.4.14}$$

The following relations are satisfied

$$H_\infty = \nu \times E_\infty \quad \text{and} \quad \nu \cdot E_\infty = \nu \cdot H_\infty = 0 \tag{4.4.15}$$

with ν being the unit outward normal on $\mathbb{S}^2$. Obviously, from (4.4.13) and (4.4.14), one can know that the radiation solution E, H to the Maxwell equations has the following asymptotic form

$$E(x) = \frac{e^{ik|x|}}{|x|}\left\{E_\infty(\hat{x}) + O\left(\frac{1}{|x|}\right)\right\}, \quad |x| \to \infty, \tag{4.4.16}$$

$$H(x) = \frac{e^{ik|x|}}{|x|}\left\{H_\infty(\hat{x}) + O\left(\frac{1}{|x|}\right)\right\}, \quad |x| \to \infty, \tag{4.4.17}$$

uniformly in all directions $\hat{x} = x/|x|$.

Again Rellich's lemma implies that the the far field E_∞ uniquely determines E as well as H by $H = (ik)^{-1}\nabla \wedge E$. Analogously H_∞ determines E and H uniquely.

The uniqueness and the existence for the EM scattering problem from obstacle can be obtained by using the representation theory introduced above. More details can be found in Colton's book. The well-posedness study of the direct electromagnetic scattering problem from medium is similar to that with the acoustic case. For the uniqueness, we use Green's formula and Rellich's theorem to arrive at $E = 0$ in the exterior of B_R. Then we apply the unique continuation principle.

For the scattering problem, the boundary values are the restriction of an analytic field to the boundary and therefore they are as smooth as the boundary. Hence, for domains D of class C^2 there exists a solution. Therefore, we can apply the Stratton-Chu formulas for the scattered field E^s , H^s and the Stratton-Chu formulas for the incident E^i, H^i. Then, adding both formulas and using the boundary condition $\nu \wedge (E^i + E^s) = 0$ on ∂D, we have the following theorem knowns as Huygens' principle. The representation for the far field pattern can be obtained also.

Overall, for the scattering of an entire electromagnetic field E^i, H^i by a perfect conductor D we have

$$E(x) = E^i(x) - \frac{1}{ik}\nabla \wedge \nabla \wedge \int_{\partial D} \nu(y) \wedge H(y)\Phi(x, y)ds(y), \tag{4.4.18}$$

$$H(x) = H^i(x) + \nabla \wedge \int_{\partial D} \nu(y) \wedge H(y)\Phi(x, y)ds(y) \tag{4.4.19}$$

for $x \in \mathbb{R}^3 \backslash \overline{D}$ where E, H is the total field. The far field pattern is given by

$$E_\infty(\hat{x}) = \frac{ik}{4\pi}\hat{x} \wedge \int_{\partial D} [\nu(y) \wedge H(y)] \wedge \hat{x}e^{-ik\hat{x}\cdot y}ds(y), \tag{4.4.20}$$

$$H_\infty(\hat{x}) = \frac{ik}{4\pi}\hat{x} \wedge \int_{\partial D} \nu(y) \wedge H(y)e^{-ik\hat{x}\cdot y}ds(y) \tag{4.4.21}$$

for $\hat{x} \in \mathbb{S}^2$.

4.5 Numerical Methods for Electromagnetic Scattering Problems

We introduce a perfectly matched layer (PML) technique for solving the time harmonic electromagnetic scattering problem with the perfectly conducting boundary condition

$$\begin{cases} \nabla \times \nabla \times E - k^2 E = 0 & \text{in } \mathbb{R}^3 \backslash \overline{D}, \\ n \times E = g & \text{on } \Gamma_D, \\ \lim_{|x| \to \infty} |x| \big[(\nabla \times E) \times \hat{x} - ikE \big] = 0, \end{cases} \tag{4.5.1}$$

where $D \subset \mathbb{R}^3$ is a bounded domain with Lipschitz polyhedral boundary Γ_D, E is the electric field, g is determined by the incoming wave, $\hat{x} = x/|x|$, and n is the unit outer normal to Γ_D.

Let D be contained in the interior of the ball $B_R = \{x \in \mathbb{R}^3, |x| < R\}$ with boundary Γ_R. We first recall the series solution of the scattering problem (4.5.1) outside the ball B_R by following the development in Monk. Let $Y_n^m(\hat{x})$, $m = -n, \ldots, n$, $n = 1, 2, \ldots$, be the spherical harmonics which satisfy

$$\Delta_{\partial B_1} Y_n^m(\hat{x}) + n(n+1)Y_n^m(\hat{x}) = 0 \qquad \text{on } \partial B_1, \tag{4.5.2}$$

where $\Delta_{\partial B_1} = \frac{1}{\sin\theta}\frac{\partial}{\partial\theta}(\sin\theta\frac{\partial}{\partial\theta}) + \frac{1}{\sin^2\theta}\frac{\partial^2}{\partial\phi^2}$ is the Laplace-Beltrami operator for the surface of the unit sphere ∂B_1. The set of all spherical harmonics $\{Y_n^m(\hat{x}) : m = -n, \ldots, n, n = 1, 2, \ldots\}$ form a complete orthonormal basis of $L^2(\partial B_1)$.

Denote the vector spherical harmonics

$$U_n^m = \frac{1}{\sqrt{n(n-1)}}\nabla_{\partial B_1} Y_n^m, \quad V_n^m = \hat{x} \times U_n^m, \tag{4.5.3}$$

where $\nabla_{\partial B_1} Y_n^m = \frac{\partial Y_n^m}{\partial\theta} e_\theta + \frac{1}{\sin\theta}\frac{\partial Y_n^m}{\partial\phi} e_\phi$, and $\{e_r, e_\theta, e_\phi\}$ are the unit vectors of the spherical coordinates. The set of all vector spherical harmonics $\{U_n^m, V_n^m : m = -n, \ldots, n, n = 1, 2, \ldots\}$ forms a complete orthormal basis of $\boldsymbol{L}_t^2(\partial B_1) = \{u \in [L^2(\partial B_1)]^3 : u \cdot \hat{x} = 0 \text{ on } \partial B_1\}$.

For any $\Phi \in H(\text{curl}, B_R)$, $\hat{x} \times \Phi|_{\Gamma_R}$ is in the trace space $H^{-1/2}(\text{div}, \Gamma_R)$, whose norm, for any $\lambda = \sum_{n=1}^{\infty}\sum_{m=-n}^{n} a_{nm}U_n^m + b_{nm}V_n^m$, is defined by

$$\|\lambda\|^2_{H^{-1/2}(\text{div},\Gamma_R)} = \sum_{n=1}^{\infty}\sum_{m=-n}^{n} \sqrt{n(n+1)}|a_{nm}|^2 + \frac{1}{\sqrt{n(n+1)}}|b_{nm}|^2. \tag{4.5.4}$$

It is also known that for $\Phi \in H(\text{curl}, B_R)$, the tangential component $(\hat{x} \times \Phi) \times \hat{x}|_{\Gamma_R}$ belongs to $H^{-1/2}(\text{curl}, \Gamma_R)$ which is the dual space of $H^{-1/2}(\text{div}, \Gamma_R)$ with respect to the scalar product in $L_t^2(\Gamma_R)$.

Let $H_n^{(1)}(z)$ be the spherical Hankel function of the first kind of order n. We introduce the vector wave functions

$$M_n^m(r, \hat{x}) = \nabla \times \{xH_n^{(1)}(kr)Y_n^m(\hat{x})\}, \quad N_n^m(r, \hat{x}) = \frac{1}{ik}\nabla \times M_n^m(r, \hat{x}),$$

which are the radiation solutions of the first Maxwell equation in (4.5.1) in $\mathbb{R}^2\backslash\{0\}$. In the domain $\mathbb{R}^3\backslash\overline{B}_R$, the solution E of (4.5.1) can be written as

$$E(r, \hat{x}) = \sum_{n=1}^{\infty}\sum_{m=-n}^{n} \frac{a_{nm}M_n^m(r, \hat{x})}{H_n^{(1)}(kR)\sqrt{n(n+1)}} + \frac{ikRb_{nm}N_n^m(r, \hat{x})}{z_n^{(1)}(kR)\sqrt{n(n+1)}}, \tag{4.5.5}$$

where $z_n^{(1)}(kR) = H_n^{(1)}(kR) + kRH_n^{(1)'}(kR)$, and a_{nm}, b_{nm} are determined by the trace of E on Γ_R through $\hat{x} \times E|_{\Gamma_R} = \sum_{n=1}^{\infty}\sum_{m=-n}^{n} a_{nm}U_n^m + b_{nm}V_n^m$. The series in (4.5.5) converges uniformly for $r > R$.

Now we turn to the introduction of the absorbing PML layer. We surround the domain $\Omega_R = B_R\backslash\overline{D}$ with a PML layer $\Omega^{PML} = \{x \in \mathbb{R}^3 : R < |x| < \rho\}$. Let $\alpha(r) = 1 + i\sigma(r)$ be the model medium property which satisfies

$$\sigma \in C(\mathbb{R}), \quad \sigma \geq 0, \quad \text{and} \;\; \sigma = 0 \;\; \text{for} \;\; r \leq R.$$

Denote by $\tilde{r}$ the complex radius defined by

$$\tilde{r} = \tilde{r}(r) = \begin{cases} r & \text{if } r \leq R \\ \displaystyle\int_0^r \alpha(t)dt = r\beta(r) & \text{if } r \geq R. \end{cases}$$

It is easy to check that the vector wave functions satisfy

$$M_n^m(r, \hat{x}) = H_n^{(1)}(kr)\nabla_{\partial B_1} Y_n^m(\hat{x}) \times \hat{x}, \tag{4.5.6}$$

$$\begin{aligned} N_n^m(r, \hat{x}) &= \frac{1}{ik}\nabla \times M_n^m \qquad (4.5.7)\\ &= \frac{\sqrt{n(n+1)}}{ikr} z_n^{(1)}(kr)U_n^m(\hat{x}) + \frac{n(n+1)}{ikr} H_n^{(1)}(kr)Y_n^m(\hat{x})\hat{x}. \end{aligned}$$

We introduce

$$\tilde{M}_n^m(\tilde{r}, \hat{x}) = H_n^{(1)}(k\tilde{r})\nabla_{\partial B_1} Y_n^m(\hat{x}) \times \hat{x}, \tag{4.5.8}$$

$$\begin{aligned} \tilde{N}_n^m(\tilde{r}, \hat{x}) &= \frac{1}{ik}\tilde{\nabla} \times \tilde{M}_n^m \qquad (4.5.9)\\ &= \frac{\sqrt{n(n+1)}}{ik\tilde{r}} z_n^{(1)}(k\tilde{r})U_n^m(\hat{x}) + \frac{n(n+1)}{ik\tilde{r}} H_n^{(1)}(k\tilde{r})Y_n^m(\hat{x})\hat{x}, \end{aligned}$$

where $\tilde{\nabla}\times$ is the curl operator with respect to the complex spherical variables $(\tilde{r}, \theta, \phi)$, that is, for $\Phi = \Phi_r e_r + \Phi_\theta e_\theta + \Phi_\phi e_\phi$,

$$\begin{aligned} \tilde{\nabla} \times \Phi = &\frac{1}{\tilde{r}\sin\theta}\left(\frac{\partial}{\partial\theta}(\sin\theta\,\Phi_\phi) - \frac{\partial\Phi_\theta}{\partial\phi}\right)e_r \\ &+ \frac{1}{\tilde{r}}\left(\frac{1}{\sin\theta}\frac{\partial\Phi_r}{\partial\phi} - \frac{\partial(\tilde{r}\Phi_\phi)}{\partial\tilde{r}}\right)e_\theta \\ &+ \frac{1}{\tilde{r}}\left(\frac{\partial(\tilde{r}\Phi_\theta)}{\partial\tilde{r}} - \frac{\partial\Phi_\phi}{\partial\theta}\right). \end{aligned}$$

It is easy to check that $\tilde{\nabla} \times \Phi = A\nabla \times B\Phi$, where $A = \mathrm{diag}(\beta^{-2}, \alpha^{-1}\beta^{-1}, \alpha^{-1}\beta^{-1})$ and $B = \mathrm{diag}(\alpha, \beta, \beta)$ are 3×3 diagonal matrices.

For any $\lambda = \sum_{n=1}^{\infty}\sum_{m=-n}^{n} a_{nm}U_n^m + b_{nm}V_n^m \in H^{-1/2}(\mathrm{div}, \Gamma_R)$, let $\mathbb{E}(\lambda)(\tilde{r}, \hat{x})$ be the PML extension given by

$$\mathbb{E}(\tilde{r}, \hat{x}) = \sum_{n=1}^{\infty} \sum_{m=-n}^{n} \frac{a_{nm} \tilde{M}_n^m(\tilde{r}\hat{x})}{H_n^{(1)}(kR)\sqrt{n(n+1)}} + \frac{ikRb_{nm} \tilde{N}_n^m(\tilde{r}, \hat{x})}{z_n^{(1)}(kR)\sqrt{n(n+1)}}, \quad r > R. \tag{4.5.10}$$

For the solution E of the scattering problem (4.5.1), let $\tilde{E} = \mathbb{E}(\hat{x} \times E|_{\Gamma_R})$ be the PML extension of $\hat{x} \times E|_{\Gamma_R}$. Since $\tilde{r} = r$ on Γ_R, we know that $\hat{x} \times \tilde{E} = \hat{x} \times E$ on Γ_R. On the other hand, since $H_n^{(1)}(z) \sim \frac{1}{z} e^{i(z - \frac{1}{2}n\pi - \frac{1}{2}\pi)}$ asymptotically as $|z| \to \infty$, heuristically $\tilde{E}(\tilde{r}, \hat{x})$ will decay exponentially for large $r > R$. It is obvious that $\tilde{E}$ satisfies

$$\tilde{\nabla} \times \tilde{\nabla} \times \tilde{E} - k^2 \tilde{E} = 0 \quad \text{in} \quad \mathbb{R}^3 \backslash \overline{B}_R,$$

which gives the desired PML equation in the spherical coordinates

$$\nabla \times B(A\nabla \times B\tilde{E}) - k^2 A^{-1} \tilde{E} = 0 \quad \text{in} \quad \mathbb{R}^3 \backslash \overline{B}_R.$$

The PML problem is then to find $\hat{E}$, which approximates E in Ω_R and $B\tilde{E}$ in

$$\begin{cases} \nabla \times BA(\nabla \times \hat{E}) - k^2 (BA)^{-1} \hat{E} = 0 & \text{in} \quad \Omega_\rho = B_\rho \backslash \overline{D}, \\ n \times \hat{E} = g & \text{on} \quad \Gamma_D, \\ \hat{x} \times \hat{E} = 0 & \text{on} \quad \Gamma_\rho. \end{cases} \tag{4.5.11}$$

The well-posedness of the PML problem (4.5.11) and the convergence of its solution to the original problem (4.5.1) can be found in [2].

Now we introduce the equivalent variational form of the scattering problem (4.5.1) and (4.5.11) on the bounded domain $\Omega_R = B_R \backslash \overline{D}$ using the Calderon operators.

Given a tangential vector λ on Γ_R, the Calderon operator $G_e : H^{-1/2}(\text{div}; \Gamma_R) \to H^{-1/2}(\text{div}; \Gamma_R)$ is the Dirichlet to Neumann to operator defined by

$$G_e(\lambda) = \frac{1}{ik} \hat{x} \times (\nabla \times E^s),$$

where E^s satisfies

$$\begin{cases} \nabla \times \nabla \times E^s - k^2 E^s = 0 \quad \text{in} \quad \mathbb{R}^3 \backslash \overline{B}_R, \\ \hat{x} \times E^s = \lambda \quad \text{on} \quad \Gamma_R, \\ \lim_{|x| \to \infty} |x| \left[(\nabla \times E^s) \times \hat{x} - ikE^s \right] = 0. \end{cases}$$

Let $\lambda = \sum_{n=1}^{\infty} \sum_{m=-n}^{n} a_{nm} U_n^m + b_{nm} V_n^m$; the function E^s is given as (4.5.5)

$$E^s(r,\hat{x}) = \sum_{n=1}^{\infty}\sum_{m=-n}^{n} \frac{a_{nm} M_n^m(r,\hat{x})}{H_n^{(1)}(kR)\sqrt{n(n+1)}} + \frac{\mathrm{i}kRb_{nm} N_n^m(r,\hat{x})}{z_n^{(1)}(kR)\sqrt{n(n+1)}}, \quad r > R.$$

Since $\frac{1}{\mathrm{i}k}\nabla \times M_n^m = N_m^n$, $-\frac{1}{\mathrm{i}k}\nabla \times N_n^m = M_n^m$, we have

$$\frac{1}{\mathrm{i}k}\sum_{n=1}^{\infty}\sum_{m=-n}^{n} \frac{a_{nm} N_n^m(r,\hat{x})}{H_n^{(1)}(kR)\sqrt{n(n+1)}} - \frac{\mathrm{i}kRb_{nm} M_n^m(r,\hat{x})}{z_n^{(1)}(kR)\sqrt{n(n+1)}}.$$

Thus by (4.5.7)-(4.5.8)

$$G_e(\lambda) = \sum_{n=1}^{\infty}\sum_{m=-n}^{n} \frac{-\mathrm{i}kRb_{nm} H_n^{(1)}(kR)}{z_n^{(1)}(kR)} U_n^m(\hat{x}) + \frac{a_{nm} z_n^{(1)}(kR)}{\mathrm{i}kRH_n^{(1)}(kR)} V_n^m(\hat{x}). \tag{4.5.12}$$

Let $a : H(\mathrm{curl}, \Omega_R) \times H(\mathrm{curl}, \Omega_R) \to \mathbb{C}$ be the sesquilinear form

$$a(E,\Phi) = \int_{\Omega_R} (\nabla \times E \cdot \nabla \times \overline{\Phi} - k^2 E \cdot \overline{\Phi})dx + \mathrm{i}k\left\langle G_e(\hat{x}\times E), (\hat{x}\times\Phi)\times\hat{x}\right\rangle_{\Gamma_R}.$$

The scattering problem (4.5.1) is equivalent to the following weak formulation: Given $g \in H^{-1/2}(\mathrm{div}, \Gamma_D)$, find $E \in H(\mathrm{curl}, \Omega_R)$ such that $\boldsymbol{n} \times E = g$ on Γ_D, and

$$a(E,\Phi) = 0, \qquad \forall\, \Phi \in H_0(\mathrm{curl}, \Omega_R), \tag{4.5.13}$$

where $H_0(\mathrm{curl}, \Omega_R) = \{\boldsymbol{v} \in H(\mathrm{curl}, \Omega_R) : \boldsymbol{n} \times \boldsymbol{v} = 0 \text{ on } \Gamma_D\}$.

The existence of a unique solution of the variational problem (4.5.13) can be found in [3]. Then the general theory in Babuska and Aziz [1] implies that there exists a constant $\mu > 0$ such that the following inf-sup conditions holds:

$$\sup_{\Phi \in H_0(\mathrm{curl},\Omega_R)} \frac{|a(\Psi,\Phi)|}{\|\Phi\|_{H(\mathrm{curl},\Omega_R)}} \ge \mu\|\Psi\|_{H(\mathrm{curl},\Omega_R)}, \qquad \forall\, \Psi \in H_0(\mathrm{curl}, \Omega_R). \tag{4.5.14}$$

For the PML problem (4.5.11), we need to reformulate it in the bounded domain Ω_R by imposing the boundary condition

$$\hat{x} \times (\nabla \times \hat{E})|_{\Gamma_R} = \mathrm{i}k\hat{G}_e(\hat{x} \times \hat{E}|_{\Gamma_R}),$$

where the approximate Calderon operator $\hat{G}_e : H^{-1/2}(\mathrm{div}, \Gamma_R) \to H^{-1/2}(\mathrm{div}, \Gamma_R)$ is defined as

$$\hat{G}_e(\lambda) := \frac{1}{\mathrm{i}k}\hat{x} \times (\nabla \times \Psi), \tag{4.5.15}$$

with Ψ satisfying

$$\nabla \times BA(\nabla \times \Psi) - k^2(BA)^{-1}\Psi = 0 \quad \text{in} \quad \Omega^{PML}, \tag{4.5.16}$$

$$\hat{x} \times \Psi = \lambda \quad \text{on} \quad \Gamma_R, \qquad \hat{x} \times \Psi = 0 \quad \text{on} \quad \Gamma_\rho. \tag{4.5.17}$$

That the approximate Calderon operator $\hat{G}_e$, let $\hat{a} : H(\text{curl}, \Omega_R) \times H(\text{curl}, \Omega_R) \to \mathbb{C}$ be the sesquilinear form

$$\hat{a}(\hat{E}, \Phi) = \int_{\Omega_R} (\nabla \times \hat{E} \cdot \nabla \times \overline{\Phi} - k^2 \hat{E} \cdot \overline{\Phi}) dx + ik \left\langle \hat{G}_e(\hat{x} \times \hat{E}), (\hat{x} \times \Phi) \times \hat{x} \right\rangle_{\Gamma_R}.$$

Then the weak formulation of (4.5.11) on the bounded domain Ω_R is:
Given $g \in H^{-1/2}(\text{div}, \Gamma_D)$, find $\hat{E} \in H(\text{curl}, \Omega_R)$, such that $\boldsymbol{n} \times \hat{E} = g$ on Γ_D, and

$$\hat{a}(\hat{E}, \Phi) = 0, \qquad \forall\, \Phi \in H_0(\text{curl}, \Omega_R). \tag{4.5.18}$$

Now we introduce the finite element method for the PML problem (4.5.11). We start by introducing the weak formulation of the PML problem (4.5.11). Let

$$b(\Psi, \Phi) = \int_{\Omega_\rho} (BA\nabla \times \Psi \cdot \nabla \times \overline{\Phi} - k^2(BA)^{-1}\Psi \cdot \overline{\Phi}) dx. \tag{4.5.19}$$

Then the weak formulation of (4.5.11) is: Given $g \in H^{-1/2}(\text{div}, \Gamma_D)$, find $\hat{E} \in H(\text{curl}, \Omega_\rho)$, such that $\boldsymbol{n} \times \hat{E} = g$ on Γ_D, $\hat{x} \times \hat{E} = 0$ on Γ_ρ, and

$$b(\hat{E}, \Phi) = 0, \qquad \forall\, \Phi \in H_0(\text{curl}, \Omega_\rho). \tag{4.5.20}$$

Let Γ_ρ^h, which consists of piecewise triangles whose vertices lie on Γ_ρ, be an approximation of Γ_ρ. Let Ω_ρ^h be the subdomain of Ω_ρ bounded by Γ_D and Γ_ρ^h. Let $\mathcal{T}_h$ be a regular triangulation of the domain Ω_ρ^h. We will use the lowest order Nédeléc edge element for which the finite element space V_h over $\mathcal{T}_h$ is defined by

$$V_h = \{\boldsymbol{v} \in H(\text{curl}, \Omega_\rho^h) : \boldsymbol{v}|_T = a_T + b_T \times x, \quad \forall\, a_T, b_T \in \mathbb{R}^3, \ \forall\, T \in \mathcal{T}_h\}.$$

Degrees of freedom of functions $\boldsymbol{v} \in V_h$ on every $T \in \mathcal{T}_h$ are $\int_{e_i} \boldsymbol{v} \cdot dl, i = 1, \ldots, 6$, where $e_1, \ldots, e_6$ are six edges of T. Denote $\widetilde{V}_h = V_h \cap H_0(\text{curl}, \Omega_\rho)$. In the following, we will always assume that the functions in $\widetilde{V}_h$ are extended to the domain Ω_ρ by zero so that any function $\boldsymbol{v} \in \widetilde{V}_h$ is also a function in $H_0(\text{curl}, \Omega_\rho)$. The finite element approximation to (4.5.20) reads as follows.

Find $E_h \in V_h$ such that $\boldsymbol{n} \times E_h = g_h$ on Γ_D, $\boldsymbol{n} \times E_h = 0$ on Γ_ρ^h, and

$$b(E_h, \Phi_h) = 0, \qquad \forall\, \Phi_h \in \widetilde{V}_h. \tag{4.5.21}$$

Here g_h is some edge element approximation of g on Γ_D. Notice that the integral in $b(E_h, \Phi_h)$ is actually over Ω_ρ^h since $\Phi_h = 0$ in $\Omega_\rho \backslash \Omega_\rho^h$ by our convention. Based on

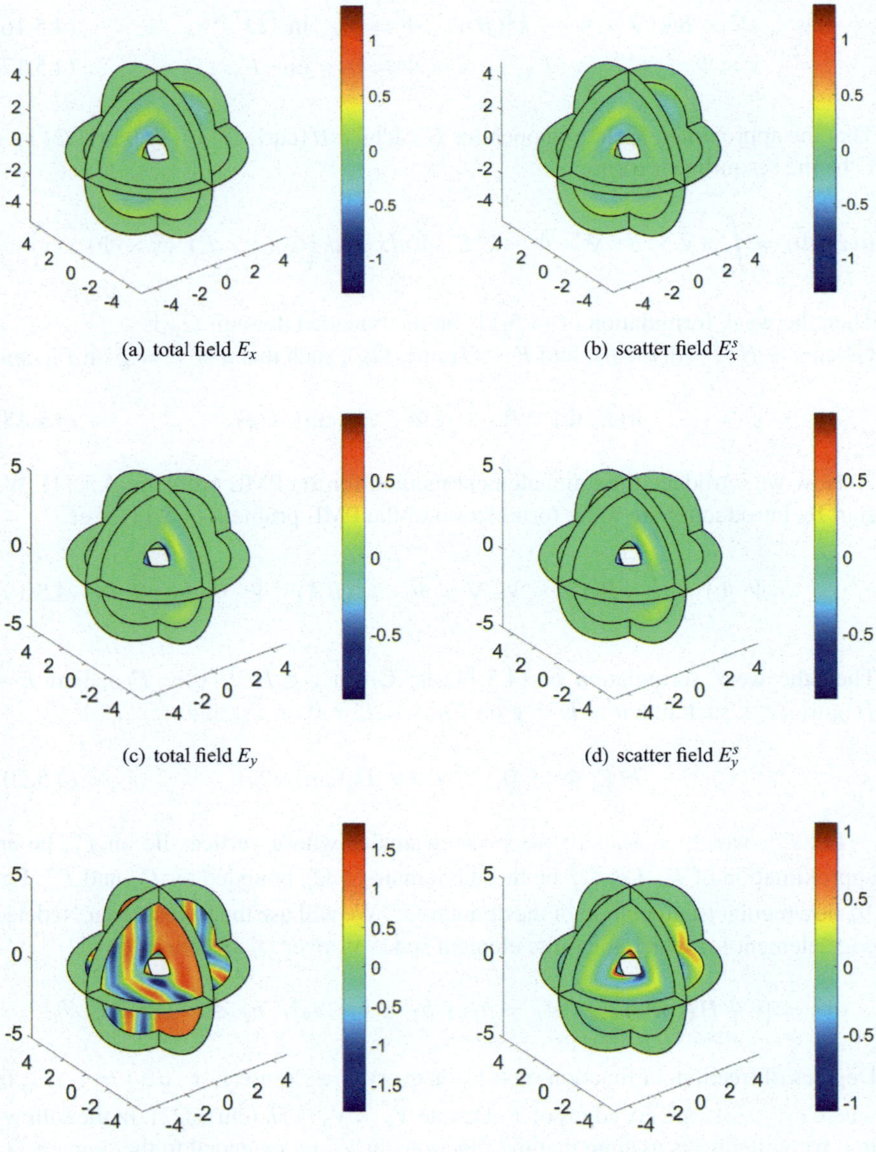

(a) total field E_x (b) scatter field E_x^s

(c) total field E_y (d) scatter field E_y^s

(e) total field E_z (f) scatter field E_z^s

Fig. 4.1 The incident direction is $d = (1, 0, 0)$ and the wavenumber is $k = 3$

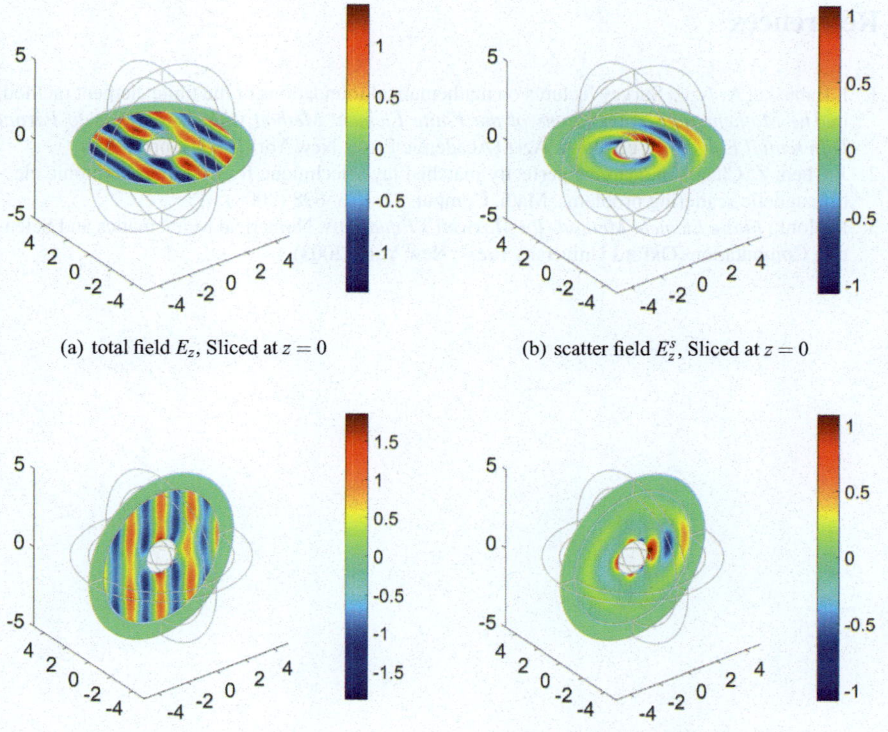

(a) total field E_z, Sliced at $z = 0$

(b) scatter field E_z^s, Sliced at $z = 0$

(c) total field E_z, Sliced at $y = 0$

(d) scatter field E_z^s, Sliced at $y = 0$

Fig. 4.2 The incident direction is $d = (1, 0)$ and the wavenumber is $k = 3$

a general argument in Schatz, the unique existence of (4.5.21) for sufficiently small mesh size $h < h^*$ can be proved by using the unique existence of the continuous problem (4.5.11).

Now we present some numerical experiments to demonstrate the theoretical results established in above.

First, the numerical experiment are conducted in $\mathbb{R}^3$. The geometry is a unit ball, the incident wave field is given by $E^i = (0, 0, \mathrm{e}^{\mathrm{i}k\mathrm{x}\cdot\mathrm{d}})^\top$.

For the setting with respect to the scattering measurement, a PML layer of width 1 is attached to the ball with $r = 5$ to truncate the whole space into a finite domain with scattering boundary conditions enforced on the outer boundary. We fix the wave number $k = 3$ and the incident direction $d = (1, 0, 0)^T$. The total field and the scatter field in 2D and 3D is given in Figs. 4.1 and 4.2, respectively.

References

1. I. Babuska, A. Aziz, Survey lectures on mathematical foundations of the finite element method, in *The Mathematical Foundations of the Finite Element Method with Application to Partial Differential Equations*, ed. by A. Aziz (Academic Press, New York, 1973), pp. 5–359
2. J. Chen, Z. Chen, An adaptive perfectly matched layer technique for 3-D time-harmonic electromagnetic scattering problems. Math. Comput. **77**, 673–698 (2008)
3. P. Monk, *Finite Element Methods for Maxwell's Equations*. Numerical Mathematics and Scientific Computation (Oxford University Press, New York, 2003)

Chapter 5
Numerical Inverse Electromagnetic Scattering Problems

5.1 Overview

The chapter is concerned with the inverse scattering problem of reconstructing an inhomogeneous scatterer located in an otherwise homogeneous space by measuring the corresponding electromagnetic (EM) wave fields far away from the scatterer produced by sending some detecting EM wave fields. The inverse electromagnetic scattering problem has been playing a central role in many areas of science and technology, such as radar and sonar, non-destructive testing, remote sensing, geophysical exploration and medical imaging to name just a few; see [7, 14, 15, 20, 26, 41, 48] and the references therein. In the current chapter, we mainly consider the reconstruction scheme for this inverse scattering problem. There are extensive studies in the literature in this aspect and many imaging schemes have been developed by various authors, and we would like to refer to [1, 5–8, 11, 17, 19, 24, 25, 27, 46, 47] and the references therein. In Sect. 5.3, we shall consider our study in a very practical setting by making use of a single electric far-field measurement. That is, we shall consider the reconstruction by measuring the far-field pattern of the electric wave field corresponding to a single pair of time-harmonic EM plane waves. From a practical viewpoint, the inverse scattering method with a single far-field measurement would be of significant interests, but is highly challenging with very limited progress in the literature; we refer to [12, 16, 20, 22, 23, 26, 31, 33–37, 37–39, 41, 48] for related discussion and surveys on some existing development. For more practical considerations, we shall work in a even more challenging setting by assuming very little a priori knowledge of the underlying scattering object, which might consist of multiple components, and both the number of the components and the physical property of each component are unknown in advance. This setting would make the study even more highly non-trivial, but on the other hand would be of significant practical importance when there is little a priori information available on the target

J. Li and H. Liu, *Numerical Methods for Inverse Scattering Problems*,
https://doi.org/10.1007/978-981-99-3772-1_5

object. In Sect. 5.4, we extend the related results in Sect. 5.3 to the case with multi-scale electromagnetic scatterers. Our discussion in this chapter follows the treatments in [30].

Our discussion in this chapter follows the treatment in [29, 30].

5.2 Strengthened Linear Sampling Methods

In this section we briefly address how to extend the theory and its resulting strengthened LSM for the inverse acoustic obstacle scattering problem studied in the previous sections to the case with inverse electromagnetic scattering.

We shall adopt the same notations below as those used in the previous sections. Then the direct electromagnetic scattering problem is to determine an electromagnetic field (E, H) governed by the system

$$\begin{cases} \mathrm{curl} E - \mathrm{i}kH = 0 & \text{in } \mathbb{R}^3 \backslash \bar{\mathbf{D}}, \\ \mathrm{curl} H + \mathrm{i}kE = 0 & \text{in } \mathbb{R}^3 \backslash \bar{\mathbf{D}}, \\ \lim_{r\to\infty}(H^s \times x - rE^s) = 0 \end{cases} \tag{5.2.1}$$

where $r = |x|$ for any $x \in \mathbb{R}^3$, (E, H) is the total field formed by the scattered field (E^s, H^s) and the incident field (E^i, H^i) given by

$$\begin{cases} E^i(x) = \frac{\mathrm{i}}{k}\mathrm{curl}\,\mathrm{curl}\, p e^{\mathrm{i}kx\cdot d} = \mathrm{i}k(d \times p) \times d e^{\mathrm{i}kx\cdot d}, \\ H^i(x) = \mathrm{curl}\, p e^{\mathrm{i}kx\cdot d} = \mathrm{i}kd \times p e^{\mathrm{i}kx\cdot d}, \end{cases} \tag{5.2.2}$$

with p being the polarization vector, and d being a unit vector along the propagation direction. System (5.2.1) is complemented by the following mixed boundary condition:

$$\nu \times E = 0 \quad \text{on } \partial\mathbf{D}_D, \quad \nu \times \mathrm{curl} E - \mathrm{i}\lambda(\nu \times E) = 0 \quad \text{on } \partial\mathbf{D}_I. \tag{5.2.3}$$

It is known that the radiating solution (E, H) to the exterior mixed boundary value problem (5.2.1)–(5.2.3) has the asymptotic behavior

$$E(x) = \frac{e^{\mathrm{i}k|x|}}{|x|}\left\{E_\infty(\hat{x}, d, p) + \mathcal{O}(\frac{1}{|x|})\right\}, \quad H(x) = \frac{e^{\mathrm{i}k|x|}}{|x|}\left\{H_\infty(\hat{x}, d, p) + \mathcal{O}(\frac{1}{|x|})\right\}, \tag{5.2.4}$$

as $|x| \to \infty$, where $E_\infty(\hat{x}, d, p)$ and $H_\infty(\hat{x}, d, p)$ are defined over the unit sphere $\mathbb{S}^2$ and known as the electric and magnetic far-field pattern, respectively.

Now the inverse problem of our interest is to recover $\mathbf{D}$ from the knowledge of the electric far-field data $E_\infty(\hat{x}, d, p)$ without any a priori information.

The LSM for the reconstruction is to use the solution g of the following far-field operator equation

$$(Fg)(\hat{x}) = E_{e,\infty}(\hat{x}, z, q) \tag{5.2.5}$$

as an indicator function to determine the boundary of the unknown scatterer **D**, where F is the far-field operator given by

$$(Fg)(\hat{x}) := \int_{\mathbb{S}^2} E_\infty(\hat{x}, d, g(d))ds(d), \quad \hat{x} \in \mathbb{S}^2, \tag{5.2.6}$$

and $E_{e,\infty}$ is a function induced by the incident field:

$$E_{e,\infty}(\hat{x}, z, q) = \frac{ik}{4\pi}(\hat{x} \times q) \times \hat{x}e^{-ik\hat{x}\cdot z}. \tag{5.2.7}$$

Similar blow-up behaviors of the indicator functions can be derived as in Theorem 3.2.1. It is straightforward to carry the reference ball technique developed in the previous sections to the LSM for the reconstruction of the boundary of the unknown scatterer **D**, using the same arguments as those in Theorems 3.2.4 and 3.2.5. Again, this reference ball technique will enable us to get rid of the interior Maxwell eigenvalue problem, and to choose the cut-off value in view of the characteristic isosurfaces of the reference ball to determine the unknown scatterer.

5.3 Single-shot Method for Multiple Multiscale Scatterers

The inverse scattering problem that we consider is to recover $(\Omega; \varepsilon, \mu, \sigma)$ from the knowledge of $A(\theta; \theta', p, \omega)$ or to recover Ω from the knowledge of $A(\theta; \theta', p, \omega)$ if $\sigma = +\infty$. We will make use of single far-field measurement, i.e., $A(\theta; \theta', p, \omega)$ for all $\theta \in \mathbb{S}^2$ but fixed θ', p and ω. Furthermore, we shall require very little a priori knowledge of the underlying scatterer, which could be composed of multiple components, and the number of the components is not required to be known in advance, and each component could be either a medium inclusion or an obstacle. Generically, in this extremely general setting, one cannot expect to recover all the details of the underlying scatterer by using only a single measurement. Instead, we would consider the recovery of the locations of the multiple components in the setting described above. Nevertheless, at this point, we would like to remark that our numerical experiments indicate that the discussed imaging scheme could also qualitatively reveal the supports/shapes of the scatterer components.

Specifically, two single-measurement locating schemes would be discussed for two separate cases depending on the size of the target scatterer. The first scheme is for locating scatterers of small size compared to the wavelength of the detecting plane wave. The multiple scatterers could be extremely general with very little a priori knowledge required. Each scatterer component could be either an impenetrable perfectly conducting obstacle or a penetrable inhomogeneous medium with an unknown content, and the number of the scatterer components is not required to be known in advance. The locating scheme is based on a indicator function $I_s(z)$ for $z \in \mathbb{R}^3$. If z happens to be the location point of a scatterer component, then z is a local maximum point for $I_s(z)$. Using the indicating behavior of $I_s(z)$, one could then locate all the scatterer components. In defining the indicator function, only inner-products of the electric far-field pattern and the vectorial spherical harmonics are involved. The indicating behavior is derived from the asymptotic expansion of the electric far-field pattern for small 'point-like' scatterers. The expansion is further based on the low frequency EM scattering asymptotics (i.e., Raleigh approximation); see [9–11, 21, 43]. But for our imaging scheme, the expansion would be formulated in terms of the vectorial spherical harmonics instead of the polarizability tensors. It is also interesting to mention that our numerical experiments show that the discussed scheme also works effectively for locating 'partially-small' scatterer, namely, the scatterer is not 'point-like' but 'segment-like'. Furthermore, in addition to finding the locations of the scatterer components, the discussed scheme also shows some very promising feature in qualitatively imaging the supports/shapes of the unknown scatterers.

The second scheme is for locating multiple perfectly conducting obstacles whose sizes are comparable to the detecting EM wavelength. For this case, we would require that the shape of each component must be from a certain known admissible class. That is, there might be multiple obstacles with an unknown number of components, but the shape of each obstacle component must be from a certain class of reference obstacles which is known in advance. Nevertheless, there could be multiple different reference obstacles. Other than the assumptions made above, no further a priori information is required for the multiple unknown scatterers. The number of the unknown scatterer components could be larger than the number of the reference obstacles; that is, some of the unknown scatterers possess the same shape. Moreover, it is not necessary for all the reference obstacles to be presented in the unknown scatterer. The setting considered would be of significant practical interests, e.g., in radar and sonar technology. The second locating scheme is based on l' indicator functions, where $l' \in \mathbb{N}$ denotes the number of the reference obstacles. Similar to the first locating scheme, in calculating the indicator functions, only inner products are involved and no inversion will be required. The discussed method is very efficient and robust to measurement noise, and could also be extended to include inhomogeneous medium components if a certain generic condition is satisfied. To our best knowledge, this is the first imaging scheme in the literature for locating multiple unknown scatterers of regular size by a single far-field measurement. Rigorous mathematical justifications are provided for both schemes. In our theoretical analysis, we would impose the sparse distribution of the multiple scatter components. However, we have conducted extensive numerical experiments, and the numerical results show that even without the sparsity assumption, the presented locating schemes still work very effectively.

The rest of the section is organized as follows. In Sect. 5.3.1, we discuss the locating schemes, respectively, for the two separate cases with small scatterers and regular-size scatterers. Section 5.3.2 is devoted to the proofs of Theorems 5.3.1 and 5.3.2 stated in Sect. 5.3.1 justifying the indicating behaviors of the indicator functions for the two locating schemes. In Sect. 5.3.3, we present extensive numerical experiments to illustrate the effectiveness of the presented methods. This section is concluded in Sect. 5.3.4 with some further discussions.

5.3.1 The Locating Schemes

In this section, we present the two schemes for locating, respectively, multiple scatterers of small size and regular size. In order to ease the exposition, throughout the rest of the section, we assume that $\omega \sim 1$, and hence the size of a scatterer can be expressed in terms of its Euclidean diameter.

5.3.1.1 Locating Small Scatterers

We first introduce the class of small EM scatterers for our current study. Let $l \in \mathbb{N}$ and let D_j, $1 \le j \le l$, be bounded C^2 domains in $\mathbb{R}^3$. It is assumed that all D_j's are simply connected and contain the origin. For $\rho \in \mathbb{R}_+$, we let $\rho D_j := \{\rho x | x \in D_j\}$ and set

$$\Omega_j = z_j + \rho D_j, \quad z_j \in \mathbb{R}^3, \quad 1 \le j \le l.$$

Each Ω_j is referred to as a scatterer component and its content is given by ε_j, μ_j and σ_j. It is assumed that $\varepsilon_j > 0$, $\mu_j > 0$ and $\sigma_j \ge 0$ are all constants, except the case that $\sigma_j = +\infty$. If σ_j is taken to be $+\infty$, then D_j would be regarded as a perfectly conducting obstacle disregarding the parameters ε_j and μ_j. In the sequel, we reserve the letter l to denote the number of components of a scatterer defined by

$$\Omega = \bigcup_{j=1}^{l} \Omega_j \quad \text{and} \quad (\Omega; \varepsilon, \mu, \sigma) = \bigcup_{j=1}^{l} (\Omega_j; \varepsilon_j, \mu_j, \sigma_j). \tag{5.3.1}$$

The parameter $\rho \in \mathbb{R}_+$ represents the relative size of the scatterer (or, more precisely, each of its component). We now make the following qualitative assumptions,

$$\rho \ll 1 \quad \text{and} \quad \text{dist}(z_j, z_{j'}) \gg 1 \ \text{ for } j \neq j', \ 1 \le j, j' \le l. \tag{5.3.2}$$

The assumption (5.3.2) means that compared to the wavelength of the detecting/incident wave, the relative size of each scatterer component is small and if there are multiple components, they must be sparsely distributed. Our numerical experiments in Sect. 5.3.3 could be more quantitative in this aspect, and it is numerically shown that if the size of the scatterer component is smaller than half a wavelength and the distance between two distinct components is bigger than half a wavelength, one could have a fine reconstruction by using the presented scheme. In this sense, the qualitative assumption (5.3.2) is needed only for our subsequent theoretical analysis of the discussed locating scheme. Furthermore, we would like to emphasize that most of the other restrictive assumptions introduced above are also mainly for the purpose of the theoretical justification, and the numerical experiments in Sect. 5.3.3 will show that the discussed imaging scheme works in a much more general setting. Specifically, the regularity of each reference component D_j is not necessarily C^2-smooth, and it could be a Lipschitz domain, and moreover, it is not necessarily simply connected, as long as ρD_j for a small $\rho \in \mathbb{R}_+$ yields an appropriate domain of small size. Moreover, the content of each medium component is not necessarily constant, and it could be variable. Some of those restrictive assumptions could be relaxed from our theoretical justification in the following. However, it is our focus and emphasis on developing the locating scheme in this section, and we would not appeal for a most general theoretical study.

In the sequel, without loss of generality, we can assume that there exists $\tilde{l} \in \mathbb{N} \cup \{0\}$, $0 \le \tilde{l} \le l$ such that

$$0 \le \sigma_j < +\infty \ \text{ for } 1 \le j \le \tilde{l} \qquad \text{and} \qquad \sigma_j = +\infty \ \text{ for } \tilde{l}+1 \le j \le l.$$

Then, we set

$$\Omega_m := \bigcup_{j=1}^{\tilde{l}} \Omega_j \qquad \text{and} \qquad \Omega_o := \bigcup_{j=\tilde{l}+1}^{l} \Omega_j \tag{5.3.3}$$

to denote, respectively, the medium component and the obstacle component of a scatterer. We emphasize that both l and $\tilde{l}$ are unknown in advance, and $\tilde{l}$ could be 0 or l, corresponding to the case that $\Omega_m = \emptyset$ or $\Omega_o = \emptyset$. Moreover, the contents and the shapes of the two components are unknown in advance either and they are not even necessarily identical to each other, i.e.,

$$(\Omega_j; \varepsilon_j, \mu_j, \sigma_j) \neq (\Omega_{j'}; \varepsilon_{j'}, \mu_{j'}, \sigma_{j'}) \quad \text{for } j \neq j'.$$

Corresponding to a single pair of excited EM plane waves (4.2.1), the electromagnetic scattering is governed by the following Maxwell system

$$\begin{cases} \nabla \wedge E - \mathrm{i}\omega \left(1 + \sum_{j=1}^{\tilde{l}} (\mu_j - 1)\chi_{\Omega_j}\right) H = 0 & \text{in } \mathbb{R}^3 \backslash \overline{\Omega}_o, \\ \nabla \wedge H + \left(\mathrm{i}\omega(1 + \sum_{j=1}^{\tilde{l}} (\varepsilon_j - 1)\chi_{\Omega_j}) - \sum_{j=1}^{\tilde{l}} \sigma_j \chi_{\Omega_j}\right) E = 0 & \text{in } \mathbb{R}^3 \backslash \overline{\Omega}_o, \\ E^- = E|_{\Omega_m},\ E^+ = (E - E^i)|_{\mathbb{R}^3 \backslash \overline{\Omega}},\ H^+ = (H - H^i)|_{\mathbb{R}^3 \backslash \overline{\Omega}}, \\ \nu \wedge E^+ = -\nu \wedge E^i \qquad \text{on } \partial\Omega_o, \\ \lim_{|x| \to +\infty} |x| \left| (\nabla \wedge E^+)(x) \wedge \frac{x}{|x|} - \mathrm{i}\omega E^+(x) \right| = 0. \end{cases} \tag{5.3.4}$$

It is known that there exists a unique pair of solutions $(E^+, H^+) \in H_{loc}(\text{curl}; \mathbb{R}^3\backslash\overline{\Omega}_0) \wedge H_{loc}(\text{curl}; \mathbb{R}^3\backslash\overline{\Omega}_0)$ to the Maxwell equations (5.3.4) (see [44]). In the following, we write

$$A(\theta; \Omega) := A(\theta; \bigcup_{j=1}^{l}(\Omega_j; \varepsilon_j, \mu_j, \sigma_j), \theta', p, \omega)$$

to denote the electric far-field pattern corresponding to the EM wave fields in (5.3.4). We also write for $1 \le j \le l$,

$$A(\theta; \Omega_j) := A(\theta; (\Omega_j; \varepsilon_j, \mu_j, \sigma_j), \theta', p, \omega)$$

to denote the far-field pattern corresponding solely to the scatterer $(\Omega_j; \varepsilon_j, \mu_j, \sigma_j)$. Both $A(\theta; \Omega)$ and $A(\theta; \Omega_j)$ are real analytic functions on the unit sphere $\mathbb{S}^2$ (cf. [20, 44]).

Next, we introduce the space of L^2 tangential fields on the unit sphere as follows,

$$T^2(\mathbb{S}^2) := \{\mathbf{a} \in \mathbb{C}^3 | \ \mathbf{a} \in L^2(\mathbb{S}^2)^3, \ \theta \cdot \mathbf{a} = 0 \ \text{a.e. on} \, \mathbb{S}^2\}.$$

Note that $T^2(\mathbb{S}^2)$ is a linear subspace of $L^2(\mathbb{S}^2)^3$, the space of vector L^2-fields on the unit sphere $\mathbb{S}^2$. In addition, we introduce the vectorial spherical harmonics (cf. [20])

$$\begin{cases} U_n^m(\theta) := \dfrac{1}{\sqrt{n(n+1)}} \text{Grad}\, Y_n^m(\theta) \\ V_n^m(\theta) := \theta \wedge U_n^m(\theta) \end{cases} \qquad n \in \mathbb{N}, \ \ m = -n, \ldots, n, \tag{5.3.5}$$

which form a complete orthonormal system in $T^2(\mathbb{S}^2)$. In (5.3.5), $Y_n^m(\theta)$ with $\theta \in \mathbb{S}^2$, $m = -n, \ldots, n$ are the spherical harmonics of order $n \ge 0$, and Grad denotes the surface gradient operator on $\mathbb{S}^2$. It is known that $A(\theta; \Omega)$ and $A(\theta; \Omega_j)$, $1 \le j \le l$ all belong to $T^2(\mathbb{S}^2)$. We define

$$K^j := \frac{\|A(\theta; \Omega_j)\|^2_{T^2(\mathbb{S}^2)}}{\|A(\theta; \Omega)\|^2_{T^2(\mathbb{S}^2)}}, \quad 1 \le j \le l, \tag{5.3.6}$$

and

$$I_s(z) := \frac{1}{\|A(\theta; \Omega)\|^2_{T^2(\mathbb{S}^2)}} \sum_{m=-1,0,1} \left(\left| \left\langle A(\theta; \Omega), e^{\mathrm{i}\omega(\theta'-\theta)\cdot z} U_1^m(\theta) \right\rangle_{T^2(\mathbb{S}^2)} \right|^2 + \left| \left\langle A(\theta; \Omega), e^{\mathrm{i}\omega(\theta'-\theta)\cdot z} V_1^m(\theta) \right\rangle_{T^2(\mathbb{S}^2)} \right|^2 \right), \tag{5.3.7}$$

where $z \in \mathbb{R}^3$, and $\langle \mathbf{u}, \mathbf{v} \rangle_{T^2(\mathbb{S}^2)} = \int_{\mathbb{S}^2} \mathbf{u} \cdot \overline{\mathbf{v}}\, ds_\theta$. Clearly, K^j is a real number, whereas $I_s(z)$ is a real-valued function depending on the point $z \in \mathbb{R}^3$. We are now ready to present the first main result of this section, whose proof will be postponed to the next section.

Theorem 5.3.1 *Let* $(\Omega; \varepsilon, \mu, \sigma)$ *be given by* (5.3.1) *satisfying* (5.3.2), *and* K^j, $1 \leq j \leq l$, *and* $I_s(z)$ *be defined in* (5.3.6) *and* (5.3.7), *respectively. Set*

$$L = \min_{1 \leq j, j' \leq l, j \neq j'} dist(z_j, z_{j'}) \gg 1.$$

Then

$$K^j = K_0^j + \mathcal{O}\left(\frac{1}{L} + \rho\right), \quad 1 \leq j \leq l, \tag{5.3.8}$$

where K_0^j *is a positive number independent of* L *and* ρ. *Moreover there exists an open neighborhood of* z_j, $neigh(z_j)$, $1 \leq j \leq l$, *such that*

$$I_s(z) \leq K_0^j + \mathcal{O}\left(\frac{1}{L} + \rho\right) \quad for\ z \in neigh(z_j), \tag{5.3.9}$$

and moreover $I_s(z)$ *achieves its maximum at* z *in* $neigh(z_j)$,

$$I_s(z_j) = K_0^j + \mathcal{O}\left(\frac{1}{L} + \rho\right), \tag{5.3.10}$$

Remark 5.3.1 Clearly, $I_s(z)$ possesses the indicating behavior which could be used to identify the location point z_j's of the scatterer components Ω_j's. Such behavior is more evident if one considers the case that Ω has only one component, i.e., $l = 1$. In the one-component case, one would have that

$$I_s(z) < 1 + \mathcal{O}(\rho) \quad \text{for all } z \in \mathbb{R}^3 \backslash \{z_1\},$$

but

$$I_s(z_1) = 1 + \mathcal{O}(\rho).$$

That is, z_1 is a global maximum point for $I(z)$.

Using Theorem 5.3.1, we can formulate our first imaging scheme of locating multiple small scatterer components as follows, which shall be referred to as *Scheme S* in the rest of the section.

The first single-shot locating method: **Scheme S**

Step 1. For an unknown EM scatterer Ω in (5.3.1), collect the far-field data by sending a single detecting EM plane wave specified by (4.2.1).

Step 2. Select a sampling region with a mesh $\mathcal{T}_h$ containing Ω.

Step 3. For each point $z \in \mathscr{T}_h$, calculate $I_s(z)$.
Step 4. Locate all the significant local maxima of $I_s(z)$ on $\mathscr{T}_h$, which represent the locations of the scatterer components.

5.3.1.2 Locating Regular-Size Scatterers

In this section, we consider the locating of multiple scatterers of regular size. We present our scheme for the case that all the scatterer components are perfectly conducting obstacles. Later, we would remark how our method could be extended to include the inhomogeneous medium components. Let $M_j \subset \mathbb{R}^3$, $1 \le j \le l$, be bounded simply connected C^2 domains that contain the origin. Let

$$Q := \bigcup_{j=1}^{l} Q_j = \bigcup_{j=1}^{l} z_j + M_j, \quad z_j \in \mathbb{R}^3, \ 1 \le j \le l. \tag{5.3.11}$$

Each $Q_j := z_j + M_j$ is a scatterer component, and it is assumed to be a perfectly conducting obstacle. Illuminated by a single pair of EM plane waves (4.2.1), the EM scattering by the scatterer Q can be described by the following Maxwell system

$$\begin{cases} \nabla \wedge E - \mathrm{i}\omega H = 0 & \text{in } \mathbb{R}^3 \backslash \overline{Q}, \\ \nabla \wedge H + \mathrm{i}\omega E = 0 & \text{in } \mathbb{R}^3 \backslash \overline{Q}, \\ E^+ = (E - E^i)|_{\mathbb{R}^3 \backslash \overline{Q}}, & \\ \nu \wedge E^+ = -\nu \wedge E^i & \text{on } \partial Q, \\ \lim\limits_{|x| \to +\infty} |x| \left| (\nabla \wedge E^+)(x) \wedge \dfrac{x}{|x|} - \mathrm{i}\omega E^+(x) \right| = 0. & \end{cases} \tag{5.3.12}$$

In the sequel, we write

$$A(\theta; Q) := A(\theta; \bigcup_{j=1}^{l} Q_j) = A(\theta; \bigcup_{j=1}^{l} z_j + M_j) \tag{5.3.13}$$

to denote the far-field pattern corresponding to the EM fields in (5.3.12). It is assumed that

$$\mathrm{diam}(Q_j) = \mathrm{diam}(M_j) \sim 1, \ \ 1 \le j \le l; \tag{5.3.14}$$

and

$$L = \min_{1 \le j, j' \le l, j \ne j'} \mathrm{dist}(z_j, z_{j'}) \gg 1. \tag{5.3.15}$$

That is, the size of the underlying scatterer components are comparable to the detecting wavelength. This is in sharp difference from our study in Sect. 5.3.1.1, where the scatterer components are of small size compared to the detecting EM wavelength. The qualitative condition (5.3.15) states that the scatterer components are sparsely distributed, and we would like to emphasize again that this is mainly needed for subsequent theoretical analysis of the discussed locating scheme. Numerical examples in Sect. 5.3.3 shows that as long as the distance between different components is bigger than half a wavelength, the presented scheme would yield a fine reconstruction. Furthermore, we introduce an admissible reference scatterer space

$$\mathscr{S} := \{\Sigma_j\}_{j=1}^{l'}, \tag{5.3.16}$$

where each $\Sigma_j \subset \mathbb{R}^3$ is a bounded simply connected C^2 domain that contains the origin and

$$\Sigma_j \neq \Sigma_{j'}, \quad \text{for } j \neq j',\ 1 \leq j, j' \leq l'. \tag{5.3.17}$$

For the present study, we assume that

$$M_j \in \mathscr{S}, \quad j = 1, 2, \ldots, l. \tag{5.3.18}$$

That is, in the practical situation, the shapes of the underlying scatterers are known in advance in the sense that each component must be of a shape from a known admissible class. But we do not know the locations of those scatterer components and intend to recover them from a single wave detection. We would like to remark that it is not necessary to have $l = l'$. It may have $l > l'$, and in this case, there must be more than one component in Q who has the same shape from $\mathscr{S}$; and it may also have $l < l'$, and in this case, there are less scatterers presented than the known admissible scatterers.

Next we introduce l' indicator functions as follows,

$$I_r^k(z) = \frac{\left|\langle A(\theta; Q), e^{\mathrm{i}\omega(\theta'-\theta)\cdot z} A(\theta; \Sigma_k)\rangle_{T^2(\mathbb{S}^2)}\right|}{\|A(\theta; \Sigma_k)\|^2_{T^2(\mathbb{S}^2)}}, \quad k = 1, 2, \ldots, l', \tag{5.3.19}$$

where $A(\theta; \Sigma_k)$ denotes the far-field pattern corresponding to the perfectly conducting obstacle Σ_k, $1 \leq k \leq l'$. In the following, we shall show that the l' indictor functions introduced in (5.3.19) can be used to locate the scatterer components Q_j of Q. To that end, we shall make the following generic assumption that

$$A(\theta; \Sigma_k) \neq A(\theta; \Sigma_{k'}), \quad \text{for } k \neq k',\ 1 \leq k, k' \leq l'. \tag{5.3.20}$$

Assumption (5.3.20) is closely related to a longstanding problem in the inverse scattering theory (cf. [18, 20, 41]): whether or not can one uniquely determine an obstacle by a single wave measurement? That is, if two obstacles produce the same far-field data corresponding to a single incident plane wave, can one conclude that they must be the same? Though such uniqueness result is widespreadly believed to be true, it still remains open, particularly for an obstacle of regular size. The uniqueness is proved for obstacles of polyhedra type in [33, 37]. Assumption (5.3.20) on the reference scatterers is of critical importance in our subsequently presented locating scheme. Nonetheless, since the admissible class $\mathscr{S}$ is known, (5.3.20) can be verified in advance. Moreover, since $\mathscr{S}$ is known in advance, we can assume by reordering if necessary that

$$\|A(\theta;\,\Sigma_k)\|_{T^2(\mathbb{S}^2)} \ge \|A(\theta;\,\Sigma_{k+1})\|_{T^2(\mathbb{S}^2)}, \quad k = 1, 2, \ldots, l' - 1. \tag{5.3.21}$$

That is, the sequence $\{\|A(\theta;\,\Sigma_k)\|_{T^2(\mathbb{S}^2)}\}_{k=1}^{l'}$ is nonincreasing. Next, we present a key theorem, which is the basis of our subsequent locating scheme.

Theorem 5.3.2 *Let Q be given in* (5.3.11), *and the obstacle components are assumed to satisfy* (5.3.15) *and* (5.3.18). *The admissible reference scatterer space $\mathscr{S}$ is assumed to satisfy* (5.3.20) *and* (5.3.21). *Consider the indicator function I_r^1 introduced in* (5.3.19). *Suppose there exists $\mathscr{J}_0 \subset \{1, 2, \ldots, l'\}$ such that $M_j = \Sigma_1$ for $j \in \mathscr{J}_0$, whereas $M_j \ne \Sigma_1$ for $j \in \{1, 2, \ldots, l\}\backslash\mathscr{J}_0$. Then for each z_j, $j = 1, 2, \ldots, l$, there exists an open neighborhood of z_j, $neigh(z_j)$, such that*

(i) if $j \in \mathscr{J}_0$, then

$$\widetilde{I}_r^1(z) := |I_r^1(z) - 1| \le \mathscr{O}\left(\frac{1}{L}\right), \quad z \in neigh(z_j), \tag{5.3.22}$$

and moreover, z_j is a local minimum point for $\widetilde{I}_r^1(z)$;

(ii) if $j \in \{1, 2, \ldots, l'\}\backslash\mathscr{J}_0$, then there exists $\varepsilon_0 > 0$ such that

$$\widetilde{I}_r^1(z) := |I_r^1(z) - 1| \ge \varepsilon_0 + \mathscr{O}\left(\frac{1}{L}\right), \quad z \in neigh(z_j). \tag{5.3.23}$$

By using Theorem 5.3.2, the presented locating scheme can be proceeded as follows, which shall be referred to as *Scheme R* in the rest of the section.

The second single-shot locating method: Scheme R

Step 1. For an unknown EM scatterer Q in (5.3.11), collect the far-field data by sending the detecting EM plane wave specified by (4.2.1).

Step 2. Select a sampling region with a mesh $\mathscr{T}_h$ containing Q.

Step 3. Collect in advance the far-field patterns associated with the admissible reference scatterer space $\mathscr{S}$ in (5.3.16), and reorder $\mathscr{S}$ if necessary to make it satisfy (5.3.21), and also verify the generic assumption (5.3.20).

Step 4. Set $k = 1$.

Step 5. For each point $z \in \mathscr{T}_h$, calculate $I_r^k(z)$ (or $\widetilde{I}_r^k(z) = |I_r^k(z) - 1|$).

Step 6. Locate all those significant local maxima of $I_r^k(z)$ such that $I_r^k(z) \sim 1$ (or the minima of $\widetilde{I}_r^k(z)$ on $\mathscr{T}_h$ such that $\widetilde{I}_r^k(z) \ll 1$), where scatterer components of the form $z + \Sigma_k$ is located.

Step 7. Trim all those $z + \Sigma_k$ found in **Step 6** from $\mathscr{T}_h$.
Step 8. If $\mathscr{T}_h = \emptyset$ or $k = l'$, then Stop; otherwise, set $k = k + 1$, and go to **Step 5**.

It can be seen that our locating scheme R progresses in a recursive manner. For a scatterer Q with multiple components, one firstly locates the sub-components of shape Σ_1, which have the most prominent scattering effect among all the scatterer components. After locating all the sub-components of shape Σ_1, one can exclude those components from the searching region, and then repeats the same procedure to locate all the sub-components of shape Σ_2, which, according to the ordering (5.3.21), have the most prominent scattering effect among all the scattering components that still remain in the searching region. Clearly, this procedure can be continued till one locates all the scatter components. Theorem 5.3.2 remains true for inhomogeneous medium scatterers if the generic condition (5.3.20) still holds and in that case, the locating **Scheme R** could be extended to include inhomogeneous medium components as well; see Remark 5.3.4 in the following for related discussions.

5.3.2 *Proofs of Theorems 5.3.1 and 5.3.2*

This section is devoted to the proofs of Theorems 5.3.1 and 5.3.2, which are the theoretical cores for our locating **Schemes S** and **R**, respectively. We first derive two key lemmas.

Lemma 5.3.1 *Let*

$$\Gamma = \bigcup_{j=1}^{l} \Gamma_j \quad and \quad (\Gamma; \varepsilon, \mu, \sigma) = \bigcup_{j=1}^{l} (\Gamma_j; \varepsilon_j, \mu_j, \sigma_j), \tag{5.3.24}$$

be a scatterer with multiple components, where each Γ_j is a bounded simply connected C^2 domain in $\mathbb{R}^3$. Assume that

$$L = \min_{1 \le j, j' \le l, j \ne j'} dist(\Gamma_j, \Gamma_{j'}) \gg 1. \tag{5.3.25}$$

Then we have

$$A(\theta; \Gamma) = \sum_{j=1}^{l} A(\theta; \Gamma_j) + \mathscr{O}(L^{-1}), \tag{5.3.26}$$

where $A(\theta; \Gamma)$ and $A(\theta; \Gamma_j)$ denote the far-field patterns, respectively, corresponding to $(\Gamma; \varepsilon, \mu, \sigma)$ and $(\Gamma_j; \varepsilon_j, \mu_j, \sigma_j)$.

Proof We only consider a specific case with $l = 2$, and Γ_1 is an obstacle component while Γ_2 is a medium component. Nevertheless, the general case can be proved following a completely similar manner. Moreover, we assume that ω is not an interior

EM eigenvalue for Γ_1 (see Remark 5.3.2). For this case, the EM scattering corresponding to Γ in (5.3.24) is governed by the Maxwell system (5.3.4) with Ω_o replaced by Γ_1 and Ω_m replaced by $(\Gamma_2; \varepsilon_2, \mu_2, \sigma_2)$. We know that there exists a unique pair of solutions $E \in (H^1_{loc}(\mathbb{R}^3\backslash\overline{\Gamma}_1))^3$ and $H \in (H^1_{loc}(\mathbb{R}^3\backslash\overline{\Gamma}_1))^3$ to the Maxwell system (cf [44]). Next, we shall make use of the integral equation method to prove the lemma. To that end, we let

$$\Phi(x, y) := \frac{1}{4\pi}\frac{e^{\mathrm{i}\omega|x-y|}}{|x-y|}, \quad x, y \in \mathbb{R}^3,\ x \neq y,$$

which is the fundamental solution to the differential operator $-\Delta - \omega^2$. Let $R \in \mathbb{R}_+$ be sufficiently large such that $\Gamma_1 \cup \Gamma_2 \Subset B_R$.

By the Stratton-Chu formula [20], we have

$$\begin{aligned} E(x) = E^i(x) &+ \nabla_x \wedge \int_{\partial\Gamma_1} \nu(y) \wedge E(y)\Phi(x, y)\, ds_y \\ &- \frac{1}{\mathrm{i}\omega}\nabla_x \wedge \nabla_x \wedge \int_{\partial\Gamma_1} \nu(y) \wedge H(y)\Phi(x, y)\, ds_y \\ &+ \nabla_x \wedge \int_{\Gamma_2} (\nabla_y \wedge E(y) - \mathrm{i}\omega H(y))\Phi(x, y)\, dy \\ &- \nabla_x \int_{\Gamma_2} \nabla_y \cdot E(y)\Phi(x, y)\, dy \\ &+ \mathrm{i}\omega \int_{\Gamma_2} (\nabla_y \wedge H(y) + \mathrm{i}\omega E(y))\Phi(x, y)\, dy, \quad x \in B_R\backslash\overline{\Gamma}_1, \end{aligned} \tag{5.3.27}$$

and

$$\begin{aligned} H(x) = H^i(x) &+ \nabla_x \wedge \int_{\partial\Gamma_1} \nu(y) \wedge H(y)\Phi(x, y)\, ds_y \\ &+ \frac{1}{\mathrm{i}\omega}\nabla_x \wedge \nabla_x \wedge \int_{\partial\Gamma_1} \nu(y) \wedge E(y)\Phi(x, y)\, ds_y \\ &+ \nabla_x \wedge \int_{\Gamma_2} (\nabla_y \wedge H(y) + \mathrm{i}\omega E(y))\Phi(x, y)\, dy \\ &- \nabla_x \int_{\Gamma_2} \nabla_y \cdot H(y)\Phi(x, y)\, dy \\ &- \mathrm{i}\omega \int_{\Gamma_2} (\nabla_y \wedge E(y) - \mathrm{i}\omega H(y))\Phi(x, y)\, dy, \quad x \in B_R\backslash\overline{\Gamma}_1. \end{aligned} \tag{5.3.28}$$

By using the Maxwell equations (5.3.4), and (5.3.27)–(5.3.28), together with the use of the mapping properties of the layer potential operators (cf. [20, 44]), we have the following system of integral equations

$$
\begin{aligned}
E(x) = E^i(x) &- \frac{1}{\mathrm{i}\omega}\nabla_x \wedge \nabla_x \wedge \int_{\partial\Gamma_1} \nu(y) \wedge H(y)\Phi(x, y)\, ds_y \\
&+ \nabla_x \wedge \int_{\Gamma_2} \mathrm{i}\omega(\mu_2 - 1)H(y)\Phi(x, y)\, dy \\
&+ \int_{\Gamma_2} \mathrm{i}\omega\left(1 - \varepsilon_2 - \mathrm{i}\frac{\sigma_2}{\omega}\right) E(y)\Phi(x, y)\, dy, \qquad x \in \Gamma_2,
\end{aligned} \tag{5.3.29}
$$

$$
\begin{aligned}
H(x) = H^i(x) &+ \nabla_x \wedge \int_{\partial\Gamma_1} \nu(y) \wedge H(y)\Phi(x, y)\, ds_y \\
&+ \nabla_x \wedge \int_{\Gamma_2} \mathrm{i}\omega\left(1 - \varepsilon_2 - \mathrm{i}\frac{\sigma_2}{\omega}\right) E(y)\Phi(x, y)\, dy \\
&- \int_{\Gamma_2} \mathrm{i}\omega(\mu_2 - 1)H(y)\Phi(x, y)\, dy, \qquad x \in \Gamma_2,
\end{aligned} \tag{5.3.30}
$$

$$
\begin{aligned}
\nu(x) \wedge H(x) = 2\nu(x) \wedge H^i(x) &+ 2\nu(x) \wedge \nabla_x \wedge \int_{\partial\Gamma_1} \nu(y) \wedge H(y)\Phi(x, y)\, ds_y \\
&+ \nu(x) \wedge \nabla_x \wedge \int_{\Gamma_2} 2\mathrm{i}\omega\left(1 - \varepsilon_2 - \mathrm{i}\frac{\sigma_2}{\omega}\right) E(y)\Phi(x, y)\, dy \\
&- \nu(x) \wedge \int_{\Gamma_2} 2\mathrm{i}\omega(\mu_2 - 1)H(y)\Phi(x, y)\, dy, \qquad x \in \partial\Gamma_1.
\end{aligned} \tag{5.3.31}
$$

Next, we introduce the following volume integral operators

$$
\begin{aligned}
(\mathscr{L}E)(x) &= \int_{\Gamma_2} \mathrm{i}\omega\left(1 - \varepsilon_2 - \mathrm{i}\frac{\sigma_2}{\omega}\right) E(y)\Phi(x, y)\, dy, \quad x \in \Gamma_2, \\
(\mathscr{L}'E)(x) &= \nabla_x \wedge (\mathscr{L}E)(x), \quad x \in \Gamma_2, \\
(\mathscr{K}H)(x) &= \int_{\Gamma_2} \mathrm{i}\omega(\mu_2 - 1)H(y)\Phi(x, y)\, dy, \quad x \in \Gamma_2, \\
(\mathscr{K}'H)(x) &= \nabla_x \wedge (\mathscr{K}H)(x), \quad x \in \Gamma_2,
\end{aligned} \tag{5.3.32}
$$

and the boundary integral operator

$$
(\mathscr{P}(\nu \wedge H))(x) = 2\nu(x) \wedge \nabla_x \wedge \int_{\partial\Gamma_1} \nu(y) \wedge H(y)\Phi(x, y)\, ds_y, \quad x \in \partial\Gamma_1. \tag{5.3.33}
$$

Moreover, we introduce the following two function spaces

$$
\begin{aligned}
TH^{-1/2}(\partial\Gamma_1) &:= \left\{\mathbf{a} \in (H^{-1/2}(\partial\Gamma_1))^3;\ \nu \wedge \mathbf{a} = 0 \quad \text{for a.e. } x \in \partial\Gamma_1\right\}, \\
TH^{-1/2}_{\mathrm{Div}}(\partial\Gamma_1) &:= \left\{\mathbf{a} \in TH^{-1/2}(\partial\Gamma_1);\ \mathrm{Div}(\mathbf{a}) \in TH^{-1/2}(\partial\Gamma_1)\right\}.
\end{aligned} \tag{5.3.34}
$$

In the sequel, we set

$$\mathbf{b}_1(x) := -\frac{1}{\mathrm{i}\omega}\nabla_x \wedge \nabla_x \wedge \int_{\partial \Gamma_1} \nu(y) \wedge H(y)\Phi(x, y)\, ds_y, \quad x \in \Gamma_2,$$

$$\mathbf{b}_2(x) := \nabla_x \wedge \int_{\partial \Gamma_1} \nu(y) \wedge H(y)\Phi(x, y)\, ds_y, \quad x \in \Gamma_2,$$

$$\mathbf{b}_3(x) := \nu(x) \wedge \nabla_x \wedge \int_{\Gamma_2} 2\mathrm{i}\omega\left(1 - \varepsilon_2 - \mathrm{i}\frac{\sigma_2}{\omega}\right) E(y)\Phi(x, y)\, dy$$
$$- \nu(x) \wedge \int_{\Gamma_2} 2\mathrm{i}\omega(\mu_2 - 1)H(y)\Phi(x, y)\, dy, \quad x \in \Gamma_1.$$

Since $L = \mathrm{dist}(\Gamma_1, \Gamma_2) \gg 1$, one readily verifies that

$$\|\mathbf{b}_l\|_{L^2(\Gamma_2)^3} \le \frac{C}{L}\|\nu \wedge H\|_{TH_{\mathrm{Div}}^{-1/2}(\partial\Gamma_1)}, \quad l = 1, 2, \tag{5.3.35}$$

and

$$\|\mathbf{b}_3\|_{TH_{\mathrm{Div}}^{-1/2}(\Gamma_1)} \le \frac{C}{L}\left(\|E\|_{L^2(\Gamma_2)^3} + \|H\|_{L^2(\Gamma_2)^3}\right), \tag{5.3.36}$$

where C is a positive constant depending only on Γ_1, Γ_2 and ω.

Next, by using the integral operators introduced in (5.3.32) and (5.3.33), the integral equations (5.3.29) and (5.3.30) can be formulated as

$$\mathbf{A} := \begin{pmatrix} I - \mathscr{L} & \mathscr{K}' \\ -\mathscr{L}' & I + \mathscr{K} \end{pmatrix}, \quad \mathbf{A}\begin{pmatrix} E(x) \\ H(x) \end{pmatrix} - \begin{pmatrix} \mathbf{b}_1(x) \\ \mathbf{b}_2(x) \end{pmatrix} = \begin{pmatrix} E^i(x) \\ H^i(x) \end{pmatrix}, \quad x \in \Gamma_2, \tag{5.3.37}$$

In a similar manner, (5.3.31) can be formulated as

$$(I - \mathscr{P})(\nu \wedge H)(x) - \mathbf{b}_3(x) = 2\nu(x) \wedge H^i(x), \quad x \in \Gamma_1. \tag{5.3.38}$$

Referring to [20, 44], we know that both $\mathbf{A} : L^2(\Gamma_2)^3 \wedge L^2(\Gamma_2)^3 \to L^2(\Gamma_2)^3 \wedge L^2(\Gamma_2)^3$ and $I - \mathscr{P} : TH_{\mathrm{Div}}^{-1/2}(\partial\Gamma_1) \to TH_{\mathrm{Div}}^{-1/2}(\partial\Gamma_1)$ are Fredholm operators of index 0, and moreover they are invertible. Using this fact and (5.3.35)–(5.3.36), one obtains from (5.3.37) that

$$\begin{pmatrix} E(x) \\ H(x) \end{pmatrix} = \mathbf{A}^{-1}\begin{pmatrix} E^i(x) \\ H^i(x) \end{pmatrix} + \mathscr{O}(L^{-1}) := \begin{pmatrix} \widetilde{E}(x) \\ \widetilde{H}(x) \end{pmatrix} + \mathscr{O}(L^{-1}), \quad x \in \Gamma_2 \tag{5.3.39}$$

and

$$(\nu \wedge H)(x) = (I - \mathscr{P})^{-1}(2\nu \wedge H^i)(x) + \mathscr{O}(L^{-1}) := (\nu \wedge \widehat{H})(x) + \mathscr{O}(L^{-1}), \quad x \in \partial\Gamma_1, \tag{5.3.40}$$

where $(\widetilde{E}, \widetilde{H})$ are actually the EM fields corresponding to $(\Gamma_2; \varepsilon_2, \mu_2, \sigma_2)$, and $\widehat{H}$ is the magnetic field corresponding to Γ_1. Finally, by using (5.3.39) and (5.3.40), and the integral representation (5.3.29), one readily has (5.3.26), which completes the proof. □

Remark 5.3.2 If ω is an interior eigenvalue for Γ_1, one can overcome the problem by using the combined electric and magnetic dipole operators technique, and we refer to [20, Chap. 6] for more details.

Lemma 5.3.2 *Let $(\Omega_1; \varepsilon_1, \mu_1, \sigma_1)$ be one component of Ω described in Sect. 5.3.1.1. Then we have*

$$\begin{aligned} A(\theta; \Omega_1) &= e^{\mathrm{i}\omega(\theta'-\theta)\cdot z_1} A(\theta; \rho D_1) \\ &= e^{\mathrm{i}\omega(\theta'-\theta)\cdot z_1} \left[(\omega\rho)^3 \left(\sum_{m=-1,0,1} a_{1,m} U_1^m(\theta) + b_{1,m} V_1^m(\theta) \right) + \mathscr{O}((\omega\rho)^4) \right], \end{aligned} \tag{5.3.41}$$

where U_1^m and V_1^m are the vectorial spherical harmonics introduced in (5.3.5), *and $a_{1,m}$ and $b_{1,m}$ ($m = -1, 0, 1$), are constants depending only on $(D_1; \varepsilon_1, \mu_1, \sigma_1)$ and p, θ', but independent of $\omega\rho$.*

Proof We first consider the case that $\sigma_1 = +\infty$, namely, Ω_1 is a perfectly conducting obstacle. It is directly verified that

$$A(\theta; \Omega_1) = A(\theta; z_1 + \rho D_1) = e^{i\omega(\theta'-\theta)\cdot z_1} A(\theta; \rho D_1). \tag{5.3.42}$$

The EM scattering corresponding to the obstacle ρD_1 is described by

$$\begin{cases} \nabla \wedge E_\rho - \mathrm{i}\omega H_\rho = 0 & \text{in } \mathbb{R}^3 \backslash \overline{\rho D_1}, \\ \nabla \wedge H_\rho + \mathrm{i}\omega E_\rho = 0 & \text{in } \mathbb{R}^3 \backslash \overline{\rho D_1}, \\ E_\rho^+ = (E_\rho - E^i)|_{\mathbb{R}^3 \backslash \overline{\rho D_1}}, & \\ \nu \wedge E_\rho^+ = -\nu \wedge E^i & \text{on } \partial(\rho D_1), \\ \lim_{|x|\to+\infty} |x| \left| (\nabla \wedge E_\rho^+)(x) \wedge \dfrac{x}{|x|} - \mathrm{i}\omega E_\rho^+(x) \right| = 0. & \end{cases} \tag{5.3.43}$$

Set

$$\widetilde{E}(x) := E_\rho(\rho x), \quad \widetilde{H}(x) := H_\rho(\rho x), \quad \widetilde{E}^i(x) = E^i(\rho x) \quad \text{for } x \in \mathbb{R}^3 \backslash \overline{D}_1. \tag{5.3.44}$$

It is verified directly that

$$\begin{cases} \nabla \wedge \widetilde{E} - \mathrm{i}\omega\rho \widetilde{H} = 0 & \text{in } \mathbb{R}^3 \backslash \overline{D_1}, \\ \nabla \wedge \widetilde{H} + \mathrm{i}\omega\rho \widetilde{E} = 0 & \text{in } \mathbb{R}^3 \backslash \overline{D_1}, \\ \widetilde{E}^+ = (\widetilde{E} - \widetilde{E}^i)|_{\mathbb{R}^3 \backslash \overline{D_1}}, & \\ \nu \wedge \widetilde{E}^+ = -\nu \wedge \widetilde{E}^i & \text{on } \partial D_1, \\ \lim_{|x|\to+\infty} |x| \left| (\nabla \wedge \widetilde{E}^+)(x) \wedge \dfrac{x}{|x|} - \mathrm{i}\omega \widetilde{E}^+(x) \right| = 0. & \end{cases} \tag{5.3.45}$$

Then by the low frequency asymptotics in [21], one has

$$\widetilde{E}^{+}(x) = \frac{e^{\mathrm{i}\omega\rho|x|}}{\mathrm{i}\omega\rho|x|}\widetilde{A}(\theta) + \mathscr{O}\left(\frac{1}{|x|^2}\right), \tag{5.3.46}$$

with

$$\widetilde{A}(\theta) = \frac{(\mathrm{i}\omega\rho)^3}{4\pi}[\theta \wedge (\theta \wedge \mathbf{a}) - \theta \wedge \mathbf{b}] + \mathscr{O}((\omega\rho)^4), \tag{5.3.47}$$

where **a** and **b** are two constant vectors, representing the electric and magnetic dipole moments, and they depend only on p, θ' and D_1, but independent of ω and ρ. By (5.3.44), one can readily see that

$$A(\theta) = \frac{1}{\mathrm{i}\omega}\widetilde{A}(\theta) = \frac{1}{\mathrm{i}\omega}\frac{(\mathrm{i}\omega\rho)^3}{4\pi}[\theta \wedge (\theta \wedge \mathbf{a}) - \theta \wedge \mathbf{b}] + \mathscr{O}((\omega\rho)^4). \tag{5.3.48}$$

Finally, by using the fact that U_n^m and V_n^m form a complete orthonormal system in $T^2(\mathbb{S}^2)$, it is straightforward to show that there exist $a_{1,m}$ and $b_{1,m}$, $m = -1, 0, 1$ such that

$$\theta \wedge (\theta \wedge \mathbf{a}) - \theta \wedge \mathbf{b} = \sum_{m=-1,0,1} a_{1,m} U_1^m(\theta) + b_{1,m} V_1^m(\theta),$$

which together with (5.3.47) and (5.3.42) implies (5.3.41).

For the case when the underlying small scatterer is an inhomogeneous medium, by using a completely same scaling argument, together with the corresponding low frequency EM asymptotics in [21], one can prove (5.3.41).

The proof is completed. □

Remark 5.3.3 The low frequency EM asymptotics and the asymptotic expansions of EM fields due to a small inclusion could also be found in [9–11, 43], where polarizability tensors are always involved. For the present study, we need asymptotic expansions in terms of the vectorial spherical harmonics.

With the technical preparations above, we are now in a position to show the proof of Theorem 5.3.1.

Proof of Theorem 5.3.1 First, by Lemmas 5.3.1 and 5.3.2, we have

$$
\begin{aligned}
A(\theta;\Omega) &= \sum_{j=1}^{l} A(\theta;\Omega_l) + \mathcal{O}\left(\frac{1}{L}\right)\\
&= \sum_{j=1}^{l} e^{\mathrm{i}\omega(\theta'-\theta)\cdot z_j}\left[(\omega\rho)^3 A^j(\theta) + \mathcal{O}((\omega\rho)^4)\right] + \mathcal{O}\left(\frac{1}{L}\right)\\
&= \sum_{j=1}^{l} e^{\mathrm{i}\omega(\theta'-\theta)\cdot z_j}\left[(\omega\rho)^3\left(\sum_{m=-1,0,1} a_{1,m}^{(j)} U_1^m(\theta) + b_{1,m}^{(j)} V_1^m(\theta)\right) + \mathcal{O}((\omega\rho)^4)\right]\\
&\quad + \mathcal{O}\left(\frac{1}{L}\right)
\end{aligned}
\tag{5.3.49}
$$

where $a_{1,m}^{(j)}$ and $b_{1,m}^{(j)}$, $m = -1, 0, 1$, are constants dependent only on $(D_j; \varepsilon_j, \mu_j, \sigma_j)$ and p, θ'. In (5.3.49), we have introduced

$$
A^j(\theta) := \sum_{m=-1,0,1}\left(a_{1,m}^{(j)} U_1^m(\theta) + b_{1,m}^{(j)} V_1^m(\theta)\right).
$$

Next, without loss of generality, we only consider the indicating behaviors of $I_s(z)$ in a small open neighborhood of z_1, i.e., $z \in neigh(z_1)$. Clearly, we have

$$
\omega|z_j - z| \geq \omega L \gg 1 \quad \text{for } z \in neigh(z_1) \text{ and } j = 2, 3, \ldots, l. \tag{5.3.50}
$$

Hence, by the Riemann-Lebesgue lemma about oscillatory integrals and (5.3.49), we have

$$
\begin{aligned}
&\left|\langle A(\theta;\Omega), e^{\mathrm{i}\omega(\theta'-\theta)\cdot z} U_1^m(\theta)\rangle_{T^2(\mathbb{S}^2)}\right|\\
&= (\omega\rho)^3\left|\langle e^{\mathrm{i}\omega(\theta'-\theta)\cdot z_1} A^1(\theta), e^{\mathrm{i}\omega(\theta'-\theta)\cdot z} U_1^m(\theta)\rangle_{T^2(\mathbb{S}^2)} + \mathcal{O}\left((\omega\rho)^3(\frac{1}{L}+\rho)\right)\right|\\
&\leq (\omega\rho)^3\left(|a_{1,m}^{(1)}| + \mathcal{O}\left(\frac{1}{L}+\rho\right)\right), \quad m = -1, 0, 1.
\end{aligned}
\tag{5.3.51}
$$

In (5.3.51), we have employed the orthogonality of vector spherical harmonics and also the Cauchy-Schwartz inequality. Moreover, by the Cauchy-Schwartz inequality, we know the strict inequality would hold in the last inequality of (5.3.51) if $z \neq z_1$, and only when $z = z_1$, the equality would hold. In a completely similar manner, one can show that

$$
\begin{aligned}
&\left|\langle A(\theta;\Omega), e^{\mathrm{i}\omega(\theta'-\theta)\cdot z} V_1^m(\theta)\rangle_{T^2(\mathbb{S}^2)}\right| \\
&= (\omega\rho)^3 \left|\langle e^{\mathrm{i}\omega(\theta'-\theta)\cdot z_1} A^1(\theta), e^{\mathrm{i}\omega(\theta'-\theta)\cdot z} V_1^m(\theta)\rangle_{T^2(\mathbb{S}^2)} + \mathscr{O}\left((\omega\rho)^3(\frac{1}{L}+\rho)\right)\right| \\
&\le (\omega\rho)^3 \left(|b_{1,m}^{(1)}| + \mathscr{O}\left(\frac{1}{L}+\rho\right)\right), \quad m=-1,0,1,
\end{aligned}
\tag{5.3.52}
$$

where strict inequality would hold for the last relation if $z \neq z_1$, and only when $z = z_1$, the equality would hold.

Hence, by (5.3.51) and (5.3.52), we have

$$
\begin{aligned}
&\sum_{m=-1,0,1} \left(\left|\left\langle A(\theta;\Omega), e^{\mathrm{i}\omega(\theta'-\theta)\cdot z}\, U_1^m(\theta)\right\rangle_{T^2(\mathbb{S}^2)}\right|^2\right. \\
&\qquad \left. + \left|\left\langle A(\theta;\Omega), e^{\mathrm{i}\omega(\theta'-\theta)\cdot z}\, V_1^m(\theta)\right\rangle_{T^2(\mathbb{S}^2)}\right|^2\right) \\
&\le (\omega\rho)^6 \left(\sum_{m=-1,0,1} |a_{1,m}^{(1)}|^2 + |b_{1,m}^{(1)}|^2 + \mathscr{O}\left(\frac{1}{L}+\rho\right)\right),
\end{aligned}
\tag{5.3.53}
$$

where the equality would hold only when $z = z_1$. On the other hand, by using (5.3.49), it is straightforward to show that

$$
\|A(\theta;\Omega)\|_{T^2(\mathbb{S}^2)}^2 = (\omega\rho)^6 \sum_{j=1}^{l} \left(\sum_{m=-1,0,1} |a_{1,m}^{(j)}|^2 + |b_{1,m}^{(j)}|^2 + \mathscr{O}\left(\frac{1}{L}+\rho\right)\right).
\tag{5.3.54}
$$

By (5.3.53) and (5.3.54), we see that for $z \in neigh(z_1)$,

$$
\begin{aligned}
I_s(z) = &\frac{1}{\|A(\theta;\Omega)\|_{T^2(\mathbb{S}^2)}^2} \sum_{m=-1,0,1} \left(\left|\left\langle A(\theta;\Omega), e^{\mathrm{i}\omega(\theta'-\theta)\cdot z}\, U_1^m(\theta)\right\rangle_{T^2(\mathbb{S}^2)}\right|^2\right. \\
&\qquad \left. + \left|\left\langle A(\theta;\Omega), e^{\mathrm{i}\omega(\theta'-\theta)\cdot z}\, V_1^m(\theta)\right\rangle_{T^2(\mathbb{S}^2)}\right|^2\right) \\
\le &\frac{\sum_{m=-1}^{1} |a_{1,m}^{(1)}|^2 + |b_{1,m}^{(1)}|^2}{\sum_{j=1}^{l}\sum_{m=-1}^{1} |a_{1,m}^{(j)}|^2 + |b_{1,m}^{(j)}|^2} + \mathscr{O}\left(\frac{1}{L}+\rho\right),
\end{aligned}
\tag{5.3.55}
$$

where only when $z = z_1$, the equality would hold in the last relation. Set

$$K_0^1 := \frac{\sum\limits_{m=-1}^{1} |a_{1,m}^{(1)}|^2 + |b_{1,m}^{(1)}|^2}{\sum\limits_{j=1}^{l}\sum\limits_{m=-1}^{1} |a_{1,m}^{(j)}|^2 + |b_{1,m}^{(j)}|^2}.$$

Using (5.3.41) and (5.3.54), it is readily seen that

$$K^1 := \frac{\|A(\theta;\Omega_1)\|^2_{T^2(\mathbb{S}^2)}}{\|A(\theta;\Omega)\|^2_{T^2(\mathbb{S}^2)}} = K_0^1 + \mathscr{O}\left(\frac{1}{L} + \rho\right).$$

Next, we present the proof of Theorem 5.3.2.

Proof of Theorem 5.3.2 First, by Lemma 5.3.1, we have

$$\begin{aligned} A(\theta;\Omega) = A(\theta;\bigcup_{j=1}^{l} Q_j) &= \sum_{j=1}^{l} A(\theta; Q_j) + \mathscr{O}\left(\frac{1}{L}\right) \\ &= \sum_{j=1}^{l} A(\theta; \bigcup_{j=1}^{l} z_j + M_j) + \mathscr{O}\left(\frac{1}{L}\right) \\ &= \sum_{j=1}^{l} e^{\mathrm{i}\omega(\theta'-\theta)\cdot z_j} A(\theta; M_j) + \mathscr{O}\left(\frac{1}{L}\right). \end{aligned} \tag{5.3.56}$$

Let us consider the indicator function $I_r^1(z)$ in (5.3.19). Without loss of generality, we assume that $M_1 = \Sigma_1$ and $M_j \neq \Sigma_1$ for $j = 2, \ldots, l$. Let $z \in neigh(z_1)$. By (5.3.15), (5.3.56) and the Riemannian-Lebesgue lemma about oscillatory integrals, we have

$$\begin{aligned} &\left|\langle A(\theta;\Omega), e^{\mathrm{i}\omega(\theta'-\theta)\cdot z} A(\theta;\Sigma_1)\rangle_{T^2(\mathbb{S}^2)}\right| \\ &= \left|\langle e^{\mathrm{i}\omega(\theta'-\theta)\cdot z_1} A(\theta;\Sigma_1), e^{\mathrm{i}\omega(\theta'-\theta)\cdot z} A(\theta;\Sigma_1)\rangle_{T^2(\mathbb{S}^2)}\right| + \mathscr{O}\left(\frac{1}{L}\right) \\ &\leq \|A(\theta;\Sigma_1)\|^2_{T^2(\mathbb{S}^2)} + \mathscr{O}\left(\frac{1}{L}\right), \end{aligned} \tag{5.3.57}$$

where in the last relation, the equality would hold only when $z = z_1$. Hence, for a sufficiently small neighborhood, $neigh(z_1)$,

$$|I_r(z) - 1| \leq \mathscr{O}\left(\frac{1}{L}\right) \quad \text{for } z \in neigh(z_1).$$

Next, we let $z \in neigh(z_2)$, with $neigh(z_2)$ sufficiently small. Then, again by (5.3.15), (5.3.56) and the Riemannian-Lebesgue lemma, we have

$$\begin{aligned}&\left|\langle A(\theta;\Omega), e^{\mathrm{i}i\omega(\theta'-\theta)\cdot z_2}A(\theta;\Sigma_1)\rangle_{T^2(\mathbb{S}^2)}\right|\\&=\left|\langle e^{\mathrm{i}\omega(\theta'-\theta)\cdot z_2}A(\theta;M_2), e^{\mathrm{i}\omega(\theta'-\theta)\cdot z_2}A(\theta;\Sigma_1)\rangle_{T^2(\mathbb{S}^2)}\right|+\mathscr{O}\left(\frac{1}{L}\right)\\&<\|A(\theta;M_2)\|_{T^2(\mathbb{S}^2)}\|A(\theta;\Sigma_1)\|_{T^2(\mathbb{S}^2)}+\mathscr{O}\left(\frac{1}{L}\right),\end{aligned}\tag{5.3.58}$$

where in the last relation, strict inequality holds in light of the assumption (5.3.20) and the fact $M_2\neq\Sigma_1$. Hence, by (5.3.58) and (5.3.21), it is readily shown that

$$I_r(z_2)<\frac{\|A(\theta;M_2)\|_{T^2(\mathbb{S}^2)}}{\|A(\theta;\Sigma_1)\|_{T^2(\mathbb{S}^2)}}\leq 1+\mathscr{O}\left(\frac{1}{L}\right).\tag{5.3.59}$$

Hence, there exists some $\varepsilon_0\in\mathbb{R}_+$ such that

$$\widetilde{I}_r^1(z)=|I_r^1(z)-1|\geq\varepsilon_0+\mathscr{O}\left(\frac{1}{L}\right)\quad\text{for } z\in neigh(z_2).$$

In a completely similar manner, one can show the indicating behaviors of $\widetilde{I}_r^1(z)$ for $z\in neigh(z_j)$, $j=3,\ldots,l$.

The proof is completed. □

Remark 5.3.4 Through our proof of Theorem 5.3.2, one can see that our locating **Scheme R** could be extended to a more general setting by including inhomogeneous medium components as follows. Suppose in $\mathscr{S}$, some reference scatterer, say, $(\Sigma_j;\varepsilon_j,\mu_j,\sigma_j)$, $1\leq j\leq\tilde{l}\leq l$, are inhomogeneous media, with the EM parameters ε_j, μ_j and σ_j known as well. Let Q be a scatterer with multiple components, such that each component is a translation of some reference scatterer. Then our locating **Scheme R** could be extended to this more general setting provided the reference scatterer space $\mathscr{S}$ satisfies the generic assumption (5.3.20). However, to our best knowledge, there is no such uniqueness result in the literature. On the other hand, as we remarked earlier that $\mathscr{S}$ is given in advance, one could verify (5.3.20) in advance, and if it is satisfied, our locating scheme could apply.

5.3.3 Numerical Experiments and Discussions

In this section, we carry out a series of numerical experiments for different benchmark problems to test the performance of the discussed locating **Schemes S** and **R**. The results achieved are consistent with our theoretical predictions in Sects. 5.3.1 and 5.3.2 in a sound manner. Besides, the numerical results reveal some very promising features of the imaging schemes that were not covered in our theoretical analysis.

We first briefly describe our experimental settings. Let $e_1=(1,0,0)^T$, $e_2=(0,1,0)^T$ and $e_3=(0,0,1)^T$ be the three canonical Cartesian bases. The single detecting/incident wave we shall employ for our numerical examples is the plane wave specified by (4.2.1) with the polarization $p = e_3$ and impinging direction $\theta' = e_1$. Moreover, we take the unitary wavelength $\lambda = 1$, namely the frequency $\omega = 2\pi$. In all the examples, the electric far-field pattern A is observed at 590 Lebedev quadrature points distributed on the unit sphere $\mathbb{S}^2$ (cf. [28] and references therein). The exact far-field data $A(\theta)$ are corrupted point-wise by the formula

$$A_\delta(\theta) = A(\theta) + \delta\zeta_1 \max_\theta |A(\theta)| \exp(\mathrm{i}2\pi\zeta_2), \tag{5.3.60}$$

where δ refers to the relative noise level, and both ζ_1 and ζ_2 follow the uniform distribution ranging from -1 to 1. The scattered electromagnetic fields are synthesized using the quadratic edge element discretization in the spherical domain centered at the origin with radius 4λ enclosed by a spherical PML layer of width λ to damp the reflection. Local adaptive refinement techniques within the inhomogeneous scatterer are adopted to enhance the resolution of the scattered field. The far-field data are approximated by the integral equation representation [45, p. 181, Theorem 3.1] on the sphere centered at the origin with radius 3.5 using the numerical quadrature. We refine the mesh successively till the relative maximum error of successive groups of far-field data is below 0.1%. The far-field patterns on the finest mesh are used as the exact data. The values of the indicator functions have been normalized between 0 and 1 to highlight the positions identified. The sampling domain is fixed to be $\mathscr{T} = [-2\lambda, 2\lambda]^3$, which is then divided into small cubes of equal width $h = 0.01\lambda$ yielding the sampling mesh $\mathscr{T}_h$. The orthogonal slices of the contours of the indicator function values will be displayed as an estimate to the profiles of the unknown scatterers. In the sequel, for brevity, we refer to the proposed single-shot locating methods of **Schemes S** and **R** as the **SSM(s)** and **SSM(r)**, respectively.

The scatterers under concern include a cube, a ball of different radii, and revolving solids from a kite and a peanut parameterized in the $x - y$ plane as follows [32]:

$$(x(t), y(t), 0) := (\cos t + 0.65\cos 2t - 0.65, 1.5\sin t, 0), \quad 0 \le t \le 2\pi,$$
$$(x(t), y(t), 0) := \sqrt{3\cos^2 t + 1}(\cos t, \sin t, 0), \quad 0 \le t \le 2\pi.$$

In the following, three groups of experiments shall be conducted. The first group of experiments is on locating point-like small-size scatterers in various scenarios by the **SSM(s)**, and the second group of experiments is on testing the **SSM(r)** for locating regular-size scatterers. In the third group of experiments, we shall test the performance of **SSM(s)** on imaging 'partially-small' segment-like scatterers.

5.3.3.1 The SSM(s) for Point-Like Scatterers

Example 1 In this example, we consider a cube scatterer of length 0.02λ located at the origin, with the EM parameters given by $\varepsilon = 4$, $\mu = 1$ and $\sigma = 0$. The orthogonal slices of the contours of the indicator function $I_s(z)$ in (5.3.7) for the **SSM(s)** are

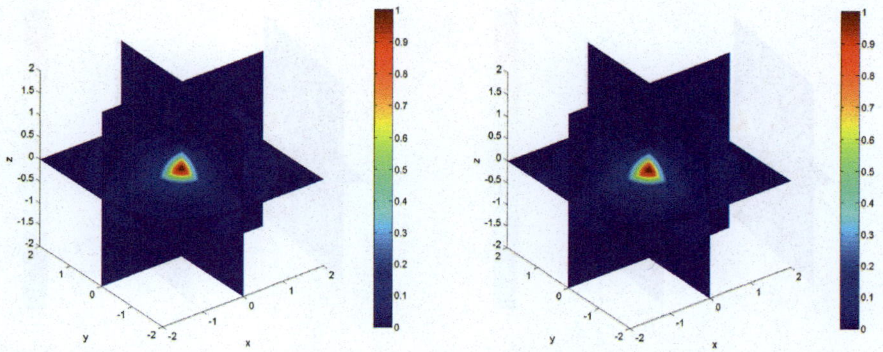

Fig. 5.1 Reconstruction results for Example 1: (Left) Exact far-field data; (Right) Noisy far-field data with $\delta = 20\%$

given in Fig. 5.1. It can be readily seen that the **SSM(s)** can locate the small scatterer in a very accurate and stable manner. Indeed, even 20% random noise is attached to the measurement data, the **SSM(s)** still yields a very robust and accurate locating.

Example 2 In this example, we consider a scatterer consisting of a ball medium component of radius 0.2λ positioned at $(1.5\lambda,\ 1.5\lambda,\ 0)$ and a ball PEC obstacle of radius 0.2λ positioned at $(-1.5\lambda,\ -1.5\lambda,\ 0)$. The EM parameters of the first scatterer component are taken to the same as the one in Example 1. For this example, the orthogonal slices of the contours of the indicator function $I_s(z)$ for the **SSM(s)** are shown in Fig. 5.2. The **SSM(s)** yields a very accurate identification of the location of both scatterers even if the measurement data is significantly perturbed to a high-level noise. This example demonstrates that **SSM(s)** can locate the multiple scatterer components without knowing the physical property of each component in advance. Moreover, we would like to note a promising feature of the **SSM(s)**: if one chooses a cut-off value to be 0.7 to separate the region where $I_s(z)$ is bigger than the cut-off value, then the rough profiles of the scatterer components would appear. Hence, in addition to finding the locations of the scatterer components, the presented scheme is also capable of qualitatively imaging the supports/shapes of the unknown scatterers.

Example 3 In this example, three scatterers are presented at $(-\lambda,\ -\lambda,\ \lambda)$, $(\lambda,\ \lambda,\ 0)$, $(\lambda,\ \lambda,\ -\lambda)$, respectively. The first two scatterers are balls of radius 0.2λ, with the EM parameters given by $\varepsilon_1 = 4 + \sin x_1$, $\mu_1 = 1$, $\sigma_1 = 0$ and $\varepsilon_2 = 4 + \cos x_2$, $\mu_2 = 1$, $\sigma_2 = 0$, respectively. Here, we have made use of $x = (x_1, x_2, x_3)$ to denote a point in $\mathbb{R}^3$. The third one is a non-convex kite-shaped revolving scatterer scaled by a relative size $\rho = 0.2$ with the EM parameters given by $\varepsilon_3 = 4 + x_3$, $\mu_3 = 1$ and $\sigma_3 = 0$. The numerical reconstruction results are shown in Fig. 5.3. It can be seen from Fig. 5.3 that the **SSM(s)** is capable of locating multiple scatterer components with variable contents.

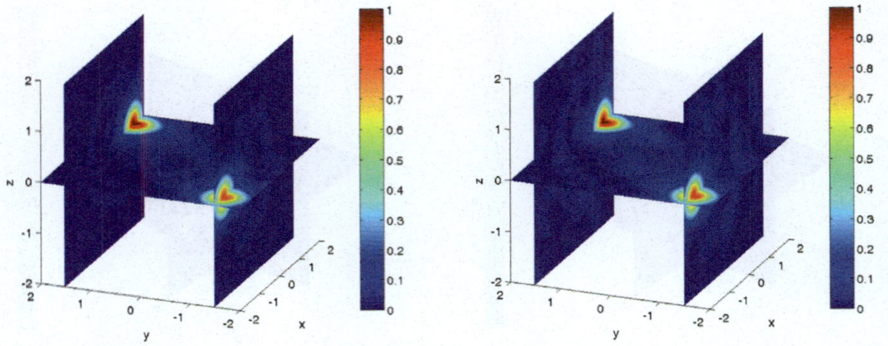

Fig. 5.2 Reconstruction results for Example 2: (Left) Exact far-field data; (Right) Noisy far-field data with $\delta = 20\%$

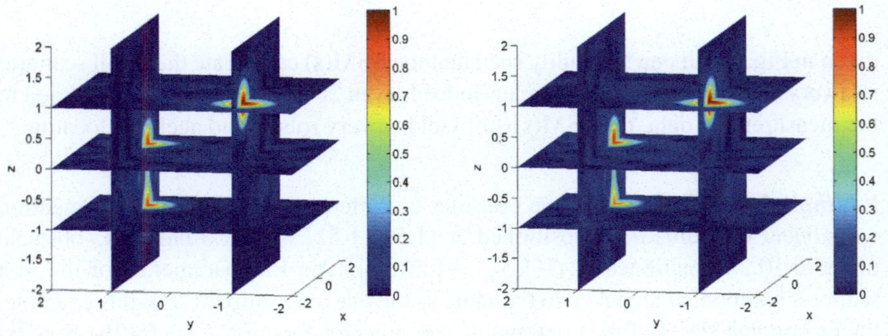

Fig. 5.3 Reconstruction results for Example 3: (Left) Exact far-field data; (Right) Noisy far-field data with $\delta = 20\%$

Example 4 We consider two ball scatterers of radius 0.2λ, located at $(-0.45\lambda, 0, 0)$ and $(0.45\lambda, 0, 0)$, respectively, with the same EM parameters $\varepsilon = 4$, $\mu = 1$ and $\sigma = 0$. We investigate the lower distance limit between the underlying separate scatterer components for the **SSM(s)**. The results are shown in Fig. 5.4. It can be seen in this case, namely the distance between the two components is of a half wavelength, the **SSM(s)** can locate both scatterer components and separate them well. If we further reduce the distance between the two components (less than a half wavelength), the **SSM(s)** can no longer separate the two scatterer components, though it can still roughly locate them.

5.3.3.2 SSM(r) for Regular-Size Scatterers

In this subsection, we consider two examples to demonstrate the capability and effectiveness of the **SSM(r)** for locating regular-size scatterers.

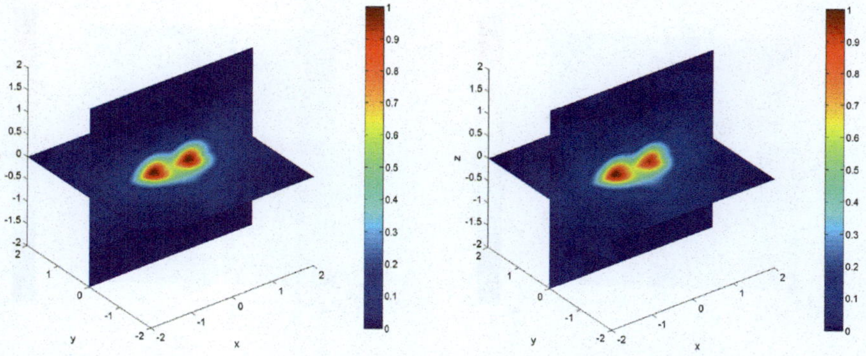

Fig. 5.4 Example 4: Reconstruction results for Example 4: (Left) Exact far-field data; (Right) Noisy far-field data with $\delta = 20\%$

Example 5 In this example, we consider a revolving peanut and a revolving kite, both are revolving solids along the x-axis. The gravitational center of the peanut is chosen at $(0,\ 0,\ 1.5\lambda)$, and the center of the kite is anchored at $(0,\ 0,\ -1.5\lambda)$. Both scatterers are taken to be PEC obstacles.

The admissible reference class is chosen to be composed of a unit ball, a revolving kite and a revolving peanut. All those admissible reference obstacles are centered at the origin and are PEC obstacles. Following the **SSM(r)** algorithm, their norms of far-field data associated with the reference obstacles in the admissible space are sorted in the descending order with the first one being the kite, the second one being the peanut and the third one being the unit ball. Using those a priori admissible known far-field data, we implement the **SSM(r)** method. The orthogonal contour slices with a certain transparency are shown in Fig. 5.5 for better visualization. For both the noise-free and noisy far-field data, the **SSM(r)** can successfully determine the location of the kite through the first indicator function; see the dark red part in the center of the kite in the left figures. Once the kite is determined and its surrounding subregion trimmed from the sampling domain, we find that the second indicator function plot shows that the center of the peanut can be identified by continuing the **SSM(r)**, see the right figures. After the peanut is determined and trimmed from the sampling domain, if one continues the **SSM(r)** by testing data of the reference ball obstacle, no significant peak value would be found for the reference unit ball obstacle. Our numerical results are nicely consistent with our theoretical predication.

Example 6 In this example, we adopt the same setting as that in Example 5, but with all the scatterer components set to be inhomogeneous media, with the EM parameters $\varepsilon = 4$, $\mu = 1$ and $\sigma = 0$. It can be verified that the far-field data for the three reference scatterers are distinct from each other, hence the generic condition (5.3.20) is satisfied and thus the **SSM(r)** applies; see Remark 5.3.4. Different from the obstacle case, the order of the norms of the far-field data in the admissible reference space in the present example is as follows, the peanut comes first, and then the kite

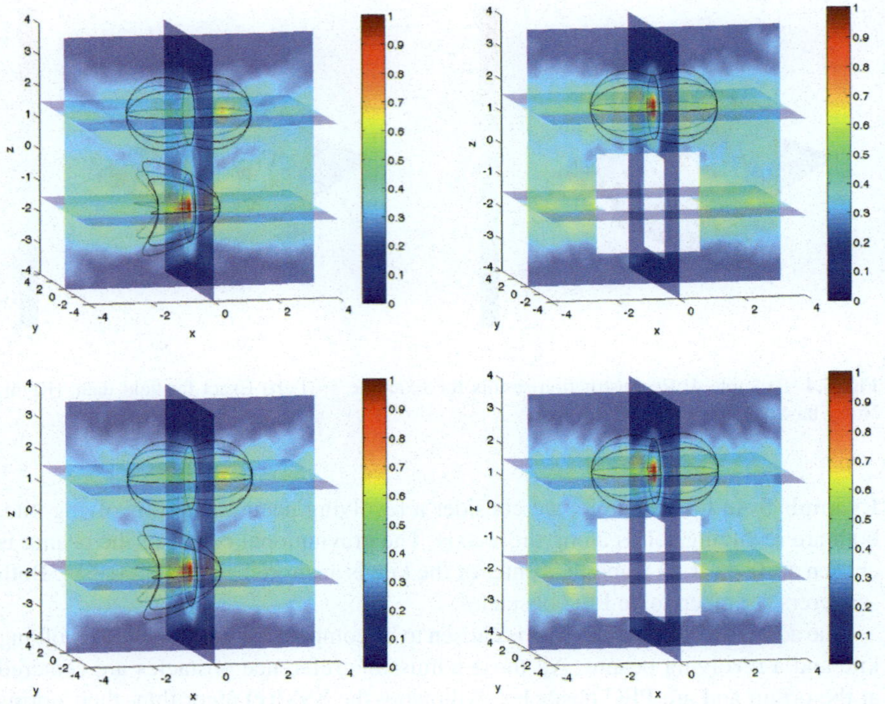

Fig. 5.5 Example 5: Reconstruction results using (top) exact far-field data, (bottom) noisy far-field data with $\delta = 5\%$

and finally is the unit ball. The numerical results are given in Fig. 5.6, from which we can see that the **SSM(r)** can also successfully determine the locations of each medium component successively.

5.3.3.3 SSM(s) for Segment-Like Scatterers

In this section, we present some numerical examples to show some very interesting and promising features of the **SSM(s)** that were not covered in our theoretical analysis. Specifically, we test the performance of the **SSM(s)** in identifying 'partially small' segment-like scatterers.

Example 7 This example considers a slender cylinder scatterer with base point anchored at $(-0.5\lambda,\ 0,\ 0)$, radius 0.1λ, height 2λ and pointing to the positive x-axis. The EM parameters inside the slender cylinder are chosen to be $\varepsilon = 4$, $\mu = 1$ and $\sigma = 0$. The numerical results are given in Fig. 5.7. In the noise-free case, the **SSM(s)** can determine successfully the location and even the length for the slender cylinder scatterer. The identified geometry is roughly a red long bar with the correct

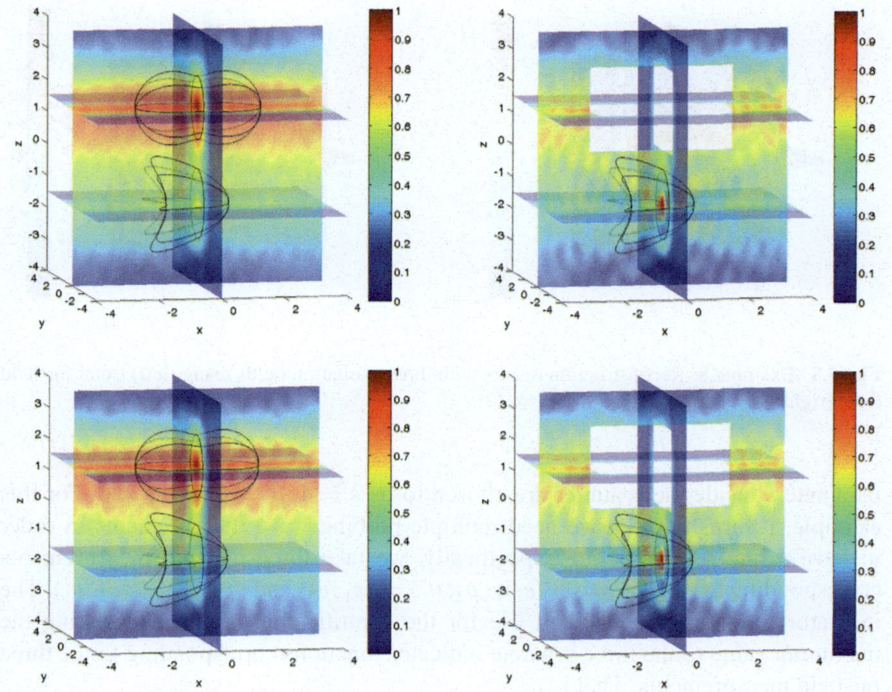

Fig. 5.6 Example 6: Reconstruction results using (top) exact far-field data, (bottom) noisy far-field data with $\delta = 5\%$

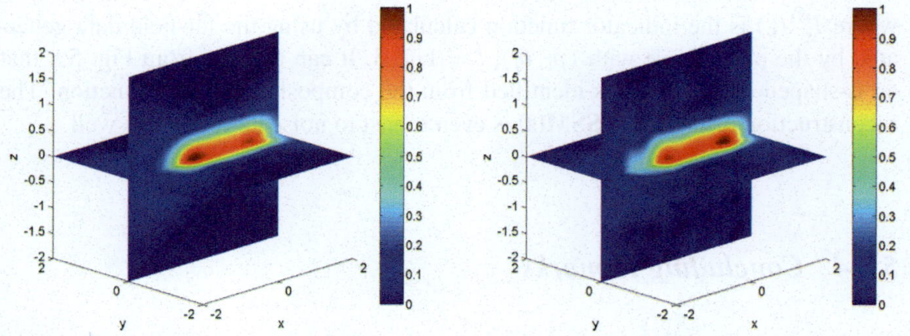

Fig. 5.7 Example 7: Reconstruction results using (left) exact far-field data, (right) noisy far-field data with $\delta = 20\%$

length. Furthermore, the **SSM(s)** also performs well for the even more challenging case with large random noise, up to $\delta = 20\%$, attached to the far-field data.

Example 8 In this example, we consider an L-shaped scatterer composed of two slender cylinders as given in Example 7, except that both base points are moved to $(-\lambda, -\lambda, 0)$ and pointing to the positive x- and y-axes, respectively. The EM

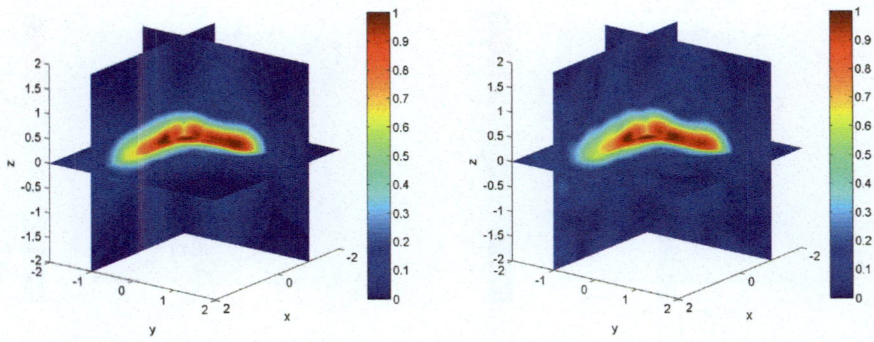

Fig. 5.8 Example 8: Reconstruction results with three excitation fields using (left) exact far-field data, (right) noisy far-field data with $\delta = 20\%$

parameters inside the scatterer are chosen to be $\varepsilon = 4$, $\mu = 1$ and $\sigma = 0$. For this example, it turns out that we need multiple probing wave measurements in order to have a fine reconstruction. Specifically, we take three far-field measurements corresponding to $(p_1, \theta_1') = (e_3, e_1)$, $(p_2, \theta_2') = (e_1, e_2)$ and $(p_3, \theta_3') = (e_2, e_3)$. The indicator function that we shall use for the identification is given by taking the maximum value of the three separate indicator functions corresponding to the three far-field measurements. That is

$$I_s(z) = \max_{1\le l\le 3} I_s^{(l)}(z), \quad z \in \mathscr{T},$$

where $I_s^{(l)}(z)$ is the indicator function calculated by using the far-field data generated by the plane wave with (p_l, θ_l'), $l = 1, 2, 3$. It can be seen from Fig. 5.8 that an L-shaped dark red bar is identified from the composite indicator function. The reconstruction result of the **SSM(s)** is even robust to noise up to 20% as well.

5.3.4 Concluding Remarks

In this section, two inverse scattering schemes, **SSM(s)** and **SSM(r)**, are discussed for locating multiple electromagnetic scatterers by a single electric far-field measurement. The locating schemes could work in an extremely general setting: the underlying scatterer might include, at the same time, obstacle components and inhomogeneous medium components; the number of the scatterer components and the physical property of each component are not required to be known in advance. The first scheme **SSM(s)** is for locating scatterers of small size compared to the detecting EM wavelength. For this scheme, the content of each medium component is not required to known in advance either. The second scheme **SSM(r)** is for locating scatterers of regular size compared to the detecting EM wavelength. For this scheme, certain a priori information would be required of each scatterer component. Specifically, if the component is an obstacle, then its shape must be from a certain known admissible reference scatterer class; and if the component is an inhomogeneous medium, then its support and content must also be from a certain known admissible reference scatterer class, and moreover, a certain generic condition must be satisfied. Nevertheless, the scheme **SSM(r)** could also work in a very general setting. The reference class may consist of multiple different reference scatterers, and some reference scatterer may not be presented as a component in the unknown scatterer, and some may be presented as components for more than one time. The setting considered would be of significant interests, e.g., in radar and sonar imaging. The locating schemes are based on some indicator functions, whose indicating behaviors could be used for identification in many applications. In calculating the indicator functions, no inversion would be involved, so the discussed methods are very efficient and robust to noise. Rigorous mathematical justifications are provided for both schemes. Extensive numerical experiments are conducted to illustrate the effectiveness and robustness of the discussed imaging schemes in various practical scenarios. The numerical results match our theoretical predications in a very sound manner. Furthermore, the numerical results also reveal some very interesting and promising features of the discussed schemes. The scheme **SSM(s)** is also capable of qualitatively imaging the supports/shapes of the unknown scatterers in addition to locating them. Second, it is also capable of qualitatively identifying ‘partially small’ segment-like scatterers.

Finally, we would like to note that in both of the discussed imaging schemes, the decoupling of the multiple scattering interaction between different scatterer components plays a critical role. On the other hand, it is known that the multiple interaction may be helpful for the locating (see, e.g. [32]). This is an interesting topic worthy of further investigation. Moreover, the issues on the relation of our locating schemes with the MUSIC algorithms (see, e.g. [5, 6]), and on how to increase the imaging resolutions are also topics worthy of further investigation.

5.4 Scatterers

This section is concerned with the numerical reconstruction algorithms for the multi-scale inverse scattering problem. The rest of the section is organized as follows. Section 5.4.1 is devoted to the description of multi-scale EM scatterers and the two locating schemes in [29]. In Sect. 5.4.2, we discuss techniques on relaxing the requirement on knowledge of the orientation and size for locating regular-size scatterers. In Sect. 5.4.3, we present the imaging schemes of locating multiple multi-scale scatterers. Finally, in Sect. 5.4.4, numerical experiments are given to demonstrate the effectiveness and the promising features of the discussed imaging schemes.

5.4.1 Multi-scale EM Scatterers and Two Locating Schemes

Throughout the rest of the subsection, we assume $k \sim 1$. That is, the wavelength of the EM plane waves is given by $\lambda = 2\pi/k \sim 1$ and hence the size of a scatterer can be expressed in terms of its Euclidean diameter.

5.4.1.1 Scheme S

We first introduce the class of small scatterers for our study. Let $l_s \in \mathbb{N}$ and D_j, $1 \le j \le l_s$ be bounded Lipschitz domains in $\mathbb{R}^3$. It is assumed that all D_j's are simply connected and contain the origin. For $\rho \in \mathbb{R}_+$, we let $\rho D_j := \{\rho x; x \in D_j\}$ and set

$$\Omega_j^{(s)} = z_j + \rho D_j, \quad z_j \in \mathbb{R}^3, \ 1 \le j \le l_s. \tag{5.4.1}$$

Each $\Omega_j^{(s)}$ is referred to as a scatterer component and its content is endowed with ε_j, μ_j and σ_j. The parameter $\rho \in \mathbb{R}_+$ represents the relative size of the scatterer (or, more precisely, each of its components). The scatterer components $(\Omega_j^{(s)}; \varepsilon_j, \mu_j, \sigma_j)$, $1 \le j \le l_s$, are assumed to satisfy: (i). if for some $j, 0 \le \sigma_j < +\infty$, then ε_j, μ_j and σ_j are all real valued C^2-smooth functions in the closure of $\Omega_j^{(s)}$; (ii). in the case of (i), the following condition is satisfied, $|\varepsilon_j(x) - 1| + |\mu_j(x) - 1| + |\sigma_j(x)| > c_0 > 0$ for all $x \in \Omega_j^{(s)}$ and some constant c_0; (iii). if for some $j, \sigma_j = +\infty$, then disregarding the parameters ε_j and μ_j, $\Omega_j^{(s)}$ is regarded as a PEC obstacle. Condition (ii) means that if $(\Omega_j^{(s)}; \varepsilon_j, \mu_j, \sigma_j)$ is a medium component, then it is inhomogeneous from the homogeneous background space. We set

$$\Omega^{(s)} := \bigcup_{j=1}^{l_s} \Omega_j^{(s)} \quad \text{and} \quad (\Omega^{(s)}; \varepsilon, \mu, \sigma) := \bigcup_{j=1}^{l_s} (\Omega_j^{(s)}; \varepsilon_j, \mu_j, \sigma_j). \tag{5.4.2}$$

and make the following qualitative assumption,

$$\rho \ll 1 \quad \text{and} \quad \operatorname{dist}(z_j, z_{j'}) \gg 1 \ \text{ for } j \ne j', \ 1 \le j, j' \le l_s. \tag{5.4.3}$$

The assumption (5.4.3) implies that compared to the wavelength of the incident plane waves, the relative size of each scatterer component is small and if there are multiple components, they are sparsely distributed. It is numerically shown in [29] that if the relative size is smaller than half a wavelength and the distance between two different components is bigger than half a wavelength, the scheme developed there works well for locating the multiple components of $\Omega^{(s)}$. Let $0 \le l_s' \le l_s$ be such that when $1 \le j \le l_s'$, $\sigma_j = +\infty$, and when $l_s' + 1 \le j \le l_s$, $0 \le \sigma_j < +\infty$. That is, if $1 \le j \le l_s'$, $\Omega_j^{(s)}$ is a PEC obstacle component, whereas if $l_s' + 1 \le j \le l_s$, $(\Omega_j^{(s)}; \varepsilon_j, \mu_j, \sigma_j)$ is a medium component. If $l_s' = 0$, then all the components of the small scatterer $\Omega^{(s)}$ are of medium type and if $l'_s = l_s$, then all the components are PEC obstacles. The EM scattering corresponding to $\Omega^{(s)}$ due to a single pair of incident waves (E^i, H^i) is governed by (4.3.1) with $O = \bigcup_{j=1}^{l'_s} \Omega_j^{(s)}$ and $(M; \varepsilon, \mu, \sigma) = \bigcup_{j=l'_s+1}^{l_s} (\Omega_j^{(s)}; \varepsilon_j, \mu_j, \sigma_j)$. We denote the electric far-field pattern by $A(\hat{x}; \Omega^{(s)})$.

In order to locate the multiple components of $\Omega^{(s)}$ in (5.4.2), the following indicator function is introduced in [29],

$$I_s(z) := \frac{1}{\|A(\hat{x}; \Omega^{(s)})\|^2_{T^2(\mathbb{S}^2)}} \sum_{m=-1,0,1} \left(\left| \left\langle A(\hat{x}; \Omega^{(s)}), e^{ik(d-\hat{x})\cdot z}\, U_1^m(\hat{x}) \right\rangle_{T^2(\mathbb{S}^2)} \right|^2 + \left| \left\langle A(\hat{x}; \Omega^{(s)}), e^{ik(d-\hat{x})\cdot z}\, V_1^m(\hat{x}) \right\rangle_{T^2(\mathbb{S}^2)} \right|^2 \right), \quad z \in \mathbb{R}^3, \tag{5.4.4}$$

where

$$T^2(\mathbb{S}^2) := \{\mathbf{a} \in \mathbb{C}^3 \,|\, \mathbf{a} \in L^2(\mathbb{S}^2)^3,\ \hat{x} \cdot \mathbf{a} = 0 \ \text{ for a.e. } \hat{x} \in \mathbb{S}^2\},$$

which is endowed with the inner product $\langle \mathbf{u}, \mathbf{v} \rangle_{T^2(\mathbb{S}^2)} = \int_{\mathbb{S}^2} \mathbf{u} \cdot \overline{\mathbf{v}}$, and

$$U_n^m(\hat{x}) := \frac{1}{\sqrt{n(n+1)}} \operatorname{Grad} Y_n^m(\hat{x}), \quad V_n^m(\hat{x}) := \hat{x} \wedge U_n^m(\hat{x}), \ n \in \mathbb{N}, \ \ m = -n, \ldots, n,$$

with $Y_n^m(\hat{x})$, $m = -n, \ldots, n$ the spherical harmonics of order $n \geq 0$ (cf. [20]). It is shown in [29] that z_j (cf. (5.4.1)), $1 \leq j \leq l_s$, is a local maximum point for $I_s(z)$. Based on such indicating behavior, the following scheme is proposed in [29] for locating the multiple components of the small scatterer $\Omega^{(s)}$.

Algorithm: Locating Scheme S

---

1. For an unknown EM scatterer $\Omega^{(s)}$ in (5.4.2), collect the far-field data by sending a single pair of detecting EM plane waves specified in (4.1.1).
2. Select a sampling region with a mesh $\mathscr{T}_h$ containing $\Omega^{(s)}$.
3. For each point $z \in \mathscr{T}_h$, calculate $I_s(z)$.
4. Locate all the significant local maxima of $I_s(z)$ on $\mathscr{T}_h$, which represent the locations of the scatterer components.

5.4.1.2 Scheme R

Next, we consider the locating of multiple obstacles of regular size. For this locating scheme, one must require the following generic uniqueness result holds for the inverse scattering problem. Let O_1 and O_2 be obstacles and both of them are assumed to be bounded simply connected Lipschitz domains in $\mathbb{R}^3$ containing the origin. Then

$$A(\hat{x}; O_1) = A(\hat{x}; O_2) \text{ if and only if } O_1 = O_2. \tag{5.4.5}$$

This result implies that by using a single far-field measurement, one can uniquely determine an obstacle. There is a widespread belief that such a uniqueness result holds, but there is only limited progress in the literature, see, e.g., [33, 37, 41]. Throughout the present section, we assume that such a generic uniqueness holds true.

We now briefly recall *Scheme R* in [29] for locating multiple regular-size obstacles. Let $l_r \in \mathbb{N}$ and let G_j, $1 \le j \le l_r$ be bounded simply connected Lipschitz domains containing the origin in $\mathbb{R}^3$. Set

$$\Omega_j^{(r)} = z_j + G_j, \quad z_j \in \mathbb{R}^3, \quad 1 \le j \le l_r. \tag{5.4.6}$$

Each $\Omega_j^{(r)}$ denotes a PEC obstacle located at the position $z_j \in \mathbb{R}^3$. It is required that

$$\mathrm{diam}(\Omega_j^{(r)}) = \mathrm{diam}(G_j) \sim 1, \ 1 \le j \le l_r; \quad L = \min_{1 \le j, j' \le l_r, j \ne j'} \mathrm{dist}(z_j, z_{j'}) \gg 1. \tag{5.4.7}$$

Furthermore, there exists an admissible reference obstacle space

$$\mathscr{S} := \{\Sigma_j\}_{j=1}^{l'}, \tag{5.4.8}$$

where each $\Sigma_j \subset \mathbb{R}^3$ is a bounded simply connected Lipschitz domain that contains the origin and

$$\Sigma_j \ne \Sigma_{j'}, \quad \text{for } j \ne j', \ 1 \le j, j' \le l', \tag{5.4.9}$$

such that

$$G_j \in \mathscr{S}, \quad j = 1, 2, \ldots, l_r. \tag{5.4.10}$$

The admissible class $\mathscr{S}$ is required to be known in advance, and by reordering if necessary, it is assumed that

$$\|A(\hat{x}; \Sigma_j)\|_{T^2(\mathbb{S}^2)} \ge \|A(\hat{x}; \Sigma_{j+1})\|_{T^2(\mathbb{S}^2)}, \quad j = 1, 2, \ldots, l' - 1. \tag{5.4.11}$$

Let

$$\Omega^{(r)} := \bigcup_{j=1}^{l_r} \Omega_j^{(r)}. \tag{5.4.12}$$

Then $\Omega^{(r)}$ denotes the regular-size scatterer for our current study, which may consist of multiple obstacle components. The second condition in (5.4.7) means that the components are sparsely distributed. It is numerically observed in [29] that if the distance is larger than a few numbers of wavelength, then Scheme R works effectively. The assumption (5.4.10) indicates that certain a priori knowledge of the target scatterer is required. It is remarked that l_r is not necessarily the same as l'. Define l' indicator functions as follows,

$$I_r^j(z) = \frac{\left|\langle A(\hat{x};\, \Omega^{(r)}),\, e^{ik(d-\hat{x})\cdot z} A(\hat{x};\, \Sigma_j)\rangle_{T^2(\mathbb{S}^2)}\right|}{\|A(\hat{x};\, \Sigma_j)\|^2_{T^2(\mathbb{S}^2)}}, \quad j = 1, 2, \ldots, l', \quad z \in \mathbb{R}^3. \tag{5.4.13}$$

The following indicating behavior of $I_r^j(z)$'s is proved in [29] and summarized below.

Theorem 5.4.1 *Consider the indicator function $I_r^1(z)$ introduced in* (5.4.13). *Suppose there exists $J_0 \subset \{1, 2, \ldots, l_r\}$ such that for $j_0 \in J_0$, $G_{j_0} = \Sigma_1$, whereas $G_j \neq \Sigma_1$ for $j \in \{1, 2, \ldots, l_r\} \backslash J_0$. Then for each z_j, $j = 1, 2, \ldots, l_r$, there exists an open neighborhood of z_j, $neigh(z_j)$, such that*

(i) if $j \in J_0$, then

$$\widetilde{I}_r^1(z) := |I_r^1(z) - 1| \leq \mathscr{O}\left(\frac{1}{L}\right), \quad z \in neigh(z_j), \tag{5.4.14}$$

and moreover, z_j is a local minimum point for $\widetilde{I}_r^1(z)$;

(ii) if $j \in \{1, 2, \ldots, l_r\} \backslash J_0$, then there exists $\varepsilon_0 > 0$ such that

$$\widetilde{I}_r^1(z) := |I_r^1(z) - 1| \geq \varepsilon_0 + \mathscr{O}\left(\frac{1}{L}\right), \quad z \in neigh(z_j). \tag{5.4.15}$$

Based on Theorem 5.4.1, the *Scheme R* for locating the multiple components in $\Omega^{(r)}$ can be successively formulated as follows.

Algorithm: Locating Scheme R

---

1. For an unknown EM scatterer $\Omega^{(r)}$ in (5.4.12), collect the far-field data by sending a single pair of detecting EM plane waves specified in (4.1.1).
2. Select a sampling region with a mesh $\mathscr{T}_h$ containing $\Omega^{(r)}$.
3. Collect in advance the far-field patterns associated with the admissible reference scatterer space $\mathscr{S}$ in (5.3.16), and reorder $\mathscr{S}$ if necessary to make it satisfy (5.4.11), and also verify the generic assumption (5.4.5).
4. Set $j = 1$.
5. For each point $z \in \mathscr{T}_h$, calculate $I_r^j(z)$ (or $\widetilde{I}_r^j(z) = |I_r^j(z) - 1|$).
6. Locate all those significant local maxima of $I_r^j(z)$ such that $I_r^j(z) \sim 1$ (or the minima of $\widetilde{I}_r^j(z)$ on $\mathscr{T}_h$ such that $\widetilde{I}_r^j(z) \ll 1$), where scatterer components of the form $z + \Sigma_j$ is located.
7. Trim all those $z + \Sigma_j$ found in (6) from $\mathscr{T}_h$.
8. If $\mathscr{T}_h = \emptyset$ or $j = l'$, then Stop; otherwise, set $j = j + 1$, and go to (5).

Remark 5.4.1 By (5.4.5) and (5.4.9), it is readily seen that

$$A(\hat{x};\, \Sigma_j) \neq A(\hat{x};\, \Sigma_{j'}), \quad j \neq j',\ 1 \leq j, j' \leq l'. \tag{5.4.16}$$

Equation (5.4.16) plays a critical role in justifying the indicating behavior of $I_r^j(z)$ in Theorem 5.4.1. Nevertheless, since the reference space (5.4.8) is given, one can verify (5.4.16) in advance. On the other hand, one can also include inhomogeneous medium components into the admissible reference space provided the relation (5.4.16) is satisfied. For the inhomogeneous medium component in $\mathscr{S}$, its content is required to be known in advance; see Remark 5.4.3 in the following.

Scheme R could find important practical applications, e.g., in radar technology in locating an unknown group of aircrafts, where one has the a priori knowledge on the possible models of the target airplanes. However, we note here some important practical situations that Scheme R does not cover. Indeed, in Scheme R, it is required that each component, say $\Omega_1^{(r)}$, is a translation of the reference obstacle Σ_1, namely $\Omega_1^{(r)} = z + \Sigma_1$. This means that, in addition to the shape of the obstacle component $\Omega_1^{(r)}$, one must also know its orientation and size in advance (here, the concept of orientation and size will be given in Sect. 5.4.2). In the radar technology, this means that in addition to the model of each aircraft, one must also know which direction the aircraft is heading to. Clearly, this limits the applicability of the locating scheme. In the next section, we shall discuss strategies to relax the limitations about the requirement on orientation and size. Furthermore, we shall consider the locating of multiple multi-scale scatterers, which may include, at the same time, small- and regular-size scatterers. To that end, we introduce the multiple multi-scale scatterer for our subsequent study as follows,

$$\Omega^{(m)} := \Omega^{(s)} \cup \Omega^{(r)} \tag{5.4.17}$$

with $\Omega^{(s)}$ and $\Omega^{(r)}$ given in (5.4.2) and (5.4.12), respectively.

5.4.2 Scheme R with Augmented Reference Spaces

In this section, we discuss an enhanced version of Scheme R with augmented reference spaces to image a regular-size scatterer with multiple components of different shapes, orientations and sizes. This goal is achieved through collecting more reference far field data of a set of a priori known components, in particular associated with their possible orientations and sizes.

Let $\Pi_{\theta,\phi,\psi}$ denote the 3D rotation whose Euler angles are θ, ϕ and ψ with the $x_1 - x_2 - x_3$ convention for $x = (x_1, x_2, x_3) \in \mathbb{R}^3$. That is, $\Pi_{\theta,\phi,\psi}x = U(\theta, \phi, \psi)x$, where $U \in SO(3)$ is given by

$$U = \begin{pmatrix} \cos\theta\cos\psi & -\cos\theta\sin\psi + \sin\phi\sin\theta\cos\psi & \sin\phi\sin\psi + \cos\phi\sin\theta\cos\psi \\ \cos\theta\sin\psi & \cos\phi\cos\psi + \sin\phi\sin\theta\sin\psi & -\sin\phi\cos\psi + \cos\phi\sin\theta\sin\psi \\ -\sin\theta & \sin\phi\cos\theta & \cos\phi\cos\theta \end{pmatrix} \tag{5.4.18}$$

with $0 \le \theta, \phi \le 2\pi$ and $0 \le \psi \le \pi$. In the sequel, we suppose there exist triplets $(\theta_j, \phi_j, \psi_j)$, $j = 1, 2, \ldots, l_r$ such that

$$\Omega_j^{(r)} = z_j + \Pi_{\theta_j,\phi_j,\psi_j} G_j, \tag{5.4.19}$$

where $G_j \in \mathscr{S}$ defined in (5.3.16). Now, we let

$$\Omega^{(r)} = \bigcup_{j=1}^{l_r} (z_j + \Pi_{\theta_j,\phi_j,\psi_j} G_j) := \bigcup_{j=1}^{l_r} (z_j + \widetilde{G}_j) \tag{5.4.20}$$

denote the regular-size target scatterer for our study. Compared to the regular-size scatterer in (5.4.12) considered in [29] (cf. (5.4.6)–(5.4.12)), the scatterer introduced in (5.4.20) possesses the new feature that each component is allowed to be rotated. In the sequel, the Euler angles $(\theta_j, \phi_j, \psi_j)$ will be referred to as the *orientation* of the scatterer component $\Omega_j^{(r)}$ in (5.4.19).

Next, we also introduce a scaling/dilation operator Λ_{τ_j}, $\tau_j \in \mathbb{R}_+$, and for $\Omega_j^{(r)} = z_j + G_j$, $G_j \in \mathscr{S}$, we set

$$\Omega_j^{(r)} := z_j + \Lambda_{\tau_j} G_j, \tag{5.4.21}$$

where $\Lambda_{\tau_j} G_j := \{\tau_j x\,;\ x \in G_j\}$. Now, for a sequence of $\{\tau_j\}_{j=1}^{l_r}$ we set

$$\Omega^{(r)} = \bigcup_{j=1}^{l_r} (z_j + \Lambda_{\tau_j} G_j). \tag{5.4.22}$$

We call τ_j the *size* or *scale* of the component $\Omega_j^{(r)}$ relative to the reference one G_j.

For our subsequent study, we would consider locating a regular-size scatterer with its components both possibly orientated and scaled,

$$\Omega^{(r)} = \bigcup_{j=1}^{l_r} (z_j + \Pi_{\theta_j,\phi_j,\psi_j} \Lambda_{\tau_j} G_j) := \bigcup_{j=1}^{l_r} (z_j + \widehat{G}_j). \tag{5.4.23}$$

Compared to the scatterer in (5.4.12) considered in [29], the scatterer introduced in (5.4.22) is scaled relatively. To that end, we first show a relation of the far-field pattern when the underlying scatterer is rotated and scaled.

Proposition 5.4.1 *Let G be a bounded simply connected Lipschitz domain containing the origin, which represents a PEC obstacle. Then, we have that*

$$A(\hat{x};\Pi_{\theta,\phi,\psi}G,d,p,k)=UA(U^T\hat{x};G,U^Tp,U^Td,k), \tag{5.4.24}$$

where $U=U(\theta,\phi,\psi)$ is the rotation matrix corresponding to $\Pi_{\theta,\phi,\psi}$; and

$$A(\hat{x};\Lambda_\tau G,d,p,k)=\tau A(\hat{x};G,d,p,k\tau) \tag{5.4.25}$$

Proof Let $E\in H^1_{loc}(\mathbb{R}^3\backslash\Pi_{\theta,\phi,\psi}G)$ and $H\in H^1_{loc}(\mathbb{R}^3\backslash\Pi_{\theta,\phi,\psi}G)$ be the solutions to the following Maxwell system

$$\begin{aligned}
&\nabla\wedge E-\mathrm{i}kH=0, \qquad \nabla\wedge H+\mathrm{i}kE=0 \quad \text{in } \mathbb{R}^3\backslash\overline{\Pi_{\theta,\phi,\psi}G},\\
&\nu\wedge E=0 \quad \text{on } \partial(\Pi_{\theta,\phi,\psi}G), \qquad E=E^i+E^+ \quad \text{in } \mathbb{R}^3\backslash\overline{\Pi_{\theta,\phi,\psi}G},\\
&\lim_{|x|\to+\infty}|x|\left|(\nabla\wedge E^+)(x)\wedge\frac{x}{|x|}-\mathrm{i}kE^+(x)\right|=0,
\end{aligned} \tag{5.4.26}$$

where $E^i(x)=pe^{\mathrm{i}kx\cdot d}$ and ν is the outward unit normal vector to $\partial(\Pi_{\theta,\phi,\psi}G)$. Set

$$\begin{aligned}
\widetilde{E}&=\Pi^*_{\theta,\phi,\psi}E:=\Pi^{-1}_{\theta,\phi,\psi}\circ E\circ\Pi_{\theta,\phi,\psi}=U^TE\circ U\\
\widetilde{H}&=\Pi^*_{\theta,\phi,\psi}H:=\Pi^{-1}_{\theta,\phi,\psi}\circ H\circ\Pi_{\theta,\phi,\psi}=U^TH\circ U
\end{aligned} \quad \text{in } \mathbb{R}^3\backslash\overline{G}, \tag{5.4.27}$$

and

$$\widetilde{E}^i(x):=(U^Tp)e^{\mathrm{i}kx\cdot(U^Td)} \tag{5.4.28}$$

Then, by the transformation properties of Maxwell's equations (see, e.g., [42]), it is straightforward to verify that

$$\begin{aligned}
&\nabla\wedge\widetilde{E}-\mathrm{i}k\widetilde{H}=0, \quad \nabla\wedge\widetilde{H}+\mathrm{i}k\widetilde{E}=0 \quad \text{in } \mathbb{R}^3\backslash\overline{G},\\
&\widetilde{\nu}\wedge\widetilde{E}=0 \quad \text{on } \partial G, \qquad \widetilde{E}=\widetilde{E}^i+\widetilde{E}^+ \quad \text{in } \mathbb{R}^3\backslash\overline{G},\\
&\lim_{|x|\to+\infty}|x|\left|(\nabla\wedge\widetilde{E}^+)(x)\wedge\frac{x}{|x|}-\mathrm{i}k\widetilde{E}^+(x)\right|=0,
\end{aligned} \tag{5.4.29}$$

where $\widetilde{\nu}$ is the outward unit normal vector to ∂G. Clearly, $A(\hat{x};\Pi_{\theta,\phi,\psi}G)$ can be read-off from the large $|x|$ asymptotics of $E(x)$ in (5.4.26),

$$E(x)=pe^{\mathrm{i}kx\cdot d}+\frac{e^{\mathrm{i}k|x|}}{|x|}A\left(\frac{x}{|x|};\Pi_{\theta,\phi,\psi}G,d,p,k\right)+\mathscr{O}\left(\frac{1}{|x|^2}\right). \tag{5.4.30}$$

Hence, by (5.4.27) and (5.4.30), we have

$$\begin{aligned}\widetilde{E}(x) &= U^T E(Ux)\\ &= U^T p e^{\mathrm{i}kUx\cdot d} + \frac{e^{\mathrm{i}k|Ux|}}{|Ux|} U^T A\left(\frac{Ux}{|Ux|}; \Pi_{\theta,\phi,\psi} G, d, p, k\right) + \mathscr{O}\left(\frac{1}{|Ux|^2}\right)\\ &= U^T p e^{\mathrm{i}kx\cdot U^T d} + \frac{e^{\mathrm{i}k|x|}}{|x|} U^T A(U\hat{x}; \Pi_{\theta,\phi,\psi} G, d, p, k) + \mathscr{O}\left(\frac{1}{|x|^2}\right).\end{aligned} \tag{5.4.31}$$

By (5.4.29) and (5.4.31), one can readily see that

$$U^T A(U\hat{x}; \Phi_{\theta,\phi,\psi} G, d, p, k) = A(\hat{x}; G, \widetilde{E}^i) = A(\hat{x}; G, U^T d, U^T p, k).$$

which immediately implies (5.4.24).

In a completely similar manner, one can show (5.4.25). The proof is complete. □

Proposition 5.4.1 suggests that in order to locate a scatterer $\Omega^{(r)}$ in (5.4.20) by using the Scheme R, one can make use of the multi-polarization and multi-incident-direction far-field data, namely $A(\hat{x}; p, d, k)$ for all $p \in \mathbb{R}^3$, $d \in \mathbb{S}^2$ and a fixed $k \in \mathbb{R}_+$. On the other hand, in order to still make use of a single far-field for the locating, one can augment the reference space $\mathscr{S}$ by letting

$$\widetilde{\mathscr{S}} = \Pi_{\theta,\phi,\psi}\mathscr{S} := \{\Pi_{\theta,\phi,\psi}\Sigma_j\}_{j=1}^{l'}, \quad (\theta,\phi,\psi) \in [0, 2\pi]^2 \times [0, \pi]. \tag{5.4.32}$$

Furthermore, from a practical viewpoint, we introduce a discrete approximation of $\widetilde{\mathscr{S}}$ and set

$$\widetilde{\mathscr{S}}_h := \{\Pi_{\theta^h,\phi^h,\psi^h}\Sigma_j\}_{j=1}^{l'} = \{\widetilde{\Sigma}_j\}_{j=1}^{\widetilde{l}_h}, \tag{5.4.33}$$

where $(\theta^h, \phi^h, \psi^h)$ denotes an equal distribution over $[0, 2\pi]^2 \times [0, \pi]$ with an angular mesh-size $h \in \mathbb{R}_+$ and its cardinality N_h, and $\widetilde{l}_h := l' \times N_h$. By reordering if necessary, we assume the non-increasing relation (5.4.11) also holds for those components. Next, based on the same single far-field data for Scheme S, one can calculate $\widetilde{l}_h$ indicator functions according to (5.4.13), but with the reference scatterers taken from $\widetilde{\mathscr{S}}_h$. We denote the $\widetilde{l}_h$ indicator functions by $I_h^j(z)$, $1 \le j \le \widetilde{l}_h$. Then, we have

Theorem 5.4.2 *Consider the multiple scatterers introduced in* (5.4.20) *and the indicator function $I_h^1(z)$ introduced above. Let $\widetilde{\Sigma}_l \in \widetilde{\mathscr{S}}_h$ be such that*

$$\widetilde{\Sigma}_1 = \Pi_{\theta_1^h,\phi_1^h,\psi_1^h}\Sigma_{m_0} \quad \textit{with} \quad \Sigma_{m_0} \in \mathscr{S}. \tag{5.4.34}$$

Suppose there exists $J_0 \subset \{1, 2, \ldots, l_r\}$ such that for $j_0 \in J_0$,

$$\Omega_{j_0}^{(r)} = z_{j_0} + \widetilde{G}_{j_0} = z_{j_0} + \Pi_{\theta_{j_0},\phi_{j_0},\psi_{j_0}} G_{j_0}$$

with

$$G_{j_0} = \Sigma_{m_0} \quad \textit{and} \quad \|(\theta_{j_0}, \phi_{j_0}, \psi_{j_0}) - (\theta_1^h, \phi_1^h, \psi_1^h)\|_{l^\infty} = \mathscr{O}(h); \tag{5.4.35}$$

whereas for the other components $\Omega_j^{(r)}$, $j \in \{1, 2, \ldots, l_r\}\backslash J_0$, either of the two conditions in (5.4.35) is violated. Then for each z_j, $j = 1, 2, \ldots, l_r$, there exists an open neighborhood of z_j, $neigh(z_j)$, such that

(i) if $j \in J_0$, then

$$\widetilde{I}_h^1(z) := |I_h^1(z) - 1| \leq \mathscr{O}\left(\frac{1}{L} + h\right), \quad z \in neigh(z_j), \tag{5.4.36}$$

and moreover, z_j is a local minimum point for $\widetilde{I}_h^1(z)$;

(ii) if $j \in \{1, 2, \ldots, l_r\}\backslash J_0$, then there exists $\varepsilon_0 > 0$ such that

$$\widetilde{I}_h^1(z) := |I_h^1(z) - 1| \geq \varepsilon_0 + \mathscr{O}\left(\frac{1}{L} + h\right), \quad z \in neigh(z_j). \tag{5.4.37}$$

Proof Let

$$\widetilde{\Gamma}_1 := \Pi_{\theta_{j_0}, \phi_{j_0}, \psi_{j_0}} \Sigma_{m_0},$$

and

$$H_r^1(z) = \frac{\left|\langle A(\hat{x}; \Omega^{(r)}), e^{ik(d-\hat{x})\cdot z} A(\hat{x}; \widetilde{\Gamma}_1)\rangle_{T^2(\mathbb{S}^2)}\right|}{\|A(\hat{x}; \widetilde{\Gamma}_1)\|_{T^2(\mathbb{S}^2)}^2}, \quad z \in \mathbb{R}^3.$$

By a completely similar argument to the proof of Theorem 2.1 in [29], one can show that $H_r^1(z)$ possesses the two indicating behaviors given in (9.118) and (9.119). Next, by Proposition 5.4.1, we have

$$A(\hat{x}; \widetilde{\Gamma}_1) = A(\hat{x}; \Pi_{\theta_0, \phi_0, \psi_0} \Sigma_{m_0}) = U_0 A(U_0^T \hat{x}; \Sigma_{m_0}, U_0^T p, U_0^T d, k), \tag{5.4.38}$$

and

$$A(\hat{x}; \widetilde{\Sigma}_1) = A(\hat{x}; \Pi_{\theta_1^h, \phi_1^h, \psi_1^h} \Sigma_{m_0}) = U_h A(U_h^T \hat{x}; \Sigma_{m_0}, U_h^T p, U_h^T d, k), \tag{5.4.39}$$

where U_0 and U_h are the rotation matrices corresponding to $\Pi_{\theta_0, \phi_0, \psi_0}$ and $\Pi_{\theta_1^h, \phi_1^h, \psi_1^h}$, respectively. By the second assumption in (5.4.35), it is straightforward to show that

$$\|A(\hat{x}; \widetilde{\Gamma}_1) - A(\hat{x}; \widetilde{\Sigma}_1)\|_{T^2(\mathbb{S}^2)} = \mathscr{O}(h). \tag{5.4.40}$$

Finally, by (5.4.40), one has by direct verification that

$$|I_h^1(z) - H_r^1(z)| = \mathscr{O}(h), \quad z \in neigh(z_j), \quad j = 1, 2, \ldots, l_r. \tag{5.4.41}$$

It is remarked that the estimate in (5.4.41) is independent of $neigh(z_j)$, $j = 1, \ldots, l_r$. By (5.4.41) and the indicating behaviors of $H_r^1(z)$, one immediately has (5.4.36) and (5.4.37). □

Based on Theorem 5.4.2, we discuss the following enhanced locating scheme for locating the multiple components of $\Omega^{(r)}$ in (5.4.20).

Algorithm: Locating Scheme AR

This scheme is the same as Scheme R in Sect. 5.4.1 with steps (3), (5), (7), respectively modified as

(3) Augment the reference space $\mathscr{S}$ to be $\widetilde{\mathscr{S}}_h$ in (5.4.33), and reorder the elements in $\widetilde{\mathscr{S}}_h$ such that

$$\|A(\hat{x};\, \widetilde{\Sigma}_j)\|_{T^2(\mathbb{S}^2)} \geq \|A(\hat{x};\, \widetilde{\Sigma}_{j+1})\|_{T^2(\mathbb{S}^2)}, \quad j = 1, 2, \ldots, \widetilde{l}_h - 1. \tag{5.4.42}$$

(5) Replace $I_r^j(z)$ by $I_{\widetilde{h}}^j(z)$.
(7) Trim all those $z + \widetilde{\Sigma}_j$ found in Step (6) from $\widetilde{\mathscr{T}}_h$.

Remark 5.4.2 We remark that in Scheme AR, if certain a priori information is available about the possible range of the orientations of the scatterer components, it is sufficient for the augmented reference space $\widetilde{\mathscr{S}}_h$ to cover that range only. Clearly, Scheme AR can not only locate the multiple components of $\Omega^{(r)}$ in (5.4.20), but can also recover the orientation of each scatterer component.

Remark 5.4.3 Similar to Remark 5.4.1, our Scheme AR can be extended to include inhomogeneous medium components as long as the relation (5.4.16) holds for the reference scatterers in $\widetilde{\mathscr{S}}_h$. Indeed, in our numerical experiments in Sect. 5.4.4, we consider the case that the reference scatterers are composed of two inhomogeneous mediums, $(\Sigma_j; \varepsilon_j, \mu_j, \sigma_j),\ j = 1, 2$ with ε_j, μ_j and σ_j all constants that are known in advance. For this case, we would like to remark that by following the same argument, Proposition 5.4.1 remains the same, which in turn guarantees that Theorem 5.4.2 remains the same as well. Furthermore, we would like to emphasize that Scheme AR could be straightforwardly extended to work in a much more general setting where there might be both inhomogeneous medium components with variable contents and PEC obstacles presented in the reference space, as long as the generic relation (5.4.16) is satisfied.

In an analogous manner, for a scatterer described in (5.4.23), Scheme AR can be modified that the reference space is augmented by the sizes of components to be

$$\widetilde{\mathscr{S}}_h := \{\widetilde{\Sigma}_j\}_{j=1}^{\widetilde{l}_{h,m}} = \cup_{h,m}\{\Pi_{\theta^h,\phi^h,\psi^h} \Lambda_{\tau^m} \Sigma_j\}_{j=1}^{l'}, \tag{5.4.43}$$

where τ^m is an equal distribution of an interval $[s_1, s_2]$ with its cardinality N_k, or some other discrete distribution depending on the availability of certain a priori information of relative sizes, and $\widetilde{l}_{h,m} = l' \times N_h \times N_m$. Here, s_1, s_2 are positive numbers such that $[s_1, s_2]$ contains the scales/sizes of all the scatterer components. With such an augmented reference space, Scheme AR can be used to locate the multiple components and also recover both orientations and relative sizes of the scatterer $\Omega^{(r)}$ in (5.4.23).

5.4.3 Locating Multiple Multi-scale Scatterers

In this section, we consider locating a multi-scale scatterer $\Omega^{(m)}$ as described in (5.4.17) with multiple components. In addition to the requirements imposed on the small component $\Omega^{(s)}$ and the regular-size component $\Omega^{(r)}$ in Sect. 5.4.1, we shall further assume that

$$L_m := \mathrm{dist}(\Omega^{(s)}, \Omega^{(r)}) \gg 1. \tag{5.4.44}$$

By Lemmas 3.1 and 3.2 in [29], one has, respectively,

$$A(\hat{x}; \Omega^{(m)}, k) = A(\hat{x}; \Omega^{(s)}, k) + A(\hat{x}; \Omega^{(r)}, k) + \mathscr{O}\left(L_m^{-1}\right), \tag{5.4.45}$$

$$A(\hat{x}; \Omega^{(s)}, k) = \mathscr{O}((k\rho)^3). \tag{5.4.46}$$

That is, if $k \sim 1$, in the far-field pattern $A(\hat{x}; \Omega^{(m)})$, the scattering information from the regular-size component $\Omega^{(r)}$ is dominant and the scattering contribution from the small component $\Omega^{(s)}$ can be taken as small perturbation. Hence, a primitive way to locate the components of $\Omega^{(m)}$ can be proceeded in two stages as follows. First, using the single far-field pattern $A(\hat{x}; \Omega^{(m)})$ as the measurement data, one utilizes Scheme AR to locate the components of the regular-size scatterer $\Omega^{(r)}$. After the recovery of the regular-size scatterer $\Omega^{(r)}$, the far-field pattern from $\Omega^{(r)}$, namely $A(\hat{x}; \Omega^{(r)})$ becomes known. By subtracting $A(\hat{x}; \Omega^{(r)})$ from $A(\hat{x}; \Omega^{(m)})$, one then has $A(\hat{x}; \Omega^{(s)})$ (approximately). Finally, by applying Scheme S with the far-field data $A(\hat{x}; \Omega^{(s)})$, one can then locate the components of $\Omega^{(s)}$. However, if the size contrast between $\Omega^{(r)}$ and $\Omega^{(s)}$ is too big, the scattering information of $\Omega^{(s)}$ will be hidden in the noisy far-field data of $\Omega^{(r)}$. Hence, in order for the above two-stage scheme to work in locating $\Omega^{(m)}$, the size contrast between $\Omega^{(s)}$ and $\Omega^{(r)}$ cannot be excessively big. But if it is this case, the scattering effect from $\Omega^{(s)}$ would be a significant constituent part to $A(\hat{x}; \Omega^{(m)})$, and this will deteriorate the recovery in the first stage and then the second-stage recovery will be deteriorated consequently as well. In order to overcome such a dilemma for this multi-scale locating, we shall show a subtle local re-sampling technique.

Algorithm: Locating Scheme M

---

1. Collect a single far-field measurements $A(\hat{x}; \Omega^{(m)}, k)$ corresponding to the multi-scale scatterers $\Omega^{(m)}$.
2. Select a sampling region with a mesh $\mathscr{T}_h$ containing $\Omega^{(m)}$.
3. Suppose that

$$\Omega^{(r)} = \bigcup_{j=1}^{l_r} (z_j + \widetilde{\Sigma}_j), \quad \widetilde{\Sigma}_j \in \mathscr{S}_h,$$

as described in (5.4.23) of Sect. 5.4.2. Using $A(\hat{x}; \Omega^{(m)}, k)$ as the measurement data, one locates the rough locations $\widetilde{z}_j \in \mathscr{T}_h$, $j = 1, 2, \ldots, l_r$, shapes and orientations of each scatterer component following Scheme AR. Here $\widetilde{z}_j$, $j = 1, 2, \ldots, l_r$, are the approximate position points to the exact ones z_j, $j = 1, 2, \ldots, l_r$.

4. Apply the *local re-sampling technique* following the next sub-steps to update $\widetilde{z}_j$'s and to locate the components of the small-size scatterer $\Omega^{(s)}$.

    (a) For each point $\widetilde{z}_j$ found in Step (3), one generates a finer local mesh $\mathscr{Q}_{h'}(\widetilde{z}_j)$ around $\widetilde{z}_j$.
    (b) For one set of sampling points, $\hat{z}_j \in \mathscr{Q}_{h'}(\widetilde{z}_j)$, $j = 1, 2, \ldots, l_r$, one calculates

$$\widetilde{A}(\hat{x}; k) = A(\hat{x}; \Omega^{(m)}, k) - \sum_{j=1}^{l_r} e^{\mathrm{i}k(d-\hat{x})\cdot\hat{z}_j} A(\hat{x}; \widetilde{\Sigma}_j, k). \tag{5.4.47}$$

    (c) Using $\widetilde{A}(\hat{x}; k)$ in Step (b) as the measurement data, one applies Scheme S to locate the significant local maximum points on $\mathscr{T}_h \backslash \cup_{j=1}^{l_r} \mathscr{Q}_{h'}(\widetilde{z}_j)$ of the corresponding indicator function.
    (d) Repeat Steps (b) and (c) by all the possible sets of sampling points from $\mathscr{Q}_{h'}(\widetilde{z}_j)$, $j = 1, 2, \ldots l_r$. The clustered local maximum points on $\mathscr{T}_h \backslash \cup_{j=1}^{l_r} \mathscr{Q}_{h'}(\widetilde{z}_j)$ are the positions corresponding to the scatterer components of $\Omega^{(s)}$.
    (e) One updates the $\widetilde{z}_j$'s to be those sampling points $\hat{z}_j$'s which generate the clustered local maximum points in Step (d).

We note that in (5.4.47), if the re-sampling points $\hat{z}_j$'s are the exact position points, namely $\hat{z}_j = z_j$, $j = 1, 2, \ldots, l_r$, then

$$\sum_{j=1}^{l_r} e^{\mathrm{i}k(d-\hat{x})\cdot\hat{z}_j} A(\hat{x}; \widetilde{\Sigma}_j, k) = A(\hat{x}; \Omega^{(r)}, k).$$

This, together with (5.4.45), implies that $\widetilde{A}(\hat{x}; k)$ calculated according to (5.4.47) is an approximation to $A(\hat{x}; \Omega^{(s)}, k)$.

Next, we discuss an enhanced Scheme M by making use of two far-field measurements which could provide a more robust and accurate locating of the multi-scale scatterers $\Omega^{(m)}$. Indeed, we assume that in $\Omega^{(m)}$, the diameters of the multiple components of $\Omega^{(r)}$ are around d_1, whereas the diameters of the multiple components of $\Omega^{(s)}$ are around d_2 such that d_1/d_2 is relatively large. We choose two wave numbers k_1 and k_2 such that for $\lambda_1 = 2\pi/k_1$ and $\lambda_2 = 2\pi/k_2$, $\lambda_1 > d_1$ with $\lambda_1 \sim d_1$, and $d_2 < \lambda_2 < d_1$ with λ_2/d_2 relatively large. Then, in $A(\hat{x}; \Omega^{(m)}, k_1)$, according to (5.4.45) and (5.4.46), $A(\hat{x}; \Omega^{(r)}, k_1)$ is more significant and this will enable Scheme AR to have a more accurate locating of $\Omega^{(r)}$. On the other hand, according to (5.4.46), $A(\hat{x}; \Omega^{(m)}, k_2)$ clearly carries more scattering information of $\Omega^{(s)}$ than that in $A(\hat{x}; \Omega^{(m)}, k_1)$. Hence, after the locating of $\Omega^{(r)}$ by using $A(\hat{x}; \Omega^{(m)}, k_1)$, one can use $A(\hat{x}; \Omega^{(m)}, k_2)$ as the measurement data for the second stage in Scheme

M to yield a more accurate reconstruction of $\Omega^{(s)}$. In summary, the enhanced Scheme M by making use of two far-field measurements can be formulated as follows.

Algorithm: Enhanced Locating Scheme M

---

1. Collect two far-field measurements $A(\hat{x}; \Omega^{(m)}, k_1)$ and $A(\hat{x}; \Omega^{(m)}, k_2)$ corresponding to the multi-scale scatterer $\Omega^{(m)}$.
2. Use $A(\hat{x}; \Omega^{(m)}, k_1)$ as the measurement data for the first stage in Scheme M, namely Steps 2) and 3).
3. Use $A(\hat{x}; \Omega^{(m)}, k_2)$ as the measurement data for the second stage in Scheme M, namely Step 4).
4. Apply the local re-sampling technique following the next sub-steps of Step 4) in Scheme M to update $\widetilde{z}_j$'s and to locate the components of the small-size scatterer $\Omega^{(s)}$. Particularly, (5.4.47) is modified to be

$$\widetilde{A}(\hat{x}; k_2) = A(\hat{x}; \Omega^{(m)}, k_2) - \sum_{j=1}^{l_r} e^{ik_2(d-\hat{x})\cdot\widetilde{z}_j} A(\hat{x}; \widetilde{\Sigma}_j, k_2). \tag{5.4.48}$$

5.4.4 Numerical Experiments and Discussions

In this section, we present some numerical results to illustrate salient features of the schemes using augmented far field data set as well as its ability to image multiple multi-scale scatterers by the Scheme M with the local re-sampling technique.

Three geometries will be considered for the scatterer components in our numerical experiments. They are given by revolving bodies through rotating the following 2D shapes in the x-y plane around the x-axis

$$\begin{aligned}
\textbf{Circle}: &\ \{(x, y) : x = \cos(s),\ y = \sin(s),\ 0 \le s \le 2\pi\},\\
\textbf{Peanut}: &\ \{(x, y) : x = \sqrt{3\cos^2(s) + 1}\cos(s),\ y = \sqrt{3\cos^2(s) + 1}\sin(s),\ 0 \le s \le 2\pi\},\\
\textbf{Kite}: &\ \{(x, y) : x = \cos(s) + 0.65\cos(2s) - 0.65,\ y = 1.5\sin(s),\ 0 \le s \le 2\pi\}.
\end{aligned}$$

In the sequel, they are denoted by **B**, **P** and **K**, respectively, for short. The candidate data set $\mathscr{S}_h$ includes far field data of all three reference components **B**, **P** and **K**, and is further lexicographically augmented by a collection of a priori known orientations and sizes. More precisely, the augmented data set is obtained by rotating **P** and **K** in the x-y plane every $\pi/4$ radian[1] as shown in Figs. 5.9 and 5.10, respectively, and by scaling **B**, **P** and **K** by one fifth, one half, one, twice and five times.

In the examples below, as assumed earlier, we set $\varepsilon_0 = \mu_0 = 1$ and $\sigma_0 = 0$ outside the scatterer, and hence the wavelength is unitary in the homogeneous background.

---

[1] There are only four different orientations for **P** due to its symmetry.

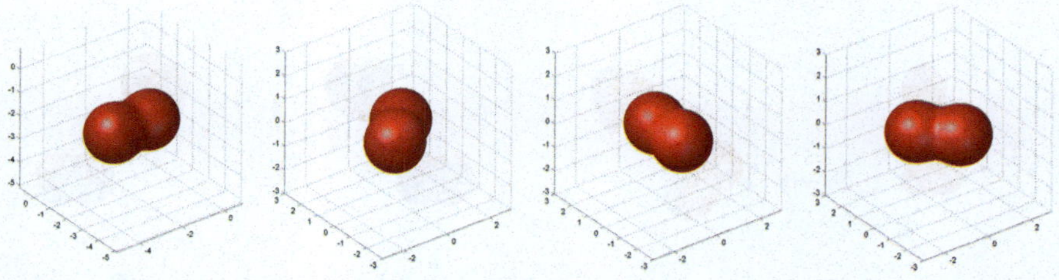

Fig. 5.9 Scatterer component Peanut with four orientations

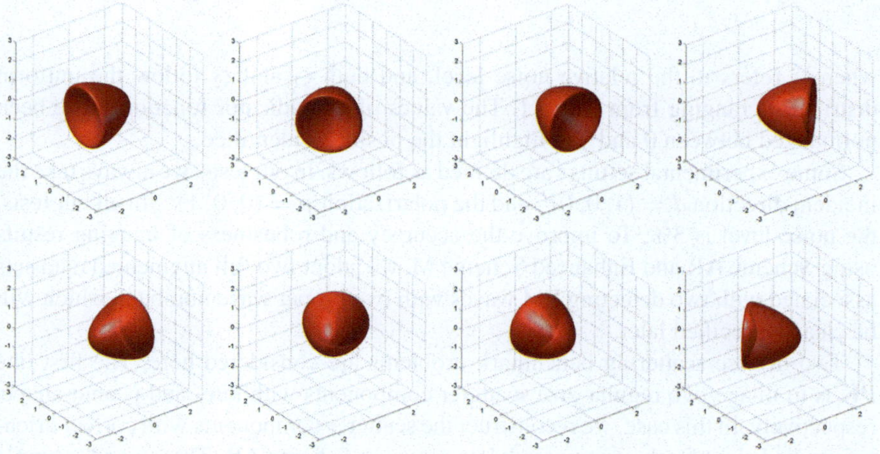

Fig. 5.10 Scatterer component Kite with eight orientations

Unless otherwise specified, all the scatterer components are either PEC conductors or inhomogeneous media with all other parameters the same as those in the homogeneous background except $\varepsilon = 4$. Our near-field data are obtained by solving the Maxwell system (4.3.1) using the quadratic $H(\text{curl})$-conforming edge element discretization in a spherical domain centered at the origin and holding inside all the scatterer components. The computational domain is enclosed by a PML layer to damp the reflection. Local adaptive refinement scheme within the inhomogeneous scatterer is adopted to enhance the accuracy of the scattered wave. The far-field data are approximated by the integral equation representation [45, p. 181, Theorem 3.1] using the spherical Lebedev quadrature (cf. [28]). We refine the mesh successively till the relative maximum error of successive groups of far-field data is below 0.1%. The far-field patterns on the finest mesh are used as the exact data. The electric far-field patterns $A(\hat{x}, \Omega)$, $\Omega = \Omega^{(r)}$ or $\Omega^{(m)}$, are observed at 590 Lebedev quadrature points distributed on the unit sphere $\mathbb{S}^2$ (cf. [28] and references therein). The exact far-field data $A(\hat{x}, \Omega)$ are corrupted point-wise by the formula

$$A_\delta(\hat{x}, \Omega) = A(\hat{x}, \Omega) + \delta\zeta_1 \max_{\hat{x}} |A(\hat{x}, \Omega)| \exp(\mathrm{i}2\pi\zeta_2), \qquad (5.4.49)$$

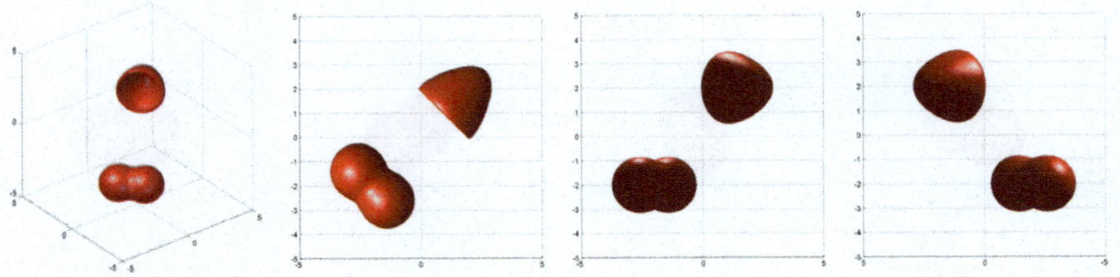

(a) (b) (c) (d)

Fig. 5.11 True scatterer for Example **PK**

where δ refers to the relative noise level, and both ζ_1 and ζ_2 follow the uniform distribution ranging from -1 to 1. The values of the indicator functions have been normalized between 0 and 1 to highlight the positions identified.

Some experimental settings are defined as follows. In our tests, we always take the incident direction $d = (1, 0, 0)^T$ and the polarization $p = (0, 0, 1)^T$. In all our tests, the noise level is 3%. To improve the accuracy and robustness of imaging results using Scheme AR and Enhanced Scheme M, we adopt two full augmented data sets associated with two detecting EM waves with two proper wave numbers, which will be clearly specified later.

Two inverse scattering benchmark problems are considered here. The first one **PK** is to image two regular-size scatterer components with kite- and peanut-shape, respectively. In this case, we reconstruct the scatterer components with correct orientations and sizes by the augmented data set using Scheme AR. The second example **KB** is to image a combined scatterer consisting of multiple multi-scale components, an enlarged kite of **K** by two times and a relatively small ball of **B** scaled to one half from the unit one. The size ratio between the two components is about six.

5.4.4.1 Scheme AR

Example PK

In this example, we try to locate with Scheme AR a kite component **K** located at $(2, 2, 2)$ with azimuthal angle $\pi/4$ radian, and a peanut component **P** located at $(-2, -2, -2)$ with azimuthal angle $3\pi/4$ radian as shown as in Fig. 5.11a and its projection on the $x - y$, $y - z$ and $z - x$ planes shown in Fig. 5.11b–d, respectively.

As remarked earlier, we choose the two scatterer components to be inhomogeneous media. There are two considerations for such a choice. First, we discussed Scheme AR in Sect. 5.4.2 mainly for locating PEC obstacles, but we also gave the extension to locate medium components if the generic situation described in Remark 5.4.3 is fulfilled. Second, we would like to illustrate the wide applicability of Scheme AR, and we refer to [29] for numerical results on recovering multiple

PEC obstacles by Scheme R. We implement Scheme AR in a two-stage imaging procedure as follows:

Scheme S

We first set $k = 1$, which amounts to sending a detecting EM wave of wavelength at least twice larger than each component of the scatterer. With the collected far-field data, we implement Scheme S to find how many components to be recovered and locate the rough positions of those scatterer components.

The imaging result at this coarse stage is shown in Fig. 5.12, indicated by the characteristic behavior of the function $I_s^j(z)$ (cf. (5.4.4)) in Scheme S. Note that no reference spaces are needed up to this stage. It can be observed that the indicator function achieves local maxima in the region where there exists a scatterer component, either kite or peanut. The rough position of the peanut is highlighted in Fig. 5.12a which indicate a possible scatterer component somewhere around the highlighted region. In Fig. 5.12b, we see that the rough position of the kite could also be found. But its dimer brightness as shown in Fig. 5.12b tells us that one cannot figure out its shape and size up to this stage.

Then we could incorporate the suspicious regions into a stack of cubes, as in Fig. 5.12c and d. And the computation of the next stage, i.e., Scheme AR, is just performed on these cubes, which are shown exclusively in Fig. 5.12e and f. It is emphasized that this preprocessing stage can be skipped and one can directly implement the Scheme AR as described in the next stage to locate the kite **K** and the peanut **P**. However, by performing this preprocessing stage, the computational costs can be significantly reduced, and the robustness and resolution can be enhanced for Scheme AR, as will be performed in the next stage.

Scheme AR

In this stage, we take $k = 5$. With the collected far-field data, we implement Scheme AR to determine the location, shape, orientation and size of each scatterer component.

When we use the far-field data of the reference peanut with $3\pi/4$ azimuthal angle and unitary scale as the test data in the indicator function $I_r^j(z)$ (cf. (5.4.13)), the distribution of the indicator function is shown in Fig. 5.13a. Then we take maximum of the indicator values and find a much precise location $(-2.1,\ -2.1,\ -2.1)$ of the peanut, as in Fig. 5.13b. Based on that position, we plot the proper shape, orientation and size based on the information carried with the far field data employed and plot the imaging result in Fig. 5.13c. Its projection on the orthogonal cut planes across its location are shown in Fig. 5.13d–f. It can be concluded that the position identified is quite good and reasonable.

After excluding the peanut component, we apply Scheme AR to the local mesh around the Kite component. When the far-field data of the reference kite with $\pi/4$ azimuthal angle and unitary scale is adopted in the indicator function $I_r^j(z)$ (cf. (5.4.13)), the value distribution of the indicator function is shown in Fig. 5.14a. Then we take maximum of the indicator values and find the the location $(2.2,\ 2.2,\ 2.2)$ of the kite, as in Fig. 5.14b. As previous, we plot the exact

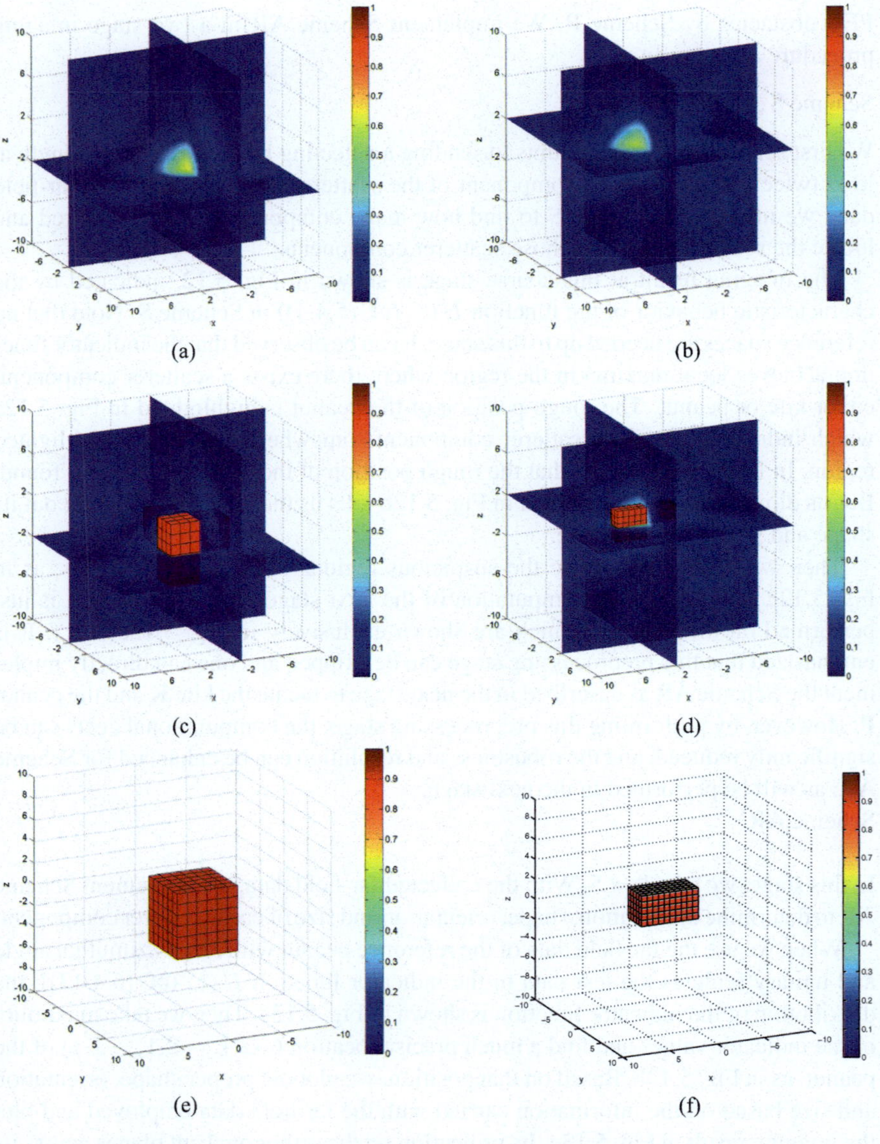

Fig. 5.12 Identification in the coarse/preprocessing stage in Example **PK**

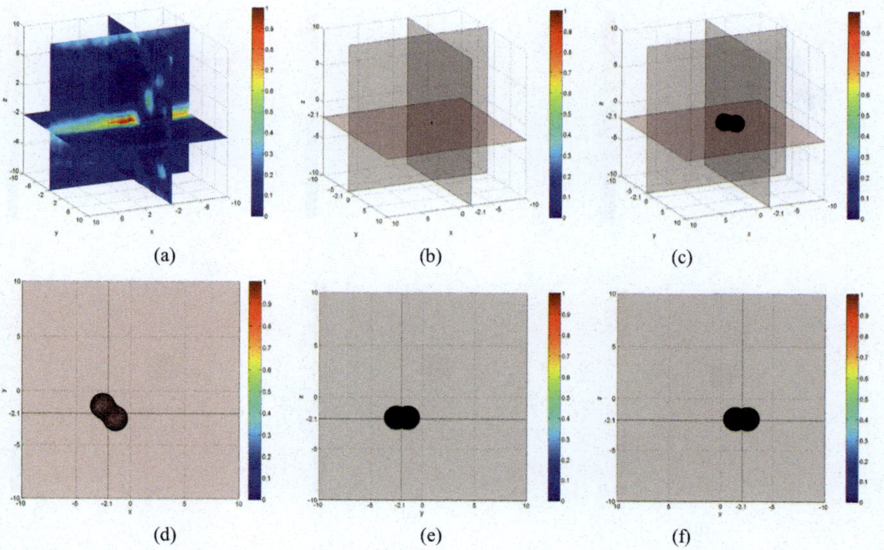

Fig. 5.13 Fine stage identification of the Peanut component in Example **PK**: **a** the multi-slice plot of the indicator function; **b** rough position by take maximum of indicator function; **c** the reconstructed component after the determination of the orientation of the peanut; **d–f** projections of the reconstruction in **c**

shape, orientation and size and show three orthogonal cut planes across the location identified in Fig. 5.14c–f. The identified location is very close to the exact position of the kite.

5.4.4.2 Enhanced Scheme M

Example KB

In this example we try to locate multiple multi-scale scattering components using Enhanced Scheme M. The exact scatterer is composed of a kite-shaped scatterer enlarged by two times from the reference one and a ball scatterer scaled by a half from the unit one. The kite is chosen to be a PEC obstacle, whereas the ball is an inhomogeneous medium. The exact scatterer is shown in Fig. 5.15, where the 3D kite-shaped component is located at $(0,\ 0,\ -4)$ and the ball component is located at $(0,\ 0,\ 9)$ with radius a half unit.

Now we employ Enhanced Scheme M to detect the unknown scatterers by applying Scheme AR first and then Scheme S with the local re-sampling technique. In the first stage of Scheme AR, the far-field data used are collected by illuminating the scatterer by an incident EM waves of $k = \pi$. In the second stage for Scheme S, the far-field data used are collected by illuminating the scatterers by an detecting EM waves of $k = 2\pi/5$. For $k = \pi$, we enrich our augmented reference space $\widetilde{\mathscr{S}}$ by the

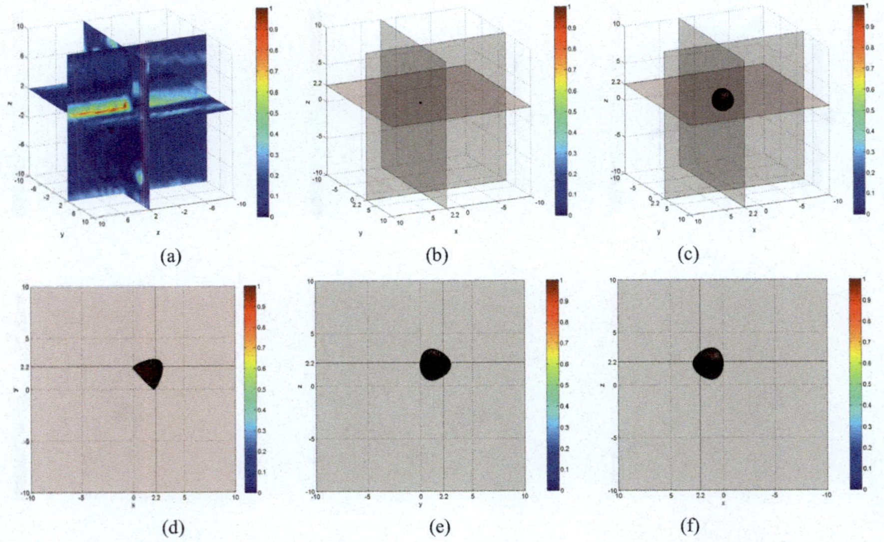

Fig. 5.14 Fine stage Identification of the Kite component in Example **PK**: **a** the multi-slice plot of the indicator function; **b** rough position by taking maximum of the indicator function; **c** the reconstructed component after the determination of the orientation and size of the kite; **d–f** projections of the reconstruction in **c**

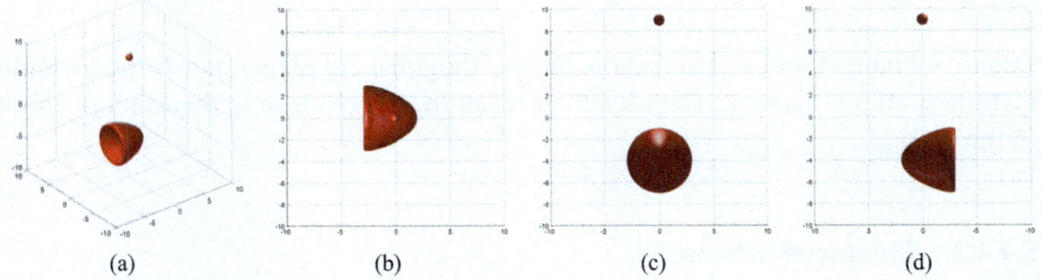

Fig. 5.15 True scatterer of Example **KB**

far-field data corresponding to each reference components with different orientations and sizes on 590 Lebedev quadrature points on the unit sphere.

It is remarked the order of Scheme AR and Scheme S cannot be interchanged. Scheme AR is first employed to detect regular components since the scattering amplitude due to the regular components is much larger than that from the small ones. After subtracting those far field data from the observed data, through the shifting formula applied to the reference data of identified regular components, the scattering amplitude of small components becomes significant among the remaining far field data. That's why it is appropriate to employ Scheme S at this stage.

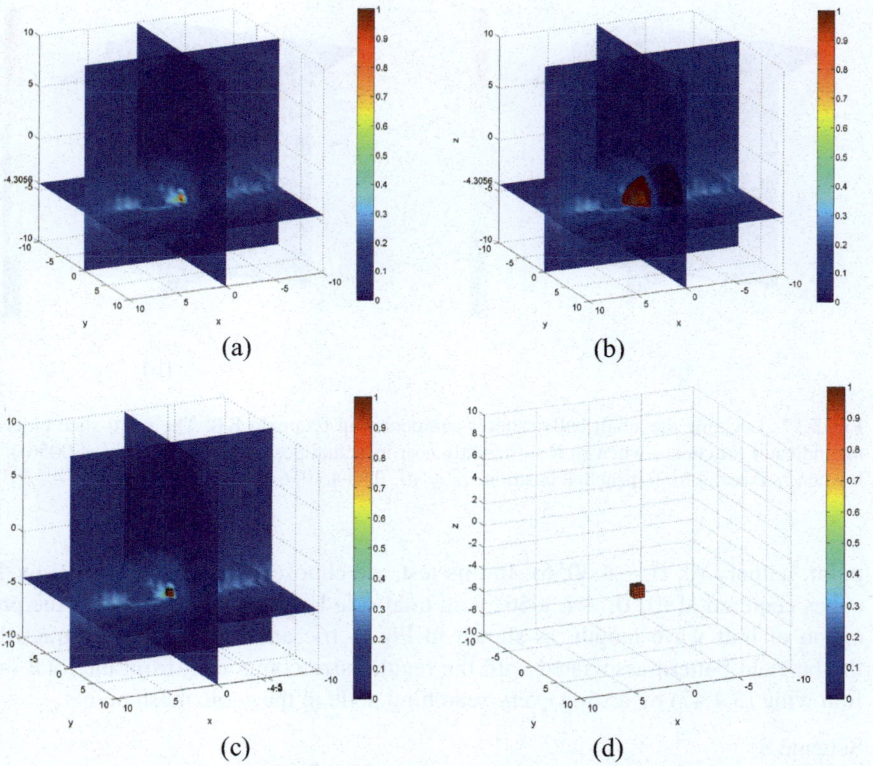

Fig. 5.16 Locating by Scheme AR in Example **KB**: **a** the multi-slice plot of the indicator function by Scheme AR; **b** the reconstructed component after the determination of the orientation and size of the kite; **c** a multi-slice plot with re-sampling cubes; **d** the isolated re-sampling cubes without the background multi-slide plot

Scheme AR

We first apply Scheme AR to the multi-scale scatterers. When the far-field data of the reference kite with vanishing azimuthal angle and double size is adopted in the indicator function $I_r^j(z)$ (cf. (5.4.13)), the local maximum behavior of the indicator function is shown in Fig. 5.16a. Using Scheme AR, we obtain a rough position of the kite component by taking the coordinates at which the indicator function achieves the maximum, namely $(0,\ 0,\ -4.3056)$ as shown in Fig. 5.16a. Its shape, orientation and size are superimposed by the message carried in the far-field data and plotted in Fig. 5.16b, where we reverse the x-axis for ease of visualization.

Local re-sampling technique

The detected position from Scheme AR in the previous step is an approximate position of the kite component due to the noise. In order to implement the local re-sampling technique, we set a local searching region around the obtained position

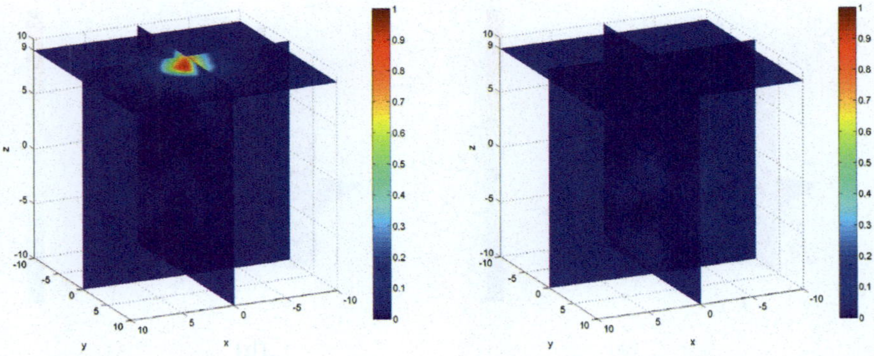

(a) (b)

Fig. 5.17 Locating the small ball scatterer component in Example **KB**. The multi-slice plots of the indicator function **a** when z_0 is sufficiently near its actual position ($z_0 = (0,\ 0,\ -4.0056)$), or **b** when z_0 is away from its actual position ($z_0 = (0,\ 0,\ -4.1056)$)

point, namely $(0,\ 0,\ -4.3056)$. In this test, we choose a stack of 10-by-10-by-10 cubes centered at $(0,\ 0,\ -4.3056)$ with total side length 1, namely within the precision of half wave length, as shown in Fig. 5.16c and d. Then we subtract the the far-field pattern associated with the regular-size component from the total one following (5.4.47) by testing every searching node in the cubic mesh points.

Scheme S

The rest of the job is to follow Step 4) in Enhanced Scheme M to test every suspicious points among the cubic grid points as shown in Fig. 5.16c. Figure 5.17 shows a gradual evolution process as we move gradually the sampling grid point from the nearly correct $z_0 = (0,\ 0,\ -4.0056)$ to a perturbed position $z_0 = (0,\ 0,\ -4.1056)$), which not only helps us update the position of the regular-size **K** component at $z_0 = (0,\ 0,\ -4.0056)$ but also determines the location of the small-size **B** component. From this example, we see that the identified position of the small ball component is no longer available if the position of the regular-size component is slightly perturbed. For the current test, the tolerance of the perturbation is within 0.05. Hence, a nice by-product from the local re-sampling technique is that it helps improve significantly the position of the regular-size component. The operation in this stage is essentially very cheap since only a few local grid points are involved and the re-sampling procedure only computes inner product of the subtracted far-field data with the test data in (5.4.4). Moreover, efficiency can be further improved by implementing the algorithm in parallel.

5.4.5 Discussions

In this section we present several variants of the one-shot method proposed in [29]. The methods can be used for the efficient numerical reconstruction of multiple multi-scale electromagnetic scatterers in a very general and practical setting. The methods are based on two types of imaging functionals in (5.4.4) and (5.4.13), respectively, for locating small-size and regular-size scatterers. A local re-sampling technique is of crucial importance to concatenate the two types of imaging functionals for locating multiple multi-scale EM scatterers.

The imaging functionals in (5.4.4) and (5.4.13) possess a similar flavor to the matched field imaging functional in finding the locations of multiple point sources (see, e.g., [13]). One of the salient features of the discussed imaging functionals is that they can be directly calculated by using the scattering data from a single far-field measurement.

The imaging functional (5.4.4) is the core for Scheme S of locating small scatterers. The amount of data used in the scheme is much small and one cannot expect to recover many details of the underlying small scatterer. It is known that frequency algorithms are efficient for detecting a conductivity inclusion with only one source and many receivers. In [25, 40], direct sampling methods is discussed for the inverse acoustic medium scattering problem by employing at most a few far-field measurements. Our method can also be extended to those important practical scenarios, but with many nontrivial and technical modifications.

Scheme R of locating regular-size scatterers is based on a series of imaging functionals in (5.4.13). For this scheme, one needs know in advance the possible shapes of the underlying scatterers. Nevertheless, Scheme R could find important applications, e.g. in radar technology as discussed at the end of Sect. 5.4.1.2. In Scheme R, we has also made use of a generic uniqueness result in (5.4.5) of determining a PEC obstacle by a single far-field measurement. It is remarked that such a unique determination result is mainly of theoretical interest. In fact, in [4] for a realistic noise model, it has been illustrated that high-frequency approximations do not provide any information in the shadow region of a scatterer and only the illuminated region can be reconstructed. This also justifies our qualitative reconstruction by using Scheme R. Nevertheless, we believe that if more measurement data could be incorporated into the Scheme R, then one could extract more information of the underlying scatterer. It is an interesting direction for further investigation. We mention in passing the other promising imaging schemes of recovering regular-size scatterers in [4] of using many measurements.

Our remark of incorporating more measurement data into Scheme R to extract more information of the regular-size scatterers also applies to Scheme S of finding small EM scatterers. In [8], resolved imaging (namely, recovering both the shape and EM parameters) on a small EM target has been obtained from the full multi-static response matrix. The configuration of the measurement data in our Scheme S corresponds to having only one column of the response matrix in [8]. It is worthwhile to investigate the connection between our Scheme S and the resolved imaging scheme in [8] of recovering small scatterers. Finally, we would like to mention the other promising schemes of imaging small scatterers by employing many measurements, including the Kirchhoff back-propagation, MUSIC, and reverse time migration among others; see [2–4] and the references therein.

References

1. R. Aramini, M. Brignone, M. Piana, The linear sampling method without sampling. Inverse Prob. **22**, 2237–2254 (2006)
2. H. Ammari, J. Garnier, H. Kang, W.K. Park, K. Solna, Imaging schemes for perfectly conducting cracks. SIAM J. Appl. Math. **71**(1), 68–91 (2011)
3. H. Ammari, J. Garnier, V. Jugnon, H. Kang, Stability and resolution analysis for a topological derivative based imaging functional. SIAM J. Control. Optim. **50**(1), 48–76 (2012)
4. H. Ammari, J. Garnier, H. Kang, M. Lim, K. Solna, Multistatic imaging of extended targets. SIAM J. Imag. Sci. **5**(2), 564–600 (2012)
5. H. Ammari, E. Iakovleva, D. Lesselier, A MUSIC algorithm for locating small inclusions buried in a half-space from the scattering amplitude at a fixed frequency. Multiscale Model. Simul. **3**, 597–628 (2005)
6. H. Ammari, E. Iakovleva, D. Lesselier, G. Perrusson, MUSIC-type electromagnetic imaging of a collection of small three-dimensional inclusions. SIAM J. Sci. Comput. **29**, 674–709 (2007)
7. H. Ammari, H. Kang, *Polarization and Moment Tensors (With Applications to Inverse Problems And Effective Medium Theory*. Applied Mathematical Sciences (Springer, New York, 2007)
8. H. Ammari, H. Kang, E. Kim, J. Lee, The generalized polarization tensors for resolved imaging. Part II: Shape and electromagnetic parameters reconstruction of an electromagnetic inclusion from multistatic measurements. Math. Comput. **81**, 839–860 (2012)
9. H. Ammari, J.-C. Nédélec, Low-frequency electromagnetic scattering. SIAM J. Math. Anal. **31**, 836–861 (2000)
10. H. Ammari, M. Vogelius and D. Volkov, Asymptotic formulas for per-turbations in the electromagnetic fields due to the presence of inhomogeneities of small diameter. II. The full Maxwell equations. Journal de Mathematiques Pures et Appliquees, (9) **80**, 769–814 (2001)
11. X. Antoine, B. Pincon, K. Ramdani, B. Thierry, Far field modeling of electromagnetic time reversal and application to selective focusing on small scatterers. SIAM J. Appl. Math. **69**, 830–844 (2008)
12. E. Blåsten, H.Y. Liu, J. Xiao, Jingni, On an electromagnetic problem in a corner and its applications. Anal. PDE **14**(7), 2207–2224 (2021)
13. J.H. Bramble, J.E. Pasciak, D. Trenev, Analysis of a finite PML approximation to the three dimensional elastic wave scattering problem. Math. Comput. **79**, 2079–2101 (2010)
14. F. Cakoni, D. Colton, *Qualitative Methods in Inverse Scattering Theory: An Introduction* (Springer, 2005)
15. F. Cakoni, D. Colton, P. Monk, *The Linear Sampling Method in Inverse Electromagnetic Scattering* (SIAM, Philadelphia, 2011)
16. X. Cao, H. Diao, H.Y. Liu, J. Zou, Two single-measurement uniqueness results for inverse scattering problems within polyhedral geometries. Inverse Prob. Imaging **16**(6), 1501–1528 (2022)
17. X. Chen, Y. Zhong, MUSIC electromagnetic imaging with enhanced resolution for small inclusions. Inverse Prob. **25**, 015008 (2009)
18. D. Colton, J. Coyle, P. Monk, Recent developments in inverse acoustic scattering theory. SIAM Rev. **42**, 369–414 (2000)
19. D. Colton, A. Kirsch, A simple method for solving inverse scattering problems in the resonance region. Inverse Prob. **12**, 383–393 (1996)

20. D. Colton, R. Kress, *Inverse Acoustic and Electromagnetic Scattering Theory*, 2nd edn. (Springer, New York, 1998)
21. G. Dassios, R. Kleinman, *Low Frequency Scattering* (Clarendon Press, Oxford, 2000)
22. H. Diao, X. Cao, H. Liu, On the geometric structures of transmission eigenfunctions with a conductive boundary condition and applications. Commun. Partial Differ. Equ. **46**(4), 630–679 (2021)
23. H. Diao, H.Y. Liu, L. Zhang, J. Zou, Unique continuation from a generalized impedance edge-corner for Maxwell's system and applications to inverse problems. Inverse Prob. **37**(3), 035004 (2021)
24. M. Ikehata, Reconstruction of obstacles from boundary measurements. Wave Motion **3**, 205–223 (1999)
25. K. Ito, B. Jin, J. Zou, A direct sampling method to an inverse medium scattering problem. Inverse Prob. **28**, 025003 (2012)
26. V. Isakov, *Inverse Problems for Partial Differential Equations*, Applied Mathematical Sciences, vol. 127, 2nd edn. (Springer, New York, 2006) Applied Mathematical Sciences, 127. (Springer-Verlag, New York, 2006)
27. A. Kirsch, N. Grinberg, *The Factorization Method for Inverse Problems* (Oxford University Press, Oxford, 2008)
28. V.I. Lebedev, D.N. Laikov, A quadrature formula for the sphere of the 131st algebraic order of accuracy. Dokl. Math. **59**, 477–481 (1999)
29. J. Li, H.Y. Liu, Z. Shang, H. Sun, Two single-shot methods for locating multiple electromagnetic scatterers. SIAM J. Appl. Math. **73**, 1721–1746 (2013)
30. J. Li, H.Y. Liu, Q. Wang, Locating multiple multiscale electromagnetic scatterers by a single far-field measurement. SIAM J. Imag. Sci. **6**(4), 2285–2309 (2013)
31. J. Li, H. Y. Liu, Y. Wang, Recovering an electromagnetic obstacle by a few phaseless backscattering measurements. Inverse Prob. **33**(3), 035011 (2017)
32. J. Li, H.Y. Liu, J. Zou, Strengthened linear sampling method with a reference ball. SIAM J. Sci. Comput. **31**, 4013–4040 (2009)
33. H.Y. Liu, A global uniqueness for formally determined inverse electromagnetic obstacle scattering. Inverse Prob. **24**, 035018 (2008)
34. H.Y. Liu, On local and global structures of transmission eigenfunctions and beyond. J. Inverse Ill-Posed Prob. **30**(2), 287–305 (2022)
35. H.Y. Liu, L. Rondi, J. Xiao, Mosco convergence for H(curl) spaces, higher integrability for Maxwell's equations, and stability in direct and inverse EM scattering problems. J. Eur. Math. Soc. **21**(10), 2945–2993 (2019)
36. H.Y. Liu, J. Xiao, On electromagnetic scattering from a penetrable corner. SIAM J. Math. Anal. **49**(6), 5207–5241 (2017)
37. H.Y. Liu, M. Yamamoto, J. Zou, Reflection principle for Maxwell's equations and its application to inverse electromagnetic scattering. Inverse Prob. **23**, 2357–2366 (2007)
38. H.Y. Liu, M. Yamamoto, J. Zou, New reflection principles for Maxwell's equations and their applications. Numer. Math.-Theory Methods Appl. **2**(1), 1–17 (2009)
39. H.Y. Liu, H. Zhang, J. Zou, Recovery of polyhedral scatterers by a single electromagnetic far-field measurement. J. Math. Phys. **50**(12), 123506 (2009)
40. K. Liu, J. Zou, A multilevel sampling algorithm for locating inhomogeneous media. Inverse Prob. **29**, 095003 (2013)
41. H.Y. Liu, J. Zou, On uniqueness in inverse acoustic and electromagnetic obstacle scattering problems. J. Phys: Conf. Ser. **124**, 012006 (2008)
42. H.Y. Liu, T. Zhou, On approximate electromagnetic cloaking by transformation media. SIAM J. Appl. Math. **71**, 218–241 (2011)
43. P.A. Martin, *Multiple Scattering: Interaction of Time-Harmonic Waves with N Obstacles* (Cambridge University Press, Cambridge, 2006)
44. J.-C. Nédélec, *Acoustic and Electromagnetic Equations: Integral Representations for Harmonic Problems* (Springer, New York, 2001)

45. R. Pike, P. Sabatier, (eds.), *Scattering : Scattering and Inverse Scattering in Pure and Applied Science* (Academic, 2002)
46. R. Potthast, A survey on sampling and probe methods for inverse problems. Inverse Prob. **22**(2), R1–R47 (2006)
47. R. Song, Y. Zhong, X. Chen, A multi-dimensional sampling method for locating small scatterers. Inverse Prob. **28**, 115004 (2012)
48. G. Uhlmann, (ed.), *Inside Out: Inverse Problems and Applications*, MSRI Publications, vol. 47 (Cambridge University Press, 2003)
49. H. Zhao, Analysis of the response matrix for an extended target. SIAM J. Appl. Math. **64**, 725–745 (2004)

Chapter 6
Direct Elastic Scattering Problems

We first introduce the Lamé system that governs the elastic wave propagation in $\mathbb{R}^n$, $n = 2, 3$. Throughout, we let $\mathscr{C}$ and ρ signify the constitutive material parameters of an elastic medium. Here, $\mathscr{C}(\mathbf{x}) = (\mathscr{C}_{ijkl}(\mathbf{x}))_{i,j,k,l=1}^n$ is a four-rank real-valued tensor satisfying the following symmetry property:

$$\mathscr{C}_{ijkl} = \mathscr{C}_{klij} \quad \text{and} \quad \mathscr{C}_{ijkl} = \mathscr{C}_{jikl} = \mathscr{C}_{ijlk}, \quad i, j, k, l = 1, 2, \ldots, n. \tag{6.0.1}$$

$\rho(\mathbf{x})$ is a bounded measurable complex-valued function with $\Re\rho > 0$ and $\Im\rho \geq 0$. Physically, $\mathscr{C}$ signifies the stiffness tensor, and $\Re\rho$ and $\Im\rho$ characterize the density and damping of an elastic medium, respectively. Let $\mathbf{u}(\mathbf{x}) = (u_j(\mathbf{x}))_{j=1}^n \in \mathbb{C}^n$ denote the displacement field in the elastic medium. In linear elasticity, one has the following Lamé system:

$$\mathscr{L}_{\mathscr{C}}\mathbf{u} + \omega^2\rho\mathbf{u} = \mathbf{0}, \quad \mathscr{L}_{\mathscr{C}}\mathbf{u} := \nabla\cdot(\mathscr{C} : \nabla\mathbf{u}) = \left(\sum_{j,k,l=1}^n \partial_j(\mathscr{C}_{ijkl}\partial_l u_k)\right)_{i=1}^n, \tag{6.0.2}$$

where $\omega \in \mathbb{R}_+$ signifies the angular frequency and $\mathscr{L}_{\mathscr{C}}$ is referred to as the Lamé operator associated with $\mathscr{C}$. In (6.0.2), the symbol ":" indicates an action of double contraction, which is defined for two matrices $\mathbf{A} = (a_{ij})_{i,j=1}^n$ and $\mathbf{B} = (b_{ij})_{i,j=1}^n$:

$$\mathbf{A} : \mathbf{B} = \sum_{i,j=1}^n a_{ij}b_{ij} \quad \text{and} \quad \mathscr{C} : \mathbf{A} = (\mathscr{C} : \mathbf{A})_{ij} = \left(\sum_{k,l=1}^n \mathscr{C}_{ijkl}a_{kl}\right).$$

Throughout we assume that the elastic tensor $\mathscr{C}$ satisfies the uniform Legendre ellipticity condition:

$$c_{\min}\|\boldsymbol{\xi}\|_2^2 \leq \boldsymbol{\xi} : \mathscr{C} : \boldsymbol{\xi}^* \leq c_{\max}\|\boldsymbol{\xi}\|_2^2, \quad \forall\boldsymbol{\xi} \in \mathbb{C}^{n\times n} \text{ being a symmetric matrix,} \tag{6.0.3}$$

J. Li and H. Liu, *Numerical Methods for Inverse Scattering Problems*,
https://doi.org/10.1007/978-981-99-3772-1_6

where $c_{\min}$ and $c_{\max}$ are two positive constants. If there exist scalar real functions $\lambda(\mathbf{x})$ and $\mu(\mathbf{x})$ such that

$$\mathscr{C}_{ijkl} = \lambda\delta_{ij}\delta_{kl} + \mu(\delta_{ik}\delta_{jl} + \delta_{il}\delta_{jk}), \tag{6.0.4}$$

where δ is the Kronecker delta function, then the elastic medium is said to be isotropic, otherwise it is said anisotropic.

6.1 Elastic Scattering from Obstacles

Consider a time-harmonic elastic plane wave $\mathbf{u}^{in}(x)$, $x \in \mathbb{R}^3$ (with the time variation of the form $e^{-i\omega t}$ being factorized out, where $\omega \in \mathbb{R}_+$ denotes the angular frequency) impinged on a rigid scatterer $D \subset \mathbb{R}^3$ embedded in an infinite isotropic and homogeneous elastic medium in $\mathbb{R}^3$. The elastic scattering is governed by the reduced Navier equation (or Lamé system)

$$(\Delta^* + \omega^2)\,\mathbf{u} = 0, \quad \text{in} \quad \mathbb{R}^3\backslash\overline{D}, \quad \Delta^* := \mu\Delta + (\lambda+\mu)\,\nabla\text{div} \tag{6.1.1}$$

where $\mathbf{u}(x) \in \mathbb{C}^3$ denotes the total displacement field, and λ, μ are the Lamé constants satisfying $\mu > 0$ and $3\lambda + 2\mu > 0$. Here, we note that the density of the background elastic medium has been normalized to be unitary. Henceforth, we suppose that $D \subset \mathbb{R}^3$ is an open bounded domain such that $\mathbb{R}^3\backslash\overline{D}$ is connected. It is emphasized that D may consist of (finitely many) multiple simply connected components. The incident elastic plane wave is of the following general form

$$\mathbf{u}^{in}(x) = \mathbf{u}^{in}(x; d, d^{\perp}, \alpha, \beta, \omega) = \alpha\, d e^{\mathrm{i}k_p x\cdot d} + \beta d^{\perp} e^{\mathrm{i}k_s x\cdot d}, \quad \alpha, \beta \in \mathbb{C}, \tag{6.1.2}$$

where $d \in \mathbb{S}^2 := \{x \in \mathbb{R}^3 : |x| = 1\}$, is the impinging direction, $d^{\perp} \in \mathbb{S}^2$ satisfying $d^{\perp}\cdot d = 0$ is the polarization direction; and $k_s := \omega/\sqrt{\mu}$, $k_p := \omega/\sqrt{\lambda + 2\mu}$ denote the shear and compressional wave numbers, respectively. If $\alpha = 1, \beta = 0$ for u^{in} in (6.1.2), then $\mathbf{u}^{in} = \mathbf{u}_p^{in} := de^{ik_p x\cdot d}$ is the (normalized) plane pressure wave; whereas if $\alpha = 0$, $\beta = 1$ for u^{in} in (6.1.2), then $\mathbf{u}^{in} = \mathbf{u}_s^{in} := d^{\perp}e^{ik_s x\cdot d}$ is the (normalized) plane shear wave. The obstacle D is a rigid body and $\mathbf{u}$ satisfies the first kind (Dirichlet) boundary condition

$$\mathbf{u} = 0 \quad \text{on} \quad \partial D. \tag{6.1.3}$$

Define $\mathbf{u}^{sc} := \mathbf{u} - \mathbf{u}^{in}$ to be the *scattered wave*, which can be easily verified to satisfy the Navier equation (6.1.1) as well. $\mathbf{u}^{sc}$ can be decomposed into the sum

$$\mathbf{u}^{sc} := \mathbf{u}_p^{sc} + \mathbf{u}_s^{sc}, \quad \mathbf{u}_p^{sc} := -\frac{1}{k_p^2}\nabla\text{div}\,\mathbf{u}^{sc}, \quad \mathbf{u}_s^{sc} := \frac{1}{k_s^2}\nabla\times\nabla\times\mathbf{u}^{sc},$$

where the vector functions $\mathbf{u}_p^{sc}$ and $\mathbf{u}_s^{sc}$ are referred to as the pressure (longitudinal) and shear (transversal) parts of $\mathbf{u}^{sc}$, respectively, satisfying

$$(\Delta + k_p^2)\,\mathbf{u}_p^{sc} = 0, \quad \nabla \times \mathbf{u}_p^{sc} = 0, \quad \text{in } \mathbb{R}^3\backslash\overline{D},$$
$$(\Delta + k_s^2)\,\mathbf{u}_s^{sc} = 0, \quad \operatorname{div}\mathbf{u}_s^{sc} = 0, \quad \text{in } \mathbb{R}^3\backslash\overline{D}.$$

Moreover, the scattered field u^{sc} is required to satisfy Kupradze's radiation condition

$$\lim_{r\to\infty}\left(\frac{\partial \mathbf{u}_p^{sc}}{\partial r} - \mathrm{i}k_p\mathbf{u}_p^{sc}\right) = 0, \quad \lim_{r\to\infty}\left(\frac{\partial \mathbf{u}_s^{sc}}{\partial r} - \mathrm{i}k_s\mathbf{u}_s^{sc}\right) = 0, \quad r = |x|, \quad (6.1.4)$$

uniformly in all directions $\hat{x} = x/|x| \in \mathbb{S}^2$ (see, e.g., [1]). The radiation conditions in (6.1.4) lead to the P-part (longitudinal part) $\mathbf{u}_p^{\infty}$ and the S-part (transversal part) $\mathbf{u}_s^{\infty}$ of the far-field pattern of $\mathbf{u}^{sc}$.

The direct elastic scattering problem (DP) is stated as follows

(DP): Given a rigid scatterer $D \subset \mathbb{R}^3$ and an incident plane wave $\mathbf{u}^{in}$ of the form (6.1.2), find the total field $\mathbf{u} = \mathbf{u}^{in} + \mathbf{u}^{sc}$ in $\mathbb{R}^3\backslash\overline{D}$ such that the Dirichlet boundary condition (6.1.3) holds on ∂D and that the scattered field $\mathbf{u}^{sc}$ satisfies Kupradze's radiation condition (6.1.4).

We refer to the monograph [8] for a comprehensive treatment of the boundary value problems of elasticity. It is well-known that the direct scattering problem admits a unique solution $\mathbf{u} \in C^2(\mathbb{R}^3\backslash\overline{D})^3 \cap C^1(\mathbb{R}^3\backslash D)^3$ if ∂D is C^2-smooth (see [8]), whereas $\mathbf{u} \in H^1_{loc}(\mathbb{R}^3\backslash\overline{D})^3$ if ∂D is Lipschitz (see [2]).

6.2 Elastic Scattering from Mediums

In this section we consider the elastic scattering from an inhomogeneous medium in $\mathbb{R}^n$, $(n = 2, 3)$. We assume that the inhomogeneity is compactly supported. Let Ω be a bounded C^2 domain in $\mathbb{R}^n$ which supports an inhomogeneous isotropic medium characterized by $\rho(x)$. We take the detecting/incident wave field to be a plane waves,

$$\mathbf{u}^{in}(x) = \mathbf{u}^{in}(x; d, d^{\perp}, \alpha, \beta, \omega) = \alpha\, d e^{\mathrm{i}k_p x\cdot d} + \beta d^{\perp} e^{\mathrm{i}k_s x\cdot d}, \quad \alpha, \beta \in \mathbb{C}, \quad (6.2.1)$$

where $d \in \mathbb{S}^2 := \{x \in \mathbb{R}^3 : |x| = 1\}$, is the impinging direction, $d^{\perp} \in \mathbb{S}^2$ satisfying $d^{\perp}\cdot d = 0$ denotes the polarization direction; and $k_s := \omega/\sqrt{\mu}$, $k_p := \omega/\sqrt{\lambda + 2\mu}$ denote the shear and compressional wave numbers, respectively. If $\alpha = 1, \beta = 0$ for $\mathbf{u}^{in}$ in (6.2.1), then $\mathbf{u}^{in} = \mathbf{u}_p^{in} := d e^{ik_p x\cdot d}$ is the (normalized) plane pressure wave; and if $\alpha = 0, \beta = 1$ for $\mathbf{u}^{in}$ in (6.2.1), then $\mathbf{u}^{in} = \mathbf{u}_s^{in} := d^{\perp} e^{ik_s x\cdot d}$ is the (normalized) plane shear wave.

The direct elastic scattering problem is described as follows: given λ, μ, ω and an incident wave u^{in} satisfying

$$(\Delta^* + \omega^2)\mathbf{u} = 0, \quad \text{in } \mathbb{R}^3, \quad \Delta^* := \mu\Delta + (\lambda + \mu)\nabla\text{div} \tag{6.2.2}$$

find a scattered field $\mathbf{u}^s$, such that the total field $\mathbf{u} = \mathbf{u}^{in} + \mathbf{u}^s$ such that

$$(\Delta^* + \omega^2\rho)\,\mathbf{u} = 0, \quad \text{in } \mathbb{R}^3. \tag{6.2.3}$$

Moreover, $\mathbf{u}^s$ satisfies a rediation condition

$$\lim_{r\to\infty}\left(\frac{\partial \mathbf{u}_p^{sc}}{\partial r} - ik_p\mathbf{u}_p^{sc}\right) = 0, \quad \lim_{r\to\infty}\left(\frac{\partial \mathbf{u}_s^{sc}}{\partial r} - ik_s\mathbf{u}_s^{sc}\right) = 0, \quad r = |x|. \tag{6.2.4}$$

6.3 Elastic Scattering from Complex Scatterers

In this section, we consider the time-harmonic elastic wave scattering from a general inhomogeneous medium with an embedded impenetrable obstacle.

We consider the medium compactly supported in a bounded domain Ω in $\mathbb{R}^n$, ($n = 2, 3$) and let $D \Subset \Omega$ be an impenetrable obstacle. To that end, we note that physically, D represents a traction-free impenetrable obstacle embedded in the elastic medium $(\Omega\backslash\overline{D}; \mathscr{C}, \rho)$. In what follows, we set $D \oplus (\Omega\backslash\overline{D}; \mathscr{C}, \rho)$ to signify such an elastic object as described above. Let λ_e, μ_e and ρ_e be real constants satisfying the strong convexity condition (induced by the ellipticity condition (6.0.3)):

$$\mu_e > 0, \quad n\lambda_e + 2\mu_e > 0 \quad \text{and} \quad \rho_e > 0. \tag{6.3.1}$$

Let $\mathscr{C}^e$ be an isotropic elastic tensor as defined in (6.0.4) with $\lambda = \lambda_e$ and $\mu = \mu_e$. Let $(\mathscr{C}, \rho)$ be extended into $\mathbb{R}^n\backslash\overline{\Omega}$ such that $(\mathscr{C}, \rho) = (\mathscr{C}^e, \rho_e)$ in $\mathbb{R}^n\backslash\overline{\Omega}$. Let $\mathbf{u}^{in}$ be an entire solution to the following Lamé system:

$$\mu_e\Delta\mathbf{u}^{in} + (\lambda_e + \mu_e)\nabla(\nabla\cdot\mathbf{u}^{in}) + \omega^2\rho_e\mathbf{u}^{in} = \mathbf{0}. \tag{6.3.2}$$

Consider the following elastic scattering system:

$$\begin{cases} \mathscr{L}_{\mathscr{C}}\mathbf{u} + \omega^2\rho\mathbf{u} = \mathbf{f} & \text{in } \mathbb{R}^n\backslash\overline{D}, \\ \mathbf{u} = \mathbf{u}^{in} + \mathbf{u}^s & \text{in } \mathbb{R}^n\backslash\overline{\Omega}, \\ \mathscr{B}(\mathbf{u}) = \mathbf{0} & \text{on } \partial D, \\ \boldsymbol{u}\big|_{\partial\Omega} = \boldsymbol{u}^s\big|_{\partial\Omega} + \boldsymbol{u}^{in}, \quad \mathscr{T}_\nu(\boldsymbol{u}) = \mathscr{T}_\nu(\boldsymbol{u}^s) + \mathscr{T}_\nu(\boldsymbol{u}^{in}) & \text{on } \partial\Omega, \\ \boldsymbol{u}^{\mathrm{p},s} = -\frac{1}{k_{\mathrm{p}}^2}\nabla(\nabla\cdot\boldsymbol{u}^s), \quad \boldsymbol{u}^{\mathrm{s},s} = \frac{1}{k_{\mathrm{s}}^2}\nabla\times(\nabla\times\boldsymbol{u}^s) & \text{in } \mathbb{R}^n\backslash\overline{\Omega}, \\ \lim_{|\mathbf{x}|\to\infty}|\mathbf{x}|^{(n-1)/2}\left(\frac{\partial\boldsymbol{u}^{\mathrm{t},s}}{\partial|\mathbf{x}|} - \imath\kappa_{\mathrm{t}}\boldsymbol{u}^{\mathrm{t},s}\right) = \mathbf{0}, & \mathrm{t} = \mathrm{p}, \mathrm{s}, \end{cases} \tag{6.3.3}$$

where $\mathscr{T}_\nu(\mathbf{u}) := \boldsymbol{\nu} \cdot (\mathscr{C} : \nabla \mathbf{u})$ with $\boldsymbol{\nu} \in \mathbb{S}^{n-1}$ signifying the exterior unit normal vector to the boundary and $\mathbf{f}(\mathbf{x})$ indicates a source and is compactly supported outside Ω, namely $\mathrm{supp}(\mathbf{f}) \subset B_{r_0} \backslash \overline{\Omega}$ for some ball B_{r_0} with center at the origin and a radius of r_0. $\iota := \sqrt{-1}$, $\kappa_s := \omega\sqrt{1/\mu_e}$ and $\kappa_p := \omega\sqrt{1/(\lambda_e + 2\mu_e)}$, and $\mathscr{B}(\mathbf{u}) = \mathbf{u}$ or $\mathscr{B}(\mathbf{u}) = \mathscr{T}_\nu(\mathbf{u})$ correspond, respectively, to the cases that D is rigid or traction-free. The system (6.3.3) describes the time-harmonic scattering due to an incident field $\mathbf{u}^{in}$ and the scatter $D \oplus (\Omega \backslash \overline{D}; \mathscr{C}, \rho)$. $\mathbf{u}^s$ is referred to as the scattered field, which characterizes the perturbation of the propagation of the incident field due to the presence of the inhomogeneous scatterer. $\mathbf{u}^{p,s}$ and $\mathbf{u}^{s,s}$ are the compressional and shear parts of $\mathbf{u}^s$, respectively. The last limit in (6.3.3) is known as the Kupradze radiation condition, which holds uniformly in the angular variable $\hat{\mathbf{x}} := \mathbf{x}/|\mathbf{x}| \in \mathbb{S}^{n-1}$.

The next lemma, which can be named as Rellich's lemma in the linear elasticity, can be proved by generalizing the arguments in [7].

Lemma 6.3.1 *Let B_r be an appropriate ball centered at origin with a radius $r \in \mathbb{R}_+$, and assume that $\boldsymbol{u}^s$ is a radiating solution to*

$$\mu \Delta \boldsymbol{u}^s + (\lambda + \mu)\nabla(\nabla \cdot \boldsymbol{u}^s) + \omega^2 \rho \boldsymbol{u}^s = \mathbf{0}, \quad \mu > 0, \quad n\lambda + 2\mu > 0, \quad \rho > 0$$

in $|\mathbf{x}| \geq r$. If

$$\Im\Big(\int_{\partial B_r} \mathscr{T}_\nu \boldsymbol{u}^s \cdot \overline{\boldsymbol{u}^s} \mathrm{d}s\Big) \leq 0, \tag{6.3.4}$$

then $\boldsymbol{u}^s = \mathbf{0}$ in $|\mathbf{x}| \geq r$.

The well-posedness of the scattering problem (6.3.3) in such a general scenario is provided as follows.

Proposition 6.3.1 *There exits a unique solution $\boldsymbol{u} \in H^1(\mathbb{R}^n \backslash \overline{D})^n$ to the scattering problem* (6.3.3). *Furthermore, it holds that*

$$\|\boldsymbol{u}\|_{H^1(\mathbb{R}^n \backslash \overline{D})^n} \leq C\Big(\|\boldsymbol{u}^{in}\|_{H^{1/2}(\partial\Omega)^n} + \|\mathscr{T}_\nu(\boldsymbol{u}^{in})\|_{H^{-1/2}(\partial\Omega)^n} + \|\mathbf{f}\|_{L^2(B_{r_0} \backslash \overline{\Omega})^n}\Big), \tag{6.3.5}$$

where C is a positive constant, $\Omega \Subset B_{r_0}$ and B_{r_0} is a ball centered at the origin with the radius $r_0 \in \mathbb{R}_+$.

Proof By using an appropriate truncation we can truncate the unbounded domain $\mathbb{R}^n \backslash \overline{D}$ in (6.3.3) into a bounded one. In what follows, we let B_r signify a central ball of radius r containing Ω and consider the following PDE system: Find $\boldsymbol{u} \in H^1(B_r \backslash \overline{D})^n$ such that

$$\begin{cases} \mathscr{L}_{\mathscr{C}}\boldsymbol{u} + \omega^2\rho(\mathbf{x})\boldsymbol{u} = \mathbf{0} & \text{in } \Omega\backslash\overline{D}, \\ \mathscr{L}_{\mathscr{C}^e}\boldsymbol{u}^s + \omega^2\rho_e\boldsymbol{u}^s = \mathbf{f} & \text{in } \mathbb{R}^n\backslash\overline{\Omega}, \\ \boldsymbol{u}^s = \boldsymbol{u}^{\mathrm{p},s} + \boldsymbol{u}^{\mathrm{s},s} & \text{in } B_r\backslash\overline{\Omega}, \\ \mathscr{B}(\mathbf{u}) = \mathbf{0} & \text{on } \partial D, \\ \boldsymbol{u}\big|_{\partial\Omega} = \boldsymbol{u}^s\big|_{\partial\Omega} + \boldsymbol{u}^{in}, \quad \mathscr{T}_\nu(\boldsymbol{u}) = \mathscr{T}_\nu(\boldsymbol{u}^s) + \mathscr{T}_\nu(\boldsymbol{u}^{in}) & \text{on } \partial\Omega, \\ \boldsymbol{u}^{\mathrm{p},s} = -\frac{1}{k_{\mathrm{p}}^2}\nabla(\nabla\cdot\boldsymbol{u}^s), \quad \boldsymbol{u}^{\mathrm{s},s} = \frac{1}{k_{\mathrm{s}}^2}\nabla\times(\nabla\times\boldsymbol{u}^s) & \text{in } B_r\backslash\overline{\Omega}, \\ \mathscr{T}_\nu(\boldsymbol{u}^s) = \Lambda\boldsymbol{u}^s & \text{on } \partial B_r. \end{cases} \tag{6.3.6}$$

where Λ is the Dirichlet-to-Newmann (DtN) map such that

$$\begin{aligned} \Lambda : H^{1/2}(\partial B_r)^n &\longrightarrow H^{-1/2}(\partial B_r)^n, \\ \widetilde{\mathbf{g}} &\longmapsto \mathscr{T}_\nu(\widetilde{\mathbf{q}}) \end{aligned} \tag{6.3.7}$$

with a radiating solution $\widetilde{\mathbf{q}}$ for Navier equation

$$\begin{cases} \mu_e\Delta\widetilde{\mathbf{q}} + (\lambda_e+\mu_e)\nabla(\nabla\cdot\widetilde{\mathbf{q}}) + \omega^2\rho_e\widetilde{\mathbf{q}} = \mathbf{0} & \text{in } \mathbb{R}^n\backslash\overline{B_r}, \\ \widetilde{\mathbf{q}} = \widetilde{\mathbf{g}} & \text{on } \partial B_r, \end{cases}$$

where λ_e, μ_e and ρ_e are real constants satisfying the strong convexity condition (6.3.1).

First, we establish the equivalence of problem (6.3.3) and problem (6.3.6). Therefore we can prove that (6.3.6) admits a unique solution and satisfies certain a priori estimates.

By applying the definition of Λ, it is easy to see that if $(\boldsymbol{u}, \boldsymbol{u}^s)$ is a solution to the scattering problem (6.3.3), then $(\boldsymbol{u}, \boldsymbol{u}^s)\big|_{B_r\backslash\overline{D}}$ solves the scattering problem (6.3.6).

On the other hand, suppose $(\boldsymbol{u}, \boldsymbol{u}^s)$ is a solution to the truncated system (6.3.6). By applying the integral representation and $\mathscr{T}_\nu\,\boldsymbol{u}^s = \Lambda\boldsymbol{u}^s$ on ∂B_r, we can derive that

$$\begin{aligned} \boldsymbol{u}^s(\mathbf{x}) = &\int_{\partial B_r}\Big\{\{\mathscr{T}_\nu^{\mathbf{y}}\Phi(\mathbf{x},\mathbf{y})\}^\top\cdot\boldsymbol{u}^s(\mathbf{y}) - \Phi(\mathbf{x},\mathbf{y})\cdot\Lambda\boldsymbol{u}^s(\mathbf{y})\Big\}\,\mathrm{d}s(\mathbf{y}) + \int_{B_r\backslash\overline{\Omega}}\Phi(\mathbf{x},\mathbf{y})\cdot\mathbf{f}(\mathbf{y})\,\mathrm{d}\mathbf{y} \\ &-\int_{\partial\Omega}\Big\{\{\mathscr{T}_\nu^{\mathbf{y}}\Phi(\mathbf{x},\mathbf{y})\}^\top\cdot\boldsymbol{u}^s(\mathbf{y}) - \Phi(\mathbf{x},\mathbf{y})\cdot\mathscr{T}_\nu^{\mathbf{y}}\boldsymbol{u}^s(\mathbf{y})\Big\}\,\mathrm{d}s(\mathbf{y}), \end{aligned} \tag{6.3.8}$$

where $\Phi(\mathbf{x},\mathbf{y})$ is the fundamental solution to the Lamé system (6.3.2) with the form

$$\Phi(\mathbf{x},\mathbf{y}) = \frac{\kappa_{\mathrm{s}}^2}{4\pi\omega^2}\frac{e^{\imath\kappa_{\mathrm{s}}|\mathbf{x}-\mathbf{y}|}}{|\mathbf{x}-\mathbf{y}|}I + \frac{1}{4\pi\omega^2}\nabla_{\mathbf{x}}\nabla_{\mathbf{x}}^\top\frac{e^{\imath\kappa_{\mathrm{s}}|\mathbf{x}-\mathbf{y}|} - e^{\imath\kappa_{\mathrm{p}}|\mathbf{x}-\mathbf{y}|}}{|\mathbf{x}-\mathbf{y}|} \tag{6.3.9}$$

and

$$\mathscr{T}_\nu^{\mathbf{y}}\Phi(\mathbf{x},\mathbf{y}) = \Big[\mathscr{T}_\nu^{\mathbf{y}}\Big(\Phi(\mathbf{x},\mathbf{y})(:,1)\Big),\ \mathscr{T}_\nu^{\mathbf{y}}\Big(\Phi(\mathbf{x},\mathbf{y})(:,2)\Big),\ \mathscr{T}_\nu^{\mathbf{y}}\Big(\Phi(\mathbf{x},\mathbf{y})(:,3)\Big)\Big]. \tag{6.3.10}$$

Here, I is the identity matrix, $\Phi(\mathbf{x}, \mathbf{y})(:, j)$ denotes the j-th column of $\Phi(\mathbf{x}, \mathbf{y})$, $j = 1, 2, 3$. $\mathscr{T}_\nu^{\mathbf{y}}$ is the exterior unit normal vector to the boundaries with respect to $\mathbf{y}$. Notice that $\Phi(\mathbf{x}, \mathbf{y}) = \Phi(\mathbf{x}, \mathbf{y})^\top$. Then, by combining the definition of Λ with the fact that each column of $\Phi(\mathbf{x}, \mathbf{y})$ satisfies the Kupradze radiation condition, we can obtain that

$$\int_{\partial B_r} \Big\{\{\mathscr{T}_\nu^{\mathbf{y}} \Phi(\mathbf{x}, \mathbf{y})\}^\top \cdot \boldsymbol{u}^s(\mathbf{y}) - \Phi(\mathbf{x}, \mathbf{y}) \cdot \Lambda \boldsymbol{u}^s(\mathbf{y})\Big\} \mathrm{d}s(\mathbf{y}) = 0. \tag{6.3.11}$$

Substituting (6.3.11) into (6.3.8) yields

$$\begin{aligned} \boldsymbol{u}^s(\mathbf{x}) = &-\int_{\partial\Omega} \Big\{\{\mathscr{T}_\nu^{\mathbf{y}} \Phi(\mathbf{x}, \mathbf{y})\}^\top \cdot \boldsymbol{u}^s(\mathbf{y}) - \Phi(\mathbf{x}, \mathbf{y}) \cdot \mathscr{T}_\nu^{\mathbf{y}} \boldsymbol{u}^s(\mathbf{y})\Big\} \mathrm{d}s(\mathbf{y}) \\ &+ \int_{B_r\backslash\overline{\Omega}} \Phi(\mathbf{x}, \mathbf{y}) \cdot \mathbf{f}(\mathbf{y}) \,\mathrm{d}\mathbf{y}. \end{aligned}$$

Clearly, $\boldsymbol{u}^s$ can be extended to a function belong to $H^1_{loc}(\mathbb{R}^n\backslash\overline{\Omega})^n$ (still denoted by $\boldsymbol{u}^s$). Since each column of $\Phi(\mathbf{x}, \mathbf{y})$ or $\mathscr{T}_\nu^{\mathbf{y}} \Phi(\mathbf{x}, \mathbf{y})$ satisfies the Kupradze radiation condition, the new function $\boldsymbol{u}^s \in H^1_{loc}(\mathbb{R}^n\backslash\overline{\Omega})^n$ also satisfies the Kupradze radiation condition. Hence, $(\boldsymbol{v}, \boldsymbol{u}^s)$ solves problem (6.3.3).

Next, we prove that there exists a unique solution to the system (6.3.3) and it is relied on the input data $\boldsymbol{u}^{in}$ and $\mathbf{f}$.

Firstly, let $\mathscr{B}(\mathbf{u}) = \mathbf{0}$ on ∂D, $\boldsymbol{u}^{in} = \mathbf{0}$, $\mathscr{T}_\nu(\boldsymbol{u}^{in}) = \mathbf{0}$, $\mathbf{f} = \mathbf{0}$. It is sufficient to show that there exists only a trivial solution to (6.3.3). Post-multiplying the first equation of (6.3.3), respectively, by $\overline{\boldsymbol{u}^{in}}$ and $\overline{\boldsymbol{u}^s}$ and using the Betti's first formula over $\mathbb{R}^n\backslash\overline{D}$ and $B_r\backslash\overline{\Omega}$ and the boundary conditions on ∂D and $\partial\Omega$, we have

$$\begin{aligned} \int_{\mathbb{R}^n\backslash\overline{D}} [\mathscr{C}(\mathbf{x}) : \nabla\overline{\boldsymbol{u}^{in}}] : \nabla\boldsymbol{v} \,\mathrm{d}\mathbf{x} = &\int_{\mathbb{R}^n\backslash\overline{D}} \omega^2\rho\, |\boldsymbol{u}^{in}|^2 \,\mathrm{d}\mathbf{x} - \int_{B_r\backslash\overline{\Omega}} [\mathscr{C}^e : \nabla\overline{\boldsymbol{u}^s}] : \nabla\boldsymbol{u}^s \,\mathrm{d}s(\mathbf{x}) \\ &+ \int_{\partial B_r} \boldsymbol{v} \cdot [\mathscr{C}^e : \nabla\boldsymbol{u}^s] \cdot \overline{\boldsymbol{u}^s} \,\mathrm{d}s(\mathbf{x}) + \int_{B_r\backslash\overline{\Omega}} \omega^2 \rho_e\, |\boldsymbol{u}^s|^2 \,\mathrm{d}\mathbf{x}. \end{aligned} \tag{6.3.12}$$

Taking the imaginary part of the equation above, we obtain

$$\Im \int_{\partial B_r} \boldsymbol{v} \cdot [\mathscr{C}^e : \nabla\boldsymbol{u}^s]\ \overline{\boldsymbol{u}^s} \,\mathrm{d}s(\mathbf{x}) = -\int_{\Omega\backslash\overline{D}} \omega^2\, \Im\rho\, |\boldsymbol{u}|^2 \,\mathrm{d}\mathbf{x} \le 0.$$

From Lemma 6.3.1 and the unique continuation principle, we know $\boldsymbol{u}^s = \mathbf{0}$ in $\Omega\backslash\overline{D}$ and $\boldsymbol{v} = \mathbf{0}$ in D. Therefore, the uniqueness of the solution to (6.3.3) is established.

We know that problems (6.3.3) and (6.3.6) are equivalent. Thus, we only need to verify the existence of solution to (6.3.6) by the variational technique. Without loss of generality, we assume $\omega^2\rho_e$ is not a Dirichlet eigenvalue in $B_r\backslash\overline{\Omega}$. It is easy to check that the vector field $\boldsymbol{w}$, which is defined by $\boldsymbol{w}(\mathbf{x}) = \boldsymbol{u}^{in}(\mathbf{x})$ in $\Omega\backslash\overline{D}$ and $\boldsymbol{w}(\mathbf{x}) = \boldsymbol{u}^s(\mathbf{x}) + \widetilde{\boldsymbol{v}}(\mathbf{x})$ in $B_r\backslash\overline{\Omega}$, satisfies

$$
\begin{cases}
\mathscr{L}_{\mathscr{C}}\boldsymbol{w}+\omega^2\rho(\mathbf{x})\boldsymbol{w}=\mathbf{f} & \text{in } B_r\backslash\overline{D},\\
\boldsymbol{w}^s=\boldsymbol{w}^{\mathrm{p},s}+\boldsymbol{w}^{\mathrm{s},s} & \text{in } B_r\backslash\overline{\Omega},\\
\mathscr{T}_\nu(\boldsymbol{w})=\mathbf{p} & \text{on } \partial D,\\
\boldsymbol{w}^-=\boldsymbol{w}^+ & \text{on } \partial\Omega,\\
\mathscr{T}_\nu(\boldsymbol{w}^-)=\mathscr{T}_\nu(\boldsymbol{w}^+)+\mathscr{T}_\nu(\boldsymbol{u}^{in})-\mathscr{T}_\nu(\widetilde{\boldsymbol{v}}) & \text{on } \partial\Omega,\\
\mathscr{T}_\nu(\boldsymbol{w}^-)=\Lambda\boldsymbol{w}^+ +\mathscr{T}_\nu(\widetilde{\boldsymbol{v}}) & \text{on } \partial B_r,
\end{cases}
\tag{6.3.13}
$$

where $\boldsymbol{w}^-$ and $\boldsymbol{w}^+$ stand for the limits from outside and inside $\partial\Omega$, respectively, Λ is the DtN operator given in (6.3.7), $\widetilde{\boldsymbol{v}}$ is a solution to the following equation:

$$
\begin{cases}
\mu_e\Delta\widetilde{\boldsymbol{v}}+(\lambda_e+\mu_e)\nabla(\nabla\cdot\widetilde{\boldsymbol{v}})+\omega^2\rho_e\widetilde{\boldsymbol{v}}=\mathbf{0} & \text{in } B_r\backslash\overline{\Omega}\\
\widetilde{\boldsymbol{v}}=\boldsymbol{u}^{in} & \text{on } \partial\Omega,\\
\widetilde{\boldsymbol{v}}=\mathbf{0} & \text{on } \partial B_r.
\end{cases}
\tag{6.3.14}
$$

By [9, Theorem 4.10], we know that $\widetilde{\boldsymbol{v}}$ is unique and $\|\widetilde{\boldsymbol{v}}\|_{H^1(B_r\backslash\overline{\Omega})^n}=O(\|\boldsymbol{u}^{in}\|_{H^{1/2}(\partial\Omega)^n})$.

Next, we introduce a bounded operator

$$
\Lambda_0 : H^{1/2}(\partial B_r)^n \longrightarrow H^{-1/2}(\partial B_r)^n
$$

which maps Φ to $\mathscr{T}_\nu(\widetilde{\boldsymbol{w}})\Big|_{\partial B_r}$ where $\widetilde{\boldsymbol{w}}\in H^1_{loc}(\mathbb{R}^n\backslash\overline{B_r})^n$ is the unique solution of the following system:

$$
\begin{cases}
\mu_e\Delta\widetilde{\boldsymbol{w}}+(\lambda_e+\mu_e)\nabla(\nabla\cdot\widetilde{\boldsymbol{w}})+\omega^2\rho_e\widetilde{\boldsymbol{w}}=\mathbf{0} & \text{in } \mathbb{R}^n\backslash\overline{B_r},\\
\widetilde{\boldsymbol{w}}=\Phi\in H^{1/2}(\partial B_r)^n & \text{on } \partial B_r.
\end{cases}
\tag{6.3.15}
$$

The operator Λ_0 has the following properties

$$
-\int_{\partial B_r}\overline{\Phi}\Lambda_0\Phi\, ds(x)\geq 0,\quad \Phi\in H^{1/2}(\partial B_r)^n,
\tag{6.3.16}
$$

and the difference $\Lambda-\Lambda_0$ is a compact operator from $H^{1/2}(\partial B_r)^n\to H^{-1/2}(\partial B_r)^n$. It is proved in [3] that these properties still hold for dyadic field by the similar analysis for the Laplace operator [5, 6]. Hence, for any $\boldsymbol{\varphi}\in H^1(B_r\backslash\overline{D})^n$, using the test function $\overline{\boldsymbol{\varphi}}$ we can easily derive the variational formulation of (6.3.13): find $\boldsymbol{w}\in H^1(B_r\backslash\overline{D})^n$ such that

$$
a_1(\boldsymbol{w},\boldsymbol{\varphi})+a_2(\boldsymbol{w},\boldsymbol{\varphi})=\mathscr{F}(\boldsymbol{\varphi}),
\tag{6.3.17}
$$

where the bilinear forms a_1, a_2 and the linear functional $\mathscr{F}(\cdot)$ are defined by

$$
\begin{aligned}
a_1(\boldsymbol{w}, \boldsymbol{\varphi}) &:= \int_{\Omega\backslash\overline{D}} (\mathscr{C}(\mathbf{x}) : \nabla\overline{\boldsymbol{\varphi}}) : \nabla\boldsymbol{w}\,\mathrm{d}\mathbf{x} + \int_{\Omega\backslash\overline{D}} \rho\,\omega^2\,\boldsymbol{w}\cdot\overline{\boldsymbol{\varphi}}\,\mathrm{d}\mathbf{x} + \int_{B_r\backslash\overline{\Omega}} (\mathscr{C}^e : \nabla\overline{\boldsymbol{\varphi}}) : \nabla\boldsymbol{w}\,\mathrm{d}\mathbf{x} \\
&\quad + \int_{B_r\backslash\overline{\Omega}} \omega^2\,\rho_e\,\boldsymbol{w}\cdot\overline{\boldsymbol{\varphi}}\,\mathrm{d}\mathbf{x} - \int_{\partial B_r} \Lambda_0\boldsymbol{w}\cdot\overline{\boldsymbol{\varphi}}\,\mathrm{d}s(\mathbf{x}), \\
a_2(\boldsymbol{w}, \boldsymbol{\varphi}) &:= -2\int_{\Omega\backslash\overline{D}} \rho\,\omega^2\,\boldsymbol{w}\cdot\overline{\boldsymbol{\varphi}}\,\mathrm{d}\mathbf{x} - 2\int_{B_r\backslash\overline{\Omega}} \omega^2\,\rho_e\,\boldsymbol{w}\cdot\overline{\boldsymbol{\varphi}}\,\mathrm{d}\mathbf{x} \\
&\quad - \int_{\partial B_r} (\Lambda - \Lambda_0)\boldsymbol{w}\cdot\overline{\boldsymbol{\varphi}}\,\mathrm{d}s(\mathbf{x}), \\
\mathscr{F}(\boldsymbol{\varphi}) &:= \int_{\partial\Omega} (\mathscr{T}_{\boldsymbol{\nu}}(\boldsymbol{u}^{in}) - \mathscr{T}_{\boldsymbol{\nu}}(\widetilde{\boldsymbol{v}}))\cdot\overline{\boldsymbol{\varphi}}\,\mathrm{d}s(\mathbf{x}) + \int_{\partial B_r} \mathscr{T}_{\boldsymbol{\nu}}(\widetilde{\boldsymbol{v}})\cdot\overline{\boldsymbol{\varphi}}\,\mathrm{d}s(\mathbf{x}) \\
&\quad - \int_{B_r\backslash\overline{D}} \mathbf{f}\cdot\overline{\boldsymbol{\varphi}}\,\mathrm{d}\mathbf{x}.
\end{aligned}
$$

By using the assumptions about $\rho(\mathbf{x})$ and $\mathscr{C}(\mathbf{x})$, Cauchy-Schwarz inequality and the definition of operator Λ_0, one can show the boundedness of the bilinear form a_1: for any $\boldsymbol{\phi}, \boldsymbol{\varphi} \in H^1(B_r\backslash\overline{D})^n$,

$$
\left|a_1(\boldsymbol{\phi}, \boldsymbol{\varphi})\right| \leq C_1 \|\boldsymbol{\phi}\|_{H^1(B_r\backslash\overline{D})^n} \|\boldsymbol{\varphi}\|_{H^1(B_r\backslash\overline{D})^n}
$$

for some constant C_1. Furthermore, by virtue of Poincaré's inequality and (6.3.16), we have the coercivity property of the bilinear form a_1: for any $\boldsymbol{\varphi} \in H^1(B_r\backslash\overline{D})^n$,

$$
a_1(\boldsymbol{\varphi}, \boldsymbol{\varphi}) \geq C_2 \|\boldsymbol{\varphi}\|^2_{H^1(B_r\backslash\overline{D})^n}
$$

for some constant C_2. According to Lax-Milgram lemma, there exists a bounded inverse operator $\mathscr{L} : H^1(B_r\backslash\overline{D})^n \longrightarrow H^1(B_r\backslash\overline{D})^n$ such that

$$
a_1(\boldsymbol{w}, \boldsymbol{\varphi}) = \langle \mathscr{L}\boldsymbol{w}, \boldsymbol{\varphi}\rangle,
$$

where $\langle\cdot,\cdot\rangle$ is the inner product in $H^1(B_r\backslash\overline{D})^n$, and the inverse of $\mathscr{L}$ is also bounded. In view of the expression of the bilinear form a_2, we introduce two bounded operators $\mathscr{K}_1$ and $\mathscr{K}_2$ given by

$$
\begin{aligned}
\langle \mathscr{K}_1\boldsymbol{w}, \boldsymbol{\varphi}\rangle &:= 2\int_{\Omega\backslash\overline{D}} \rho\,\omega^2\,\boldsymbol{w}\cdot\overline{\boldsymbol{\varphi}}\,\mathrm{d}\mathbf{x} + 2\int_{B_r\backslash\overline{\Omega}} \omega^2\,\rho_e\,\boldsymbol{w}\cdot\overline{\boldsymbol{\varphi}}\,\mathrm{d}\mathbf{x}, \qquad (6.3.18) \\
\langle \mathscr{K}_2\boldsymbol{w}, \boldsymbol{\varphi}\rangle &:= \int_{\partial B_r} (\Lambda - \Lambda_0)\boldsymbol{w}^+ \cdot \overline{\boldsymbol{\varphi}}\,\mathrm{d}s(\mathbf{x}).
\end{aligned}
$$

We claim that the operators $\mathscr{K}_1$ and $\mathscr{K}_2$ are both compact. In fact, let $\{\boldsymbol{w}_n\}_{n=1}^{\infty}$ be a bounded sequence in $H^1(B_r\backslash\overline{D})^n$ and weakly converge to $\boldsymbol{w}_*$ in the sense of $\|\cdot\|_{H^1(B_r\backslash\overline{D})^n}$ (denoted by $\boldsymbol{w}_n \rightharpoonup \boldsymbol{w}_*$). Since $\mathscr{I} : H^1(B_r\backslash\overline{D})^n \longrightarrow L^2(B_r\backslash\overline{D})^n$ is a compact embedding operator, we get

$$\langle \mathcal{K}_1(\boldsymbol{w}_n - \boldsymbol{w}_*), \boldsymbol{\varphi}\rangle = 2\int_{\Omega\setminus\overline{D}} \rho\,\omega^2\,(\boldsymbol{w}_n - \boldsymbol{w}_*)\cdot\overline{\boldsymbol{\varphi}}\,\mathrm{d}\mathbf{x} + 2\int_{B_r\setminus\overline{\Omega}} \omega^2\,\rho_e\,(\boldsymbol{w}_n - \boldsymbol{w}_*)\cdot\overline{\boldsymbol{\varphi}}\,\mathrm{d}\mathbf{x}$$

and thus

$$\begin{aligned}
\left\|\mathcal{K}_1(\boldsymbol{w}_n - \boldsymbol{w}_*)\right\|^2_{H^1(B_r\setminus\overline{D})^n} &= \langle \mathcal{K}_1(\boldsymbol{w}_n - \boldsymbol{w}_*),\ \mathcal{K}_1(\boldsymbol{w}_n - \boldsymbol{w}_*)\rangle \\
&= 2\int_{\Omega\setminus\overline{D}} \rho\,\omega^2\,(\boldsymbol{w}_n - \boldsymbol{w}_*)\cdot\overline{\mathcal{K}_1(\boldsymbol{w}_n - \boldsymbol{w}_*)}\,\mathrm{d}\mathbf{x} \\
&\quad + 2\int_{B_r\setminus\overline{\Omega}} \rho_e\,\omega^2\,(\boldsymbol{w}_n - \boldsymbol{w}_*)\cdot\overline{\mathcal{K}_1(\boldsymbol{w}_n - \boldsymbol{w}_*)}\,\mathrm{d}\mathbf{x} \\
&\le 2\,C\,\omega^2\,\max\left\{\|\rho(\mathbf{x})\|_{\mathbf{L}^\infty(\Omega\setminus\overline{D})},\,\rho_e\right\}\left\|\boldsymbol{w}_n - \boldsymbol{w}_*\right\|^2_{L^2(B_r\setminus\overline{D})^n},
\end{aligned}$$

which implies that $\mathcal{K}_1$ is compact. Similarly, we can verify the compactness of $\mathcal{K}_2$. Since $\boldsymbol{w}_n \rightharpoonup \boldsymbol{w}_*$ in $H^1(B_r\setminus\overline{D})^n$, we have $\boldsymbol{w}_n\big|_{\partial B_r} \rightharpoonup \boldsymbol{w}_*\big|_{\partial B_r}$ in $H^{1/2}(\partial B_r)^n$ by the trace operator. Together with the compactness of $\Lambda - \Lambda_0$, it is easy to obtain that

$$(\Lambda - \Lambda_0)\boldsymbol{w}_n\big|_{\partial B_r} \longrightarrow (\Lambda - \Lambda_0)\boldsymbol{w}_*\big|_{\partial B_r}$$

in $H^{-1/2}(\partial B_r)^n$. For any $\boldsymbol{\varphi} \in H^1(B_r\setminus\overline{D})^n$, it holds that

$$\langle \mathcal{K}_2(\boldsymbol{w}_n - \boldsymbol{w}_*), \boldsymbol{\varphi}\rangle = \int_{\partial B_r} (\Lambda - \Lambda_0)(\boldsymbol{w}_n - \boldsymbol{w}_*)\cdot\overline{\boldsymbol{\varphi}}\,\mathrm{d}s(x).$$

Therefore we have

$$\begin{aligned}
\left\|\mathcal{K}_2(\boldsymbol{w}_n - \boldsymbol{w}_*)\right\|^2_{H^1(B_r\setminus\overline{D})^n} &= \langle \mathcal{K}_2(\boldsymbol{w}_n - \boldsymbol{w}_*), \mathcal{K}_2(\boldsymbol{w}_n - \boldsymbol{w}_*)\rangle \\
&= \int_{\partial B_r} (\Lambda - \Lambda_0)(\boldsymbol{w}_n - \boldsymbol{w}_*)\cdot\overline{\mathcal{K}_2(\boldsymbol{w}_n - \boldsymbol{w}_*)}\,\mathrm{d}s(x) \\
&\le \|(\Lambda - \Lambda_0)(\boldsymbol{w}_n - \boldsymbol{w}_*)\|_{H^{-1/2}(\partial B_r)^n}\,\|\mathcal{K}_2(\boldsymbol{w}_n - \boldsymbol{w}_*)\|_{H^{1/2}(\partial B_r)^n} \\
&\le C\left\|(\Lambda - \Lambda_0)(\boldsymbol{w}_n - \boldsymbol{w}_*)\right\|_{H^{-1/2}(\partial B_r)^n}\left\|\boldsymbol{w}_n - \boldsymbol{w}_*\right\|_{L^2(B_r\setminus\overline{D})^n},
\end{aligned}$$

which implies that $\mathcal{K}_2$ is compact.

Since $\mathcal{L}$ is bounded and $\mathcal{K}_1 + \mathcal{K}_2$ is compact, we know that $\mathcal{L} - (\mathcal{K}_1 + \mathcal{K}_2)$ is a Fredholm operator of index zero. According to the Fredholm alternative theorem, Riesz representation theory and the uniqueness of (6.3.3), we know there must exist a solution to (6.3.3). Since the inverse of $\mathcal{L} - (\mathcal{K}_1 + \mathcal{K}_2)$ is bounded, by applying the Lax-Milgram lemma to

$$\big\langle(\mathcal{T} - \mathcal{K}_1 - \mathcal{K}_2)\,\boldsymbol{w}, \boldsymbol{\varphi}\big\rangle = \mathcal{F}(\boldsymbol{\varphi}),$$

we get

$$\|\boldsymbol{w}\|_{H^1(B_R\setminus\overline{D})^n} \le C\|\mathcal{F}\|.$$

On the other hand, it is straightforward to verify that

$$|\mathscr{F}(\boldsymbol{\varphi})| \le C\left(\|\boldsymbol{u}^{in}\|_{H^{1/2}(\partial\Omega)^n} + \|\mathscr{T}_\nu(\boldsymbol{u}^{in})\|_{H^{-1/2}(\partial\Omega)^n} + \|\mathbf{f}\|_{L^2(B_{r_0}\setminus\overline{\Omega})^n}\right)\|\boldsymbol{\varphi}\|_{H(B_r\setminus\overline{D})^n},$$

which can directly imply the inequality (6.3.5).

The proof is complete. □

6.4 Green's Theorems and Representation Formulas

From now on we suppose that the given real constants ω, λ and μ satisfy

$$\omega > 0, \quad \mu > 0, \quad 2\mu + \lambda > 0.$$

Moreover, for a smooth vector field U, $\Delta^* U := \mu U + (\lambda + \mu)\nabla(\nabla \cdot U)$. If $\nu \in \mathbb{C}^3$ is a vector and U is a smooth vector field in a neighborhood of a point $x \in \mathbb{R}^3$, we denote by $[T(U, \nu)](x)$ the vector

$$[T(U,\nu)](x) := (\beta_1 + \mu)\frac{\partial U}{\partial \nu}(x) + \beta_2(\nabla \cdot U)(x)\nu(x) + \beta_1 \nu(x) \wedge [\nabla \wedge U(x)].$$

Here $\beta_1, \beta_2 \in \mathbb{R}$ are arbitrary constants satisfying $\beta_1 + \beta_2 = \lambda + \mu$. Note that there is no indication that T dependent on β_1 and β_2. For $\beta_1 = \mu$ and $\beta_2 = \lambda$, $[T(U, \nu)](x)$ is the tranction vector on a surface containing x with normal vector ν at x. Notice that the notation $(TU)(x)$ is shorten for $[T(U, \nu(x))](x)$ which stands for the integration on a surface having the normal vector ν.

If $D \subset \mathbb{R}^3$ is a C^2-smooth, bounded, open set and if $U, V : \overline{D} \to \mathbb{C}^3$ denote $C^2(\overline{D})$-smooth vector fields, then Gauss' theorem implies

$$\begin{aligned}
\int_{\partial D} (TU) \cdot V ds &= \int_D (\beta_1 + \mu)\Big(\Delta U \cdot V + \sum_{j,k=1}^{3} \frac{\partial U_j}{\partial x_k}\frac{\partial V_j}{\partial x_k}\Big)dx \\
&\quad + \int_D \beta_2[\nabla(\nabla \cdot U) \cdot V + (\nabla \cdot U)(\nabla \cdot V)]dx \\
&\quad + \int_D \beta_1[(\nabla \wedge \nabla \wedge U) \cdot V - (\nabla \wedge U) \cdot (\nabla \wedge V)]dx \qquad (6.4.1) \\
&= \int_D \Big\{(\beta_1 + \mu)\sum_{j,k=1}^{3} \frac{\partial U_j}{\partial x_k}\frac{\partial V_j}{\partial x_k} + \beta_2(\nabla \cdot U)(\nabla \cdot V) \\
&\quad - \beta_1(\nabla \wedge U) \cdot (\nabla \wedge V)\Big\}dx + \int_D (\Delta^* U) \cdot V dx,
\end{aligned}$$

which we call the first Betti formula.

It is possible to weaken the regularity assumptions on U and V. $U, V \in C^1(\overline{D}) \cap C^2(D)$ and $\Delta^* U \in C(\overline{D})$ are sufficient for the above first Betti formula.

Interchanging the roles of U and V in (6.4.1) and subtracting yields the second Betti formula

$$\int_{\partial D} \{(TU) \cdot V - (TV) \cdot U\} ds = \int_D \{(\Delta^* U + \omega^2 U) \cdot V - (\Delta^* V + \omega^2 V) \cdot U\} dx, \tag{6.4.2}$$

for any $\omega > 0$.

In order to state representation theorems we need a fundamental solution for the operator $\Delta^* + \omega I$. This will be a matrix valued function $\Pi : \mathbb{R}^3 \setminus \{0\} \to \mathbb{C}^{3\times 3}$. Denoting by d_1, d_2, d_3 the cartesian unit vectors in $\mathbb{R}^3$ we define for $x \in \mathbb{R}^3$, $x \neq 0$, the jth column of $\Pi(x)$ by

$$\Pi(x) d_j := \frac{e^{\mathrm{i}\kappa_s |x|}}{4\pi\mu|x|} d_j + \frac{1}{\omega^2} \nabla\nabla \cdot \left\{ \frac{e^{\mathrm{i}\kappa_s |x|} - e^{\mathrm{i}\kappa_p |x|}}{4\pi|x|} d_j \right\}, \quad j = 1, 2, 3.$$

This matrix is called Kupradze's matrix. We denote its entry in the jth row and kth column by Π_{jk}. From its definition we can infer that it is an even function of x satisfying $\Pi(x) = \Pi(x)^T$, i.e., it coincide with its transpose. In addition, we see with the help of $\nabla \wedge \nabla \wedge \cdot = -\Delta + \nabla(\nabla \cdot)$ that

$$\Pi(x) d_j = \frac{1}{\omega^2} \nabla \wedge \nabla \wedge \left\{ \frac{e^{\mathrm{i}\kappa_s |x|}}{4\pi|x|} d_j \right\} - \frac{1}{\omega^2} \nabla\nabla \cdot \left\{ \frac{e^{\mathrm{i}\kappa_p |x|}}{4\pi|x|} d_j \right\}, \quad j = 1, 2, 3. \tag{6.4.3}$$

We have to study some more properties of Π, especially its behavior for $|x| \to 0$. To this end we expand $e^{i\kappa|x|}/(4\pi|x|)$ in a power series and obtain

$$\frac{e^{i\kappa|x|}}{4\pi|x|} = \frac{\cos(\kappa|x|)}{4\pi|x|} + \mathrm{i}\frac{\sin(\kappa|x|)}{4\pi|x|} \tag{6.4.4}$$

$$= \frac{1}{4\pi|x|} - \frac{\kappa^2}{8\pi}|x| + \kappa^4|x|^3 f_1(\kappa^2|x|^2) + \mathrm{i}\kappa f_2(\kappa^2|x|^2) \tag{6.4.5}$$

with two entire functions f_1 and f_2. Inserting these expressions in the definition of Π, collecting the terms having a $1/|x|$-singularity, and using $(\kappa_p^2 - \kappa_s^2)/\omega^2 = -(\lambda + \mu)/[\mu(2\mu + \lambda)]$ motivates the definition of

$$\Pi_{jk}^{(0)}(x) := \frac{\delta_{jk}}{4\pi\mu|x|} - \frac{\lambda + \mu}{8\pi\mu(2\mu + \lambda)} \frac{\partial^2 |x|}{\partial x_j \partial x_k}, \quad j, k = 1, 2, 3,$$

and of the matrix $\Pi^{(0)}(x) := \left(\Pi_{jk}^{(0)}(x)\right)$ for $x \neq 0$. For the Δ^*-operator Kelvin's matrix $\Pi^{(0)}$ has the same role that $1/(4\pi|\cdot|)$ has for the operator Δ-operator.

Now we provide the representation theorems for the elastic scattering problem. To this end we also need the analogues of the double-layer potentials and we define for a vector $\nu \in \mathbb{C}^3$ the matrix valued functions Ξ and $\Xi^{(0)}$ by

$$\Xi(x, y, \nu)^T d_j := T_y\big(\Pi(x-y)d_j, \nu\big), \quad x, y \in \mathbb{R}^3,\ x \neq y,$$
$$\Xi^{(0)}(x, y, \nu)^T d_j := T_y\big(\Pi^{(0)}(x-y)d_j, \nu\big), \quad x, y \in \mathbb{R}^3,\ x \neq y,$$

i.e., the jth row of of $\Xi(x, y, \nu)$ consists of the pseudostress vector of the jth column of Ξ and similarly for $\Xi^{(0)}$. Since in the sequel ν is always the unit normal vector at a point y lying on a surface, we omit the dependence on ν and write $\Xi(x, y)$ instead of $\Xi\big(x, y, \nu(y)\big)$ and similarly for $\Xi^{(0)}$.

Theorem 6.4.1 *Let $D \subset \mathbb{R}^3$ be a bounded, open, C^2-smooth set with exterior unit normal vector ν. For a vector field $U \in C^1(\overline{D}) \cap C^2(D)$ with $\Delta^* U \in C(\overline{D})$ we have the representation formulas*

$$U(x) = \int_{\partial D} \{\Pi(x-y)(TU)(y) - \Xi(x, y)U(y)\} ds(y) - \int_D \Pi(x-y)(\Delta^* U + \omega^2 U)(y) dy, \tag{6.4.6}$$

and

$$U(x) = \int_{\partial D} \{\Pi^{(0)}(x-y)(TU)(y) - \Xi^{(0)}(x, y)U(y)\} ds(y) - \int_D \Pi^{(0)}(x-y)(\Delta^* U)(y) dy, \tag{6.4.7}$$

for all $x \in D$.

The above representation formulas imply that solutions to $\Delta^* U + \omega^2 U = 0$ or $\Delta^* U = 0$ are analytic.

We then consider the representation formula for the solution U of $\Delta^* U + \omega^2 U = 0$ in an exterior domain. To this end we have to impose an additional requirement on U, namely a radiation condition. The radiation condition and the fundamental solution must match. There are two ways to obtain a radiation condition. One can study the behavior of the fundamental solution for large $|x|$ and then formulate a radiation condition accordingly. A second possibility is to require an integral relation. We choose the later approach now.

Let $U \in C^2(\mathbb{R}^3 \backslash B_R)$ be a solution to $\Delta^* U + \omega^2 U = 0$. U is a radiation solution, if for all $r > R$ and for all $|x| < r$ the identity

$$\int_{|y|=r} \{\Pi(x-y)(TU)(y) - \Xi(x, y)U(y)\} ds(y) = 0 \tag{6.4.8}$$

holds true.

The radiation condition and the representation formula applied in the spherical shell $\{R < |x| < r\}$ to a radiating solution U to $\Delta^* U + \omega^2 U = 0$ immediately yields the following theorem.

Theorem 6.4.2 *Let $U \in C^2(\mathbb{R}^3 \backslash B_R)$ be a radiation to $\Delta^* U + \omega^2 U = 0$ in $\mathbb{R}^3 \backslash \overline{B_R}$. Then we have*

$$U(x) = \int_{\partial B_R} \{\Xi(x, y)U(y) - \Pi(x - y)(TU)(y)\} ds(y), \quad |x| > R. \tag{6.4.9}$$

Now we check whether the columns of the fundamental solution $\Pi(y - z)$ regarded as vector fields of the variable y are radiating solutions.

Lemma 6.4.1 *Fix $x \in \mathbb{R}^3$, $k = \{1, 2, 3\}$, and $R > |z|$. Then, $U := \Pi(\cdot - z)d_k$ is a radiating solution to $\Delta^* U + \omega^2 U = 0$ in $\mathbb{R} \backslash \{z\}$.*

We first show that a solution of the above scattering problem is also a solution to a Lippmann-Schwinger type integral equation and vice versa.

Lemma 6.4.2 *If $u^s \in C^2(\mathbb{R}^3)$ is a solution to the above elastic scattering problem, then $u = u^i + u^s$ is a solution to*

$$u(x) = u^i(x) - \omega^2 \int_{B_R} (1 - \rho(y))\Pi(x - y)u(y)dy, \quad x \in \mathbb{R}^3. \tag{6.4.10}$$

If $\varphi \in C(\overline{B_R})$ is a solution to (6.4.10) *in $\overline{B_R}$ and if u^s is defined by*

$$u^s(x) := -\omega^2 \int_{B_R} (1 - \rho(y))\Pi(x - y)\varphi(y)dy, \quad x \in \mathbb{R}^3,$$

then, u^s is a solution to the elastic scattering problem with incident wave u^i.

Theorem 6.4.3 *For any incident wave $u^i \in C^2(\mathbb{R}^3)$, the integral equation* (6.4.10) *and the direct elastic scattering problem both have the same unique solution.*

Proof The equivalence of the scattering problem and the integral equation implies that it suffices to show that the scattering problem has at most one solution in order to establish the existence of a solution.

Let u be a solution to the scattering problem with incident wave $u^i = 0$. We pick $R_1 > R$ and compute

$$\begin{aligned}
&\mathscr{I}\left(\int_{\partial B_{R_1}} u \cdot \{-\mu\nu \wedge \nabla \wedge \overline{u} + (\lambda + 2\mu)(\nabla \cdot \overline{u})\nu\} ds\right) \\
&\quad = \mathscr{I}\left(\int_{B_{R_1}} u \cdot \Delta^* \overline{u} dx + \int_{B_{R_1}} \{\mu|\nabla \wedge u|^2 + (\lambda + 2\mu)|\nabla \cdot u|^2\} dx\right) = 0
\end{aligned}$$

Hence, u vanishes in the exterior of B_R.

Also, we can represent u as

$$u(x) = -\omega^2 \int_{B_R} (1 - \rho(y))\Pi(x - y)u(y)dy, \quad x \in \mathbb{R}^3.$$

Then, the mapping properties of the volume potential imply that u is C^3-smooth, i.e., $u \in C_0^3(B_R)$. Next, we define $v := \nabla \cdot u \in C_0^2(B_R)$ and we obtain the following system of differential equations for u and v:

$$\Delta u + \frac{\lambda + \mu}{\mu} \nabla v + \kappa_s^2 \rho u = 0,$$
$$\Delta v + \kappa_p^2 \rho v + \kappa_p^2 \nabla \rho \cdot u = 0.$$

The first equation is the elasticity equation $\mu \Delta u + (\lambda + \mu)\nabla(\nabla \cdot u) + \omega^2 \rho u = 0$ and the second one arises when taking the divergence of the elasticity equation.

Hence, there is a constant c_1 such that

$$|\Delta u(x)| \leq c_1 \big(|u(x)|^2 + |\nabla v(x)|^2\big)^{1/2},$$
$$|\Delta v(x)| \leq c_1 \big(|u(x)|^2 + |v(x)|^2\big)^{1/2}, \quad x \in \mathbb{R}^3.$$

We choose $R' > R$ and $t \geq 1$ sufficiently large to ensure

$$\frac{c_1^2 R'^2}{\pi^2 t^2} \left[2c^2 c_1^2 + 4 \frac{c_1^2 R'^2}{\pi^2} + 2 \right] < 1. \tag{6.4.11}$$

Here, c denotes the constant from the previous lemma. Furthermore, we define $\xi := (t, it, 0) \in \mathscr{C}^3$ and

$$\begin{aligned} V(x) &= \big(V_1(x), V_2(x), V_3(x), V_4(x)\big) \\ &= e^{-\mathrm{i}\xi \cdot x} \big(U_1(x), U_2(x), U_3(x), v(x)\big), x \in C := (-R', R')^3. \end{aligned}$$

If we are able to show $V = 0$, we can conclude $U(x) = 0, x \in C$, whence U vanishes identically in $\mathbb{R}^3$.

Now, we estimate

$$\begin{aligned} \sum_{j=1}^{3} |(\Delta + 2\mathrm{i}\xi \cdot \nabla) V_j(x)|^2 &= |e^{-\mathrm{i}\xi \cdot x} \Delta U(x)|^2 \\ &\leq c_1^2 \Big(\sum_{j=1}^{3} |V_j(x)|^2 + |e^{-\mathrm{i}\xi \cdot x} \nabla v(x)|^2 \Big) \qquad (6.4.12) \\ &\leq c_1^2 \Big(\sum_{j=1}^{3} |V_j(x)|^2 + 2|\nabla V_4(x)|^2 + 4t^2 |V_4(x)|^2 \Big) \end{aligned}$$

where we have used $e^{-\mathrm{i}\xi \cdot x} \nabla v(x) = \nabla V_4(x) + \mathrm{i} V_4(x) \xi$ in the last line, and similarly

$$|(\Delta + 2\mathrm{i}\xi \cdot)V_4(x)|^2 = |e^{-\mathrm{i}\xi\cdot x}\Delta v(x)|^2 \le c_1^2\Big(\sum_{j=1}^{4}|V_j(x)|^2\Big) \tag{6.4.13}$$

Also, for $V_4 \in C_0^2(C)$ we have $V_4 = -G_\xi\big((\Delta + 2\mathrm{i}\xi\cdot)V_4\big)$, whence by the above inequality

$$\|V_4\|_{L^2}^2 \le \frac{c_1^2 R'^2}{\pi^2 t^2}\sum_{j=1}^{4}\|V_j\|_{L^2}^2. \tag{6.4.14}$$

Due to the preceding lemma we also know

$$\|\nabla V_4\|_{L^2}^2 \le c^2 c_1^2 \sum_{j=1}^{4}\|V_j\|_{L^2}^2. \tag{6.4.15}$$

The same reasoning applied to V_1, V_2, $V_3 \in C_0^2(C)$ leads to

$$\begin{aligned}\sum_{j=1}^{3}\|V_j\|_{L^2}^2 &= \sum_{j=1}^{3}\|G_\xi\big((\Delta + 2\mathrm{i}\xi\cdot\nabla)V_j\big)\|_{L^2}^2\\ &\le \frac{c_1^2 R'^2}{\pi^2 t^2}\Big(\sum_{j=1}^{3}\|V_j\|_{L^2}^2 + 2\|\nabla V_4\|_{L^2}^2 + 4t^2\|V_4\|_{L^2}^2\Big)\\ &\le \frac{c_1^2 R'^2}{\pi^2 t^2}\Big(\sum_{j=1}^{3}\|V_j\|_{L^2}^2 + 2c^2c_1^2\sum_{j=1}^{4}\|V_j\|_{L^2}^2 + 4t^2\frac{c_1^2 R'^2}{\pi^2 t^2}\sum_{j=1}^{4}\|V_j\|_{L^2}^2\Big)\\ &\le \frac{c_1^2 R'^2}{\pi^2 t^2}\Big(\Big[2c^2c_1^2 + 4\frac{c_1^2 R'^2}{\pi^2} + 1\Big]\sum_{j=1}^{4}\|V_j\|_{L^2}^2\Big).\end{aligned} \tag{6.4.16}$$

Adding (6.4.14) and (6.4.16) finally yields the inequality

$$\sum_{j=1}^{4}\|V_j\|_{L^2}^2 \le \frac{c_1^2 R'^2}{\pi^2 t^2}\Big[2c^2c_1^2 + 4\frac{c_1^2 R'^2}{\pi^2} + 2\Big]\sum_{j=1}^{4}\|V_j\|_{L^2}^2,$$

whence $V = 0$.

This means that the scattering problem with $U^i = 0$ only has the trivial solution and the proof of the theorem is complete. □

We conclude this section with a discussion of the asymptotic behavior of U^s. Since the solution U^s of the elastic scattering problem has the form

$$
\begin{aligned}
U^s(x) &= -\omega^2 \int_{B_R} (1-\rho(y))\Pi(x-y)U(y)dy \\
&= -\nabla\wedge\nabla\wedge \int_{B_R} \Phi_{\kappa_s}(x,y)(1-\rho(y))U(y)dy \\
&\quad + \nabla\nabla\cdot \int_{B_R} \Phi_{\kappa_p}(x,y)(1-\rho(y))U(y)dy,
\end{aligned}
$$

we obtain from the asymptotic behavior of Φ_κ that

$$
\begin{aligned}
U^s(x) &= -\frac{\kappa_s^2 e^{\mathrm{i}\kappa_s|x|}}{4\pi|x|}\int_{B_R} e^{-\mathrm{i}\kappa_s\hat{x}\cdot y}(1-\rho(y))\hat{x}\wedge(U(y)\wedge\hat{x})dy \qquad (6.4.17)\\
&\quad -\frac{\kappa_p^2 e^{\mathrm{i}\kappa_p|x|}}{4\pi|x|}\int_{B_R} e^{-\mathrm{i}\kappa_p\hat{x}\cdot y}(1-\rho(y))\hat{x}\cdot U(y)dy\hat{x} + O\left(\frac{1}{|x|^2}\right), \quad |x|\to\infty.
\end{aligned}
$$

Hence, we know

$$
U^s(x) = \frac{e^{\mathrm{i}\kappa_s|x|}}{|x|}a(\hat{x}) + \frac{e^{\mathrm{i}\kappa_p|x|}}{|x|}u(\hat{x})\hat{x} + O\left(\frac{1}{|x|^2}\right), \quad |x|\to\infty,
$$

with a smooth function u and a smooth tangential vector field a on S^2. We call $U^s_\infty(\hat{x}) := a(\hat{x}) + u(\hat{x})\hat{x}$, $\hat{x}\in S^2$, the far field of U^s.

The formulas

$$
\begin{aligned}
\nabla\wedge U^s(x) &= -\kappa_s^2\nabla\wedge\int_{B_R}\Phi_{\kappa_s}(x,y)(1-\rho(y))U(y)dy,\\
\nabla\cdot U^s(x) &= -\kappa_p^2\nabla\cdot\int_{B_R}\Phi_{\kappa_p}(x,y)(1-\rho(y))U(y)dy,
\end{aligned}
$$

show that $\nabla\wedge U^s$ and $\nabla\cdot U^s$ are both radiating solutions to a Helmholtz equation in the exterior of B_R. Furthermore, using the asymptotic behavior of Φ_κ again, we can compute their far field patterns and compare them with (6.4.17). This yields $[\nabla\wedge U^s]_\infty(\hat{x}) = i\kappa_s\hat{x}\wedge a(\hat{x})$ and $(\nabla\cdot U)_\infty(\hat{x}) = i\kappa_p u(\hat{x})$, $\hat{x}\in S^2$. We can now infer from the one-to-one correspondence between far field patterns and radiating solutions to the Helmholtz equation, i.e., from Rellich's lemma, that any solution U^s to the elastic scattering problem, which has a vanishing far field $U^s_\infty = 0$, must vanish identically in the exterior of B_R. This follows immediately from $(\nabla\cdot U^s)_\infty(\hat{x}) = i\kappa_p\hat{x}\cdot U^s_\infty(\hat{x}) = 0$, $[\nabla\wedge U^s]_\infty(\hat{x}) = i\kappa_s\hat{x}\wedge U^s_\infty(\hat{x}) = 0$, whence $\nabla\wedge U^s = 0$ and $\nabla\cdot U^s = 0$ in $\mathbb{R}^3\backslash B_R$. Relation

$$
U^s = \frac{\mu}{\omega^2}\nabla\wedge\nabla\wedge U^s - \frac{\lambda+2\mu}{\omega^2}\nabla(\nabla\cdot U^s)
$$

now implies $U^s = 0$. Let us summarize this one-to-one correspondence between radiating solutions which are the scattered part of a solution to the elastic scattering problem and its far field patterns in the following theorem.

Theorem 6.4.4 *Let U^s be the scattered part of a solution $U = U^i + U^s$ to the elastic scattering problem. Then, the far field U^s_∞ of U^s uniquely determines U^s in the exterior of B_R.*

In elastic scattering a plane incident wave is defined by

$$U^i(x, d, p) = -\frac{1}{\omega^2}\nabla_x(\nabla_x \cdot [pe^{\mathrm{i}\kappa_p d\cdot x}]) + \frac{1}{\omega^2}\nabla_x \wedge \nabla_x \wedge [pe^{\mathrm{i}\kappa_s d\cdot x}],\ \ x \in \mathbb{R}^3,$$

where $d \in S^2$ is its direction of propagation and $p \in \mathbb{C}^3$ controls its polarization. Straightforward calculations show that U^i is a solution to $\Delta^* U^i + \omega^2 U^i = 0$. Note, that for $d \cdot p = 0$ the first term vanishes and we have a pure shear wave, whereas for $d \wedge p = 0$ the second term vanishes and we have a pure pressure wave.

Denoting by $U^s(\cdot, d, p), U(\cdot, d, p)$ the scattered wave and the total wave belonging to the elastic scattering problem with incident wave $U^i(\cdot, d, p)$, we define the far field pattern belonging to the density ρ to be the matrix valued function $U_\infty : S^2 \times S^2 \to \mathbb{C}^{3\times 3}$, having as its jth column

$$U_\infty(\hat{x}, d)d_j = [U^s(\cdot, d, d_j)]_\infty(\hat{x}),\quad j = 1, 2, 3, \hat{x}, d \in S^2.$$

Our considerations from above imply that

$$\begin{aligned} U_\infty(\hat{x}, d)d_j = &-\frac{\kappa_s^2}{4\pi}\int_{B_R} e^{-\mathrm{i}\kappa_s\hat{x}\cdot y}(1-\rho(y))\hat{x} \wedge (U(y, d, d_j) \wedge \hat{x})dy \\ &-\frac{\kappa_p^2}{4\pi}\int_{B_R} e^{-\mathrm{i}\kappa_p\hat{x}\cdot y}(1-\rho(y))\hat{x} \cdot U(y, d, d_j)dy\hat{x}. \end{aligned} \tag{6.4.18}$$

6.5 Numerical Methods for Elastic Scattering Problems

The PML can be thought of as the introduction of a fictitious material designed to absorb energy away from the region of interest (usually close to the scatterer). A perfectly matched layer is one which absorbs all energy sent into it without creating any reflected waves. In this subsection we study the perfectly matched layer (PML) method for solving the elastic wave scattering problems as follows

$$\begin{cases} \nabla \cdot \tau(\boldsymbol{u}) + \gamma^2\boldsymbol{u} = -\boldsymbol{q} & \text{in } \mathbb{R}^3\backslash\overline{D}, \\ \tau(\boldsymbol{u})\boldsymbol{n}_D = \boldsymbol{g} & \text{on } \Gamma_D. \end{cases} \tag{6.5.1}$$

Here $D \subset \mathbb{R}^3$ is a bounded domain with Lipschitz boundary Γ_D, $\boldsymbol{q} \in ([H^1(\mathbb{R}^3\backslash\overline{D})]^3)'$ has support inside $B_l := \{\boldsymbol{x} = (x_1, x_2, x_3)^T \in \mathbb{R}^3 : |x_i| < l_i, i = 1, 2, 3\}$ for some constants $l_i > 0$, $i = 1, 2, 3$, $\boldsymbol{g} \in [H^{-1/2}(\Gamma_D)]^3$ is determined by the traction on the boundary, $\boldsymbol{n}_D$ is the unit outer normal to Γ_D, and $\gamma = \sqrt{\rho_0}\omega > 0$ with the angular frequency $\omega > 0$ and the constant density $\rho_0 > 0$.

In the region outside D, the medium is assumed to be linear, homogeneous, and isotropic with constant Lamé constant λ and μ. The stress tensor $\tau(\boldsymbol{u})$ relates to the displacement vector $\boldsymbol{u} = (u_1, u_2, u_3)^T$ by the generalized Hooke law:

$$\tau(\boldsymbol{u}) = 2\mu\varepsilon(\boldsymbol{u}) + \lambda tr(\varepsilon(\boldsymbol{u}))I, \quad \varepsilon(\boldsymbol{u}) = \frac{1}{2}(\nabla\boldsymbol{u} + (\nabla\boldsymbol{u})^T), \tag{6.5.2}$$

where $I \in \mathbb{R}^{3\times 3}$ is the identity matrix and $\nabla\boldsymbol{u}$ is the displacement gradient tensor whose elements are $(\nabla\boldsymbol{u})_{ij} = \partial u_i \backslash \partial x_j$, $i, j = 1, 2, 3$. We remark that the results in this subsection can be extended to solve the scattering problems with other boundary conditions such as Dirichlet or mixed boundary conditions on Γ_D.

It is known that under the constitutive relation (6.5.2), (6.5.1) can be rewritten to the following equation:

$$\boldsymbol{u} + \frac{1}{k_p^2}\nabla(\nabla\cdot\boldsymbol{u}) - \frac{1}{k_s^2}\nabla\times\nabla\times\boldsymbol{u} = 0 \quad \text{in } \mathbb{R}^3\backslash\overline{B}_l, \tag{6.5.3}$$

where $k_p = \frac{\gamma}{\sqrt{\lambda+2\mu}}$ and $k_s = \frac{\gamma}{\sqrt{\mu}}$ are respectively the wave numbers of compressional and shear waves. Let $\boldsymbol{u}_p = -\frac{1}{k_p^2}\nabla(\nabla\cdot\boldsymbol{u})$ be the compressional part and $\boldsymbol{u}_s = \frac{1}{k_s^2}\nabla\times\nabla\times\boldsymbol{u}$ be the shear part of the wave field. They satisfy the Helmholtz equations

$$\Delta\boldsymbol{u}_p + k_p^2\boldsymbol{u}_p = 0, \qquad \Delta\boldsymbol{u}_s + k_s^2\boldsymbol{u}_s = 0 \quad \text{in } \mathbb{R}^3\backslash\overline{B}_l.$$

It is clear that $\boldsymbol{u} = \boldsymbol{u}_p + \boldsymbol{u}_s$ in $\mathbb{R}^3\backslash\overline{B}_l$. The Kupradze-Sommerfeld radiation condition is given by requirement that $\boldsymbol{u}_p$ and $\boldsymbol{u}_s$ should satisfy the Sommerfeld radiation condition

$$\lim_{|\boldsymbol{x}|\to\infty}|\boldsymbol{x}|\left(\frac{\partial\boldsymbol{u}_p}{\partial|\boldsymbol{x}|} - \mathrm{i}k_p\boldsymbol{u}_p\right) = 0, \quad \lim_{|\boldsymbol{x}|\to\infty}|\boldsymbol{x}|\left(\frac{\partial\boldsymbol{u}_s}{\partial|\boldsymbol{x}|} - \mathrm{i}k_s\boldsymbol{u}_s\right) = 0.$$

Now we derive the PML equation. Let $B_l := \{\boldsymbol{x} = (x_1, x_2, x_3)^T \in \mathbb{R}^3 : |x_i| < l_i, i = 1, 2, 3\}$ contain the scatter D and the support of $\boldsymbol{q}$. Let $\Gamma_l = \partial B_l$ and $\boldsymbol{n}_l$ the unit outer normal to Γ_l. We start by introducing the Dirichlet-to-Neumann operator $\mathbb{T} : [H^{1/2}(\Gamma_l)]^3 \to [H^{-1/2}(\Gamma_l)]^3$. Given $\boldsymbol{f} \in [H^{1/2}(\Gamma_l)]^3$, we define $\mathbb{T}\boldsymbol{f} = \tau(\boldsymbol{\xi})\boldsymbol{n}_l$ with $\boldsymbol{\xi}$ being the solution of the following exterior Dirichlet problem:

$$\begin{cases} \nabla \cdot \tau(\boldsymbol{\xi}) + \gamma^2 \boldsymbol{\xi} = 0 & \text{in } \mathbb{R}^3 \backslash \overline{B}_l, \\ \boldsymbol{\xi} = \boldsymbol{f} & \text{on } \Gamma_l, \\ \boldsymbol{\xi} \text{ satisfies the Kupradze-Sommerfeld radiation conditions at infinity.} \end{cases} \tag{6.5.4}$$

Since (6.5.4) has a unique solution $\boldsymbol{\xi} \in [H^1_{\text{loc}}(\mathbb{R}^3 \backslash \overline{B}_l)]^3$, $\mathbb{T} : [H^{1/2}(\Gamma_l)]^3 \to [H^{-1/2}(\Gamma_l)]^3$ is well-defined and is continuous linear operator.

Let $a : [H^1(\Omega_l)]^3 \times [H^1(\Omega_l)]^3 \to \mathbb{C}$, where $\Omega_l = B_l \backslash \overline{D}$, be the sesquilinear form

$$a(\boldsymbol{\phi}, \boldsymbol{\psi}) = \int_{\Omega_l} \left(\tau(\boldsymbol{\phi}) : \nabla \overline{\boldsymbol{\psi}} - \gamma^2 \boldsymbol{\phi} \cdot \overline{\boldsymbol{\psi}}\right) d\boldsymbol{x} - \langle \mathbb{T}\boldsymbol{\phi}, \boldsymbol{\psi} \rangle_{\gamma_l}. \tag{6.5.5}$$

The weak formulation of the scattering problem (6.5.1) is: Given $\boldsymbol{q} \in ([H^1(\mathbb{R}^3 \backslash \overline{D})]^3)'$ and $\boldsymbol{g} \in [H^{-1/2}(\Gamma_D)]^3$, find $\boldsymbol{u} \in [H^1(\Omega_l)]^3$ such that

$$a(\boldsymbol{u}, \boldsymbol{v}) = (\boldsymbol{q}, \boldsymbol{v})_{\Omega_l} + \langle \boldsymbol{g}, \boldsymbol{v} \rangle_{\Gamma_D}, \qquad \boldsymbol{v} \in [H^1(\Omega_l)]^3. \tag{6.5.6}$$

The existence of a unique of the scattering problem (6.5.6) is a direct consequence of the following lemma [4].

Lemma 6.5.1 *For any $\boldsymbol{q} \in ([H^1(\mathbb{R}^3 \backslash \overline{D})]^3)'$ with compact support and $\boldsymbol{g} \in [H^{-1/2}(\Gamma_D)]^3$, the problem (6.5.1) with the Kupradze-Sommerfeld radiation condition has a unique solution $\boldsymbol{u} \in [H^1_{loc}(\mathbb{R}^3 \backslash \overline{D})]^3$ such that for any bounded open set $\mathscr{O} \subset \mathbb{R} \backslash \overline{D}$ that contains the support of $\boldsymbol{q}$,*

$$\|\boldsymbol{u}\|_{H^1(\mathscr{O} \backslash \overline{D})} \leq C\left(\|\boldsymbol{q}\|_{H^1(\mathbb{R}^3 \backslash \overline{D})'} + \|\boldsymbol{g}\|_{H^{-1/2}(\Gamma_D)}\right). \tag{6.5.7}$$

For the sesquilinear form $a(\cdot, \cdot)$, we associated with a bounded linear operator $\hat{A} : [H^1(\Omega_l)]^3 \to ([H^1(\Omega_l)]^3)'$ such that

$$(\hat{A}\boldsymbol{\phi}, \boldsymbol{\psi})_{\Omega_l} = a(\boldsymbol{\phi}, \boldsymbol{\psi}), \qquad \forall \boldsymbol{\phi}, \boldsymbol{\psi} \in [H^1(\Omega_l)]^3.$$

By Lemma 6.5.1, $\hat{A}$ is surjective and one-to-one. Thus, by the open mapping theorem, we know that there exists a constant $C > 0$ such that the following inf-sup condition is satisfied

$$\sup_{0 \neq \boldsymbol{\psi} \in [H^1(\Omega_l)]^3} \frac{|a(\boldsymbol{\phi}, \boldsymbol{\psi})|}{\|\boldsymbol{\psi}\|_{H^1(\Omega_l)}} \geq C \|\boldsymbol{\phi}\|_{H^1(\Omega_l)}, \qquad \forall \boldsymbol{\phi} \in [H^1(\Omega_l)]^3. \tag{6.5.8}$$

The PML method is based on the complex coordinate stretching outside B_l. Let $\alpha_j(x_j) = 1 + \zeta \sigma_j(x_j) + i \sigma_j(x_j)$, $j = 1, 2, 3$, be the model medium property. We should assume that $\zeta \geq \sqrt{(\lambda + 2\mu)/\mu}$ on the parameter ζ to guarantee the ellipticity of the PML equation.

For $t \in \mathbb{R}$, $\sigma_j(t) \in C^1(\mathbb{R})$, $j = 1, 2, 3$, is an even function such that

$$\sigma_j'(t) \geq 0 \quad \text{for } t \geq 0, \quad \sigma_j = 0 \quad \text{for } |t| \leq l_j, \quad \text{and } \sigma_j = \sigma_0 \quad \text{for } |t| \geq \bar{l}_j, \quad (6.5.9)$$

where $\bar{l}_j > l_j$ is fixed and $\sigma_0 > 0$ is a constant. The requirement that the medium property $\sigma_j(t)$ is constant for $|t| \geq \bar{l}_j$ is essential for using a reflection argument to prove the inf-sup condition for the PML problem in the truncated domain.

For $\boldsymbol{x} \in \mathbb{R}^3$, denote by $\tilde{\boldsymbol{x}}(\boldsymbol{x}) = \left(\tilde{x}_1(x_1), \tilde{x}_2(x_2), \tilde{x}_3(x_3)\right)^T$ the complex coordinate, where

$$\tilde{x}_j(x_j) = \int_0^{x_j} \alpha_j(t)dt = x_j + (\zeta + \mathrm{i}) \int_0^{x_j} \sigma_j(t)dt, \quad j = 1, 2, 3.$$

Note that $\tilde{x}_j(x_j)$ depends only on x_j. For any $z \in \mathbb{C}_{++} := \{z \in \mathbb{C} : Re(z) \geq 0, Im(z) \geq 0\}$, denote

$$\tilde{x}_j^z(x_j) = x_j + z \int_0^{x_j} \sigma_j(t)dt, \quad j = 1, 2, 3. \quad (6.5.10)$$

Write $\tilde{\boldsymbol{x}}_z = \left(\tilde{x}_1^z(x_1), \tilde{x}_2^z(x_2), \tilde{x}_3^z(x_3)\right)^T$ and $\tilde{\boldsymbol{y}}_z = \left(\tilde{y}_1^z(y_1), \tilde{y}_2^z(y_2), \tilde{y}_3^z(y_3)\right)^T$. We define the complex distance

$$d(\tilde{\boldsymbol{x}}_z, \tilde{\boldsymbol{y}}_z) = [(\tilde{x}_1^z(x_1) - \tilde{y}_1^z(y_1))^2 + (\tilde{x}_2^z(x_2) - \tilde{y}_2^z(y_2))^2 + (\tilde{x}_3^z(x_3) - \tilde{y}_3^z(y_3))^2]^{1/2}.$$

Here and in the following, for any $z \in \mathbb{C}$, $z^{1/2}$ is the analytic branch of $\sqrt{z}$ such that $R(z^{1/2}) > 0$ for any $z \in \mathbb{C}\backslash(-\infty, 0]$. It is obvious that $\tilde{\boldsymbol{x}}_{z_0} = \tilde{\boldsymbol{x}}$, where $z_0 = \zeta + i$.

Lemma 6.5.2 *For any* $z \in U := \{z \in \mathbb{C} : Re(z) > |Im(z)|\}$, *we have*

$$|\boldsymbol{x} - \boldsymbol{y}| \leq d(\tilde{\boldsymbol{x}}_z, \tilde{\boldsymbol{y}}_z) \leq (1 + |z|\sigma_0)|\boldsymbol{x} - \boldsymbol{y}|, \quad \forall \boldsymbol{x}, \boldsymbol{y} \in \mathbb{R}^3.$$

Now we are ready to introduce the PML equation based on the method of complex coordinate stretching. By Betti formula, the solution $\boldsymbol{\xi}$ of the exterior Dirichlet problem (6.5.4) satisfies:

$$\boldsymbol{\xi} = -\boldsymbol{\Psi}_{SL}(\mathbb{T}\boldsymbol{f}) + \boldsymbol{\Psi}_{DL}(\boldsymbol{f}) \quad \text{in } \mathbb{R}^3 \backslash \overline{B}_l, \quad (6.5.11)$$

where $\boldsymbol{\Psi}_{SL}$, $\boldsymbol{\Psi}_{DL}$ are respectively the single and double layer potentials. For $n = 1, 2, 3$, the n-th component of the potentials are, for $\boldsymbol{\lambda} \in [H^{-1/2}(\Gamma_l)]^3$, $\boldsymbol{f} \in [H^{1/2}(\Gamma_l)]^3$,

$$\boldsymbol{\Psi}_{SL}(\boldsymbol{\lambda})(\boldsymbol{x}) \cdot \boldsymbol{e}_n = \left\langle \boldsymbol{\lambda}, \overline{\Gamma(\boldsymbol{x}, \cdot)\boldsymbol{e}_n} \right\rangle_{\Gamma_l}, \quad \boldsymbol{\Psi}_{DL}(\boldsymbol{\lambda})(\boldsymbol{f}) \cdot \boldsymbol{e}_n = \left\langle \mathbb{T}[\Gamma(\boldsymbol{x}, \cdot)\boldsymbol{e}_n], \overline{\boldsymbol{f}} \right\rangle_{\Gamma_l}.$$

Here $\boldsymbol{e}_n$ is the unit vector in the in the x_n direction and $\Gamma((x), (y))\boldsymbol{e}_n$ is the n-th column of the fundamental solution matrix $\Gamma((x), (y))$ of the time harmonic elastic wave equation satisfying the Kupradze-Sommerfeld radiation condition. The (j, k)-element of $\Gamma((x), (y))$ is

$$\Gamma_{jk}((x),(y)) = \frac{1}{\gamma^2}\left[k_s^2 G_{k_s}(\boldsymbol{x},\boldsymbol{y})\delta_{jk} - \frac{\partial^2}{\partial x_j \partial x_k}(G_{k_p}(\boldsymbol{x},\boldsymbol{y}) - G_{k_s}(\boldsymbol{x},\boldsymbol{y}))\right],$$

where $G_k(\boldsymbol{x},\boldsymbol{y}) = f_k(|\boldsymbol{x}-\boldsymbol{y}|)$, $f_k(r) = \frac{e^{ikr}}{4\pi r}$ for $r > 0$, is the fundamental solution of the Helmholtz equation of wave number k. It is known that $\boldsymbol{\Psi}_{SL}(\boldsymbol{\lambda}) \in [H^1_{\text{loc}}(\mathbb{R}^3)]^3$ for $\boldsymbol{\lambda} \in [H^{-1/2}(\Gamma_l)]^3$ and $\boldsymbol{\Psi}_{DL}(\boldsymbol{f}) \in [H^{-1/2}(\Gamma_l)]^3$ for $\boldsymbol{f} \in [H^{1/2}(\Gamma_l)]^3$.

Straightforward calculation shows that

$$\Gamma_{jk}(\boldsymbol{x},\boldsymbol{y}) = \Gamma_1(|\boldsymbol{x}-\boldsymbol{y}|)\delta_{jk} + \Gamma_2(|\boldsymbol{x}-\boldsymbol{y}|)\frac{(x_j - y_j)(x_k - y_k)}{|\boldsymbol{x}-\boldsymbol{y}|^2}, \tag{6.5.12}$$

where, for $r > 0$,

$$\Gamma_1(r) = \frac{1}{\gamma^2}\left[k_s^2 f_{k_s}(r) - \frac{f'_{k_p} - f'_{k_s}}{r}\right], \tag{6.5.13}$$

$$\Gamma_2(r) = \frac{1}{\gamma^2}\left[3\frac{f'_{k_p} - f'_{k_s}}{r} + (k_p^2 f'_{k_p}(r) - k_s^2 f'_{k_s}(r))\right]. \tag{6.5.14}$$

The functions Γ_1 and Γ_2 can be extended to be analytic functions defined in $\mathbb{C}\backslash\{0\}$.

Lemma 6.5.3 *For $j = 1, 2$, $\Gamma_j(z)$ is analytic in $\mathbb{C}\backslash\{0\}$. Moreover, $|\Gamma_j(z)| \le C|z|^{-1}$, $|\Gamma'_j(z)| \le C|z|^{-2}$, and $|\Gamma''_j(z)| \le C|z|^{-3}$ uniformly for $z \in \mathbb{C}\backslash\{0\}$, $|z| \le 1$.*

For any $z \in U = \{z \in \mathbb{C} : Re(z) > |Im(z)|\}$ defined in Lemma 6.5.2, we define the modified single and double layer potentials $(\tilde{\Psi})^z_{SL}$ and $\tilde{\boldsymbol{Psi}}^z_{DL}$ as follows. For $\boldsymbol{\lambda} \in [H^{-1/2}(\Gamma_l)]^3$, $\boldsymbol{f} \in [H^{1/2}(\Gamma_l)]^3$, the n-th component, $n = 1, 2, 3$, of the modified potentials are

$$(\tilde{\Psi})^z_{SL}(\boldsymbol{\lambda})(\boldsymbol{x}) \cdot \boldsymbol{e}_n = \left\langle \boldsymbol{\lambda}, \overline{\tilde{\Gamma}_z(\boldsymbol{x},\cdot)\boldsymbol{e}_n} \right\rangle_{\Gamma_l}, \quad (\tilde{\Psi})^z_{DL}(\boldsymbol{f})(\boldsymbol{x}) \cdot \boldsymbol{e}_n = \left\langle \mathbb{T}[\tilde{\Gamma}_z(\boldsymbol{x},\cdot)\boldsymbol{e}_n], \overline{\boldsymbol{f}} \right\rangle_{\Gamma_l},$$

where the (j,k)-element of the matrix $\tilde{\Gamma}_z(\boldsymbol{x},\boldsymbol{y})$ is

$$\tilde{\Gamma}^{jk}_j(\boldsymbol{x},\boldsymbol{y}) = \Gamma_1(d(\tilde{\boldsymbol{x}}_z,\tilde{\boldsymbol{y}}_z))\delta_{jk} + \Gamma_2(d(\tilde{\boldsymbol{x}}_z,\tilde{\boldsymbol{y}}_z))\frac{(\tilde{x}^z_j - \tilde{y}^z_j)(\tilde{x}^z_k - \tilde{y}^z_k)}{d(\tilde{\boldsymbol{x}}_z,\tilde{\boldsymbol{y}}_z)^2}. \tag{6.5.15}$$

In the following, for $z_0 = \zeta + \mathrm{i}$, we denote $\tilde{\Gamma}(\boldsymbol{x},\boldsymbol{y}) = \tilde{\Gamma}_{z_0}(\boldsymbol{x},\boldsymbol{y})$, $\tilde{\Gamma}_{jk}(\boldsymbol{x},\boldsymbol{y}) = \tilde{\Gamma}^{jk}_{z_0}(\boldsymbol{x},\boldsymbol{y})$, and, for any $\boldsymbol{\lambda} \in [H^{-1/2}(\Gamma_l)]^3$, $\boldsymbol{f} \in [H^{1/2}(\Gamma_l)]^3$,

$$(\tilde{\Psi})_{SL}(\boldsymbol{\lambda}) = (\tilde{\Psi})^{z_0}_{SL}(\boldsymbol{\lambda}), \qquad (\tilde{\Psi})_{DL}(\boldsymbol{f}) = (\tilde{\Psi})^{z_0}_{DL}(\boldsymbol{f}).$$

Lemma 6.5.4 *Let the assumption $\zeta \ge \sqrt{(\lambda + 2\mu)/\mu}$ be satisfied. For $j, k = 1, 2, 3$, we have for any $\boldsymbol{x}, \boldsymbol{y} \in \mathbb{R}^3$ such that $Imd(\tilde{x},\tilde{y}) > 0$,*

$$|\tilde{\Gamma}_{jk}(\boldsymbol{x},\boldsymbol{y})| \le C(1+|z_0|\sigma_0)^2|\boldsymbol{x}-\boldsymbol{y}|^{-1}e^{-k_p Imd(\tilde{x},\tilde{y})},$$
$$|\nabla_{\boldsymbol{x}}\tilde{\Gamma}_{jk}(\boldsymbol{x},\boldsymbol{y})| \le C(1+|z_0|\sigma_0)^4(|\boldsymbol{x}-\boldsymbol{y}|^{-1}+|\boldsymbol{x}-\boldsymbol{y}|^{-2})e^{-k_p Imd(\tilde{x},\tilde{y})},$$
$$|\nabla_{\boldsymbol{y}}\tilde{\Gamma}_{jk}(\boldsymbol{x},\boldsymbol{y})| \le C(1+|z_0|\sigma_0)^4(|\boldsymbol{x}-\boldsymbol{y}|^{-1}+|\boldsymbol{x}-\boldsymbol{y}|^{-2})e^{-k_p Imd(\tilde{x},\tilde{y})},$$
$$|\nabla_{\boldsymbol{x}}\nabla_{\boldsymbol{y}}\tilde{\Gamma}_{jk}(\boldsymbol{x},\boldsymbol{y})| \le C(1+|z_0|\sigma_0)^6(|\boldsymbol{x}-\boldsymbol{y}|^{-1}+|\boldsymbol{x}-\boldsymbol{y}|^{-3})e^{-k_p Imd(\tilde{x},\tilde{y})}.$$

Lemma 6.5.5 *For any $z_i = a_i + ib_i$ with $a_i, b_i \in \mathbb{R}$, $i = 1, 2, 3$, such that $a_1b_1 + a_2b_2 + a_3b_3 \ge 0$ and $a_1^2 + a_2^2 + a_3^2 > 0$, we have*

$$Im(z_1^2 + z_2^2 + z_3^2)^{1/2} \ge \frac{a_1b_1 + a_2b_2 + a_3b_3}{\sqrt{a_1^2 + a_2^2 + a_3^2}}.$$

Let $z_j = \tilde{x}_j - \tilde{y}_j = (x_j - y_j) + (\zeta + i)\int_{y_j}^{x_j} \sigma_j(t)dt$, $j = 1, 2, 3$. By the above Lemma, $d(\tilde{\boldsymbol{x}}, \tilde{\boldsymbol{y}}) = (z_1^2 + z_2^2 + z_3^2)^{1/2}$ satisfies

$$Imd(\tilde{\boldsymbol{x}}, \tilde{\boldsymbol{y}}) \ge \frac{\sum_{j=1}^{3}\left(|x_j - y_j|\left|\int_{y_j}^{x_j}\sigma_j(t)dt\right| + \zeta\left|\int_{y_j}^{x_j}\sigma_j(t)dt\right|^2\right)}{(1+\zeta\sigma_0)|\boldsymbol{x}-\boldsymbol{y}|}. \tag{6.5.16}$$

The following lemma shows that $Imd(\tilde{\boldsymbol{x}}, \tilde{\boldsymbol{y}})$ is bounded below by $|\boldsymbol{x} - \boldsymbol{y}|$ if $\boldsymbol{x}, \boldsymbol{y}$ are far away.

Lemma 6.5.6 *Let $\beta > 1$ be a fixed number. If $|\boldsymbol{x} - \boldsymbol{y}| \ge 2\sqrt{3}\beta\bar{l}_{max}$, where $\bar{l}_{max} = \max_{j=1,2,3}\bar{l}_j$, where $\bar{l}_j$, $j = 1, 2, 3$, are defined in (6.4.3), we have $Imd(\tilde{\boldsymbol{x}}, \tilde{\boldsymbol{y}}) \ge \frac{1}{3}(1-\beta^{-1})^2\sigma_0|\boldsymbol{x}-\boldsymbol{y}|$.*

For any $\boldsymbol{f} \in [H^{1/2}(\Gamma_l)]^3$, let $\mathbb{E}(\boldsymbol{f})(\boldsymbol{x})$ be the PML extension:

$$\mathbb{E}(\boldsymbol{f})(\boldsymbol{x}) = -\tilde{\boldsymbol{\Psi}}_{SL}(\mathbb{T}\boldsymbol{f}) + \tilde{\boldsymbol{\Psi}}_{DL}(\boldsymbol{f}), \quad \forall \boldsymbol{x} \in \mathbb{R}^3\backslash\overline{B}_l. \tag{6.5.17}$$

By (6.5.11) we know that $\mathbb{E}(\boldsymbol{f}) = \boldsymbol{f}$ on Γ_l for any $\boldsymbol{f} \in [H^{1/2}(\Gamma_l)]^3$. By Lemma 6.5.2, $|d(\tilde{\boldsymbol{x}}, \tilde{\boldsymbol{y}})| \ge |\boldsymbol{x} - \boldsymbol{y}|$ for $\boldsymbol{x} \in \mathbb{R}^3\backslash\overline{B}_l$, $\boldsymbol{y} \in \Gamma_l$. Thus since $\sigma_j \in C^1(\mathbb{R})$, $j = 1, 2, 3$, we have $\mathbb{E}(\boldsymbol{f}) \in C^2(\mathbb{R}^3\backslash\overline{B}_l)$. Moreover, by Lemmas 6.5.4 and 6.5.6 we know that $\mathbb{E}(\boldsymbol{f})$ decays exponentially as $|\boldsymbol{x}| \to \infty$.

For the solution $\boldsymbol{u}$ of the scattering problem (6.5.6), let $\tilde{\boldsymbol{u}} = \mathbb{E}(\boldsymbol{u}|_{\Gamma_l})$ be the PML extension of $\boldsymbol{u}|_{\Gamma_l}$. It satisfies $\tilde{\boldsymbol{u}} = \boldsymbol{u}|_{\Gamma_l}$ on Γ_l and the equation

$$\tilde{\nabla}\cdot\tilde{\tau}(\tilde{\boldsymbol{u}}) + \gamma^2\tilde{\boldsymbol{u}} = 0 \quad \text{in } \mathbb{R}^3\backslash\overline{B}_l, \tag{6.5.18}$$

where

$$\tilde{\tau}(\tilde{\boldsymbol{u}}) = 2\mu\tilde{\varepsilon}(\tilde{\boldsymbol{u}}) + \lambda\mathrm{tr}(\tilde{\varepsilon}(\tilde{\boldsymbol{u}}))I, \quad \tilde{\varepsilon}(\tilde{\boldsymbol{u}}) = \frac{1}{2}(\tilde{\nabla}\tilde{\boldsymbol{u}} + (\tilde{\nabla}\tilde{\boldsymbol{u}})^T).$$

Here $\tilde{\nabla}\tilde{\boldsymbol{u}} \in \mathbb{C}^{3\times 3}$ whose elements are $(\partial \tilde{u}_i)/\partial \tilde{x}_j, i, j = 1, 2, 3$. For $\boldsymbol{x} \in \mathbb{R}^3$, let $\boldsymbol{F}(\boldsymbol{x}) = \big(F_1(x_1), F_2(x_2), F_3(x_3)\big)^T$ with $F_j(x_j) = \tilde{x}_j(x_j), j = 1, 2, 3$. Then $\tilde{\boldsymbol{x}}(\boldsymbol{x}) = \boldsymbol{F}(\boldsymbol{x})$. Denote by $\nabla \boldsymbol{F}$ the Jacobi matrix of $\boldsymbol{F}$, then

$$\tilde{\nabla}\cdot = J^{-1}\nabla \cdot J(\nabla \boldsymbol{F})^{-1}, \qquad J = \det(\nabla \boldsymbol{F}). \tag{6.5.19}$$

By (6.5.19) we easily obtain from (6.5.18) the desired PML equation

$$\nabla \cdot (\tilde{\tau}(\tilde{\boldsymbol{u}})A) + \gamma^2 J\tilde{\boldsymbol{u}} = 0 \quad \text{ in } \mathbb{R}^3\backslash \overline{B}_l.$$

Here

$$\tilde{\tau}(\tilde{\boldsymbol{u}}) = 2\mu\tilde{\varepsilon}(\tilde{\boldsymbol{u}}) + \lambda \mathrm{tr}(\tilde{\varepsilon}(\tilde{\boldsymbol{u}}))I, \quad \tilde{\varepsilon}(\tilde{\boldsymbol{u}}) = \frac{1}{2}\big(\tilde{\nabla}\tilde{\boldsymbol{u}}B^T + B(\tilde{\nabla}\tilde{\boldsymbol{u}})^T\big).$$

where $B = (\nabla \boldsymbol{F})^{-T} = \mathrm{diag}\big(\alpha_1(x_1)^{-1}, \alpha_2(x_2)^{-1}, \alpha_3(x_3)^{-1}\big) \in \mathbb{C}^{3\times 3}$ is a diagonal matrix and $A = J(\nabla \boldsymbol{F})^{-T} = JB$. We notice that $\tilde{\tau}(\boldsymbol{\phi}) = \tilde{\tau}(\boldsymbol{x}, \boldsymbol{\phi})$, $\tilde{\varepsilon}(\boldsymbol{\phi}) = \tilde{\varepsilon}(\boldsymbol{x}, \boldsymbol{\phi})$ which satisfies $\tilde{\tau}(\boldsymbol{x}, \boldsymbol{\phi}) = \tau(\boldsymbol{\phi})$, $\tilde{\varepsilon}(\boldsymbol{x}, \boldsymbol{\phi}) = \varepsilon(\boldsymbol{\phi})$ for $\boldsymbol{x} \in B_l$.

Let $B_L = \{\boldsymbol{x} \in \mathbb{R}^3 : |x_i| < L_i, i = 1, 2, 3\}$ be the domain containing B_l. The PML solution $\hat{\boldsymbol{u}}$ in $\Omega_L = B_L\backslash \overline{D}$ is defined as the weak solution of the following problem:

$$\begin{cases} \nabla \cdot (\tilde{\tau}(\tilde{\boldsymbol{u}})A) + \gamma^2 J\tilde{\boldsymbol{u}} = -\boldsymbol{q} & \text{in } \Omega_L, \\ \tilde{\tau}(\tilde{\boldsymbol{u}})A\boldsymbol{n}_D = -\boldsymbol{g} & \text{on } \Gamma_D, \\ \hat{\boldsymbol{u}}\cdot \boldsymbol{n} = 0, \quad \tilde{\tau}(\tilde{\boldsymbol{u}})A\boldsymbol{n}\times \boldsymbol{n} = 0 & \text{on } \Gamma_L := \partial B_L. \end{cases} \tag{6.5.20}$$

The well-posedness of the above PML problem the the convergence of its solution to the solution of the original scattering problem was studied in [4].

We then can introduce the finite element approximation of the PML problem (6.5.20). We assume $\boldsymbol{q} \in [L^2(\Omega_l)]^3$, $\boldsymbol{g} \in [L^2(\Gamma_D)]^3$. Let $\mathcal{T}_h$ be a regular triangulation of the domain Ω_L. We assume the elements $T \in \mathcal{T}_h$ may have one curved side align with Γ_D so that $\Omega_L = \cup_{T\in\mathcal{T}_h} T$. Let $V_h \in [H^1(\Omega_L)]^3$ be the conforming quadratic finite element space over Ω_L, and $V_{0,h} = \{\boldsymbol{v}_h \in V_h : \boldsymbol{v}_h \cdot \boldsymbol{n} = 0 \text{ on } \Gamma_L\}$. The finite element approximation to the PML problem (6.5.20) reads as follows: Find $\boldsymbol{u}_h \in V_{0,h}$ such that

$$b(\boldsymbol{u}_h, \boldsymbol{\psi}_h) = (\boldsymbol{q}, \boldsymbol{\psi}_h)_{\Omega_L} + \langle \boldsymbol{g}, \boldsymbol{\psi}_h\rangle_{\Gamma_D}, \quad \forall \boldsymbol{\psi}_h \in V_{0,h}. \tag{6.5.21}$$

Let $\omega = 5, d = (1, 0, 0)^\top, \lambda = \mu = 1, \alpha = 1, \beta = 0$, then the incident wave is $u^i = (\mathrm{e}^{\mathrm{i}k_p \mathrm{x}\cdot \mathrm{d}}, 0, 0)^\top$. The numerical results are shown in (Fig. 6.1).

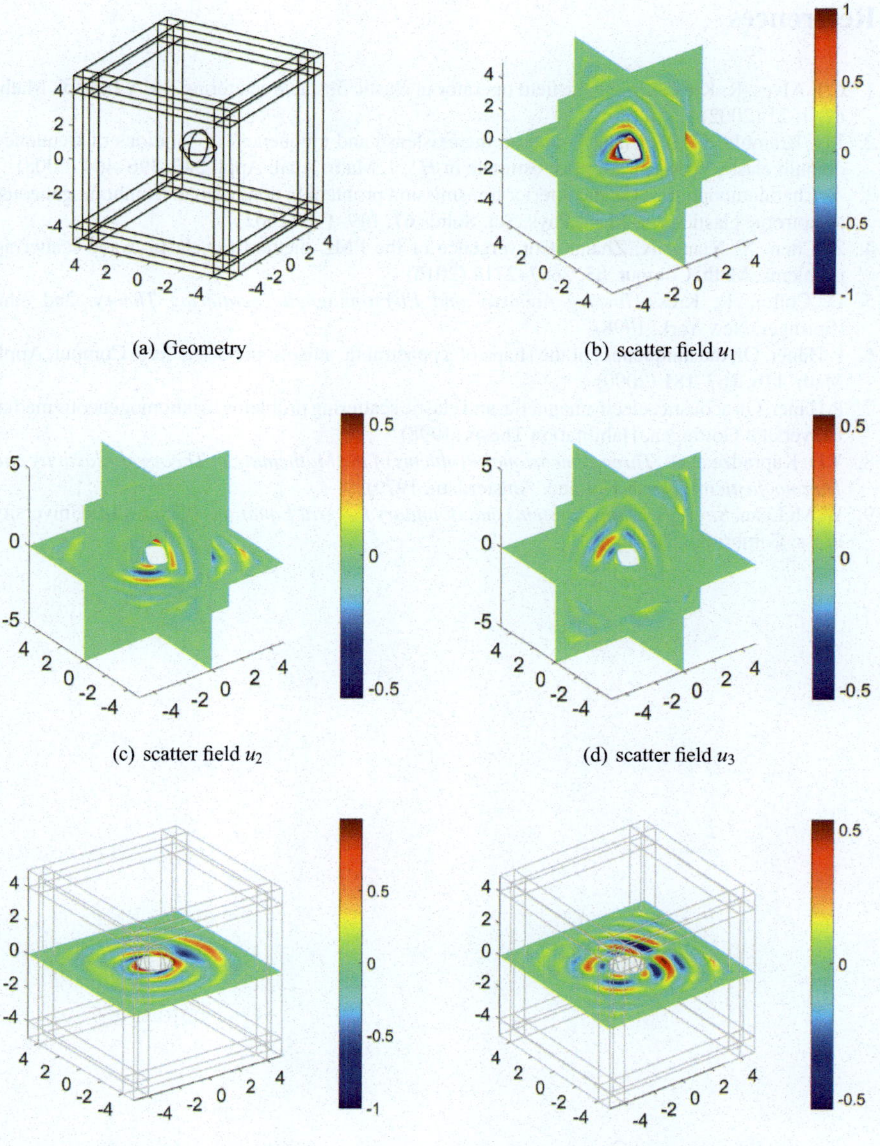

(a) Geometry

(b) scatter field u_1

(c) scatter field u_2

(d) scatter field u_3

(e) scatter field u_1, slice at $z = 0$

(f) scatter field u_2, slice at $z = 0$

Fig. 6.1 The incident direction is $d = (1, 0, 0)$

References

1. C.J. Alves, R. Kress, On the far-field operator in elastic obstacle scattering. IMA J. Appl. Math. **67**, 1–21 (2002)
2. J.H. Bramble, J.E. Pasciak, A note on the existense and uniqueness of solutions of frequency domain elastic problems: A prior estimate in H^1. J. Math. Anal. Appl. **345**, 396–404 (2008)
3. A. Charalambopoulos, On the interior transmission problem in nondissipative, inhomogeneous, anisotropic elasticity. J. Elast. Phys. Sci. Solids **67**, 149–170 (2002)
4. Z. Chen, X. Xiang, X. Zhang, Convergence of the PML method for elastic wave scattering problems. Math. Comput. **85**, 2687–2714 (2016)
5. D. Colton, R. Kress, *Inverse Acoustic and Electromagnetic Scattering Theory*, 2nd edn. (Springer, New York, 1998)
6. P. Häner, On the uniqueness of the shape of a penetrable, anisotropic obstacle. J. Comput. Appl. Math. **116**, 167–181 (2000)
7. P. Häner, On acoustic, electromagnetic, and elastic scattering problems in inhomogeneous media. Universität Göttingen, Habilitation Thesis (1998)
8. V.D. Kupradze et al., *Three-dimensional Problems of the Mathematical Theory of Elasticity and Thermoelasticity* (North-Holland, Amsterdam, 1979)
9. W. McLean, *Strongly Elliptic Systems and Boundary Integral Equations* (Cambridge University Press, Cambridge, 2000)

Chapter 7
Numerical Inverse Elastic Scattering Problems

7.1 Overview

The elastic wave propagation problems have a wide range of applications, particularly in geophysics, nondestructive testing and seismology. The associated inverse problems arise from the use of transient elastic waves to identify the elastic properties as well as to detect flaws and cracks of solid speciments, especially in the nondestructive evaluation of concrete structures (see e.g. [55, 58]). Moreover, the problem of elastic pulse transmission and reflection through the earth is fundamental to both the investigation of earthquakes and the utility of seismic waves in search for oil and ore bodies (see, e.g., [1, 20, 21, 36, 57] and the references therein). The scattering of elastic waves are very complicated due to the coexistence of compressional and shear waves propagating at different speeds.

There is a vast literature on the inverse elastic scattering problem as described above. We refer to the theoretical uniqueness results proved in [24, 47, 49–54] and, the sampling-type reconstruction methods for impenetrable elastic bodies developed in [3, 7] and those for penetrable ones in [13, 56]. Note that in the above works, both u_p^∞ and u_s^∞ are needed for infinitely many incident plane waves, namely infinitely many far-field patterns are needed. In [2, 4–6] for reconstructing small elastic inclusions and in [22] for reconstructing extended rigid scatterers, one may implement one type of elastic waves, but still with multiple or even infinitely many far-field patterns. Using a single set of boundary data, an extraction formula of an unknown linear crack or the convex hull of an unknown polygonal cavity in $\mathbb{R}^2$ was established in [30, 31] by means of the enclosure method introduced by Ikehata [32]. The inverse elastic scattering problems by a single measurement is highly challenging. We refer to [8, 16–18, 35, 42, 43] for the theoretical uniqueness and stability results, and [27, 29, 37] for related numerical studies on this intriguing topic.

This chapter follows the treatment in [28, 29].

J. Li and H. Liu, *Numerical Methods for Inverse Scattering Problems*,
https://doi.org/10.1007/978-981-99-3772-1_7

7.2 Single-Shot Method for Multiple Multiscale Scatterers

The rest of the section is organized as follows. In Sect. 7.2.1, we first describe Scheme S of locating multiple small scatterers, and then present the theoretical justification. In Sect. 7.2.2, we first describe Scheme R of locating multiple extended scatterers and then present the corresponding theoretical justification. Section 7.2.3 is devoted to Scheme M of locating multiple multiscale scatterers. In Sect. 7.2.4, numerical experiments are given to demonstrate the effectiveness and the promising features of the discussed inverse scattering schemes. We conclude our study in Sect. 7.2.5 with several remarks.

7.2.1 *Locating Multiple Small Scatterers*

Throughout the rest of the subsection, we assume the angular frequency of incidence is $\omega = 1$. Then, the wavelength of the pressure wave is $2\pi/k_p = 2\pi\sqrt{\lambda + 2\mu}/\omega = \mathscr{O}(1)$, whereas the wavelength of the shear wave is $2\pi/k_s = 2\pi\sqrt{\mu}/\omega = \mathscr{O}(1)$. Hence, the size of a scatterer can be expressed in terms of its Euclidean diameter. In the sequel we write $u_\tau^\infty(\hat{x}; D, d, d^\perp, \omega)$ ($\tau = p, s$ or $\emptyset$) to signify the dependence of far-field pattern on the rigid scatterer D, incident direction d, polarization direction $d^\perp$ and incidence frequency ω. In certain situations we only indicate the dependence of the far-field pattern on D or ω, but the notation shall be clear from the context. Unless otherwise stated, the space L^2 always signifies $L^2(\mathbb{S}^2)^3$.

Next, we first describe Scheme S of locating multiple small rigid elastic scatterers and then present the corresponding theoretical justifications.

7.2.1.1 Description of Scheme S

We first introduce the class of small elastic rigid scatterers. For $l_s \in \mathbb{N}$, let M_j, $1 \le j \le l_s$, be bounded Lipschitz simply-connected domains in $\mathbb{R}^3$. It is supposed that all M_j's contain the origin and their diameters are comparable with the S-wavelength or P-wavelength, i.e., $\mathrm{diam}(M_j) \sim \mathscr{O}(1)$ for all $j = 1, 2, \ldots, l_s$. For $\rho \in \mathbb{R}_+$, we introduce a scaling/dilation operator Λ_ρ by

$$\Lambda_\rho M_j := \{\rho x : x \in M_j\} \tag{7.2.1}$$

and set

$$D_j := z_j + \Lambda_\rho M_j, \quad z_j \in \mathbb{R}^3, \quad 1 \le j \le l_s. \tag{7.2.2}$$

Each D_j is referred to as a scatterer component located at z_j with the shape M_j. The number ρ represents the *relative size/scale* of each component. In the sequel, we reserve the letter l_s to denote the number of components of a small scatterer given by

$$D = \bigcup_{j=1}^{l_s} D_j. \tag{7.2.3}$$

For technical purpose, we next make the following qualitative assumption that $\rho \ll 1$ and

$$L_s = \min_{j \neq j', 1 \leq j, j' \leq l_s} \operatorname{dist}(z_j, z_{j'}) \gg 1. \tag{7.2.4}$$

The above assumption means that the size of each scatterer component is small compared to detecting wavelength, and if there are multiple components, they are sparsely distributed. In our numerical experiments in Sect. 7.2.4, we could speak a bit more about the qualitative assumption (7.2.4). Indeed, it is shown that as long as the size of the target scatterer is smaller than half a wavelength, and if there are multiple components presented, the distance between different components is bigger than half a wavelength, the discussed locating methods in this section work in an effective manner.

In order to present Scheme S of locating the multiple components of D in (7.2.3), we introduce the following three indicator functions $I_m(z)$, $z \in \mathbb{R}^3$, $m = 1, 2, 3$, relying on the availability of different types of far-field patterns. Define

$$I_1(z) = \frac{1}{\|u_p^\infty(\hat{x}; D)\|_{L^2}^2} \sum_{j=1}^{3} \left| \left\langle u_p^\infty(\hat{x}; D), (\hat{x} \otimes \hat{x})\mathbf{e}_j \, e^{-ik_p \hat{x}\cdot z} \right\rangle \right|^2,$$

$$I_2(z) = \frac{1}{\|u_s^\infty(\hat{x}; D)\|_{L^2}^2} \sum_{j=1}^{3} \left| \left\langle u_s^\infty(\hat{x}; D), (\mathbf{I} - \hat{x} \otimes \hat{x})\mathbf{e}_j \, e^{-ik_s \hat{x}\cdot z} \right\rangle \right|^2,$$

$$I_3(z) = \frac{1}{\|u^\infty(\hat{x}; D)\|_{L^2}^2} \sum_{j=1}^{3} \left| \left\langle u^\infty(\hat{x}; D), (\hat{x} \otimes \hat{x})\mathbf{e}_j \, e^{-ik_p \hat{x}\cdot z} + (\mathbf{I} - \hat{x} \otimes \hat{x})\mathbf{e}_j \, e^{-ik_s \hat{x}\cdot z} \right\rangle \right|^2,$$

where and in the following, the notation $\langle \cdot, \cdot \rangle$ denotes the inner product in $L^2 = L^2(\mathbb{S}^2)^3$ with respect to the variable $\hat{x} \in \mathbb{S}^2$; the symbol $\hat{x} \otimes \hat{x} := \hat{x}^\top \hat{x} \in \mathbb{R}^{3\times 3}$ stands for the tensor product; $\mathbf{I}$ denotes the 3×3 identity matrix; and

$$\mathbf{e}_1 = (1, 0, 0)^\top, \quad \mathbf{e}_2 = (0, 1, 0)^\top, \quad \mathbf{e}_3 = (0, 0, 1)^\top,$$

are the three Euclidean base vectors in $\mathbb{R}^3$. Obviously, I_m ($m = 1, 2, 3$) are all non-negative functions and they can be obtained, respectively, by using a single P-part far-field pattern ($m = 1$), S-part far-field pattern ($m = 2$), or the full far-field pattern ($m = 3$). The functions introduced above possess certain indicating behavior, which lies in the core of Scheme S. Before stating the theorem of the indicating behavior for those imaging functions, we introduce the real numbers

$$K_1^j := \frac{\|u_p^\infty(\hat{x}; D_j)\|_{L^2}^2}{\|u_p^\infty(\hat{x}; D)\|_{L^2}^2}, \quad K_2^j := \frac{\|u_s^\infty(\hat{x}; D_j)\|_{L^2}^2}{\|u_s^\infty(\hat{x}; D)\|_{L^2}^2}, \quad K_3^j := \frac{\|u^\infty(\hat{x}; D_j)\|_{L^2}^2}{\|u^\infty(\hat{x}; D)\|_{L^2}^2}, \tag{7.2.5}$$

for $1 \le j \le l_s$.

Theorem 7.2.1 *For a rigid elastic scatterer D described in* (7.2.1)–(7.2.4), *and* K_m^j, $m = 1, 2, 3$, *defined in* (7.2.5), *we have*

$$K_m^j = \widetilde{K}^j + \mathcal{O}(L_s^{-1} + \rho), \quad 1 \le j \le l_s, \ m = 1, 2, 3, \tag{7.2.6}$$

where $\widetilde{K}^j$*'s are positive numbers independent of* L_s, ρ *and* m. *Moreover, there exists an open neighborhood of* z_j, neigh(z_j), *such that*

$$I_m(z) \le \widetilde{K}^j + \mathcal{O}(L_s^{-1} + \rho) \ \ \textit{for all} \ \ z \in \text{neigh}(z_j), \tag{7.2.7}$$

and $I_m(z)$ *achieves its maximum at* z_j *in* neigh(z_j), *i.e.,*

$$I_m(z_j) = \widetilde{K}^j + \mathcal{O}(L_s^{-1} + \rho). \tag{7.2.8}$$

Remark 7.2.1 The local maximizing behavior of $I_m(z)$ clearly can be used to locate the positions of the scatterer components D, namely z_j, $1 \le j \le l_s$. Such indicating behavior is more evident if one considers the case that D has only one component, i.e., $l_s = 1$. In the one-component case, one has that

$$\widetilde{K}^j = 1, \quad I_m(z) < 1 + \mathcal{O}(\rho) \ \ \text{for all} \ \ m = 1, 2, 3, \ z \ne z_1,$$

but

$$I_m(z_1) = 1 + \mathcal{O}(\rho), \quad m = 1, 2, 3.$$

That is, z_1 is a global maximizer for $I_m(z)$.

Based on Theorem 7.2.1 we can formulate Scheme S to locate the multiple small scatterer components of D in (7.2.3) as follows.

Scheme S	Locating small scatterers of D in (7.2.3)
Step 1	For an unknown rigid scatterer D in (7.2.3), collect the P-part ($m = 1$), S-part ($m = 2$) or the full far-field data ($m = 3$) by sending a single detecting plane wave
Step 2	Select a sampling region with a mesh $\mathcal{T}_h$ containing D
Step 3	For each sampling point $z \in \mathcal{T}_h$, calculate $I_m(z)$ ($m = 1, 2, 3$) according to the measurement data
Step 4	Locate all the local maximizers of $I_m(z)$ on $\mathcal{T}_h$, which represent the locations of the scatterer components

Remark 7.2.2 In practice, the compressional wave number $k_p = \omega/\sqrt{\lambda + 2\mu}$ is smaller than the shear wave number $k_s = \omega/\sqrt{\mu}$. Hence, the P-wavelength $2\pi/k_p$ is usually larger than the S-wavelength $2\pi/k_s$. This suggests that using the shear wave measurement would yield better reconstruction than using the compressional wave measurement for locating the multiple small scatterers. That is, the indicator function I_2 would work better than I_1 for the reconstruction purpose, especially when the Láme constant λ is very large compared to μ. This also suggests that the reconstruction using the indicator function I_3 with the full far-field pattern will be more stable (w.r.t. noise) and reliable than the other two; see also Sect. 7.2.4.

7.2.1.2 Proof of Theorem 7.2.1

In this section, we provide the proof for Theorem 7.2.1. First, we recall the fundamental solution (Green's tensor) to the Navier equation (6.1.1) given by

$$\Pi(x, y) = \Pi^{(\omega)}(x, y) = \frac{k_s^2}{4\pi\omega^2}\frac{e^{ik_s|x-y|}}{|x-y|}\mathbf{I} + \frac{1}{4\pi\omega^2}\nabla_x\nabla_x^{\top}\left[\frac{e^{ik_s|x-y|}}{|x-y|} - \frac{e^{ik_p|x-y|}}{|x-y|}\right], \tag{7.2.9}$$

for $x, y \in \mathbb{R}^3$, $x \neq y$. In order to prove Theorem 7.2.1 we need the following critical lemma on the asymptotic behavior of the elastic far-field patterns due to small scatterers.

Lemma 7.2.1 *Let the incident be the plane wave and D be given in* (7.2.1)–(7.2.4). *The P-part and S-part far-field patterns have the following asymptotic expressions as* $\rho/L_s \to +0$*:*

$$\begin{aligned} u_p^\infty(\hat{x}; D) &= \frac{\rho}{4\pi(\lambda + 2\mu)}(\hat{x} \otimes \hat{x})\left[\sum_{j=1}^{l_s} e^{-ik_p\hat{x}\cdot z_j}\left(C_{p,j}\,\alpha\, e^{ik_p z_j\cdot d} + C_{s,j}\,\beta\, e^{ik_s z_j\cdot d}\right)\right] \\ &\quad + \mathcal{O}\left(\rho^2 l_s(1 + L_s^{-1})\right), \\ u_s^\infty(\hat{x}; D) &= \frac{\rho}{4\pi\mu}(\mathbf{I} - \hat{x} \otimes \hat{x})\left[\sum_{j=1}^{l_s} e^{-ik_s\hat{x}\cdot z_j}\left(C_{p,j}\,\alpha\, e^{ik_p z_j\cdot d} + C_{s,j}\beta\, e^{ik_s z_j\cdot d}\right)\right] \\ &\quad + \mathcal{O}\left(\rho^2 l_s(1 + L_s^{-1})\right), \end{aligned}$$

where $C_{p,j}, C_{s,j} \in \mathbb{C}^3$ *are constant vectors independent of* ρ, l_s, L_s *and* z_j.

The proof of Lemma 7.2.1 relies essentially on the asymptotic expansions of u_p^∞ and u_s^∞ in the work [12], where the Lax-Foldy formulations for the Lamé system was justified without the condition (7.2.4). The other references on the asymptotic expansions associated with small inclusions can be found in a series of works by H. Ammari and H. Kang and their collaborators using integral equation methods; see e.g. [4–6]. We also mention the monographs [46] by P. Martin where the multiple scattering issues are well treated and [15] for analysis of acoustic, electromagnetic and elastic scattering problems at low frequencies. For the reader's convenience, we present a proof of Lemma 7.2.1 under the sparsity assumption (7.2.4).

Proof of Lemma 7.2.1 By [12, Remark 1.3], there exists a small number $\varepsilon > 0$ such that for $(l_s - 1)\rho/L_s < \varepsilon$

$$u_p^\infty(\hat{x}; D) = \frac{1}{4\pi(\lambda+2\mu)}(\hat{x}\otimes\hat{x})\left[\sum_{j=1}^{l_s} c_j e^{-\mathrm{i}k_p\hat{x}\cdot z_j} Q_j\right] + \mathscr{O}\left(\rho^2 l_s(1+L_s^{-1})\right),$$
$$u_s^\infty(\hat{x}; D) = \frac{1}{4\pi\mu}(\mathbf{I}-\hat{x}\otimes\hat{x})\left[\sum_{j=1}^{l_s} c_j e^{-\mathrm{i}k_s\hat{x}\cdot z_j} Q_j\right] + \mathscr{O}\left(\rho^2 l_s(1+L_s^{-1})\right), \tag{7.2.10}$$

where the vector coefficients $Q_j \in \mathbb{C}^3$, $j = 1, 2, \ldots, l_s$ are the unique solutions to the linear algebraic system

$$C_j^{-1} Q_j = -u^{in}(z_j) - \sum_{m=1, m\neq j}^{l_s} \Pi^{(\omega)}(z_j, z_m)\, Q_m, \tag{7.2.11}$$

with $\Pi^{(\omega)}(z_j, z_m)$ denoting the Kupradze matrix (7.2.9) and

$$C_j := \int_{\partial D_j} \Theta_j(y) ds(y) \in \mathbb{C}^{3\times 3}.$$

Here, Θ_j is the solution matrix of the first kind integral equation

$$\int_{\partial D_j} \Pi^{(0)}(x, y)\Theta_j(y) ds(y) = \mathbf{I}, \quad x \in \partial D_j, \tag{7.2.12}$$

where the matrix $\Pi^{(0)}(x, y)$, which denotes the the Kelvin matrix of the fundamental solution of the Lamé system with $\omega = 0$, takes the form (see, e.g., [34, Chapter 2] or [25, Chapter 2.2])

$$\Pi^{(0)}(x, y) := \frac{\lambda+3\mu}{8\pi\mu(\lambda+2\mu)}\frac{1}{|x-y|}\mathbf{I} + \frac{\lambda+\mu}{8\pi\mu(\lambda+2\mu)}\frac{1}{|x-y|^3}\left((x-y)\otimes(x-y)\right) \tag{7.2.13}$$

Since $\Pi^{(0)}(x, y) \sim |x-y|^{-1}$ as $x \to y$, it follows from (7.2.12) that $\Theta_j(y) \sim \rho^{-3}$ for $y \in \partial D_j$, from which we get $C_j \sim \rho^{-1}$ for $j = 1, 2\ldots, l_s$ as $\rho \to +0$. Now, inserting the estimate of C_j into (7.2.11) and taking into account the fact that

$$\Pi^{(\omega)}(z_j, z_m) = \mathscr{O}(L_s^{-1}) \quad \text{for} \quad j \neq m,$$

we obtain

$$Q_j = \rho\, \mathbf{H}_j\, u^{in}(z_j) + \mathscr{O}(L_s^{-1} + \rho^2) \quad \text{as} \quad \rho/L_s \to +0, \quad j = 1, 2 \ldots l_s,$$

where $\mathbf{H}_j \in \mathbb{C}^{3\times 3}$ are some constant matrices independent of ρ and L. Therefore, Lemma 7.2.1 is proved by taking $C_{p,j} = \mathbf{H}_j\, d$, $C_{s,j} = \mathbf{H}_j\, d^{\perp}$. □

We are in a position to present the proof of Theorem 7.2.1.

Proof of Theorem 7.2.1 We first consider the indicating function $I_1(z)$, $z \in \mathbb{R}^3$. For notational convenience we write

$$A_j = A_j(z_j, \alpha, \beta) := C_{p,j}\, \alpha\, e^{ik_p z_j \cdot d} + C_{s,j}\, \beta\, e^{ik_s z_j \cdot d} \in \mathbb{C}^3, \quad j = 1, 2, \ldots, l_s,$$

with $C_{p,j}$, $C_{s,j}$ given as in Lemma 7.2.1. Then, it is seen from Lemma 7.2.1 that

$$\|u_p^\infty(\hat{x}; D)\|_{L^2}^2 = \frac{\rho^2}{4(\lambda + 2\mu)^2} \sum_{j=1}^{l_s} |A_j|^2 + \mathscr{O}\left(\rho^3 + \rho^2 L^{-1}\right),$$

$$\|u_p^\infty(\hat{x}; D_j)\|_{L^2}^2 = \frac{\rho^2}{4(\lambda + 2\mu)^2} |A_j|^2 + \mathscr{O}\left(\rho^3 + \rho^2 L_s^{-1}\right).$$

Hence,

$$K_1^j = \frac{\|u_p^\infty(\hat{x}; D_j)\|_{L^2}^2}{\|u_p^\infty(\hat{x}; D)\|_{L^2}^2} = \widetilde{K}^j + \mathscr{O}(\rho + L_s^{-1}), \quad \widetilde{K}^j := \frac{A_j^2}{\sum_{j=1}^{l_s} |A_j|^2}. \tag{7.2.14}$$

This proves (7.2.6) for $m = 1$. The case of using the S-part of the far-field pattern (i.e., $m = 2$) can be treated in an analogous way.

For the full-wave scenaio, namely when $m = 3$, the orthogonality of u_p^∞ and u_s^∞ should be used in the treatment. Since $\langle \mathbf{I} - \hat{x} \otimes \hat{x}, \hat{x} \otimes \hat{x} \rangle = 0$, by applying Lemma 7.2.1 again to D and D_j, we have

$$\|u^\infty(\hat{x}; D)\|_{L^2}^2 = \frac{\rho^2}{4} \left(\frac{1}{(\lambda + 2\mu)^2} + \frac{1}{\mu^2} \right) \sum_{j=1}^{l_s} |A_j|^2 + \mathscr{O}\left(\rho^3 + \rho^2 L_s^{-1}\right),$$

$$\|u^\infty(\hat{x}; D_j)\|_{L^2}^2 = \frac{\rho^2}{4} \left(\frac{1}{(\lambda + 2\mu)^2} + \frac{1}{\mu^2} \right) |A_j|^2 + \mathscr{O}\left(\rho^3 + \rho^2 L_s^{-1}\right).$$

Hence, the equality (7.2.6) with $m = 3$ is proved with the same $\widetilde{K}^j$ given in (7.2.14).

To verify (7.2.7) and (7.2.8), without loss of generality we only consider the indicating behavior of $I_1(z)$ in a small neighborhood of z_j for some fixed $1 \le j \le l_s$, i.e., $z \in \text{neigh}(z_j)$. Clearly, under the assumption (7.2.4) we have

$$\omega |z_{j'} - z| \sim \omega\, L_s \gg 1, \quad \text{for all } z \in \text{neigh}(z_j), \quad j' \neq j.$$

By using the Reimann-Lebesgue lemma about oscillating integrals and Lemma 7.2.1 we can obtain

$$\left|\left\langle u_p^\infty(\hat{x}; D), \sum_{j=1}^{3}(\hat{x}\otimes\hat{x})\mathbf{e}_j e^{-\mathrm{i}k_p\hat{x}\cdot z}\right\rangle\right|^2 \tag{7.2.15}$$

$$= \frac{\rho^2\,|A_j|^2}{16\pi^2(\lambda+2\mu)^2}\left\langle e^{-\mathrm{i}k_p\hat{x}\cdot z_j}, e^{-\mathrm{i}k_p\hat{x}\cdot z}\right\rangle + \mathscr{O}(\rho^3+\rho^2 L_s^{-1}) \tag{7.2.16}$$

$$\leq \frac{\rho^2\,|A_j|^2}{4(\lambda+2\mu)^2} + \mathscr{O}(\rho^3+\rho^2 L_s^{-1}), \tag{7.2.17}$$

where the last inequality follows from the Cauchy-Schwartz inequality. Moreover, the strict inequality in (7.2.17) holds if $z \neq z_j$ and the equal sign holds only when $z = z_j$. Therefore, by the definition of I_1,

$$I_1(z) \leq \widetilde{K}^j + \mathscr{O}(\rho + L_s^{-1}),$$

and only when $z = z_j$ the equality holds. This proves (7.2.7) and (7.2.8). The indicating behavior of I_2 and I_3 can be treated in the same manner.

The proof is completed. □

7.2.2 Locating Multiple Extended Scatterers

In this section we consider the locating of multiple rigid scatterers of regular size by using a single incident plane wave. As discussed earlier in Introduction, it is extremely challenging to recover a generic rigid elastic scatterer by using a single far-field pattern. The scheme that we shall discuss for locating multiple extended (namely, regular-size) scatterers requires a certain a priori knowledge of the underlying target objects; that is, their shapes must be from a certain known class. In what follows, we first describe the multiple extended scatterers for our study and then present the corresponding locating Scheme R.

For $j = 1, 2, \ldots, l_e$, set $r_j \in \mathbb{R}_+$ such that

$$r_j \in [R_0, R_1], \quad 0 < R_0 < R_1 < +\infty, \quad R_0 \sim \mathscr{O}(1).$$

Let $E_j \subset \mathbb{R}^3, 1 \leq j \leq l_e$ denote a bounded simply-connected Lipschitz domain containing the origin. Throughout, we assume that $\mathrm{diam}(E_j) \sim 1$, $1 \leq j \leq l_e$. Define the scaling operator $\Lambda_r E_j$ with $r \in \mathbb{R}_+$ to be the same one as that given in (7.2.1). Denote by $\mathscr{R}_j := \mathscr{R}(\theta_j, \phi_j, \psi_j) \in SO(3), 1 \leq j \leq l_e$, the 3D rotation matrix around the origin whose Euler angels are $\theta_j \in [0, 2\pi]$, $\phi_j \in [0, 2\pi]$ and $\psi_j \in [0, \pi]$; and define $\mathscr{R}_j E := \{\mathscr{R}_j x : x \in E\}$. For $z_j \in \mathbb{R}^3$, we let

$$\Omega = \bigcup_{j=1}^{l_e} \Omega_j, \quad \Omega_j := z_j + \mathscr{R}_j\, \Lambda_{r_j}\, E_j, \tag{7.2.18}$$

denote the extended target scatterer for our current study. Obviously, Ω is a collection of scatterer components Ω_j that obtained by scaling, rotating and translating E_j with the parameters r_j, $(\theta_j, \phi_j, \psi_j)$ and z_j, respectively. In the sequel, the parameter z_j, Euler angles $(\theta_j, \phi_j, \psi_j)$, number r_j and the reference scatterer E_j will be respectively referred to as the *position*, *orientation*, *size* and *shape* of the scatterer component Ω_j in Ω. For technical purpose, we impose the following sparsity assumption on the extended scatterer Ω introduced in (7.2.18),

$$L_e = \min_{j \neq j', 1 \leq j, j' \leq l_e} \mathrm{dist}(\Omega_j, \Omega_{j'}) \gg 1. \tag{7.2.19}$$

Furthermore, it is assumed that there exists an admissible reference scatterer space

$$\mathscr{A} := \{\Sigma_j\}_{j=1}^{l'} \tag{7.2.20}$$

where each $\Sigma_j \subset \mathbb{R}^3$ is a bounded simply-connected Lipschitz domain containing the origin, such that for Ω in (7.2.18),

$$E_j \in \mathscr{A}. \tag{7.2.21}$$

For the admissible reference space $\mathscr{A}$ introduced in (7.2.20), we require that

$$\Sigma_j \neq \Sigma_{j'} \quad \text{for} \quad j \neq j', \quad 1 \leq j, j' \leq l', \tag{7.2.22}$$

and it is known in advance. The number $l' \in \mathbb{N}$ in (7.2.20) is not necessarily equal to l_e in (7.2.18). Condition (7.2.21) implies that the shapes of target scatterer components must be known in advance. Nevertheless, it may happen that more than one scatterer component possesses the same shape, or some shapes from the admissible class $\mathscr{A}$ may not appear in the target scatterer components.

In the following, we present Scheme R by using a single far-field pattern to locate the multiple components of the scatterer Ω described above. The inverse problem could find important practical applications in the real world. For instance, in locating an unknown group of plastic-cased land mines or pipelines buried in dry soils, one has the a priori knowledge on the possible shapes of the target objects.

7.2.2.1 Description of Scheme R

For $h \in \mathbb{R}_+, h \ll 1$, let $\mathscr{N}_1$ be a suitably chosen finite index set such that $\{\mathscr{R}_j\}_{j \in \mathscr{N}_1} = \{\mathscr{R}(\theta_j, \phi_j, \psi_j)\}_{j \in \mathscr{N}_1}$ is an h-net of $SO(3)$. That is, for any rotation matrix $\mathscr{R} \in SO(3)$, there exists $j \in \mathscr{N}_1$ such that $\|\mathscr{R}_j - \mathscr{R}\| \leq h$. For a simply-connected domain Σ containing the origin, we define

$$\mathscr{R}_h\Sigma := \{\mathscr{R}_j\Sigma\}_{j\in\mathscr{N}_1}. \tag{7.2.23}$$

In an analogous manner, for Λ_r with $r \in [R_0, R_1]$, we let $\mathscr{N}_2$ be a suitably chosen finite index set such that $\{r_j\}_{j\in\mathscr{N}_2}$ is an h-net of $[R_0, R_1]$. Define

$$\Lambda_h\Sigma := \{\Lambda_{r_j}\Sigma\}_{j\in\mathscr{N}_2}. \tag{7.2.24}$$

Next, we augment the admissible reference space $\mathscr{A}$ to be

$$\mathscr{A}_h = \mathscr{R}_h\Lambda_h\mathscr{A} = \bigcup_{j=1}^{l'}\{\mathscr{R}_h\Lambda_h\Sigma_j\} := \{\widetilde{\Sigma}_j\}_{j=1}^{l''}, \tag{7.2.25}$$

where l'' denotes the cardinality of the discrete set $\mathscr{A}_h$. Indeed, $\mathscr{A}_h$ can be taken as an h-net of $\mathscr{A}$ in the sense that for any $\Sigma \in \mathscr{A}$, there exists $\widetilde{\Sigma} \in \mathscr{A}_h$ such that $d_H(\overline{\Sigma}, \overline{\widetilde{\Sigma}}) \le Ch$, where d_H denotes the Hausdorff distance and C is a positive constant depending only on $\mathscr{A}$. We make the following two assumptions about the augmented admissible reference space $\mathscr{A}_h$:

(i) $u_\tau^\infty(\hat{x}; \widetilde{\Sigma}_j) \neq u_\tau^\infty(\hat{x}; \widetilde{\Sigma}_{j'})$ for $\tau = s, p$ or $\emptyset$, and $j \neq j'$, $1 \le j, j' \le l''$.
(ii) $\|u_\tau^\infty(\hat{x}; \widetilde{\Sigma}_j)\|_{L^2} \ge \|u_\tau^\infty(\hat{x}; \widetilde{\Sigma}_{j'})\|_{L^2}$ for $\tau = s, p$ or $\emptyset$, and $j < j'$, $1 \le j, j' \le l''$.

Assumption (ii) can be fulfilled by reordering the elements in $\mathscr{A}_h$ if necessary. For assumption (i), we recall the following well-known conjecture in the inverse elastic scattering theory:

$$u_\tau^\infty(\hat{x}; D_1) = u_\tau^\infty(\hat{x}; D_2) \quad \text{for all} \quad \hat{x} \in \mathbb{S}^2 \quad \text{if and only if} \quad D_1 = D_2, \tag{7.2.26}$$

where D_1 and D_2 are two rigid elastic scatterers. Equation (7.2.26) states that one can uniquely determine an elastic rigid scatterer by using a single far-field pattern. There is a widespread belief that (7.2.26) holds true, but there is very limited progress in the literature, and still largely remains open. We refer to [22, 24, 47] for uniqueness results established by using infinitely many far-field measurements. Nevertheless, since $\mathscr{A}_h$ is known, assumption (i) can be verified in advance.

In order to identify the multiple extended scatterers of Ω in (7.2.18), we introduce the following $l'' \times 3$ indicator functions:

$$W_1^j(z) = \frac{1}{\|u_p^\infty(\hat{x}; \widetilde{\Sigma}_j)\|_{L^2}^2}\left|\left\langle u_p^\infty(\hat{x}; \Omega),\ e^{-ik_p\hat{x}\cdot z}u_p^\infty(\hat{x}; \widetilde{\Sigma}_j)\right\rangle\right|^2, \tag{7.2.27}$$

$$W_2^j(z) = \frac{1}{\|u_s^\infty(\hat{x}; \widetilde{\Sigma}_j)\|_{L^2}^2}\left|\left\langle u_s^\infty(\hat{x}; \Omega),\ e^{-ik_s\hat{x}\cdot z}u_s^\infty(\hat{x}; \widetilde{\Sigma}_j)\right\rangle\right|^2, \tag{7.2.28}$$

$$W_3^j(z) = \frac{1}{\|u^\infty(\hat{x}; \widetilde{\Sigma}_j)\|_{L^2}^2}\left|\left\langle u^\infty(\hat{x}; \Omega),\ e^{-ik_p\hat{x}\cdot z}u_p^\infty(\hat{x}; \widetilde{\Sigma}_j) + e^{-ik_s\hat{x}\cdot z}u_s^\infty(\hat{x}; \widetilde{\Sigma}_j)\right\rangle\right|^2, \tag{7.2.29}$$

where $z \in \mathbb{R}^3$ and $\widetilde{\Sigma}_j \in \mathscr{A}_h$ for $j = 1, 2, \ldots, l''$.

Next, we present a key theorem on the indicating behavior of these indicator functions, which forms the basis of our Scheme R. Recall that α, β are the coefficients attached to u_p^{in} and u_s^{in}, respectively.

Theorem 7.2.2 *Suppose that $\alpha\,\beta = 0$ and that $\widetilde{\Sigma}_1 \in \mathscr{A}_h$ is of the following form*

$$\widetilde{\Sigma}_1 = \mathscr{R}_{j_\sigma} \Lambda_{r_{j_\tau}} \Sigma_{j_0}, \quad \Sigma_{j_0} \in \mathscr{A}, \; j_\sigma \in \mathscr{N}_1, \; j_\tau \in \mathscr{N}_2. \tag{7.2.30}$$

Suppose that in Ω given by (7.2.18), there exists $J_0 \subset \{1, 2, \ldots, l_e\}$ such that for $j \in J_0$, the component $\Omega_j = \mathscr{R}_j \Lambda_{r_j} E_j$ satisfies

$$(i)\; E_j = \Sigma_{j_0}; \quad (ii)\; \|\mathscr{R}_j - \mathscr{R}_{j_\sigma}\| \le h; \quad (iii)\; \|r_j - r_{j_\tau}\| \le h; \tag{7.2.31}$$

whereas for $j \in \{1, 2, \ldots, l_e\} \backslash J_0$, at least one of the conditions in (7.2.31) is not fulfilled by the scatterer component Ω_j. Then for each z_j, $1 \le j \le l_e$, there exists an open neighborhood of z_j, $neigh(z_j)$, such that

1. *if $j \in J_0$, then*

$$W_m^1(z) \le 1 + \mathscr{O}\left(\frac{1}{L_e} + h\right), \quad \forall z \in \text{neigh}(z_j), \quad m = 1, 2, 3. \tag{7.2.32}$$

   Moreover, the equality relation holds in (7.2.32) only when $z = z_j$. That is, z_j is a local maximum point for $W_m^1(z)$.

2. *if $j \in \{1, 2, \ldots, l\} \backslash J_0$, then there exists $\varepsilon_0 \in \mathbb{R}_+$ such that*

$$W_m^1(z) \le 1 - \varepsilon_0 + \mathscr{O}\left(\frac{1}{L_e} + h\right), \quad \forall z \in \text{neigh}(z_j), \quad m = 1, 2, 3. \tag{7.2.33}$$

Remark 7.2.3 The condition $\alpha\beta = 0$ implies that W_m^j's in (7.2.28) are valid for incident plane pressure or shear waves only. Following the proof of Theorem 7.2.2, one can formulate the indicator functions for general elastic plane waves; see Remark 7.2.6 at the end of this section.

In Theorem 7.2.2, it may happen that $J_0 = \emptyset$. In this case, there is no scatterer in Ω which is of the shape Σ_{j_0}. Clearly, by using the indicating behavior of the functional $W_m^1(z)$ presented in Theorem 7.2.2, one can locate all the scatterer components possessing the shape Σ_{j_0}. After the locating of those scatterers of the shape Σ_{j_0}, one can exclude them from the searching region. Moreover, by Lemmas 7.2.2 and 7.2.3 in the following, one can calculate the far-field pattern generated by the remaining scatterer components. With the updated far-field pattern, one can then use $\widetilde{\Sigma}_2$ as the reference, and proceed as before to locate all the scatterer components of Ω possessing the same shape as $\widetilde{\Sigma}_2$. Clearly, this procedure can be carried out till we find all the scatterer components of Ω. In summary, Scheme R is read as follows.

Scheme R	Locating extended scatterers of Ω in (7.2.18)
Step 1	For the admissible reference scatterer class $\mathscr{A}$ in (7.2.20), formulate the augmented admissible class $\mathscr{A}_h$ in (7.2.25)
Step 2	Collect in advance the P-part ($m = 1$), S-part ($m = 2$) or the full far-field data ($m = 3$) associated with the admissible reference scatterer class $\mathscr{A}_h$ corresponding to a single incident plane wave. Reorder $\mathscr{A}_h$ if necessary to make it satisfy assumption (ii), and also verify the generic assumption (i)
Step 3	For an unknown rigid scatterer Ω in (7.2.18), collect the P-part, S-part or the full far-field data by sending the same detecting plane wave as specified in **Step 2**
Step 4	Select a sampling region with a mesh $\mathscr{T}_h$ containing Ω
Step 5	Set $j = 1$
Step 6	For each sampling point $z \in \mathscr{T}_h$, calculate $W_m^j(z)$ ($m = 1, 2, 3$) according to available far-field data for Ω
Step 7	Locate all those significant local maximum points of $W_m^j(z)$ satisfying $W_m^j(z) \approx 1$ for the scatterer components of the form $z + \widetilde{\Sigma}_j$. Let z_η, $\eta = 1, \ldots, \eta_0$ be the local maximum points found this step
Step 8	Remove all those $z + \widetilde{\Sigma}_j$ found in **Step 6** from the mesh $\mathscr{T}_h$
Step 9	Update the far-field patterns according to the following formulae $u_p^{\infty,new} = u_p^\infty(\hat{x}; d, d^\perp, \alpha, \beta, \Omega) - u_p^\infty(\hat{x}; d, d^\perp, \alpha, 0, \widetilde{\Sigma}_j) \sum_{\eta=1}^{\eta_0} e^{\mathrm{i}k_p(d-\hat{x})\cdot z_\eta}$ $-u_p^\infty(\hat{x}; d, d^\perp, 0, \beta, \widetilde{\Sigma}_j) \sum_{\eta=1}^{\eta_0} e^{\mathrm{i}(k_s d - k_p \hat{x})\cdot z_\eta}$, $u_s^{\infty,new} = u_s^\infty(\hat{x}; d, d^\perp, \alpha, \beta, \Omega) - u_s^\infty(\hat{x}; d, d^\perp, \alpha, 0, \widetilde{\Sigma}_j) \sum_{\eta=1}^{\eta_0} e^{\mathrm{i}(k_p d - k_s \hat{x})\cdot z_\eta}$ $-u_s^\infty(\hat{x}; d, d^\perp, 0, \beta, \widetilde{\Sigma}_j) \sum_{\eta=1}^{\eta_0} e^{\mathrm{i}k_s(d-\hat{x})\cdot z_\eta}$, $u^{\infty,new} = u_p^{\infty,new} + u_s^{\infty,new}$
Step 10	If $j = l''$, namely, the maximum number of the reference scatterers reaches, then stop the reconstruction; otherwise set $j = j + 1$, and go to **Step 6**

7.2.2.2 Proof of Theorem 7.2.2

Throughout the present section, we let ν denote the unit normal vector to $\partial\Omega$ pointing into $\mathbb{R}^3\backslash\overline{\Omega}$. Denote the linearized strain tensor by

$$\varepsilon(u) \;:=\; \frac{1}{2}\left(\nabla u + \nabla u^\top\right) \;\in\; \mathbb{R}^{3\times 3}, \tag{7.2.34}$$

where $\nabla u \in \mathbb{R}^{3\times 3}$ and $\nabla u^\top$ stand for the Jacobian matrix of u and its adjoint, respectively. By Hooke's law the strain tensor is related to the stress tensor via the identity

$$\sigma(u) \;=\; \lambda\,(\operatorname{div} u)\,\mathbf{I} \;+\; 2\mu\,\varepsilon(u) \;\in\; \mathbb{R}^{3\times 3}. \tag{7.2.35}$$

The surface traction (or the stress operator) on $\partial\Omega$ is given by

$$T_\nu u := \sigma(u)\nu = (2\mu\nu\,\cdot\nabla + \lambda\,\nu\,\operatorname{div} + \mu\nu\,\times\operatorname{curl}\,)\,u. \tag{7.2.36}$$

We next present several auxiliary lemmas.

Lemma 7.2.2 *Let Ω be a scatterer with multiple components given in (7.2.18). Under the assumption (7.2.4), we have*

$$u^\infty(\hat{x};\Omega) = \sum_{j=1}^{l_e} u^\infty(\hat{x};\Omega_j) + \mathscr{O}(L_e^{-1}).$$

Proof For simplicity we assume $l_e = 2$. We begin with the single- and double-layer potential operators in elasticity. For $j = 1, 2$, let

$$(S_j\varphi)(x) := 2\int_{\partial\Omega_j} \Pi(x,y)\varphi(y)ds(y), \quad \varphi \in C(\partial\Omega_j), \quad x \in \partial\Omega_j, \tag{7.2.37}$$

$$(K_j\varphi)(x) := 2\int_{\partial\Omega_j} \Xi(x,y)\varphi(y)ds(y), \quad \varphi \in C(\partial\Omega_j), \quad x \in \partial\Omega_j, \tag{7.2.38}$$

where $\Xi(x, y)$ is a matrix-valued function whose j-th column vector is defined by

$$\Xi(x,y)^\top \mathbf{e}_j := T_{\nu(y)}(\Pi(x,y)\,\mathbf{e}_j) \quad \text{on} \quad \partial\Omega_j, \quad \text{for } x \neq y, \quad j = 1, 2, 3.$$

Recall that the superscript $(\cdot)^\top$ denotes the transpose, $\mathbf{e}_j \in \mathbb{C}^{3\times 1}$ the usual cartesian unit vectors and $T_{\nu(y)}$ the stress operator. Under the regularity assumption $\partial\Omega_j \in C^2$, it was proved in [24] that the scattered field $u^{sc}(x;\Omega_j)$ corresponding to Ω_j can be represented as

$$u^{sc}(x;\Omega_j) = \int_{\partial\Omega_j} \Xi(x,y)\varphi_j(y)ds(y) + i\int_{\partial\Omega_j} \Pi(x,y)\varphi_j(y)ds(y), \quad x \in \mathbb{R}^3\backslash\overline{\Omega}_j,$$

where the density function $\varphi_j \in C(\partial\Omega_j)$ is given by

$$\varphi_j = -2(I + K_j + iS_j)^{-1}u^{in}|_{\partial\Omega_j}, \quad j = 1, 2.$$

To prove the lemma for the scatterer $\Omega = \Omega_1 \cup \Omega_2$, we make use of the ansatz

$$u^{sc}(x;\Omega) = \sum_{j=1,2}\left\{\int_{\partial\Omega_j} \Xi(x,y)\phi_j(y)ds(y) + i\int_{\partial\Omega_j} \Pi(x,y)\phi_j(y)ds(y)\right\}, \quad x \in \mathbb{R}^3\backslash\overline{\Omega},$$

with $\phi_j \in C(\partial\Omega_j)$. Using the Dirichlet boundary condition $u^{sc} + u^{in} = 0$ on each $\partial\Omega_j$, we obtain the integral equations

$$\begin{pmatrix} I + K_1 + iS_1 & J_2 \\ J_1 & I + K_2 + iS_2 \end{pmatrix}\begin{pmatrix}\phi_1 \\ \phi_2\end{pmatrix} = -2\begin{pmatrix} u^{in}|_{\partial\Omega_1} \\ u^{in}|_{\partial\Omega_2}\end{pmatrix}, \tag{7.2.39}$$

where the operators $J_1 : C(\partial\Omega_1) \to C(\partial\Omega_2)$, $J_2 : C(\partial\Omega_2) \to C(\partial\Omega_1)$ are defined respectively by

$$J_1\phi_1 = 2\left\{\int_{\partial\Omega_1} \Xi(x,y)\phi_1(y)ds(y) + i\int_{\partial\Omega_1} \Pi(x,y)\phi_1(y)ds(y)\right\}, \quad x \in \partial\Omega_2,$$

$$J_2\phi_2 = 2\left\{\int_{\partial\Omega_2} \Xi(x,y)\phi_2(y)ds(y) + i\int_{\partial\Omega_2} \Pi(x,y)\phi_2(y)ds(y)\right\}, \quad x \in \partial\Omega_1.$$

Since $L_e \gg 1$ (cf. (7.2.19)), using the fundamental solution (7.2.9), it is readily to estimate

$$\|J_1\phi_1\|_{C(\partial\Omega_2)} \le C_1 L_e^{-1}\|\phi_1\|_{C(\partial\Omega_1)}, \quad \|J_2\phi_2\|_{C(\partial\Omega_1)} \le C_2 L_e^{-1}\|\phi_2\|_{C(\partial\Omega_2)}, \quad C_1, C_2 > 0.$$

Hence, it follows from (7.2.39) and the invertibility of $I + K_j + iS_j : C(\partial\Omega_j) \to C(\partial\Omega_j)$ that

$$\begin{aligned}\begin{pmatrix}\phi_1\\ \phi_2\end{pmatrix} &= \begin{pmatrix}(I+K_1+iS_1)^{-1} & 0\\ 0 & (I+K_2+\mathrm{i}S_2)^{-1}\end{pmatrix}\begin{pmatrix}-2u^{in}|_{\partial\Omega_1}\\ -2u^{in}|_{\partial\Omega_2}\end{pmatrix} + \mathscr{O}(L_e^{-1})\\ &= \begin{pmatrix}\varphi_1\\ \varphi_2\end{pmatrix} + \mathscr{O}(L_e^{-1}).\end{aligned}$$

This implies that

$$u^{sc}(x;\Omega) = u^{sc}(x;\Omega_1) + u^{sc}(x;\Omega_2) + \mathscr{O}(L_e^{-1}) \quad \text{as} \quad L_e \to \infty,$$

which finally leads to

$$u^{\infty}(\hat{x};\Omega) = u^{\infty}(\hat{x};\Omega_1) + u^{\infty}(\hat{x};\Omega_2) + \mathscr{O}(L_e^{-1}).$$

□

Remark 7.2.4 In the proof of Lemma 7.2.2, we require that the boundary $\partial\Omega$ is C^2 continuous. This is mainly due to the requirements of the mapping properties of the single- and double-layer potential operators (cf. (7.2.37) and (7.2.38)) in the proof. This regularity assumption can be relaxed to be Lipschitz continuous by using a similar argument, together with the mapping properties of the layer potential operators defined on Lipschitz surfaces (cf. [48]).

In what follows, we establish the relation between far-field patterns for translated, rotated and scaled elastic bodies. For $D \subset \mathbb{R}^3$ and $\mathbf{a} = (a_1, a_2, a_3) \in \mathbb{R}^3$, we write $D_{\mathbf{a}} = \mathbf{a} + D$ for simplicity.

Lemma 7.2.3 *Assume ∂D is Lipschitz. If $\alpha = 1$, $\beta = 0$, then*

$$u_p^{\infty}(\hat{x}; D_{\boldsymbol{a}}) = u_p^{\infty}(\hat{x}; D)\, e^{\mathrm{i}k_p(d-\hat{x})\cdot\boldsymbol{a}}, \quad u_s^{\infty}(\hat{x}; D_{\boldsymbol{a}}) = u_s^{\infty}(\hat{x}; D)\, e^{\mathrm{i}(k_p d - k_s\hat{x})\cdot\boldsymbol{a}}. \tag{7.2.40}$$

If $\alpha = 0$, $\beta = 1$, then

$$u_p^{\infty}(\hat{x}; D_{\boldsymbol{a}}) = u_p^{\infty}(\hat{x}; D)\, e^{\mathrm{i}(k_s d - k_p\hat{x})\cdot\boldsymbol{a}}, \quad u_s^{\infty}(\hat{x}; D_{\boldsymbol{a}}) = u_s^{\infty}(\hat{x}; D)\, e^{\mathrm{i}k_s(d-\hat{x})\cdot\boldsymbol{a}}. \tag{7.2.41}$$

Proof We first consider the case of incident plane pressure waves, i.e., $\alpha = 1, \beta = 0$. Denote by $u^\infty(\hat{x}; D) = u^\infty(\hat{x}; D, d)$ the far-field pattern corresponding to the rigid scatter D with the incident direction $d \in \mathbb{S}^2$. For any $y = z + \mathbf{a} \in \partial D_\mathbf{a}$ with $z \in \partial D$, we have

$$u^{sc}(y; D_\mathbf{a}) = -d\, e^{\mathrm{i}k_p z\cdot d}\, e^{\mathrm{i}k_p \mathbf{a}\cdot d} = u^{sc}(z; D)\, e^{\mathrm{i}k_p \mathbf{a}\cdot d} = u^{sc}(y - \mathbf{a}; D)\, e^{\mathrm{i}k_p \mathbf{a}\cdot d}.$$

From the uniqueness of the exterior problem of the Navier equation for rigid scatterers (cf. [24]), it follows that

$$u^{sc}(y; D_\mathbf{a}) = u^{sc}(y - \mathbf{a}; D)\, e^{\mathrm{i}k_p \mathbf{a}\cdot d} = u^{sc}(z; D)\, e^{\mathrm{i}k_p \mathbf{a}\cdot d}, \quad \text{for all } y \in \mathbb{R}^3\backslash\overline{D}. \tag{7.2.42}$$

This implies that for any $y = z + \mathbf{a} \in \partial D_\mathbf{a}$ with $z \in \partial D$,

$$T_{\nu(y)} u^{sc}(y; D_\mathbf{a}) = T_{\nu(z)} u^{sc}(z; D)\, e^{\mathrm{i}k_p \mathbf{a}\cdot d}. \tag{7.2.43}$$

Recall that, the P-part and S-part far-field patterns of $u^\infty(\hat{x}; D_\mathbf{a})$ can be respectively characterized as follows (cf. [3]):

$$\begin{aligned} u_p^\infty(\hat{x}; D_\mathbf{a}) = \int_{\partial D_\mathbf{a}} \Big\{ [T_{\nu(y)}\, \{\hat{x}\otimes\hat{x}\, e^{-\mathrm{i}k_p\hat{x}\cdot y}\}]^\top u^{sc}(y; D_\mathbf{a}) \\ -\hat{x}\otimes\hat{x}\, e^{-\mathrm{i}k_p\hat{x}\cdot y}\; T_{\nu(y)} u^{sc}(y; D_\mathbf{a})\Big\}\, ds(y), \end{aligned}$$

$$\begin{aligned} u_s^\infty(\hat{x}; D_\mathbf{a}) = \int_{\partial D_\mathbf{a}} \Big\{ [T_{\nu(y)}\{(\mathbf{I} - \hat{x}\otimes\hat{x}) e^{-\mathrm{i}k_s\hat{x}\cdot y}\}]^\top u^{sc}(y; D_\mathbf{a}) \\ -(\mathbf{I} - \hat{x}\otimes\hat{x}) e^{-\mathrm{i}k_s\hat{x}\cdot y} T_{\nu(y)} u^{sc}(y; D_\mathbf{a})\Big\} ds(y). \end{aligned}$$

Changing the variable $y = z + \mathbf{a}$ in the previously two expressions and making use of (7.2.42) and (7.2.43), we obtain

$$\begin{aligned} u_p^\infty(\hat{x}; D_\mathbf{a}) &= \int_{\partial D} \Big\{ [T_{\nu(z)}\, \{\hat{x}\otimes\hat{x}\, e^{-\mathrm{i}k_p\hat{x}\cdot z}\}]^\top u^{sc}(z; D) \\ &\quad -\hat{x}\otimes\hat{x} e^{-\mathrm{i}k_p\hat{x}\cdot z}\; T_{\nu(y)} u^{sc}(z; D)\Big\}\, ds(z)\, e^{\mathrm{i}k_p(d-\hat{x})\cdot\mathbf{a}} \\ &= u_p^\infty(\hat{x}; D)\, e^{\mathrm{i}k_p(d-\hat{x})\cdot\mathbf{a}}, \end{aligned}$$

and

$$\begin{aligned} u_s^\infty(\hat{x}; D_\mathbf{a}) &= \int_{\partial D} \Big\{ [T_{\nu(z)}\{(\mathbf{I} - \hat{x}\otimes\hat{x}) e^{-\mathrm{i}k_s\hat{x}\cdot z}\}]^\top u^{sc}(z; D) \\ &\quad -(\mathbf{I} - \hat{x}\otimes\hat{x}) e^{-\mathrm{i}k_s\hat{x}\cdot z} T_{\nu(z)} u^{sc}(z; D)\Big\} ds(z)\, e^{\mathrm{i}(k_p d - k_s\hat{x})\cdot\mathbf{a}} \\ &= u_s^\infty(\hat{x}; D)\, e^{\mathrm{i}(k_p d - k_s\hat{x})\cdot\mathbf{a}}, \end{aligned}$$

from which the relations in (7.2.40) follow. The case for incident plane shear waves can be treated in the same manner. □

Remark 7.2.5 For general plane waves, one can obtain the corresponding relations between translated elastic bodies by supposition, giving rise to the updating formulae in Step 9 of Scheme R. Note that the identities in (7.2.40) (resp. (7.2.41)) are valid for incident plane shear (resp. pressure) wave only.

As an application of the relations established in Lemma 7.2.3, we prove uniqueness in locating the position of translated elastic bodies with a single plane pressure or shear wave.

Lemma 7.2.4 *Let $d, d^\perp \in \mathbb{S}^2$ and $\omega \in \mathbb{R}_+$ be fixed. Assume $\alpha\,\beta = 0$. Then the relation $u_\tau^\infty(\hat{x}, D_a) = u_\tau^\infty(\hat{x}, D)$ with $\tau = p$ or $\tau = s$ for all $\hat{x} \in \mathbb{S}^2$ implies $|\boldsymbol{a}| = 0$.*

Proof Without loss of generality we assume $\beta = 0$. This implies that the incident wave is a plane pressure wave. If $u_p^\infty(\hat{x}; D_{\mathbf{a}}) = u_p^\infty(\hat{x}; D)$ for all $\hat{x} \in \mathbb{S}^2$, then it follows from the first identity in (7.2.40) that $(d - \hat{x}) \cdot \mathbf{a} = 0$ for all $\hat{x} \in \mathbb{S}^2$. Since the set $\{d - \hat{x} : \hat{x} \in \mathbb{S}^2\}$ contains three linearly independent vectors of $\mathbb{R}^3$, it follows that $|\mathbf{a}| = 0$. By arguing similarly we deduce from $u_s^\infty(\hat{x}; D_{\mathbf{a}}) = u_s^\infty(\hat{x}; D)$ and the second identity in (7.2.40) that $(k_p d - k_s\hat{x}) \cdot \mathbf{a} = 0$ for all $\hat{x} \in \mathbb{S}^2$, which also leads to $|\mathbf{a}| = 0$. The proof for the case with $\alpha = 0$ can be shown in the same way by using (7.2.41). □

Let $\mathscr{R}$ be a rotation matrix in $\mathbb{R}^3$. The following relation between $u^\infty(\hat{x}; D)$ and $u^\infty(\hat{x}; \mathscr{R}D)$ was mentioned in [47, Sect. 5].

Lemma 7.2.5

$$\mathscr{R}\,u^\infty(\hat{x}; D, d, d^\perp) = u^\infty(\mathscr{R}\hat{x}; \mathscr{R}D, \mathscr{R}d, \mathscr{R}d^\perp), \quad \textit{for all} \;\; \hat{x}, d, d^\perp \in \mathbb{S}^2, d \cdot d^\perp = 0. \tag{7.2.44}$$

Now, we recall the scaling operator $\Lambda_\rho D = \{\rho\, x : x \in D\}$ for $\rho > 0$. Given the incident plane wave u^{in}, we write $u^\infty(\hat{x}) = u^\infty(\hat{x}; D, \omega)$ and $u^{sc}(\hat{x}) = u^{sc}(\hat{x}; D, \omega)$ to indicate the dependance on the obstacle D and the frequency of incidence ω. There holds

Lemma 7.2.6

$$u_\tau^\infty(\hat{x}; \Lambda_\rho D, \omega) = \rho\, u_\tau^\infty(\hat{x}; D, \rho\,\omega), \quad \hat{x} \in \mathbb{S}^2, \quad \tau = p, s \textit{ or } \emptyset.$$

Proof We see that

$$u^{sc}(x; \Lambda_\rho D, \omega) = \frac{\exp(ik_p|x|)}{4\pi(\lambda+\mu)|x|}\, u_p^\infty(\hat{x}; \Lambda_\rho D, \omega) + \frac{\exp(ik_s|x|)}{4\pi\mu|x|}\, u_s^\infty(\hat{x}; \Lambda_\rho D, \omega) + \mathscr{O}(\frac{1}{|x|^2}) \tag{7.2.45}$$

as $|x| \to +\infty$. Define $\widetilde{u}^{sc}(y) := u^{sc}(\rho y; \Lambda_\rho D, \omega)$ for $y \in \mathbb{R}^3\backslash\overline{D}$. It is readily seen that

$$\begin{cases} (\Delta^* + \rho^2\omega^2)\widetilde{u}^{sc} = 0 & \text{in } \mathbb{R}^3\backslash\overline{D}, \\ \widetilde{u}^{sc}(y) = -u^{in}(y) & \text{on } \partial D. \end{cases}$$

Moreover, $\widetilde{u}^{sc}(y)$ is still a radiating solution with the asymptotic behavior

$$\widetilde{u}^{sc}(y; D, \rho\omega) = \frac{\exp(\mathrm{i}\rho k_p|y|)}{4\pi(\lambda+\mu)|y|}\, u_p^\infty(\hat{y}; D, \rho\omega) + \frac{\exp(\mathrm{i}\rho d_s|y|)}{4\pi\mu|y|}\, u_s^\infty(\hat{y}; D, \rho\omega) + \mathscr{O}(\frac{1}{|y|^2}),$$

as $|y| \to \infty$. Changing the variable $y = x/\rho$, we deduce from the above expression that

$$\begin{aligned} u^{sc}(x) = \widetilde{u}^{sc}(x/\rho) = {} & \frac{\rho\exp(\mathrm{i}k_p|y|)}{4\pi(\lambda+\mu)|x|}\, u_p^\infty(\hat{x}; D, \rho\omega) + \frac{\rho\exp(\mathrm{i}k_s|x|)}{4\pi\mu|x|}\, u_s^\infty(\hat{x}; D, \rho\omega) \\ & + \mathscr{O}(\frac{1}{|x|^2}) \end{aligned} \tag{7.2.46}$$

as $|x| \to \infty$. Comparing (7.2.47) with (7.2.47) yields

$$u_\tau^\infty(\hat{x}; \Lambda_\rho D, \omega) = \rho\, u_\tau^\infty(\hat{x}; D, \rho\omega), \quad \tau = p, s,$$

and thus

$$u^\infty(\hat{x}; \Lambda_\rho D, \omega) = \rho\, u^\infty(\hat{x}; D, \rho\omega).$$

The proof is completed. □

We are in a position to present the proof of Theorem 7.2.2.

Proof of Theorem 7.2.2 Without loss of generality, we assume $\alpha = 1, \beta = 0$. Let the scatterer component $\Omega_j = \Omega_j(z_j, \mathscr{R}_j, r_j, E_j)$ fulfill (7.2.18) and (7.2.31). Combining Lemmas 7.2.3, 7.2.5 and 7.2.6, we obtain

$$\begin{aligned} u_p^\infty(\hat{x}; d, \omega, \Omega_j) &= u_p^\infty(\hat{x}; d, \omega, \mathscr{R}_j\, \Lambda_{r_j}\, E_j)\, e^{\mathrm{i}k_p(d-\hat{x})\cdot z_j} \\ &= \mathscr{R}_j\, u_p^\infty(\mathscr{R}_j^{-1}\hat{x}; \mathscr{R}_j^{-1}d,\, \omega,\, \Lambda_{r_j}\, E_j)\, e^{\mathrm{i}k_p(d-\hat{x})\cdot z_j} \\ &= r_j\, \mathscr{R}_j\, u_p^\infty(\mathscr{R}_j^{-1}\hat{x}; \mathscr{R}_j^{-1}d,\, r_j\,\omega,\, E_j)\, e^{\mathrm{i}k_p(d-\hat{x})\cdot z_j}. \end{aligned} \tag{7.2.47}$$

Using (7.2.31) and again Lemmas 7.2.5 and 7.2.6, we have for $j \in J_0$

$$\begin{aligned} & r_j\, \mathscr{R}_j\, u_p^\infty(\mathscr{R}_j^{-1}\hat{x}; \mathscr{R}_j^{-1}d,\, r_j\,\omega,\, E_j) \\ &= r_{j_\tau}\, \mathscr{R}_{j_\sigma}\, u_p^\infty(\mathscr{R}_{j_\sigma}^{-1}\hat{x}; \mathscr{R}_{j_\sigma}^{-1}d, r_{j_\tau}\,\omega,\, \Sigma_{j_0}) + \mathscr{O}(h) \\ &= u_p^\infty(\hat{x}; d, \omega, \mathscr{R}_{j_\sigma}\Lambda_{r_{j_\tau}}\Sigma_{j_0},) + \mathscr{O}(h) \\ &= u_p^\infty(\hat{x}; d, \omega,\, \widetilde{\Sigma}_1) + \mathscr{O}(h), \end{aligned} \tag{7.2.48}$$

where $\widetilde{\Sigma}_1$ is given as in (7.2.30). Inserting (7.2.49) into (7.2.48), it follows from Lemma 7.2.2 that

$$\begin{aligned} u_p^\infty(\hat{x};\Omega) &= \sum_{j=1}^{l_e} u_p^\infty(\hat{x};\Omega_j) + \mathscr{O}(L_e^{-1}) \\ &= \sum_{j\in J_0} u_p^\infty(\hat{x};\widetilde{\Sigma}_1) e^{\mathrm{i}k_p(d-\hat{x})\cdot z_j} + \sum_{j\in\{1,\ldots l_e\}\backslash J_0} u_p^\infty(\hat{x};\Omega_j) + \mathscr{O}(L_e^{-1}+h). \end{aligned}$$

Hence, for $z \in \text{neigh}(z_j)$ with some $j \in J_0$ we have

$$\begin{aligned} &|\langle u_p^\infty(\hat{x};\Omega), u_p^\infty(\hat{x};\widetilde{\Sigma}_1)e^{-\mathrm{i}k_p\hat{x}\cdot z}\rangle| \qquad (7.2.49) \\ &= |\langle u_p^\infty(\hat{x};\widetilde{\Sigma}_1)e^{\mathrm{i}k_p(d-\hat{x})\cdot z_j}, u_p^\infty(\hat{x};\widetilde{\Sigma}_1)e^{-\mathrm{i}k_p\hat{x}\cdot z}\rangle| + \mathscr{O}(L_e^{-1}+h) \\ &\le \|u_p^\infty(\hat{x};\widetilde{\Sigma}_1)\|_{L^2} + \mathscr{O}(L_e^{-1}+h). \end{aligned}$$

The equality in (7.2.49) follows from the the Riemann-Lebesgue lemma about oscillatory integrals by noting $|z_{j'} - z| \sim L_e \gg 1$ for $j' \neq j$, $1 \le j' \le l_e$ and $z \in \text{neigh}(z_j)$. For the inequality in (7.2.50), we have applied the Cauchy-Schwartz inequality, and it is easily seen that the equality holds only at $z = z_j$. Therefore, from the definition of the indicator function W_1^1,

$$W_1^1(z) \le 1 + \mathscr{O}(L_e^{-1}+h), \quad \text{for} \quad z \in \text{neigh}(z_j).$$

On the other hand, by a similar argument, together with assumption (i) on $\widetilde{\Sigma}_j$ and the equality (7.2.47) , we can directly verify that

$$W_1^1(z) < 1 + \mathscr{O}(L_e^{-1}+h), \quad z \in \text{neigh}(z_j), \quad j \in \{1,2,\ldots,l_e\}\backslash J_0.$$

This proves Theorem 7.2.2 with $m = 1$ for an incident pressure wave. In a completely similar manner, our argument can be extended to show the indicating behavior of $W_2^1(z)$ $(m = 2)$ by using the first equality in (7.2.41). Regarding $W_3^1(z)$ $(m = 3)$ where the full far-field pattern data are involved, we apply the orthogonality of u_p^∞ and u_s^∞ to obtain

$$W_3^j = \frac{\left|\langle u_p^\infty(\hat{x};\Omega),\, e^{-\mathrm{i}k_p\hat{x}\cdot z}\, u_p^\infty(\hat{x};\widetilde{\Sigma}_j)\rangle + \langle u_s^\infty(\hat{x};\Omega), e^{-\mathrm{i}k_s\hat{x}\cdot z}\, u_s^\infty(\hat{x};\widetilde{\Sigma}_j)\rangle\right|^2}{\|u_p^\infty(\hat{x};\widetilde{\Sigma}_j)\|_{L^2}^2 + \|u_s^\infty(\hat{x};\widetilde{\Sigma}_j)\|_{L^2}^2}.$$

Thus, the behavior of $W_3^1(z)$ follows from those of $W_1^1(z)$ and $W_2^1(z)$.

In the case of an incident shear wave, the indicating behavior of $W_m^1(z)$ $(m = 1, 2, 3)$ can be shown similarly . The proof of Theorem 7.2.2 is complete. □

Remark 7.2.6 For a general incident plane wave, following a similar argument to the proof of Theorem 7.2.2, one can show that Theorem 7.2.2 still holds with the indicator functions replaced, respectively, by

$$W_1^j(z) = \frac{\left|\langle u_p^\infty(\hat{x}; \Omega),\ A_1^j(\hat{x}; z)\rangle\right|^2}{\|u_p^\infty(\hat{x}; \widetilde{\Sigma}_j)\|_{L^2}^2}, \quad W_2^j(z) = \frac{\left|\langle u_s^\infty(\hat{x}; \Omega),\ A_2^j(\hat{x}; z)\rangle\right|^2}{\|u_s^\infty(\hat{x}; \widetilde{\Sigma}_j)\|_{L^2}^2},$$

$$W_3^j(z) = \frac{\left|\langle u^\infty(\hat{x}; \Omega),\ A_1^j(\hat{x}; z) + A_2^j(\hat{x}; z)\rangle\right|^2}{\|u^\infty(\hat{x}; \widetilde{\Sigma}_j)\|_{L^2}^2},$$

where for $j = 1, 2, \ldots, l''$,

$$A_1^j(\hat{x}; z) := e^{\mathrm{i}k_p(d-\hat{x})\cdot z} u_p^\infty(\hat{x}; d, d^\perp, \alpha, 0, \widetilde{\Sigma}_j) + e^{\mathrm{i}(k_s d - k_p\hat{x})\cdot z} u_p^\infty(\hat{x}; d, d^\perp, 0, \beta, \widetilde{\Sigma}_j),$$

$$A_2^j(\hat{x}; z) := e^{\mathrm{i}(k_p d - k_s\hat{x})\cdot z} u_s^\infty(\hat{x}; d, d^\perp, \alpha, 0, \widetilde{\Sigma}_j) + e^{\mathrm{i}k_s(d-\hat{x})\cdot z} u_s^\infty(\hat{x}; d, d^\perp, 0, \beta, \widetilde{\Sigma}_j).$$

7.2.3 *Locating Multiple Multiscale Scatterers*

In this section, we consider the recovery of a scatterer consisting of multiple multiscale components given by

$$G = D \cup \Omega, \tag{7.2.50}$$

where D is as described in (7.2.1)–(7.2.4) denoting the union of the small components, and Ω is as described in (7.2.18)–(7.2.22) denoting the union of the extended components. As before, we assume that the shapes of the extended components are from a known admissible class, as described in (7.2.20)–(7.2.22). In addition, we require that

$$L_m := \mathrm{dist}(D, \Omega) \gg 1. \tag{7.2.51}$$

Next, we show Scheme M to locate the $l_s + l_e$ multiscale scatterer components of G in (7.2.50) by using a single far-field pattern. Our treatment follows the one in [40] of locating multiscale acoustic scatterers. More specifically, we concatenate Schemes S and R of locating small and extended scatterers, respectively, by a *local tuning* technique, to form Scheme M of locating the multiscale scatterers.

Definition 7.2.1 Let $\mathscr{A}_h$ be the augmented admissible class in (7.2.25) with the two sets $\{\mathscr{R}_j\}_{j\in\mathscr{N}_1}$ and $\{r_j\}_{j\in\mathscr{N}_2}$ of rotations and scalings respectively, and $\mathscr{T}_h$ be the sampling mesh in Scheme R. Suppose that $\widehat{\Omega}_j = \widehat{z}_j + \widehat{\mathscr{R}}_j \Lambda_{\widehat{r}_j} \Sigma_j$, $j = 1, 2, \ldots, l_e$, are the reconstructed images of $\Omega_j = z_j + \mathscr{R}_j \Lambda_{r_j} \Sigma_j$, $j = 1, 2, \ldots, l_e$. For a properly chosen $\delta \in \mathbb{R}_+$, let $\mathscr{O}_1^j$, $\mathscr{O}_2^j$ and $\mathscr{O}_3^j$ be, respectively, δ-neighborhoods of $\widehat{z}_j$, $\widehat{\mathscr{R}}_j$ and $\widehat{r}_j$, $j = 1, 2, \ldots, l_e$. Then let $\{\mathscr{T}_{h_l'}, \{\mathscr{R}_j\}_{j\in\mathscr{P}_l}, \{r_j\}_{j\in\mathscr{Q}_l}\}$ be a refined mesh of $\{\mathscr{T}_h \cap \mathscr{O}_1^l, \{\mathscr{R}_j\}_{j\in\mathscr{N}_1} \cap \mathscr{O}_2^l, \{r_j\}_{j\in\mathscr{N}_2} \cap \mathscr{O}_3^l\}, l = 1, 2, \ldots, l_e$.

Define

$$\widehat{\widehat{\Omega}}_l(\widehat{\widehat{z}}, \widehat{\widehat{\mathscr{R}}}, \widehat{\widehat{r}}) := \widehat{\widehat{z}} + \widehat{\widehat{\mathscr{R}}} \Lambda_{\widehat{\widehat{r}}} \Sigma_l \quad \text{for } \widehat{\widehat{z}} \in \mathscr{T}_{h_l'},\ \widehat{\widehat{\mathscr{R}}} \in \{\mathscr{R}_j\}_{j\in\mathscr{P}_l},\ \widehat{\widehat{r}} \in \{r_j\}_{j\in\mathscr{Q}_l}, \tag{7.2.52}$$

a *local tune-up* of $\widehat{\Omega}_l = \widehat{z}_l + \widehat{\mathscr{R}}_l \Lambda_{\widehat{r}_l} \Sigma_l$ relative to $\{\mathscr{T}_{h'_l}, \{\mathscr{R}_j\}_{j\in\mathscr{P}_l}, \{r_j\}_{j\in\mathscr{Q}_l}\}$, $1 \le l \le l_e$.

Define

$$\widehat{\widehat{\Omega}} := \bigcup_{l=1}^{l_e} \widehat{\widehat{\Omega}}_l, \tag{7.2.53}$$

with each $\widehat{\widehat{\Omega}}_l$, $1 \le l \le l_e$, a local tune-up in (7.2.52) relative to $\{\mathscr{T}_{h'_l}, \{\mathscr{R}_j\}_{j\in\mathscr{P}_l}, \{r_j\}_{j\in\mathscr{Q}_l}\}$, a *local tune-up* of $\widehat{\Omega} := \bigcup_{j=1}^{l_e} \widehat{\Omega}_j$, relative to the *local tuning mesh*

$$\mathscr{L} := \bigcup_{l=1}^{l_e} \{\mathscr{T}_{h'_l}, \{\mathscr{R}_j\}_{j\in\mathscr{P}_l}, \{r_j\}_{j\in\mathscr{Q}_l}\}. \tag{7.2.54}$$

According to Definition 7.2.1, $\widehat{\Omega}$ is the reconstructed image of the extended scatterer Ω, whereas $\widehat{\widehat{\Omega}}$ is an adjustment of $\widehat{\Omega}$ by locally adjusting the position, orientation and size of each component of $\widehat{\Omega}$.

With the above preparation, we are ready to present Scheme M to locate the multiple components of G in (7.2.50), which can be first sketched as follows. First, by Lemmas 7.2.2 and 7.2.1, we know

$$u_\tau^\infty(\hat{x}; G) \approx u_\tau(\hat{x}; \Omega), \quad \tau = s, p \text{ or } \emptyset, \tag{7.2.55}$$

where $u_\tau^\infty(\hat{x}; G)$ and $u_\tau^\infty(\hat{x}; \Omega)$ are, respectively, the far-field patterns of G and Ω corresponding to a single incident plane wave. Hence, one can use $u_\tau^\infty(\hat{x}; G)$ as the far-field data for Scheme R to locate the extended scatterer components of Ω (approximately). We suppose the reconstruction in the above step yields $\widehat{\Omega}$, which is an approximation to Ω. Then, according to Lemma 7.2.2 again, we have

$$u_\tau(\hat{x}; D) \approx u_\tau^\infty(\hat{x}; G) - u_\tau^\infty(\hat{x}; \Omega) \approx u_\tau^\infty(\hat{x}; G) - u_\tau^\infty(\hat{x}; \widehat{\Omega}). \tag{7.2.56}$$

With the above calculated far-field data, one can then use Scheme S to locate the small scatterer components of D. However, the error introduced in (7.2.56) might be even more significant than the scattering data of D, hence the second-stage reconstruction cannot be expected to yield some reasonable result. In order to tackle this problem, a *local tuning technique* can be implemented by replacing $\widehat{\Omega}$ in (7.2.56) by a local tune-up $\widehat{\widehat{\Omega}}$. Clearly, a more accurate recovery of the extended scatterer Ω is included in the local tune-ups relative to a properly chosen local tuning mesh. Hence, one can repeat the second-stage reconstruction as described above by running through all the local tune-ups, and then locate the "clustered" local maximum points which denote the positions of the small scatterers. Meanwhile, one can also achieve much more accurate reconstruction of the extended scatterers. In summary, Scheme M can be proceeded as follows.

Scheme M	Locating multiple multi-scale scatterers of G in (7.2.50)
Step 1	For an unknown scatterer G, collect the P-part ($u_p^\infty(\hat{x}; G)$), S-part ($u_s^\infty(\hat{x}; G)$) or the full far-field ($u^\infty(\hat{x}; G)$) patterns, by sending a single detecting plane wave
Step 2	Select a sampling region with a mesh $\mathscr{T}_h$ containing Ω
Step 3	Apply Scheme M with $u_\tau^\infty(\hat{x}; G)$, $\tau = s, p$ or $\emptyset$, as the far-field data, to reconstruct approximately the extended scatterer Ω, denoted by $\widehat{\Omega}$. Clearly, $\widehat{\Omega}$ is as described in Definition 7.2.1
Step 4	For $\widehat{\Omega}$ obtained in **Step 3**, select a local-tuning mesh $\mathscr{L}$ of the form (7.2.54)
Step 5	For a tune-up $\widehat{\widehat{\Omega}}$ relative to the local tuning mesh $\mathscr{L}$ in **Step 4**, calculate $\widetilde{u}_\tau^\infty(\hat{x}) := u_\tau^\infty(\hat{x}; G) - u_\tau^\infty(\hat{x}; \widehat{\widehat{\Omega}})$ (7.2.57) Apply Scheme S with $\widetilde{u}_\tau^\infty(\hat{x})$ as the far-field data to locate the significant local maximum points on $\mathscr{T}_h \backslash \mathscr{L}$
Step 6	Repeat **Step 5** by running through all the local tune-ups relative to $\mathscr{L}$. Locate the clustered local maximum points on $\mathscr{T}_h \backslash \mathscr{L}$, which correspond to the small scatterer components of D
Step 7	Update $\widehat{\Omega}$ to the local tune-up $\widehat{\widehat{\Omega}}$ which generates the clustered local maximum points in **Step 6**

7.2.4 *Numerical Examples*

In this section, three numerical tests are presented to verify the applicability of the discussed schemes (Scheme S, R and M) in inverse elastic scattering problems for rigid bodies in three dimensions. Either plane pressure wave or shear wave can be used as the detecting field incident on the rigid scatterer and it generates coexisting scattering P- and S-waves coupled by the rigid body boundary condition. However, for brevity, we only present the numerical results where the plane shear wave is employed for the locating schemes.

In the sequel, the exact far field data are synthesized by a forward solver using quadratic finite elements for each displacement field component on a truncated spherical (3D) domain centered at the origin and enclosed by a PML layer following [11]. The computation is carried out on a sequence of successively refined meshes till the relative error of two successive finite element solutions between the two adjacent meshes is below 0.1%. The synthetic far field data are computed via the integral representation formulae [3, Eqs. (2.12) and (2.13)] and taken as the exact one.

In all the experiments, we always take the Lamé constants $\lambda = 2$ and $\mu = 1$, the incident direction $d = (0,\ 0,\ 1)$, the perpendicular direction $d^\perp = (1,\ 0,\ 0)$ and the angular frequency $\omega = 2$. In such a way, we know that the two wavenumbers $k_p = 1$

and $k_s = 2$ and the incident S-wavelength is π. For scatterers of small size or regular size, we always add to the exact far field data a uniform noise of 5% and use it as the measurement data in our numerical tests. While for multiscale scatterers, a uniform noise of 3% is added to the exact far field data.

Five revolving bodies will be considered for the scatterer components in our numerical tests. They are characterized by revolving the following 2D parametric curves along the x-axis. Some geometries are adjusted to their upright positions if necessary.

$$\begin{aligned}
\textbf{Ball}: &\ \{(x, y) : x = \cos(s),\ y = \sin(s),\ 0 \le s \le 2\pi\},\\
\textbf{Peanut}: &\ \{(x, y) : x = \sqrt{3\cos^2(s) + 1}\cos(s),\ y = \sqrt{3\cos^2(s) + 1}\sin(s),\ 0 \le s \le 2\pi\},\\
\textbf{Kite}: &\ \{(x, y) : x = \cos(s) + 0.65\cos(2s) - 0.65,\ y = 1.5\sin(s),\ 0 \le s \le 2\pi\},\\
\textbf{Acorn}: &\ \{(x, y) : x = (1 + \cos(\pi s)\cos(2\pi s)/3)\cos(\pi s),\\
&\qquad y = (1 + \cos(\pi s)\cos(2\pi s)/3)\sin(\pi s),\ 0 \le s \le 2\pi\},\\
\textbf{UFO}: &\ \{(x, y) : x = (1 + 0.2\cos(4\pi s))\cos(\pi s),\\
&\qquad y = (1 + 0.2\cos(4\pi s))\sin(\pi s),\ 0 \le s \le 2\pi\}.
\end{aligned}$$

They will be denoted for short by **B**, **P**, **K**, **A** and **U**, respectively, and shown in Figs. 7.1b, c, d and 7.3b, c.

Example 1 (*Scatterer of three small components*) The scatterer consists of three components **B**, **P** and **K**, all of which are scaled by one tenth so that their sizes are much smaller than the incident wave length. As shown in Fig. 7.1a, one small ball is located at $(-2,\ 3,\ -2)$, a small peanut at $(3,\ -2,\ -2)$ and a small kite at $(3,\ 3,\ 3)$. With resort to Scheme S, the reconstruction results of the small components are shown in Fig. 7.2 based on the indicator functions $I_1(z)$, $I_2(z)$ and $I_3(z)$ using the P-wave, S-wave and full-wave far field data, respectively. It is clearly seen from Fig. 7.2 that all the indicator functions I_m $(m = 1, 2)$ in Scheme S can identify the scatterer with the correct positions of its three components. As emphasized in Remark 7.2.2 that the resolution of the S-wave reconstruction in Fig. 7.2b is much sharper than its P-wave counterpart in Fig. 7.2a due to the shorter wavelength of S-wave. However, the full-wave imaging result in Fig. 7.2c exhibits the most accurate and stable reconstruction compared with the other two in that $I_3(z)$, by combining the highlighted ball and kite positions (lower two components in Fig. 7.2a) from I_1 and the highlighted peanut position (upper component in Fig. 7.2b) from I_1, yields the best indicating behavior and meanwhile retains the resolution as in the S-wave scenario. Thus to avoid redundancy and for better resolution, we always take the full-wave indicator function for later examples.

Example 2 (*Multiple extended scatterers*) The scatterer is composed of a UFO and an acorn. Their sizes are around 3 and is comparable with the incident plane shear wave. The UFO is located at $(-2,\ 0,\ -2)$, and the acorn is located at $(2,\ 0,\ 2)$ as shown in Fig. 7.3a.

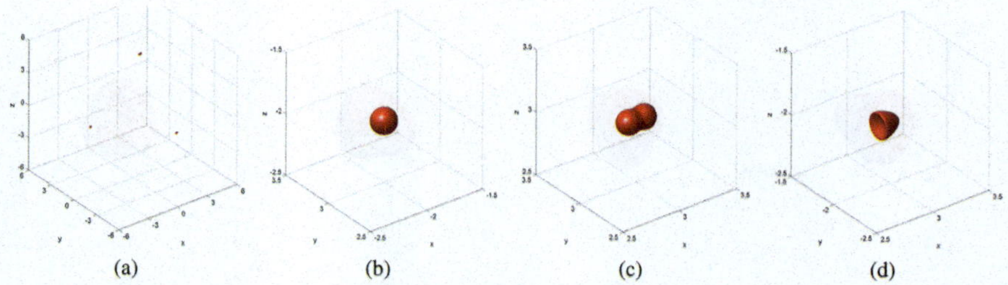

Fig. 7.1 True scatter and its components before scaling in Example 1

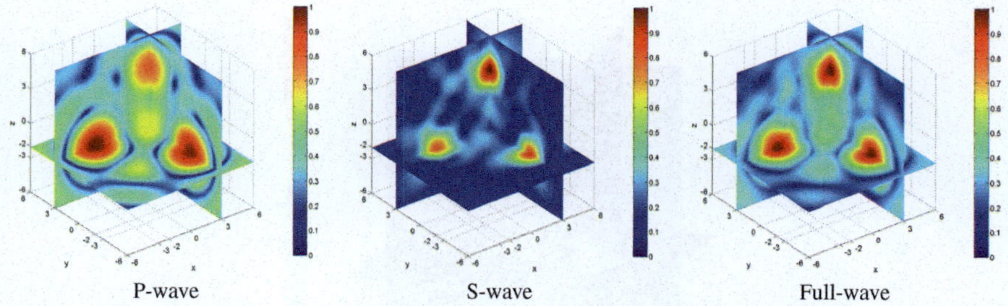

Fig. 7.2 From left to right : Reconstruction results based on the indicator functions $I_1(z)$, $I_2(z)$ and $I_3(z)$ using u_p^∞, u_s^∞ and u^∞, respectively, in Example 1

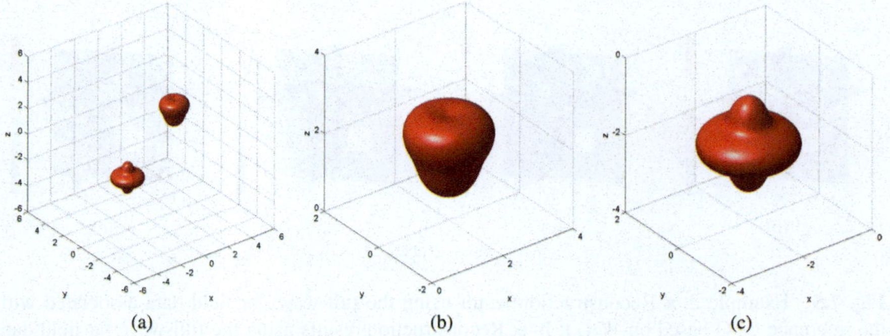

Fig. 7.3 True scatter and its components in Example 2

The candidate data set $\mathscr{A}_h$ includes far-field data of both reference components **U** and **A**, and is further lexicographically augmented by a collection of a priori known orientations and sizes. More precisely, the augmented data set is obtained by rotating **U** and **A** in the x-z plane every 90°, see, e.g., the four orientations of **A** in Fig. 7.4, and by scaling **U** and **A** by one half, one and twice.

The indicator function $W_3(z)$ is adopted to locate regular-size scatterer components. By the increasing magnitude of the far field patterns, the UFO reference data is first employed for locating purpose. Figure 7.5a tells us that the first unknown component is a UFO and its position is highlighted. What's interesting in Fig. 7.5a

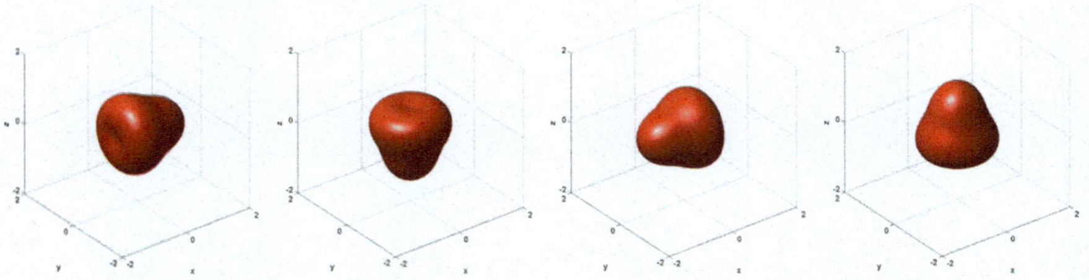

(a) (b) (c) (d)

Fig. 7.4 Example 2: Basic scatterer components : a reference acorn with four orientations

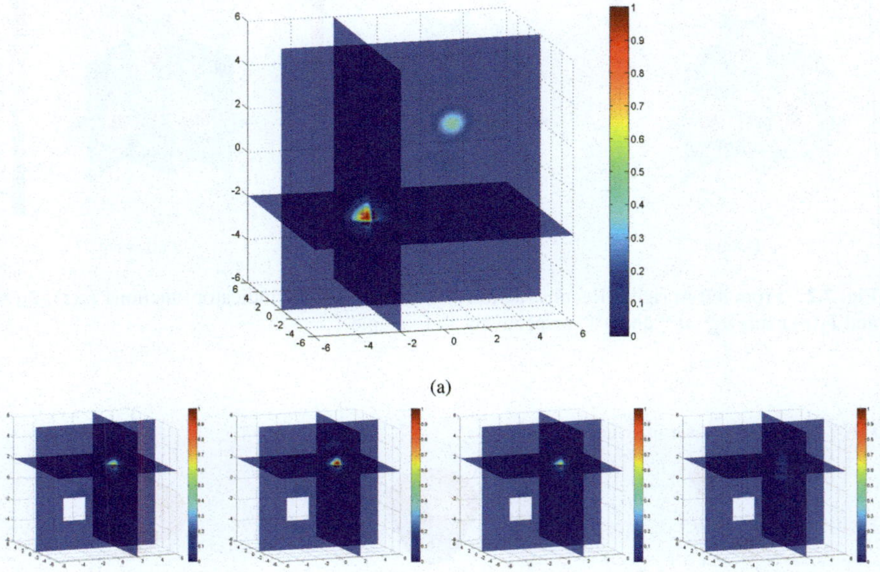

(a)

(b) (c) (d) (e)

basis_2(pear)

Fig. 7.5 Example 2. **a** Reconstruction result using the full-wave far field data associated with the reference UFO based on $W_3(z)$; **b–e**: Reconstruction results using the full-wave far field data associated with the reference acorn and its four orientations based on $W_3(z)$

is that it also indicates a ghost highlight which is close to the position of the acorn, which is due to the similarity between the UFO and acorn geometries. In the next stage, by subtracting the UFO contribution from the total far field data through Step 9 in Scheme R, we try further the far field data associated with the reference acorn geometry and all its possible orientations. For example, associated with the four orientations in Fig. 7.4, the corresponding reconstruction results are plotted in Fig. 7.5b–e. It is found in Fig. 7.5c the most prominent indicating behavior which identifies the acorn shape, its location and upside-down configuration of the second unknown component.

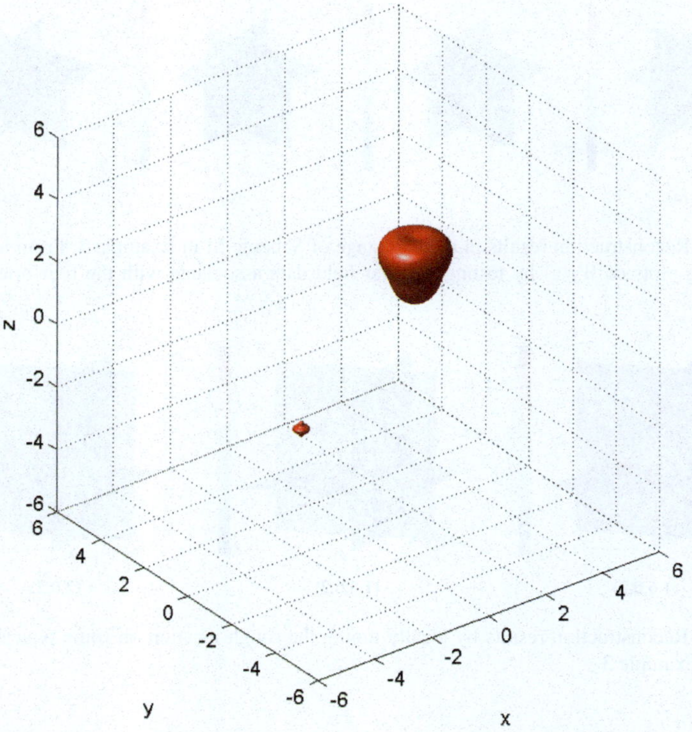

Fig. 7.6 True scatterer in Example 3

Example 3 (*Multi-scale scatter of multiple components*) In this example, we test further a multi-scale imaging problem using Scheme M. The true scatterer is composed of a small UFO scaled by one fifth and an acorn of unitary size. The small UFO is located at $(-2, 0, -2)$, and the big pear located at $(2, 0, 2)$ as shown in Fig. 7.6. As for each reference component of **A** and **U**, we rotate it every 90° in the $x - y$, $y - z$ and $z - x$ planes. Three different sizes of the reference components are tested, namely scaled by 0.2, 1 and 1.5.

In the first stage, we extract the information of the regular-size component using the indicator function $W_3(z)$ of Scheme R by computing the inner product with a priori known far-field data associated with those reference scatterer component with different orientations and sizes. We plot in Fig. 7.7 the indicator function values of $W_3(z)$ in one-to-one correspondence with the four orientations of the reference acorn as shown in Fig. 7.4. It can be observed in Fig. 7.7b that the highlighted part tells us that the first regular-size unknown component is the approximate location of an acorn with no scaling and upside-down configuration. By testing other regular-size components, no significant maxima are found and it is now safe to undergo the second stage for detecting the possible remaining small-size components.

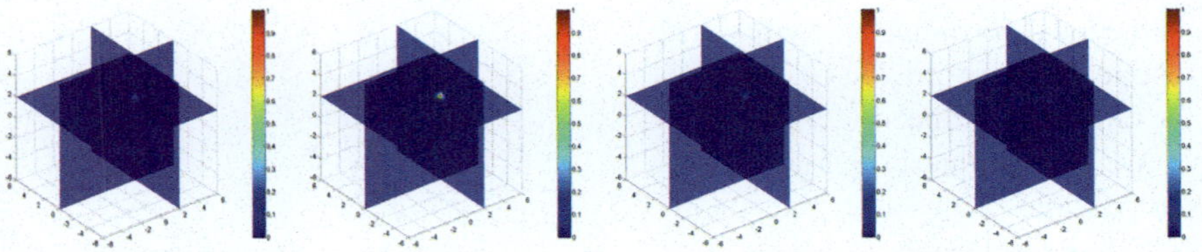

(a) (b) (c) (d)

Fig. 7.7 Reconstruction results in the first stage of Scheme M in Example 3. From left to right: Iindicating plots of $W_3(z)$ by testing with far field data associated with the four orientations in Fig. 7.4

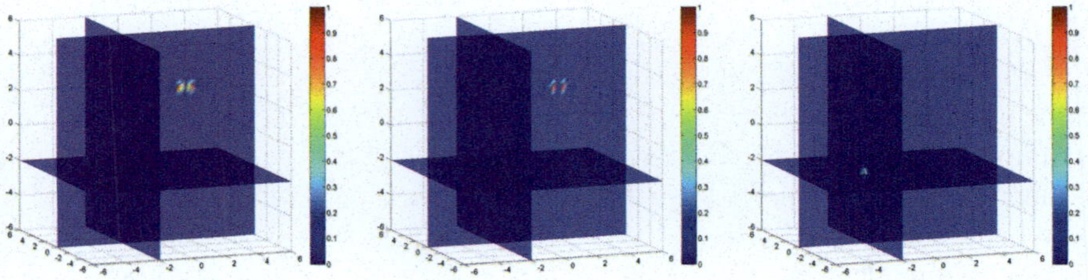

(1.6,0,2) (1.8,0,2) (2,0,2)

Fig. 7.8 Reconstruction results by locally tuning the rough location on some typical local grid points in Example 3

In the next stage, we adopt the local tuning technique pby performing a local search over a small cubic mesh around the rough position of the acorn determined by the highlighted local maximum in Fig. 7.7b. In Fig. 7.8, as the search grid points approach gradually from $(1.6\,,0\,,2)$ to $(2\,,0\,,2)$ (from left to right), the value distribution of the indicator function in Scheme S displays an gradual change of the highlighted position. In Fig. 7.8c, the red dot indicates the approximate position of the smaller UFO component, which agrees with the exact one very well. In such a way, the small UFO component could be correctly identified and positioned, and it helps us fine tune the position of the acorn and update it to be $(2\,,0\,,2)$.

7.2.5 Concluding Remarks

In this section, three imaging schemes S, R and M are presented to identify respectively, multiple small, extended and multiscale rigid elastic scatterers from the far-field pattern corresponding to a single incident plane wave with fixed incident direction and frequency. The incident elastic wave is allowed to be a plane pressure wave, a plane shear wave or a general linear combination of P- and S-waves. Relying on the availability of the far-field data, we discuss three indicating functions in each scheme by using the P-part, S-part or the full far-field pattern. Our locating schemes are based on the local *maximum* behaviors of the indicating functions. Rigorous mathematical justifications are provided and several benchmark examples are presented to illustrate the efficiency of the schemes.

We remark that in Scheme R, if certain a priori information is available about the possible range of the orientations and sizes of the scatterer components, it is sufficient for the augmented reference space $\mathscr{A}_h$ in (7.2.25) to cover that range only. In Lemma 7.2.4, we have shown uniqueness in locating the position of a translated elastic body from either the P-part or S-part of the far-field pattern corresponding to a single plane pressure or shear wave. However, we do not know if analogous uniqueness results hold for rotated and scaled elastic bodies; that is, wether or not a single far-field pattern can uniquely determine a rotating or scaling operator acting on the scatterer.

Although only the rigid scatterers are considered in the current study, the discussed schemes can be generalized to locating multiple multiscale cavities modeled by the traction-free boundary condition on the surface, the Robin-type impenetrable elastic scatterers as well as inhomogeneous penetrable elastic bodies with variable densities and Lamé coefficients inside. To achieve this, one only needs to investigate the analogous asymptotic expansions of the far-field pattern to Lemma 7.2.1, which will be used to design the locating functionals for small scatterers. The results in Sect. 7.2.2.2 remain valid for extended elastic scatterers of different physical natures. Hence, the schemes of locating extended scatterers can be straightforwardly extended to the cases mentioned above. Our approach can be also extended to the case where only limited-view measurement data are available. Further, the extension to the use of time dependent measurement data would be nontrivial and poses interesting challenges for further investigation.

7.3 Traction Free Case

This section concerns the time-harmonic elastic scattering from cavities (e.g., empty or fluid-filled cracks and inclusions) and rigid bodies, which has its origin in industrial and engineering applications; see, e.g., [10, 34, 36, 57] and the references therein. In seismology and geophysics, it is important to understand how anomalies diffract the detecting elastic waves and to characterize them from the surface measurement data. This leads to the inverse problem of determining the position and shape of an elastic scatterer; see, e.g., [1, 20, 21]. The inverse elastic scattering problem also plays a key role in many other science and technology such as petroleum and mine exploration, nondestructive testing of concrete structures etc. The inverse problem is nonlinear and ill-posed and far from well understood. In this section, we discuss several qualitative inverse elastic scattering schemes in an extremely general and practical scenario. In what follows, we first present the mathematical formulations of the forward and inverse elastic scattering problems for the present study, and then we briefly discuss the results obtained.

The rest of the section is organized as follows. In Sect. 7.3.1, we derive the asymptotic expansion of the scattered wave field from multiple small elastic scatterers. In Sect. 7.3.2, we present the inverse scattering schemes of locating multiple small, extended and multiscale elastic scatterers with the corresponding theoretical justifications.

7.3.1 Elastic Scattering From Multiscale Scatterers

In this section, we consider the elastic scattering from multiple multiscale scatterers. To that end, we first recall the fundamental solution (Green's tensor) to the Navier equation given by

$$\Pi(x,y) = \Pi^{(\omega)}(x,y) = \frac{k_s^2}{4\pi\omega^2}\frac{e^{ik_s|x-y|}}{|x-y|}\mathbf{I} + \frac{1}{4\pi\omega^2}\nabla_x\nabla_x^{\top}\left[\frac{e^{ik_s|x-y|}}{|x-y|} - \frac{e^{ik_p|x-y|}}{|x-y|}\right], \quad (7.3.1)$$

for $x, y \in \mathbb{R}^3$, $x \neq y$. Let D_j, $j \in \mathbb{N}$ be a bounded simply connected domain in $\mathbb{R}^3$ with C^2 boundary ∂D_j. Define the single and double layer potential operators, respectively, by

$$(S_j\varphi)(x) = (S_{D_j}\varphi)(x) := 2\int_{\partial D_j} \Pi(x,y)\varphi(y)ds(y), \quad \varphi \in C(\partial D_j),\ x \in \partial D_j, \quad (7.3.2)$$

$$(K_j\varphi)(x) = (K_{D_j}\varphi)(x) := 2\int_{\partial D_j} \frac{\partial \Pi(x,y)}{\partial \nu(y)}\varphi(y)ds(y), \quad \varphi \in C(\partial D_j),\ x \in \partial D_j, \quad (7.3.3)$$

where $\partial_{\nu(y)}\Pi(x,y)$ is a matrix function whose l-th column vector is given by

$$\left[\frac{\partial \Pi(x,y)}{\partial \nu(y)}\right]^{\top}\mathbf{e}_l = T_{\nu(y)}\left[\Pi(x,y)\mathbf{e}_l\right] = \nu(y)\cdot\left[\sigma(\Pi(x,y)\,\mathbf{e}_l)\right] \quad \text{on } \partial D_j,$$

for $x \neq y$, $l = 1, 2, 3$. Here, $\mathbf{e}_l$, $1 \leq l \leq 3$ are the standard Euclidean base vectors in $\mathbb{R}^3$, and $T_{\nu(y)}$ is the stress operator. The adjoint operator K_j' of K_j is given by

$$(K_j'\varphi)(x) = (K_{D_j}'\varphi)(x) := 2\int_{\partial D_j} \frac{\partial \Pi(x,y)}{\partial \nu(x)}\varphi(y)ds(y), \quad \varphi \in C(\partial D_j),\ x \in \partial D_j. \quad (7.3.4)$$

As seen by interchanging the order of integration, K_j' and K_j are adjoint with respect to the dual system $(C(\partial D_j), C(\partial D_j))$ defined by

$$(f, g) := \int_{\partial D_j} f\, g\, ds, \qquad f, g \in C(\partial D_j).$$

Using Taylor series expansion for exponential functions, one can rewrite the matrix $\Pi^{(\omega)}(x,y)$ as the series (see, e.g., [5])

$$\Pi^{(\omega)}(x, y) = \frac{1}{4\pi}\sum_{n=0}^{\infty} \frac{(n+1)(\lambda+2\mu)+\mu}{\mu(\lambda+2\mu)} \frac{(\mathrm{i}\omega)^n}{(n+2)n!}|x-y|^{n-1}\mathbf{I} \tag{7.3.5}$$

$$-\frac{1}{4\pi}\sum_{n=0}^{\infty} \frac{\lambda+\mu}{\mu(\lambda+2\mu)} \frac{(\mathrm{i}\omega)^n(n-1)}{(n+2)n!}|x-y|^{n-3}(x-y)\otimes(x-y),$$

from which it follows that

$$\Pi^{(\omega)}(x, y) = \frac{\lambda+3\mu}{8\pi\mu(\lambda+2\mu)}\frac{1}{|x-y|}\mathbf{I} + \mathrm{i}\omega\frac{2\lambda+5\mu}{12\pi\mu(\lambda+2\mu)}\mathbf{I} \tag{7.3.6}$$

$$+\frac{\lambda+\mu}{8\pi\mu(\lambda+2\mu)}\frac{1}{|x-y|^3}(x-y)\otimes(x-y) + o(1)\omega^2 \tag{7.3.7}$$

as $x \to y$. Taking $\omega \to +0$ in (7.3.7), we obtain the fundamental tensor of the Lamé system with $\omega = 0$

$$\widetilde{\Pi}(x, y) = \Pi^{(0)}(x, y) := \frac{\lambda+3\mu}{8\pi\mu(\lambda+2\mu)}\frac{1}{|x-y|}\mathbf{I} + \frac{\lambda+3\mu}{8\pi\mu(\lambda+2\mu)}\frac{1}{|x-y|^3}(x-y)\otimes(x-y). \tag{7.3.8}$$

Similar to the definitions of S_j, K_j, K_j', we define the operators $\widetilde{S}_j$, $\widetilde{K}_j$, $\widetilde{K}_j'$ in the same way as (7.3.2), (7.3.3) and (7.3.4), but with the tensor $\Pi^{(\omega)}(x, y)$ replaced by $\Pi^{(0)}(x, y)$. By comparing (7.3.6), (7.3.7) and (7.3.8), we obtain

$$\Pi^{(\omega)}(x, y) - \Pi^{(0)}(x, y) = \mathrm{i}\omega\frac{2\lambda+5\mu}{12\pi\mu(\lambda+2\mu)}\mathbf{I} + \omega^2\mathscr{O}\left(|x-y|\,\mathbf{I} - \frac{(x-y)\otimes(x-y)}{|x-y|}\right),$$

which together with (7.2.35) yields

$$|\sigma([\Pi^{(\omega)}(x, y) - \Pi^{(0)}(x, y)]\mathbf{e}_j)| \le C(\lambda, \mu)\,\omega^2, \quad C > 0, \tag{7.3.9}$$

uniformly in $j = 1, 2, 3$ as $x \to y$. The operators K_j, K_j', $\widetilde{K}_j$ and $\widetilde{K}_j'$ all have weakly singular kernels and therefore are compact on $C(\partial D_j)$; see, e.g., [25, 34].

Throughout the rest of the subsection, in order to simplify the exposition, we assume that $\omega \sim 1$. Hence, the size of a scatterer can be interpreted in terms of its Euclidean diameter. Next, we first consider the scattering from multiple sparsely distributed scatterers. Let $l \in \mathbb{N}$ and D_j, $1 \le j \le l$ be bounded simply-connected domains in $\mathbb{R}^3$ with C^2-smooth boundaries. Set

$$D = \bigcup_{j=1}^{l} D_j \quad \text{and} \quad L = \min_{j\ne j', 1\le j, j'\le l} \mathrm{dist}(\overline{D}_j, \overline{D}_{j'}). \tag{7.3.10}$$

Lemma 7.3.1 *Consider an elastic scatterer with multiple components given in (7.3.10), where each component D_j, $1 \le j \le l$ is either traction-free or rigid. For L sufficiently large, we have*

$$u^{\infty}(\hat{x}; D) = \sum_{j=1}^{l} u^{\infty}(\hat{x}; D_j) + \mathcal{O}(L^{-1}). \tag{7.3.11}$$

Proof The case that all the components of D are rigid was considered in [29]. In what follows, for simplicity we first assume that $l = 2$, and moreover we assume that both D_1 and D_2 are traction-free and ω^2 is not an eigenvalue for $-\Delta^*$ in D_j associated with the homogeneous traction-free boundary condition on ∂D_j, $j = 1, 2$.

The scattered field $u^{sc}(x; D_j)$ corresponding to D_j can be represented as the single layer potential

$$u^{sc}(x; D_j) = \int_{\partial D_j} \Pi(x, y)\varphi_j(y)ds(y), \quad x \in \mathbb{R}^3 \backslash \overline{D}_j,$$

where the density function $\varphi_j \in C(\partial D_j)$ is uniquely determined from the traction-free boundary condition on ∂D_j, and is implied in the boundary integral equation

$$\varphi_j = 2(I - K_j')^{-1}(Tu^{in}|_{\partial D_j}), \quad j = 1, 2.$$

The uniqueness and existence of φ_j follow from the Fredholm alternative applied to the operator $I - K_j'$. To prove the lemma for the scatterer $D = D_1 \cup D_2$, we make the ansatz

$$u^{sc}(x; D) = \sum_{j=1,2} \left\{ \int_{\partial D_j} \Pi(x, y)\phi_j(y)ds(y) \right\}, \quad x \in \mathbb{R}^3 \backslash \overline{D},$$

with $\phi_j \in C(\partial D_j)$. By using the boundary condition $T(u^{sc} + u^{in}) = 0$ on each ∂D_j, we obtain the system of integral equations

$$\begin{pmatrix} I - K_1' & J_2 \\ J_1 & I - K_2' \end{pmatrix} \begin{pmatrix} \phi_1 \\ \phi_2 \end{pmatrix} = 2 \begin{pmatrix} Tu^{in}|_{\partial D_1} \\ Tu^{in}|_{\partial D_2} \end{pmatrix}, \tag{7.3.12}$$

where the operators $J_1 : C(\partial D_1) \to C(\partial D_2)$, $J_2 : C(\partial D_2) \to C(\partial D_1)$ are defined respectively by

$$(J_1\phi_1)(x) := -2 \int_{\partial D_1} [T_{\nu(x)}\Pi(x, y)]\phi_1(y)ds(y), \quad x \in \partial D_2,$$

$$(J_2\phi_2)(x) := -2 \int_{\partial D_2} [T_{\nu(x)}\Pi(x, y)]\phi_2(y)ds(y), \quad x \in \partial D_1.$$

Since $L \gg 1$, using the fundamental solution (7.3.1) one readily estimates

$$\|J_1\phi_1\|_{C(\partial D_2)} \leq C_1 L^{-1}\|\phi_1\|_{C(\partial D_1)}, \quad \|J_2\phi_2\|_{C(\partial D_1)} \leq C_2 L^{-1}\|\phi_2\|_{C(\partial D_2)}, \quad C_1, C_2 > 0.$$

Hence, it follows from (7.3.12) and the invertibility of $I - K_j' : C(\partial D_j) \to C(\partial D_j)$ that

$$\begin{aligned}\begin{pmatrix}\phi_1\\ \phi_2\end{pmatrix} &= \begin{pmatrix}(I-K_1')^{-1} & 0\\ 0 & (I-K_2')^{-1}\end{pmatrix}\begin{pmatrix}2Tu^{in}|_{\partial D_1}\\ 2Tu^{in}|_{\partial D_2}\end{pmatrix} + \mathscr{O}(L^{-1})\\ &= \begin{pmatrix}\varphi_1\\ \varphi_2\end{pmatrix} + \mathscr{O}(L^{-1}).\end{aligned}$$

This implies that

$$u^{sc}(x; D) = u^{sc}(x; D_1) + u^{sc}(x; D_2) + \mathscr{O}(L^{-1}),$$

which further leads to

$$u^{\infty}(\hat{x}; D) = u^{\infty}(\hat{x}; D_1) + u^{\infty}(\hat{x}; D_2) + \mathscr{O}(L^{-1}).$$

The case that D has more than two components or D has mixed type components can be proved in a similar manner by making use of the integral equation method. In the argument above, there is a technical assumption that ω^2 is not an eigenvalue for $-\Delta^*$ in D_j with the traction-free boundary condition. If the eigenvalue problem happens, one can make use of the combined layer potentials (cf. [25, 34]) and then by a completely similar argument as above, one can show (7.3.11).

The proof is completed. □

Next, we consider the scattering from multiple small scatterers. Let $l_s \in \mathbb{N}$ and let M_j, $1 \leq j \leq l_s$, be bounded simply-connected domains in $\mathbb{R}^3$ with C^2-smooth boundaries. It is supposed that M_j, $1 \leq j \leq l_s$, contains the origin and its diameter is comparable with the S-wavelength or P-wavelength, i.e., $\mathrm{diam}(M_j) \sim \mathscr{O}(1)$. For $\rho \in \mathbb{R}_+$, we introduce a scaling/dilation operator Λ_ρ by

$$\Lambda_\rho M_j := \{\rho x : x \in M_j\} \tag{7.3.13}$$

and set

$$D_j^\rho := z_j + \Lambda_\rho M_j, \ \ z_j \in \mathbb{R}^3, \ \ 1 \leq j \leq l_s. \tag{7.3.14}$$

Let

$$D^\rho := \bigcup_{j=1}^{l_s} D_j^\rho. \tag{7.3.15}$$

Theorem 7.3.1 *Consider an elastic scatterer D^ρ given in (7.3.15). Assume that $\rho \ll 1$, $\omega \sim 1$ and*

$$L_s = \min_{j \neq j', 1 \leq j, j' \leq l_s} \operatorname{dist}(z_j, z_{j'}) \gg 1. \tag{7.3.16}$$

Moreover, we assume that D_j^ρ, $j = 1, 2, \ldots, l_s$, are all traction-free cavities. Then we have

$$u_p^\infty(\hat{x}; D^\rho) = -\rho^3 \left[(\hat{x} \otimes \hat{x}) \sum_{j=1}^{l_s} e^{-ik_p \hat{x} \cdot z_j} U_j(\hat{x}; \alpha, \beta, d, d^\perp) + \mathscr{O}(\rho + L_s^{-1}) \right], \tag{7.3.17}$$

$$u_s^\infty(\hat{x}; D^\rho) = -\rho^3 \left[(\boldsymbol{I} - \hat{x} \otimes \hat{x}) \sum_{j=1}^{l_s} e^{-ik_s \hat{x} \cdot z_j} U_j(\hat{x}; \alpha, \beta, d, d^\perp) + \mathscr{O}(\rho + L_s^{-1}) \right] \tag{7.3.18}$$

where

$$U_j(\hat{x}; \alpha, \beta, d, d^\perp) := \sum_{n=0}^{1} \sum_{m=-n}^{n} \left(\beta\, e^{ik_s z_j \cdot d}\, C_{n,j}^{m,s} + \alpha\, e^{ik_p z_j \cdot d}\, C_{n,j}^{m,p} \right) Y_n^m(\hat{x}).$$

Here, α, β are the coefficients attached to the incident plane wave, $Y_n^m(\hat{x})$ are the spherical harmonics and $C_{n,j}^{m,p}, C_{n,j}^{m,s} \in \mathbb{C}^3$ are complex-valued vectors independent of ρ, l_s, L_s and z_j.

Proof By Lemma 7.3.1, it suffices to analyze the asymptotics of the far-field patterns for only one single scatterer component. For notational convenience, we employ $\Omega = z + \Lambda_\rho M$ to denote $D_j^\rho = z + \Lambda_\rho M_j$ with any fixed $j \in \{1, 2, \ldots, l_s\}$. For $f \in C(\partial\Omega)$ and $g \in C(\partial M)$, we introduce the transforms

$$\hat{f}(\xi) = f^\wedge = f(\rho\xi + z), \quad \xi \in \partial M, \qquad \check{g}(x) = g^\vee := g((x - z)/\rho), \quad x \in \partial\Omega.$$

Using change of variables it is not difficult to verify that (see, e.g., [12, 23])

$$K_\Omega \varphi = (K_M \hat{\varphi})^\vee, \quad (I - K_\Omega)\varphi = \big((I - K_M)\hat{\varphi}\big)^\vee, \quad (I - K_\Omega)^{-1}\psi = \big((I - K_M)^{-1}\hat{\psi}\big)^\vee,$$

and similarly

$$K'_\Omega \varphi = (K'_M \hat{\varphi})^\vee, \quad (I - K'_\Omega)\varphi = \big((I - K'_M)\hat{\varphi}\big)^\vee, \quad (I - K'_\Omega)^{-1}\psi = \big((I - K'_M)^{-1}\hat{\psi}\big)^\vee.$$

These identities also hold for $\widetilde{K}$ and $\widetilde{K}'$ defined via the tensor $\widetilde{\Pi}(x, y) = \Pi^{(0)}(x, y)$. Hence, using (7.3.9), there hold

$$
\begin{aligned}
(I - K'_{\Omega})\varphi - \big((I - \widetilde{K}'_M)\hat{\varphi}\big)^{\vee} &= (I - K'_{\Omega})\varphi - (I - \widetilde{K}'_{\Omega})\varphi \\
&= (\widetilde{K}'_{\Omega} - K'_{\Omega})\varphi \\
&= 2\int_{\Omega} \nu(x)\cdot[\sigma(\Pi^{(0)}(x,y) - \Pi^{(\omega)}(x,y))]\varphi(y)\,ds(y) \\
&\leq C(\lambda,\mu)\rho^2\,||\varphi||_{C(\partial\Omega)},
\end{aligned}
$$

as $\rho \to +0$. Since $\rho \ll 1$, by the Neumann series we have

$$
(I - K'_{\Omega})^{-1}\varphi = \big((I - \widetilde{K}'_M)^{-1}\hat{\varphi}\big)^{\vee} + \mathscr{O}(\rho^2) \quad \text{as} \quad \rho \to 0. \tag{7.3.19}
$$

It is worth pointing out that the operator $I - \widetilde{K}'_M$ is bijective over the space (see, e.g., [25])

$$
\left\{ \psi \in C(\partial M) : \int_{\partial M} \psi(x)\cdot(\mathbf{a} + \mathbf{b}\times x)\,ds(x) = 0 \quad \text{for all} \quad \mathbf{a}, \mathbf{b} \in \mathbb{C}^3 \right\}.
$$

To proceed with the proof, we represent the scattered field $u^{sc}(x,\Omega)$ as the single layer potential

$$
u^{sc}(x,\Omega) = \int_{\partial\Omega} \Pi(x,y)\varphi(y)\,ds(y), \quad x \in \mathbb{R}^3\backslash\overline{\Omega},
$$

with the density function $\varphi \in C(\partial\Omega)$ given by

$$
\varphi = 2(I - K'_{\Omega})^{-1}(Tu^{in}|_{\partial\Omega}).
$$

Then, the P-part and S-part far-field patterns of u^{∞} are given respectively by (after the normalization)

$$
u_p^{\infty}(\hat{x},\Omega) = 2(\hat{x}\otimes\hat{x})\int_{\partial\Omega} e^{-ik_p\hat{x}\cdot y}\,[(I - K'_{\Omega})^{-1}\varphi](y)\,ds(y), \tag{7.3.20}
$$

$$
u_s^{\infty}(\hat{x},\Omega) = 2(\mathbf{I} - \hat{x}\otimes\hat{x})\int_{\partial\Omega} e^{-ik_s\hat{x}\cdot y}\,[(I - K'_{\Omega})^{-1}\varphi](y)\,ds(y), \tag{7.3.21}
$$

where $\varphi := Tu^{in}|_{\partial\Omega} = \nu\cdot\sigma(u^{in})|_{\partial\Omega}$. In the rest of the proof, we only justify the asymptotic behavior of $u_p^{\infty}(\hat{x},\Omega)$ as $\rho \to +0$. Our argument can be readily adapted to the case of the S-part far-field pattern.

Changing the variable $y = z + \rho\xi$ with $\xi \in \partial M$ in (7.3.20) and making use of the estimate (7.3.19), we find

$$
u_p^{\infty}(\hat{x},\Omega) = 2(\hat{x}\otimes\hat{x})\rho^2\int_{\partial M} e^{-ik_p\hat{x}\cdot(z+\rho\xi)}\,\big[(I - \widetilde{K}'_M)^{-1}\hat{\varphi}(\xi) + \mathscr{O}(\rho^2)\big]\,ds(\xi). \tag{7.3.22}
$$

Expanding the exponential function $\xi \to \exp(-ik_p\hat{x}\cdot(z+\rho\xi))$ around z yields

$$\exp(-ik_p\hat{x}\cdot(z+\rho\xi)) = \exp(-ik_p\hat{x}\cdot z) - \mathrm{i}k_p\rho(\hat{x}\cdot\xi)\exp(-ik_p\hat{x}\cdot z) + \mathscr{O}(k_p^2\rho^2) \tag{7.3.23}$$

as $\rho \to +0$. Inserting (7.3.23) into (7.3.22) gives

$$u_p^\infty(\hat{x},\Omega) = 2(\hat{x}\otimes\hat{x})\rho^2 e^{-\mathrm{i}k_p\hat{x}\cdot z}\left(\int_{\partial M}(I-\widetilde{K}'_M)^{-1}\hat{\varphi}(\xi)ds(\xi)\right) \tag{7.3.24}$$

$$-2\mathrm{i}(\hat{x}\otimes\hat{x})k_p\rho^3 e^{-\mathrm{i}k_p\hat{x}\cdot z}\left(\int_{\partial M}(\hat{x}\cdot\xi)\,(I-\widetilde{K}'_M)^{-1}\hat{\varphi}(\xi)ds(\xi)\right) \tag{7.3.25}$$

$$+\mathscr{O}(\rho^4). \tag{7.3.26}$$

To estimate the integrals on the right hand side of (7.3.26), we will investigate the incident plane pressure and shear waves, respectively. The asymptotics for general plane waves can be derived by linear superposition.

Case (i): $\beta = 1, \alpha = 0$, i.e., $u^{in} = u_s^{in} = d^\perp e^{\mathrm{i}k_s x\cdot d}$ is an incident plane shear wave.

Since $\operatorname{div} u_s^{in} = 0$, by (7.2.35) we have $\sigma(u_s^{in}) = \mu(\nabla u_s^{in} + \nabla u_s^{in\,\top})$. Expanding the function $\xi \to (\nabla u_s^{in\,\wedge})(\xi) = \nabla u_s^{in}(\rho\xi + z)$ around z,

$$(\nabla u_s^{in\,\wedge})(\xi) = \mathrm{i}k_s(d^\perp\otimes d)\left[e^{\mathrm{i}k_s z\cdot d} + \mathrm{i}k_s\rho\, e^{\mathrm{i}k_s z\cdot d}(d\cdot\xi) + \mathscr{O}(k_s^2\rho^2)\right] \quad \text{as} \quad \rho \to +0.$$

Hence,

$$(\sigma(u_s^{in})^\wedge)(\xi) = \mathrm{i}\mu k_s e^{\mathrm{i}k_s z\cdot d}\,\mathbb{H}(d)\left[1 + \mathrm{i}k_s\rho\,(d\cdot\xi) + \mathscr{O}(k_s^2\rho^2)\right], \tag{7.3.27}$$

where $\mathbb{H}(d) := (d^\perp\otimes d) + (d^\perp\otimes d)^\top$. Recalling that (which can actually be proved by using the jump relations for the double layer potential with constant density, see, e.g., [33, Example 6.14])

$$\widetilde{K}_M 1 = 2\int_{\partial M}\frac{\widetilde{\Pi}(x,y)}{\partial\nu(y)}ds(y) = -1,$$

we see $(I - \widetilde{K}_M)^{-1}\,1 = 1/2$, and thus

$$\int_{\partial M}(I-\widetilde{K}'_M)^{-1}\hat{\varphi}(\xi)ds(\xi) = \int_{\partial M}\hat{\varphi}(\xi)(I-\widetilde{K}_M)^{-1}\,1\;ds(\xi) \tag{7.3.28}$$

$$= \frac{1}{2}\int_{\partial M}\nu(\xi)\cdot(\sigma(u_s^{in})^\wedge)(\xi)\;ds(\xi). \tag{7.3.29}$$

Inserting (7.3.27) into (7.3.29) and applying Gauss's theorem yield

$$\int_{\partial M} (I - \widetilde{K}'_M)^{-1} \hat{\varphi}(\xi) ds(\xi) \tag{7.3.30}$$

$$= \mathrm{i}\mu k_s e^{\mathrm{i}k_s z\cdot d}/2 \left\{ \int_M \operatorname{div}_\xi [\mathbb{H}(d)(1 + \mathrm{i}k_s \rho d \cdot \xi)] \, ds(\xi) \right\} + \mathscr{O}(\rho^2) \tag{7.3.31}$$

$$= -\rho d^\perp e^{\mathrm{i}k_s z\cdot d} \, |M|/2 + \mathscr{O}(\rho^2). \tag{7.3.32}$$

Note that $|M|$ denotes the volume of M and that the last equality follows from the relation

$$\operatorname{div}_\xi [\mathbb{H}(d)(1 + \mathrm{i}k_s \rho d \cdot \xi)] = \mathrm{i}k_s \rho d \cdot \mathbb{H}(d) = \mathrm{i}k_s \rho \, d^\perp.$$

Again using (7.3.27) we can evaluate the second integral over ∂M on the right hand of (7.3.26) as follows:

$$\int_{\partial M} (\hat{x} \cdot \xi) \, (I - \widetilde{K}'_M)^{-1} \hat{\varphi}(\xi) ds(\xi) \tag{7.3.33}$$

$$= \mathrm{i}\mu k_s e^{\mathrm{i}k_s z\cdot d} \left\{ \int_{\partial M} (\hat{x} \cdot \xi) \, (I - \widetilde{K}'_M)^{-1} (\nu(\xi) \cdot \mathbb{H}(d)) ds(\xi) \right\} + \mathscr{O}(\rho) \tag{7.3.34}$$

$$= -\mathrm{i}\mu k_s e^{\mathrm{i}k_s z\cdot d} (\hat{x} \cdot \mathbb{M}) \cdot \mathbb{H}(d) + \mathscr{O}(\rho), \tag{7.3.35}$$

where the polarization tensor $\mathbb{M}$ depending only on M is defined as

$$\mathbb{M} = -\int_{\partial M} \xi \otimes (I - \widetilde{K}'_M)^{-1} \nu(\xi) \, ds(\xi). \tag{7.3.36}$$

Now, combining (7.3.35), (7.3.32) and (7.3.26) gives the asymptotics

$$u_p^\infty(\hat{x}, \Omega) = -(\hat{x} \otimes \hat{x}) \rho^3 e^{\mathrm{i}z\cdot(k_s d - k_p \hat{x})} \left[d^\perp \, |M| + 2(\hat{x} \cdot \mathbb{M}) \cdot \mathbb{H}(d) \right] + \mathscr{O}(\rho^4), \tag{7.3.37}$$

as $\rho \to +0$. Recalling that the spherical harmonics $Y_0^0 = \sqrt{1/4\pi}$ and that each Cartesian component of the vector $\hat{x} \in \mathbb{R}^3$ can be expressed in terms of Y_1^{-1}, Y_1^0, Y_1^1, we may reformulate the previous identity as

$$u_p^\infty(\hat{x}, \Omega) = -(\hat{x} \otimes \hat{x}) \rho^3 e^{\mathrm{i}z\cdot(k_s d - k_p \hat{x})} \left[\sum_{n=0}^{1} \sum_{m=-n}^{n} C_n^m \, Y_n^m(\hat{x}) \right] + \mathscr{O}(\rho^4), \tag{7.3.38}$$

with $C_0^0 = d^\perp \, |M| \, 2\sqrt{\pi}$ and $C_1^m \in \mathbb{C}^3$ for $m = -1, 0, 1$. This proves (7.3.17) when $\alpha = 0, \beta = 1$ and $l_s = 1$. In the case $\alpha = 0, \beta \in \mathbb{C}$ and $l_s > 1$, there holds

$$u_p^\infty(\hat{x}, \Omega) = -(\hat{x} \otimes \hat{x}) \rho^3 \sum_{j=1}^{l_s} e^{-\mathrm{i}k_p z_j \cdot \hat{x}} \left[\sum_{n=0}^{1} \sum_{m=-n}^{n} \beta \, e^{\mathrm{i}k_s z_j \cdot d} \, C_{n,j}^{m,s} \, Y_n^m(\hat{x}) \right] + \mathscr{O}(\rho^4 + L_s^{-1}),$$

with $C_{0,j}^{0,s} = d^{\perp}|M_j|2\sqrt{\pi}$ and $C_{1,j}^{m,s} \in \mathbb{C}^3$ for $m = -1, 0, 1$, and $j = 1, 2, \ldots, l_s$. Note that the constant $C_{n,j}^{m,s}$ are independent of z_j, ρ, l_s and L_s.

Case (ii): $\beta = 0, \alpha = 1$, i.e., $u^{in} = u_p^{in} = de^{ik_p x\cdot d}$ is an incident plane pressure wave.

We sketch the proof, since it can be carried out analogously to Case (i). The corresponding expansion of $(\sigma(u_p^{in}))^\wedge$ to (7.3.27) reads as follows:

$$(\sigma(u_p^{in})^\wedge)(\xi) = \mathrm{i}(\lambda + 2\mu)k_p e^{\mathrm{i}k_p z\cdot d}\,\mathbb{L}(\lambda, \mu, d)\left[1 + \mathrm{i}k_p\rho\,(d\cdot\xi) + \mathscr{O}(k_p^2\rho^2)\right] \tag{7.3.39}$$

when $\rho \to +0$, where $\mathbb{L}(\lambda, \mu, d) := (\lambda\mathbf{I} + 2\mu(d\otimes d))/(\lambda + 2\mu)$. As a consequence of (7.3.29), we have for $\varphi = \nu\cdot\sigma(u_p^{in})|_{\partial\Omega}$,

$$\int_{\partial M}(I - \widetilde{K}_M')^{-1}\hat{\varphi}(\xi)ds(\xi) \tag{7.3.40}$$

$$= i\mathrm{i}(\lambda + 2\mu)k_p e^{\mathrm{i}k_p z\cdot d}/2\left\{\int_M \mathrm{div}_{\xi}\left[\mathbb{L}(\lambda, \mu, d)(1 + \mathrm{i}k_p\rho d\cdot\xi)\right]ds(\xi)\right\} + \mathscr{O}(\rho^2) \tag{7.3.41}$$

$$= -\rho d e^{\mathrm{i}k_p z\cdot d}\,|M|/2 + \mathscr{O}(\rho^2). \tag{7.3.42}$$

Similar to (7.3.35), one has

$$\int_{\partial M}(\hat{x}\cdot\xi)\,(I - \widetilde{K}_M')^{-1}\hat{\varphi}(\xi)ds(\xi) = -\mathrm{i}(\lambda + 2\mu)k_p\,e^{\mathrm{i}k_p z\cdot d}(\hat{x}\cdot\mathbb{M})\cdot\mathbb{L}(\lambda, \mu, d) + \mathscr{O}(\rho), \tag{7.3.43}$$

where the polarization tensor $\mathbb{M}$ is given as the same in (7.3.36). Therefore, the insertion of (7.3.42) and (7.3.43) into (7.3.26) yields

$$u_p^{\infty}(\hat{x}, \Omega) = -(\hat{x}\otimes\hat{x})\rho^3 e^{\mathrm{i}k_p\hat{x}\cdot(d-z)}\left[d\,|M| + 2(\hat{x}\cdot\mathbb{M})\cdot\mathbb{L}(\lambda, \mu, d)\right] + \mathscr{O}(\rho^4), \tag{7.3.44}$$

as $\rho \to +0$, and it further leads to the asymptotic behavior in (7.3.17) when $\alpha = 1, \beta = 0$ and $l_s = 1$.

The proof is completed. □

The asymptotic expansions of the far-field patterns corresponding to small rigid bodies was considered in [12, 29], and for completeness and also the subsequent use, we include it in the following theorem.

Theorem 7.3.2 *Consider an elastic scatterer D^ρ given in (7.3.15). Assume that $\rho \ll 1$ and $L_s = \min_{j\neq j', 1\le j,j'\le l_s}\mathrm{dist}(z_j, z_{j'}) \gg 1$. Moreover, we assume that D_j^ρ, $j = 1, 2, \ldots, l_s$, are all traction-free cavities. Then we have*

$$u_p^\infty(\hat{x}; D^\rho) = \frac{\rho}{4\pi(\lambda+2\mu)}(\hat{x}\otimes\hat{x})\left[\sum_{j=1}^{l_s} e^{-\mathrm{i}k_p\hat{x}\cdot z_j}\,(C_{p,j}\,\alpha\, e^{\mathrm{i}k_p z_j\cdot d} + C_{s,j}\,\beta\, e^{\mathrm{i}k_s z_j\cdot d})\right]$$
$$+\mathscr{O}\left(\rho^2\, l_s(1+L_s^{-1})\right),$$
$$u_s^\infty(\hat{x}; D^\rho) = \frac{\rho}{4\pi\mu}(\mathbf{I}-\hat{x}\otimes\hat{x})\left[\sum_{j=1}^{l_s} e^{-\mathrm{i}k_s\hat{x}\cdot z_j}\,(C_{p,j}\,\alpha\, e^{\mathrm{i}k_p z_j\cdot d} + C_{s,j}\beta\, e^{\mathrm{i}k_s z_j\cdot d})\right]$$
$$+\mathscr{O}\left(\rho^2\, l_s(1+L_s^{-1})\right),$$

where $C_{p,j}, C_{s,j} \in \mathbb{C}^3$ are constant vectors independent of ρ, l_s, L_s and z_j.

Finally, for our subsequent study on the inverse scattering problem, we also need some results on the scattering from extended elastic bodies. Let Σ be a bounded simply-connected set that contains the origin. Denote by $\mathscr{R} := \mathscr{R}(\theta, \phi, \psi) \in SO(3)$ the 3D rotation matrix around the origin whose Euler angles are $\theta \in [0, 2\pi]$, $\phi \in [0, 2\pi]$ and $\psi \in [0, \pi]$; and define $\mathscr{R}\Sigma := \{\mathscr{R}x : x \in \Sigma\}$. We introduce

$$\mathscr{A} := \{\Sigma_j\}_{j=1}^{l'}, \quad l' \in \mathbb{N} \tag{7.3.45}$$

where each $\Sigma_j \subset \mathbb{R}^3$ is a bounded simply-connected C^2 domain containing the origin. $\mathscr{A}$ is called a *base scatterer class*, and each base scatterer Σ_j, $1 \le j \le l'$, could be either rigid or traction-free. Next, we introduce the multiple extended scatterers for our study via the base class $\mathscr{A}$ in (7.3.45). Let $l_e \in \mathbb{N}$ and for $j = 1, 2, \ldots, l_e$, set $r_j \in \mathbb{R}_+$ such that

$$r_j \in [R_0, R_1], \quad 0 < R_0 < R_1 < +\infty, \quad R_0 \sim \mathscr{O}(1),$$

and moreover, let $(\theta_j, \phi_j, \psi_j) \in [0, 2\pi]^2 \times [0, \pi]$, $j = 1, 2 \ldots, l_e$, be l_e Euler angles. For $z_j \in \mathbb{R}^3$, we let

$$D = \bigcup_{j=1}^{l_e} D_j, \quad D_j := z_j + \mathscr{R}_j\, \Lambda_{r_j}\, E_j, \quad E_j \in \mathscr{A}, \quad \mathscr{R}_j := \mathscr{R}(\theta_j, \phi_j, \psi_j). \tag{7.3.46}$$

The physical property of D_j is inherited from that of the base scatterer E_j; namely, if E_j is traction-free (resp. rigid), then D_j is also traction-free (resp. rigid). For technical purpose, we impose the following sparsity assumption on the extended scatterer D introduced in (7.3.46),

$$L_e = \min_{j\neq j', 1\le j, j'\le l_e} \mathrm{dist}(D_j, D_{j'}) \gg 1. \tag{7.3.47}$$

Theorem 7.3.3 *Consider an elastic scatterer D given in (7.3.46). Assume that the sparsity condition (7.3.47) is satisfied. If* $\alpha = 1$ *and* $\beta = 0$*, then*

$$u_\tau(\hat{x}; D, d, d, \omega) = \sum_{j=1}^{l_e} \kappa(z_j) r_j \mathscr{R}_j u_\tau^\infty(\mathscr{R}_j^\perp \hat{x}; E_j, \mathscr{R}_j^\perp d, \mathscr{R}_j^\perp, r_j\omega) + \mathscr{O}(L_e^{-1}), \quad \tau = p, s \tag{7.3.48}$$

where

$$\kappa(z_j) = e^{\mathrm{i}k_p(d-\hat{x})\cdot z_j} \text{ if } \tau = p; \quad e^{\mathrm{i}(k_p d - k_s\hat{x})\cdot z_j} \text{ if } \tau = s. \tag{7.3.49}$$

If $\alpha = 0$ *and* $\beta = 1$*, then one has a similar expansion as that in (7.3.48) but with*

$$\kappa(z_j) = e^{\mathrm{i}(k_s d - k_p\hat{x})\cdot z_j} \text{ if } \tau = p; \quad e^{\mathrm{i}k_s(d-\hat{x})\cdot z_j} \text{ if } \tau = s. \tag{7.3.50}$$

Proof We only consider the first case with $\alpha = 1$ and $\beta = 0$, and the second case with $\alpha = 0$ and $\beta = 1$ can be proved in a similar manner. If E_j is a rigid elastic body, the following identities were proved in [29]:

$$u_\tau^\infty(\hat{x}; z_j + E_j) = \kappa(z_j) u_\tau^\infty(\hat{x}; E_j) \quad \text{where } \kappa(z_j) \text{ is given in (7.3.49)}, \tag{7.3.51}$$

and

$$\mathscr{R} u_\tau^\infty(\hat{x}; E_j, d, d^\perp) = u_\tau^\infty(\mathscr{R}\hat{x}; \mathscr{R}E_j, \mathscr{R}d, \mathscr{R}d^\perp), \tag{7.3.52}$$

and

$$u_\tau^\infty(\hat{x}; \Lambda_{r_j} E_j, \omega) = r_j u_\tau^\infty(\hat{x}; E_j, r_j\omega). \tag{7.3.53}$$

By following a completely similar argument, one can show that the above identities also hold when E_j is a traction-free cavity. Finally, by using Lemma 7.3.1, and (7.3.51)–(7.3.53), one can show (7.3.49), which completes the proof. □

7.3.2 Locating Multiple Multiscale Elastic Scatterers

In this section, we consider the inverse scattering problem of recovering multiple elastic scatterers. We first consider the locating of multiple small scatterers introduced in (7.3.15), and then we consider the locating of multiscale scatterers with both small components of those in (7.3.15) and extended components of those described in (7.3.46). The key ingredients of the developed inverse scattering schemes are some indicator functions, whose local maximum behaviors can be used to identify the multiple elastic bodies in an effective and efficient manner.

7.3.2.1 Locating Small Scatterers

Let D^ρ be a small elastic scatterer consisting of multiple components as introduced in (7.3.15). In order to present the scheme of locating the multiple components of D^ρ, we introduce the following three indicator functions

$$I_1(z) = \frac{1}{\|u_p^\infty(\hat{x}; D^\rho)\|_{L^2}^2} \sum_{n=0}^{1} \sum_{m=-n}^{n} \sum_{l=1}^{3} \left| \left\langle u_p^\infty(\hat{x}; D^\rho), (\hat{x} \otimes \hat{x}) Y_n^m(\hat{x}) \mathbf{e}_l \, e^{-\mathrm{i}k_p \hat{x} \cdot z} \right\rangle \right|^2, \tag{7.3.54}$$

$$I_2(z) = \frac{1}{\|u_s^\infty(\hat{x}; D^\rho)\|_{L^2}^2} \sum_{n=0}^{1} \sum_{m=-n}^{n} \sum_{l=1}^{3} \left| \left\langle u_s^\infty(\hat{x}; D^\rho), (\mathbf{I} - \hat{x} \otimes \hat{x}) Y_n^m(\hat{x}) \mathbf{e}_l \, e^{-\mathrm{i}k_s \hat{x} \cdot z} \right\rangle \right|^2, \tag{7.3.55}$$

$$I_3(z) = \frac{1}{\|u^\infty(\hat{x}; D^\rho)\|_{L^2}^2} \sum_{n=0}^{1} \sum_{m=-n}^{n} \sum_{l=1}^{3} |f_{n,m,l}(z)|^2, \tag{7.3.56}$$

where

$$f_{n,m,l}(z) := \left\langle u^\infty(\hat{x}; D^\rho), \ \left[(\hat{x} \otimes \hat{x}) \, e^{-\mathrm{i}\mathrm{i}k_p \hat{x} \cdot z} + (\mathbf{I} - \hat{x} \otimes \hat{x}) \, e^{-\mathrm{i}k_s \hat{x} \cdot z} \right] Y_n^m(\hat{x}) \mathbf{e}_l \right\rangle.$$

Here and in what follows, the notation $\langle \cdot, \cdot \rangle$ denotes the inner product in $L^2 := L^2(\mathbb{S}^2)^3$ with respect to the variable $\hat{x} \in \mathbb{S}^2$, defined as $\langle \mathbf{u}, \mathbf{v} \rangle := \int_{\mathbb{S}^2} \mathbf{u}(\hat{x}) \cdot \overline{\mathbf{v}(\hat{x})} \, ds(\hat{x})$. Clearly, I_m $(m = 1, 2, 3)$ are all nonnegative functions and they can be obtained, respectively, by using a single P-part far-field pattern $(m = 1)$, S-part far-field pattern $(m = 2)$, or the full far-field pattern $(m = 3)$. The functions introduced above possess certain indicating behaviors, which lies in the essence of our inverse scattering schemes. Before stating the theorem of the indicating behaviors for those functions, we introduce the following real numbers

$$K_1^j := \frac{\|u_p^\infty(\hat{x}; D_j^\rho)\|_{L^2}^2}{\|u_p^\infty(\hat{x}; D^\rho)\|_{L^2}^2}, \quad K_2^j := \frac{\|u_s^\infty(\hat{x}; D_j^\rho)\|_{L^2}^2}{\|u_s^\infty(\hat{x}; D^\rho)\|_{L^2}^2}, \quad K_3^j := \frac{\|u^\infty(\hat{x}; D_j^\rho)\|_{L^2}^2}{\|u^\infty(\hat{x}; D^\rho)\|_{L^2}^2}, \tag{7.3.57}$$

for $1 \le j \le l_s$.

Theorem 7.3.4 *Consider the elastic scatterer D^ρ described in (7.3.15), and assume that D^ρ is traction-free. For K_m^j, m=1,2,3, defined in (7.3.57), we have*

$$K_m^j = \widetilde{K}^j + \mathscr{O}(L_s^{-1} + \rho), \quad 1 \le j \le l_s, \ m = 1, 2, 3, \tag{7.3.58}$$

where $\widetilde{K}^j$'s are positive numbers independent of L_s, ρ and m. Moreover, there exists an open neighborhood of z_j, neigh(z_j), such that

$$I_m(z) \le \widetilde{K}^j + \mathscr{O}(L_s^{-1} + \rho) \ \textit{ for all } \ z \in \mathrm{neigh}(z_j), \tag{7.3.59}$$

and $I_m(z)$ achieves its maximum value at z_j in neigh(z_j)*, i.e.,*

$$I_m(z_j) = \widetilde{K}^j + \mathscr{O}(L_s^{-1} + \rho). \tag{7.3.60}$$

Proof For notational convenience we write

$$A_j =: \sum_{n=0}^{1} \sum_{m=-n}^{n} |B_{n,m,j}|^2, \quad B_{n,m,j} = \beta\, e^{\mathrm{i}k_s z_j \cdot d}\, C_{n,j}^{m,s} + \alpha\, e^{\mathrm{i}k_p z_j \cdot d}\, C_{n,j}^{m,p}, \tag{7.3.61}$$

where the constants $C_{n,j}^{m,p}$, $C_{n,j}^{m,s}$ are those given in (7.3.17). Then, it is seen from Theorem 7.3.1 and the orthogonality of Y_n^m that

$$\|u_p^\infty(\hat{x};\, D_j^\rho)\|_{L^2}^2 = \rho^6 A_j + \mathscr{O}\left(\rho^7\right) \quad \text{as} \quad \rho \to +0.$$

Under the sparsity assumption (7.3.16), by using the Riemann-Lebesgue lemma about oscillating integrals, we can obtain

$$\|u_p^\infty(\hat{x};\, D^\rho)\|_{L^2}^2 = \rho^6 \sum_{j=1}^{l_s} A_j + \mathscr{O}(\rho^7) + \mathscr{O}(L_s^{-1}).$$

Hence,

$$K_1^j = \frac{\|u_p^\infty(\hat{x};\, D_j^\rho)\|_{L^2}^2}{\|u_p^\infty(\hat{x};\, D^\rho)\|_{L^2}^2} = \widetilde{K}^j + \mathscr{O}(\rho + L_s^{-1}), \quad \widetilde{K}^j := \frac{A_j}{\sum_{j=1}^{l_s} A_j}. \tag{7.3.62}$$

This proves (7.3.58) for $m = 1$. The case of using the S-part far-field pattern (i.e., $m = 2$) can be treated analogously. To treat the case $m = 3$, we use the orthogonality of u_p^∞ and u_s^∞. Since $\langle \mathbf{I} - \hat{x} \otimes \hat{x}, \hat{x} \otimes \hat{x} \rangle = 0$, again applying Theorem 7.3.1 to D^ρ and D_j^ρ yields

$$\|u^\infty(\hat{x};\, D^\rho)\|_{L^2}^2 = 2\rho^6 \sum_{j=1}^{l_s} A_j + \mathscr{O}\left(\rho^7\right) + \mathscr{O}(L_s^{-1}),$$
$$\|u^\infty(\hat{x};\, D_j^\rho)\|_{L^2}^2 = 2\rho^6 A_j + \mathscr{O}\left(\rho^7\right).$$

Hence, (7.3.58) is proved with $\widetilde{K}^j$ defined as in (7.3.62).

To verify (7.3.59) and (7.3.60), without loss of generality we only consider the indicating behavior of $I_1(z)$ in a small neighborhood of z_j for some fixed $1 \le j \le l_s$, i.e., $z \in$ neigh(z_j). We assume further that $|z - z_j| < \rho$. Clearly, under the assumption (7.3.16),

$$\omega |z_{j'} - z| \sim \omega\, L_s \gg 1, \quad \text{for all} \quad z \in \text{neigh}(z_j), \quad j' \neq j.$$

By using the Riemann-Lebesgue lemma and Theorem 7.3.1, one can obtain

$$\sum_{l'=1}^{3}\left|\left\langle u_p^\infty(\hat{x}; D^\rho), (\hat{x}\otimes\hat{x})Y_{n'}^{m'}(\hat{x})\mathbf{e}_{l'}e^{-\mathrm{i}k_p\hat{x}\cdot z}\right\rangle\right|^2 \tag{7.3.63}$$

$$= \rho^6\sum_{l'=1}^{3}\left|\left\langle e^{-\mathrm{i}k_p\hat{x}\cdot z_j}\sum_{n=0}^{1}\sum_{m=-n}^{n} B_{n,m,j}\, Y_n^m(\hat{x}),\ Y_{n'}^{m'}(\hat{x})\mathbf{e}_{l'}e^{-\mathrm{i}k_p\hat{x}\cdot z}\right\rangle\right|^2 + \mathscr{O}(\rho^7 + L_s^{-1}) \tag{7.3.64}$$

$$\leq \rho^6\sum_{l'=1}^{3}|B_{n',m',j}\cdot e_{l'}|^2 + \mathscr{O}(\rho^7 + L_s^{-1}) \tag{7.3.65}$$

$$= \rho^6|B_{n',m',j}|^2 + \mathscr{O}(\rho^7 + L_s^{-1}), \tag{7.3.66}$$

where the inequality (7.3.65) follows from the Cauchy-Schwarz inequality and $B_{n',m',j} \in \mathbb{C}^3$ are given in (7.3.61). Moreover, the strict inequality in (7.3.65) holds if $z \neq z_j$ and the equal sign holds only when $z = z_j$. Therefore, by the definitions of I_1, A_j and $\widetilde{K}^j$,

$$I_1(z) \leq \frac{\rho^6\sum_{n'=0}^{1}\sum_{m'=-n'}^{n'}|B_{n',m',j}|^2 + \mathscr{O}(\rho^7 + L_s^{-1})}{\rho^6\sum_{j=1}^{l_s}A_j + \mathscr{O}(\rho^7 + L_s^{-1})} = \widetilde{K}^j + \mathscr{O}(\rho + L_s^{-1}),$$

where the equality holds only when $z = z_j$. This proves (7.3.59) and (7.3.60). The indicating behavior of I_2 and I_3 can be verified in the same way. □

Remark 7.3.1 The local maximum behavior of $I_m(z)$ can be used to locate the scatterer components of D^ρ, namely z_j, $1 \leq j \leq l_s$. Such indicating behavior is much evident if one considers the case that D^ρ has only one component, i.e., $l_s = 1$. In the one-component case, one has that

$$\widetilde{K}^j = 1, \quad I_m(z) < 1 + \mathscr{O}(\rho) \quad \text{for all} \quad m = 1, 2, 3,\ z \neq z_1,$$

and

$$I_m(z_1) = 1 + \mathscr{O}(\rho), \quad m = 1, 2, 3.$$

That is, z_1 is a global maximizer for $I_m(z)$.

Remark 7.3.2 In Theorem 7.3.4, we only consider that D^ρ is a traction-free scatterer. If D^ρ is a rigid scatterer, by using Theorem 7.3.2 and following a similar argument, one can show that Theorem 7.3.4 remains valid. Moreover, by Theorem 7.3.2, it is easily seen that in the rigid case the terms with the index $n = 1$ in $I_m(z)$ are high-order terms and hence can be eliminated. Eliminating the terms with the index $n = 1$ in (7.3.54), (7.3.55) and (7.3.56) actually gives the indicator functions proposed in [29]. However, it is clear that the indicator functions proposed in [29] work only for locating rigid bodies. The indicator functions proposed in (7.3.54)–(7.3.56) works for locating both rigid and traction-free cavities. Furthermore, we can consider a even more general case by assuming $D_s = D^{\rho_1} \cup D^{\rho_2}$ with D^{ρ_j}, $j = 1, 2$, both of the form (7.3.15). D^{ρ_1} and D^{ρ_2}, respectively, contain the rigid bodies and traction-free cavities. It is assumed that $\rho_1 \sim \rho_2^3 \ll 1$. This means that both D^{ρ_1} and D^{ρ_2} are small scatterers, and by Theorems 7.3.1 and 7.3.2, the scattering strengths from the components of D^{ρ_1} and D^{ρ_2} are comparable. Then it is straightforward to show that Theorem 7.3.4 remains valid for the scatterer D_s described above.

Based on Theorem 7.3.4, it is ready to formulate a reconstruction scheme of locating the multiple scatterers of D^ρ in (7.3.15) as follows.

Scheme I	Locating small scatterers of D^ρ in (7.3.15)
Step 1	For an unknown scatterer D^ρ with multiple components in (7.3.15), collect the P-part ($m = 1$), S-part ($m = 2$) or the full far-field data ($m = 3$) by sending a single detecting plane wave
Step 2	Select a sampling region with a mesh $\mathcal{T}_h$ containing D^ρ
Step 3	For each sampling point $z \in \mathcal{T}_h$, calculate $I_m(z)$ ($m = 1, 2, 3$) according to the measurement data
Step 4	Locate all the local maximizers of $I_m(z)$ on $\mathcal{T}_h$, which represent locations of the scatterer components of D^ρ

7.3.2.2 Locating Multiscale Scatterers

Consider an elastic scatterer with multiscale components of the following form

$$D_m := D^\rho \cup D, \tag{7.3.67}$$

where D^ρ given in (7.3.15) and D given in (7.3.45)–(7.3.47) represent, respectively, the collections of small-size and extended-size scatterers. For D_m introduced above, we assume that $\mathrm{dist}(\overline{D^\rho}, \overline{D}) \gg 1$. Next, we consider the recovery of the multiple multiscale scatterer components of D_m, under the a priori knowledge that the base scatterer class $\mathscr{A}$ in (7.3.45) is known in advance. In the present section, $\mathscr{A}$ is also referred to as an *admissible class*. If D_m consists of only rigid bodies, the recovery was considered in [29]. By using Scheme I developed for locating small scatterers, together with the help of Lemma 7.3.1 and Theorem 7.3.3, and some slight necessary modifications, the inverse scattering scheme developed in [29] for locating multiscale rigid bodies can be readily extended to the locating of the more general multiscale scatterers contained in D_m. In what follows, for completeness and self-containedness, we sketch the reconstruction procedure.

First, for the admissible class $\mathscr{A}$ and a sufficiently small $\varepsilon \in \mathbb{R}_+$, we introduce an ε-net $\mathscr{A}_\varepsilon := \{\widetilde{\Sigma}_j\}_{j=1}^{l''}$ of $\mathscr{A}$, such that for any $\Sigma \in \mathscr{A}$, there exists $\widetilde{\Sigma} \in \mathscr{A}_\varepsilon$ satisfying $d_H(\Sigma, \widetilde{\Sigma}) \le \varepsilon$, where d_H denotes the Hausdorff distance. It is assumed that

(a) $u_\tau^\infty(\hat{x}; \widetilde{\Sigma}_j) \neq u_\tau^\infty(\hat{x}; \widetilde{\Sigma}_{j'})$ for $\tau = s, p$ or $\emptyset$, and $j \neq j', 1 \le j, j' \le l''$.
(b) $\|u_\tau^\infty(\hat{x}; \widetilde{\Sigma}_j)\|_{L^2} \ge \|u_\tau^\infty(\hat{x}; \widetilde{\Sigma}_{j'})\|_{L^2}$ for $\tau = s, p$ or $\emptyset$, and $j < j', 1 \le j, j' \le l''$.

Assumption (a) is the generic uniqueness for the inverse elastic scattering problem, whereas assumption (b) can be achieved by reordering if necessary. Next, for simplicity, we only consider the case by making use of the P-part far-field pattern. However, all the results presented below still hold when the S-part or full far-field patterns are employed, if one replaces the locating functional by the corresponding functionals developed in [29]. Let either α or β be taken to be zero in the detecting plane wave and define

$$J_j(z) = \frac{1}{\|u_p^\infty(\hat{x}; \widetilde{\Sigma}_j)\|_{L^2}^2} \left|\left\langle u_p^\infty(\hat{x}; D_m),\ e^{-ik_p\hat{x}\cdot z} u_p^\infty(\hat{x}; \widetilde{\Sigma}_j)\right\rangle\right|^2, \quad z \in \mathbb{R}^3. \quad (7.3.68)$$

Since $\widetilde{\Sigma}_j \in \mathscr{A}$ is known in advance, $J_j(z)$ is actually obtained by projecting the scattering measurement data into a space generated by the scattering data from the admissible base scatterers. Then, one starts with the indicator function $J_1(z)$ to locate all the local maximum points on a sampling mesh $\mathscr{T}$ containing the target scatterer. We denote the obtained local maximum points by $z_1^1, z_1^2, \ldots, z_l^{l_1}$, which represent the approximate locations of scatterer components of the form $z_1^j + \widetilde{\Sigma}_1$, $j = 1, 2, \ldots, l_1$. With the located scatterer components $z_1^j + \widetilde{\Sigma}_1$, one updates the P-part of the far-field pattern according to the following formula,

$$u_p^\infty(\hat{x}) := u_p^\infty(\hat{x}; D_m) - \sum_{j=1}^{l_1} \kappa(z_1^j) u_p^\infty(\hat{x}; \widetilde{\Sigma}_1),$$

where $\kappa(z_j)$ is given in (7.3.49) and (7.3.50). Using the updated far-field pattern as the measurement data, one continues the locating procedure with the indicator function $J_2(z)$ and finds the corresponding local maximum points on $\mathscr{T}$, say, $z_2^1, z_2^2, \ldots, z_2^{l_2}$, which represent the approximate locations of scatterer components of the form $z_2^j + \widetilde{\Sigma}_2$, $j = 1, 2, \ldots, l_2$. By continuing the above procedure, one can find $z_j^1, z_j^2, \ldots, z_j^{l_j}$, $j = 3, \ldots, l''$, which represent the approximate locations of the scatterer components of the form $z_j^m + \widetilde{\Sigma}_j, m = 1, 2, \ldots, l_j$. It is emphasized that it may happen that $l_j = 0$ for some $1 \le j \le l''$, which means that the scatterer components obtained from the base scatterer $\widetilde{\Sigma}_j$ does not appear in the target elastic scatterer D_m.

In the above step, one finds $\bigcup_{j=1}^{l''} \bigcup_{m=1}^{l_j} \{z_j^m\}$, and from which one recovers the extended scatterer components of D in (7.3.67) in an approximate manner. Next, one proceeds to the recovery of the small scatterer components of D^ρ. To that end, we let $U(z_j^m)$ denote an open neighborhood of z_j^m and V_j^m be an h-net of $U(z_j^m)$ with $h \ll 1$. Each set

$$\bigcup_{j=1}^{l''} \bigcup_{m=1}^{l_j} \{\widetilde{z_j^m}\}, \quad z_j^m \in V_j^m, \quad (7.3.69)$$

is a called a *local tuneup* of $\bigcup_{j=1}^{l''} \bigcup_{m=1}^{l_j} \{z_j^m\}$ relative to $\bigcup_{j=1}^{l''} \bigcup_{m=1}^{l_j} V_j^m$. For a local tuneup in (7.3.69), let

$$u_p^\infty(\hat{x};\, D^\rho) := u_p^\infty(\hat{x};\, D_m) - \sum_{j=1}^{l''}\sum_{m=1}^{l_j} \kappa(z_j^m) u_p^\infty(\hat{x};\, \widetilde{\Sigma}_j), \tag{7.3.70}$$

where $\kappa(z_j)$ is given in (7.3.49) and (7.3.50). Applying $u_p(\hat{x};\, D^\rho)$ as the measurement data to Scheme I developed at the end of Sect. 7.3.2.1, and then locate all the local maximum points of the corresponding indicator function. By running through all the possible local tuneups and repeating the above procedure, one can locate the clustered local maximum points, which represent the locations of the small scatterer components of D^ρ in (7.3.67).

7.4 Reconstructing Multiple Small Scatterers

In this section, we consider the reconstruction of multiple small scatterers. Throughout the rest of the section, we assume that $\omega \sim 1$. That is, the wavelength of the elastic waves is given by $2\pi/\omega \sim 1$ and hence the size of a scatterer can be expressed in terms of its Euclidean diameter.

We first introduce the class of small scatterers for our study. Let $l_s \in \mathbb{N}$ and $D_j \subset \mathbb{R}^3$, $1 \le j \le l_s$ be bounded simply-connected C^2 domains containing the origin. For $\rho \in \mathbb{R}_+$, we introduce a scaling operator

$$\Lambda_\rho D_j := \left\{\rho x;\ x \in D_j\right\} \tag{7.4.1}$$

and set

$$\Omega_j^{(s)} = z_j + \Lambda_\rho D_j, \quad z_j \in \mathbb{R}^3, \quad 1 \le j \le l_s, \tag{7.4.2}$$

where each $\Omega_j^{(s)}$ is referred to as a scatterer component. The parameter $\rho_j \in \mathbb{R}_+$ represents the relative size of the scatterer (or, more precisely, each of its components). Then a scatterer component $\Omega_j^{(s)}$ is said to be *small* if $\rho \ll 1$. For a collection of small scatterers, we set

$$\Omega^{(s)} := \bigcup_{j=1}^{l_s} \Omega_j^{(s)}. \tag{7.4.3}$$

$\Omega^{(s)}$ is called a *multiple small scatterer* if $l_s > 1$ and the following qualitative sparsity assumption is satisfied

$$L_s := \mathrm{dist}(z_i,\, z_j) \gg 1 \quad \text{for } i \ne j,\ 1 \le i,\ j \le l_s. \tag{7.4.4}$$

In order to recover the multiple scatterers in $\Omega^{(s)}$ in (7.4.3), we present the following three imaging functionals,

$$I_1(z) = \frac{1}{\left\|u_p^\infty(\hat{x},\, \Omega^{(s)})\right\|_{L^2}^2} \sum_{n=0}^{1} \sum_{m=-n}^{n} \sum_{l=1}^{3} \left|\left\langle u_p^\infty(\hat{x},\, \Omega^{(s)}),\, (\hat{x} \otimes \hat{x}) Y_n^m(\hat{x}) \mathbf{e}_l e^{-\mathrm{i}k_p \hat{x}\cdot z} \right\rangle\right|^2, \tag{7.4.5}$$

$$I_2(z) = \frac{1}{\left\|u_s^\infty(\hat{x},\, \Omega^{(s)})\right\|_{L^2}^2} \sum_{n=0}^{1} \sum_{m=-n}^{n} \sum_{l=1}^{3} \left|\left\langle u_s^\infty(\hat{x},\, \Omega^{(s)}),\, (\hat{x} \otimes \hat{x}) Y_n^m(\hat{x}) \mathbf{e}_l e^{-\mathrm{i}k_s \hat{x}\cdot z} \right\rangle\right|^2, \tag{7.4.6}$$

$$I_3(z) = \frac{1}{\left\|u^\infty(\hat{x},\, \Omega^{(s)})\right\|_{L^2}^2} \sum_{n=0}^{1} \sum_{m=-n}^{n} \sum_{l=1}^{3} \left| f_{n,m,l}(z) \right|^2, \tag{7.4.7}$$

where $z \in \mathbb{R}^3$ and

$$f_{n,m,l}(z) := \left\langle u^\infty(\hat{x};\, \Omega^{(s)}),\, \left[\left(\hat{x} \otimes \hat{x}\right) e^{-\mathrm{i}k_p \hat{x}\cdot z} + \left(\mathbf{I} - \hat{x} \otimes \hat{x}\right) e^{-\mathrm{i}k_s \hat{x}\cdot z} \right] Y_n^m(\hat{x}) \mathbf{e}_l \right\rangle.$$

Here and in what follows, the notation $\langle \cdot,\, \cdot \rangle$ denotes the inner product in $L^2 := L^2\left(\mathbb{S}^2\right)^3$ with respect to the variable $\hat{x} \in \mathbb{S}^2$, defined as $\langle \mathbf{u},\, \mathbf{v} \rangle := \int_{\mathbb{S}^2} \mathbf{u}\left(\hat{x}\right) \cdot \overline{\mathbf{v}\left(\hat{x}\right)} ds\left(\hat{x}\right)$. In (7.4.5)–(7.4.7), the constant vectors $\mathbf{e}_l$, $1 \le l \le 3$ are the standard Euclidean base vectors in $\mathbb{R}^3$ and $Y_n^m(\hat{x})$, $n = 0, 1, \ldots$, $m = -n, \ldots, n$, are the spherical harmonics (cf. [14]). The imaging functionals $I_m(z)$, $m = 1, 2, 3$, possess a certain local maximum behavior which can be used to locate the scatterer components of $\Omega^{(s)}$, namely z_j, $1 \le j \le l_s$. To be more specific, the values of $I_m(z)$ $(m = 1,\, 2,\, 3)$ would become larger when approaching each z_j $(1 \le j \le l_s)$, and be close to zero when z is far away from each z_j $(1 \le j \le l_s)$. Here, we would like to note that $k_s/k_p = \sqrt{(\lambda + 2\mu)/\mu}$, and hence $k_s > k_p$ if $\lambda > 0$. Generically, one would have $\lambda > 0$, and therefore, one can expect that the imaging functionals I_2 and I_3 would produce better reconstructions than the imaging functional I_1. This is actually numerically confirmed in our experiments. The scheme for locating the multiple small scatterers is formulated as Algorithm 1.

Algorithm 1 Locating multiple small scatterers $\Omega^{(s)}$ in (7.4.3) with I_m, $m = 1, 2, 3$.

Step 1 For an unknown scatterer $\Omega^{(s)}$ with multiple components in (7.4.3), collect $u_p^\infty (m = 1)$, $u_s^\infty (m = 2)$, $u^\infty (m = 3)$ by sending a single detecting plane wave 6.1.2.

Step 2 Select a sampling region with a mesh $\mathscr{T}_h$ containing $\Omega^{(s)}$;

Step 3 For each sampling point $z \in \mathscr{T}_h$, calculate I_m according to the measurement data.

Step 4 Locate all the local maximizers of $I_m(z)$ on $\mathscr{T}_h$, which represent locations of the scatterer components of $\Omega^{(s)}$.

We proceed to present a numerical example to illustrate the effectiveness of Algorithm 1. Throughout the rest of the section, for the numerical examples, the synthetic far-field data are obtained by computing the corresponding direct elastic scattering problem. To be more specific, we solve the Lamé system (6.1.1) by the standard Finite Element Method (FEM), where the unbounded computed region is truncated by the Perfect Matched Layers (PMLs). Furthermore, we have refined the mesh several times until a sufficiently accurate solution is achieved. Then the P-part and S-part far-field could be directly derived by implementing

$$u_p^\infty(\hat{x}) = \frac{k_p^2}{4\pi\omega^2}\int_{\partial D} [T_{\nu(y)}\hat{x}\hat{x}^T e^{-\mathrm{i}k_p\hat{x}\cdot y}]^T u(y) - \hat{x}\hat{x}^T e^{-\mathrm{i}k_p\hat{x}\cdot y} T_{\nu(y)}u(y), \quad \hat{x}\in\Omega, \tag{7.4.8}$$

and

$$u_s^\infty(\hat{x}) = \frac{k_s^2}{4\pi\omega^2}\int_{\partial D} [T_{\nu(y)}[\mathbf{I}-\hat{x}\hat{x}^T]e^{-\mathrm{i}k_s\hat{x}\cdot y}]^T u(y) - [\mathbf{I}-\hat{x}\hat{x}^T]e^{-\mathrm{i}k_s\hat{x}\cdot y} T_{\nu(y)}u(y), \quad \hat{x}\in\Omega, \tag{7.4.9}$$

respectively. Unless otherwise stated, we always collect the far-field pattern $u^\infty(\hat{x}_i)\,(i = 1, ..., 590)$ on 590 Lebdev quadrature points, which is important in evaluating the integral in the indicator functions. And to show the stability of the locating scheme, we have also added a 5% Gaussian noise on the obtained far-field data. That is,

$$u^\infty_{\tau,\text{noise}}(\hat{x}_i) = u^\infty_\tau(\hat{x}_i) + 0.05\xi \max_i \left|u^\infty_\tau(\hat{x}_i)\right|, \quad \tau = s,\ p \text{ or } \emptyset, \tag{7.4.10}$$

where ξ obeys a gaussian distribution with expectation 0 and variance 1.

Example 4 Let a multiple small scatterer consist of three different components, i.e., $\Omega^{(s)} = \cup_{j=1}^3 \Omega_j^{(s)}$, which is illustrated in Fig. 7.9. Here, $\Omega_1^{(s)}$ is a rigid ball located at $(-2, 0, 0)$, whose radius is 0.01; $\Omega_2^{(s)}$ is a traction-free ellipsoid located at $(2, 0, -1.5)$, the lengths of whose semi-principal axes are 0.2, 0.2, 0.1, respectively; $\Omega_3^{(s)}$ is a traction-free cube of side length 0.2 located at $(0, 0, 1)$. Moreover, we set $\lambda = 2$ and $\mu = 1$. Impinge an incident wave of the form (6.1.2) with $\omega = 2\pi$, $d = (1, 0, 0)$, $\alpha = 1$ and $\beta = 0$, then we collect the corresponding far-field $u^\infty(\hat{x})$ on the unit sphere $\mathbb{S}^1$.

From $k_s = \omega/\sqrt{\mu}$, $k_p = \omega/\sqrt{\lambda+2\mu}$ we have $k_s = 2\pi$ and $k_p = \pi$. We present the slice $y = 0$ of I_m $(m = 1, 2, 3)$ in Fig. 7.10. It is clear that all of the three imaging functionals correctly locate the unknown scatterer components. In addition, the decay speed is faster in Fig. 7.10b and c than that in Fig. 7.10a, which verifies our expectation made earlier.

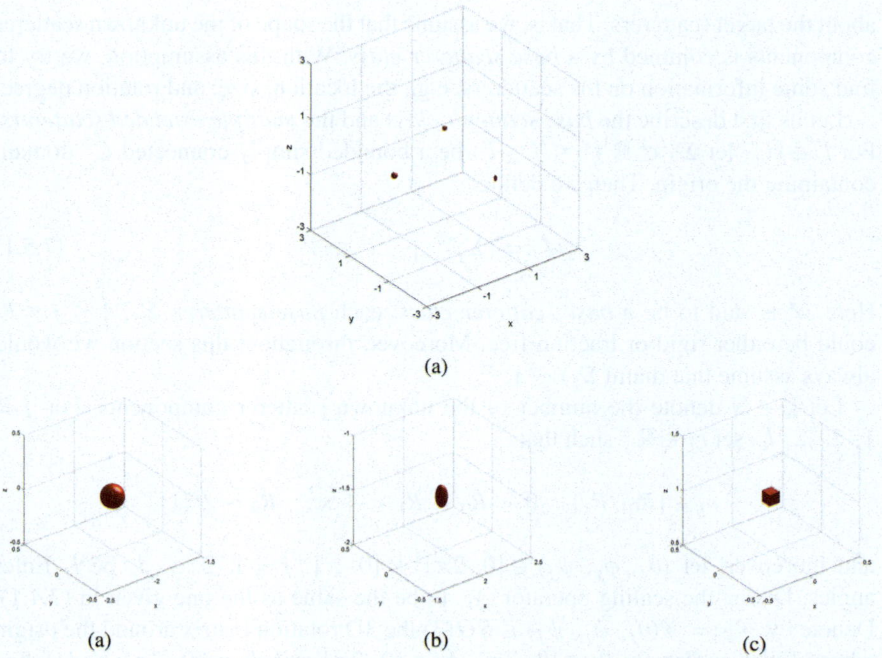

Fig. 7.9 Locating multiple small scatterers. **a** The true scatterer with 3 components; **b** The true scatterer component 1 (a rigid ball); **c** The true scatterer component 2 (a traction-free ellipsoid); **b** The true scatterer component 3 (a traction-free cube)

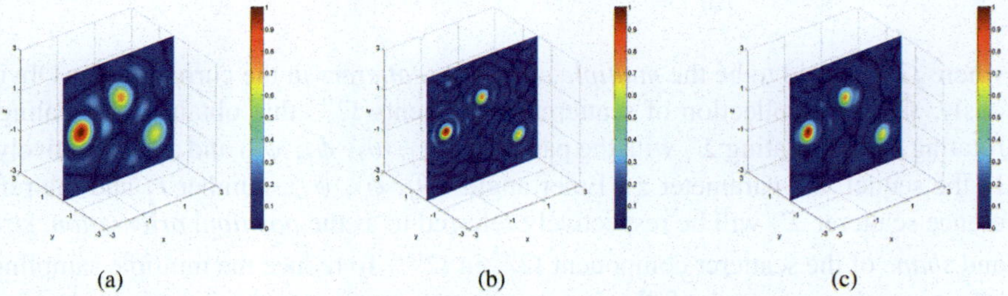

Fig. 7.10 Locating multiple small scatterers. **a** The configuration of $I_1(z)$; **b** The configuration of $I_2(z)$; **c** The configuration of $I_3(z)$

7.5 Reconstructing Multiple Extended Scatterers

In this section, we consider locating multiple extended elastic scatterers with a single incident plane wave. Reconstructing a general elastic scatterer with a single far-field measurement is known to be a challenging issue, both theoretically and numerically. In order to overcome the difficulties involved with the generality and complexity of the scatterers, the scheme mentioned in this section requires some a priori knowledge about the target scatterers. That is, we assume that the shape of the unknown scatterer components is confined by a *base scatterer class*. With this assumption, we try to find some information on the scatterers, e.g., the location, size, and rotation degree.

Let us first describe the *base scatterer class* and the *multiple extended scatterers*. For $l' \in \mathbb{N}_+$, let $\Sigma_j \subset \mathbb{R}^3 (1 \le j \le l')$ be a bounded simply-connected C^2 domain containing the origin. Then we define

$$\mathscr{A} := \{\Sigma_j\}_{j=1}^{l'}, \quad l' \in \mathbb{N}. \tag{7.5.1}$$

Here $\mathscr{A}$ is said to be a *base scatterer class*; each *base scatterer* Σ_j, $1 \le j \le l'$, could be either rigid or traction-free. Moreover, throughout this section we would always assume that $\operatorname{diam}(\Sigma_j) \sim 1$.

Let $l_e \in \mathbb{N}$ denote the number of the unknown scatterer components. For $j = 1, 2, \ldots, l_e$, set $r_j \in \mathbb{R}_+$ such that

$$r_j \in [R_0, R_1], \quad 0 < R_0 < R_1 < +\infty, \quad R_0 \sim \mathscr{O}(1),$$

and moreover, let $(\theta_j, \phi_j, \psi_j) \in [0, 2\pi]^2 \times [0, \pi]$, $j = 1, 2, \ldots, l_e$ be l_e Euler angles. Define the scaling operator Λ_{r_j} to be the same as the one given in (7.4.1). Denote by $\mathscr{R}_j := \mathscr{R}(\theta_j, \phi_j, \psi_j) \in SO(3)$ the 3D rotation matrix around the origin whose Euler angles are $\theta_j \in [0, 2\pi]$, $\phi_j \in [0, 2\pi]$ and $\psi_j \in [0, 2\pi]$; and define $\mathscr{R}_j\Sigma := \{\mathscr{R}_j x : x \in \Sigma, \ \Sigma \in \mathscr{A}\}$. For $z_j \in \mathbb{R}^3$, we let

$$\Omega^{(e)} = \bigcup_{j=1}^{l_e} \Omega_j^{(e)}, \quad \Omega_j^{(e)} := z_j + \mathscr{R}_j \Lambda_{r_j} \Sigma_j, \quad \Sigma_j \in \mathscr{A}, \tag{7.5.2}$$

where $\Omega^{(e)}$ is said to be the *multiple extended scatterers* in the current study. Obviously, $\Omega^{(e)}$ is a collection of scatterer components $\Omega_j^{(e)}$ that obtained by scaling, rotating and translating Σ_j with the parameters r_j, $(\theta_j, \phi_j, \psi_j)$ and z_j, respectively. In the sequel, the parameter z_j, Euler angles $(\theta_j, \phi_j, \psi_j)$, number r_j and the reference scatterer Σ_j will be respectively referred to as the *position*, *orientation*, *size* and *shape* of the scatterer component $\Omega_j^{(e)}$ in $\Omega^{(e)}$. To reduce the multiple sampling effect, we also impose the following sparsity assumption on the extended scatterer,

$$L_e = \min_{j \ne j', 1 \le j, j' \le l_e} \operatorname{dist}(\Omega_j^{(e)}, \Omega_{j'}^{(e)}) \gg 1. \tag{7.5.3}$$

Next, we introduce the h-net ($h \in \mathbb{R}_+, h \ll 1$) for the base scatterer class $\mathscr{A}$. Let $\mathscr{N}_1$ be a suitably chosen finite index set such that $\{\mathscr{R}_j\}_{j \in \mathscr{N}_1} = \{R(\theta_j, \phi_j, \psi_j)\}_{j \in \mathscr{N}_1}$ is an h-net of $SO(3)$. That is, for any rotation matrix $\mathscr{R} \in SO(3)$, there exists $j \in \mathscr{N}_1$ such that $\left\|\mathscr{R}_j - \mathscr{R}\right\| \le h$. For a simply-connected domain Σ_h containing the origin, we define

$$\mathscr{R}_h \Sigma := \{\mathscr{R}_j \Sigma\}_{j \in \mathscr{N}_1}. \tag{7.5.4}$$

In a similar manner, for Λ_r with $r \in [R_0,\ R_1]$, we let $\mathscr{N}_2$ be a suitably chosen finite index set such that $\{r_j\}_{j\in\mathscr{N}_2}$ is an h-net of $[R_0,\ R_1]$. Define

$$\Lambda_h \Sigma := \{\Lambda_{r_j} \Sigma\}_{j\in\mathscr{N}_2}. \tag{7.5.5}$$

Then we augment the admissible reference space $\mathscr{A}$ to be

$$\mathscr{A}_h = \mathscr{R}_h \Lambda_h \mathscr{A} = \bigcup_{j=1}^{l'} \{\mathscr{R}_h \Lambda_h \Sigma_j\} := \{\widetilde{\Sigma}_j\}_{j=1}^{l''}, \tag{7.5.6}$$

where l'' denotes the cardinality of the discrete set $\mathscr{A}_h$. Indeed, $\mathscr{A}_h$ can be taken as an h-net of $\mathscr{A}$ in the sense that for any $\Sigma \in \mathscr{A}$, there exists $\widetilde{\Sigma} \in \mathscr{A}_h$ such that $d_H(\overline{\Sigma},\ \overline{\widetilde{\Sigma}}) \le Ch$, where d_H denotes the Hausdorff distance and C is a positive constant depending only on $\mathscr{A}$. For the augmented admissible reference space $\mathscr{A}_h$, two assumptions should be made:

(i) $u_\tau^\infty(\hat{x},\ \widetilde{\Sigma}_j) \ne u_\tau^\infty$ for $\tau = s,\ p$ or $\emptyset$, and $j \ne j'$, $1 \le j,\ j' \le l''$.
(ii) $\|u_\tau^\infty(\hat{x},\ \widetilde{\Sigma}_j)\|_{L^2} \ge \|u_\tau^\infty(\hat{x},\ \widetilde{\Sigma}_{j'})\|_{L^2}$ for $\tau = s,\ p$, or $\emptyset$, and $j < j'$, $1 \le j$, $j' \le l''$.

Assumption (i) states that one can uniquely determine an elastic scatterer by using a single far-field pattern, which is a well-known conjecture in the inverse scattering theory. Since $\mathscr{A}_h$ is known, assumption (i) can be verified in advance. Assumption (ii) gives the recovered order of the scatterer components, which can be fulfilled by reordering the elements in $\mathscr{A}_h$ if necessary.

For an incident plane wave of the form (6.1.2) with $\alpha = 1, \beta = 0$ or $\alpha = 0, \beta = 1$, we introduce the following indicator functions,

$$W_1^j(z) = \frac{\left|\left\langle u_p^\infty(\hat{x};\ \Omega^{(m)}),\ A_1^j(\hat{x};\ z)\right\rangle\right|}{\|u_p^\infty(\hat{x},\ \Omega_j^{(e)})\|_{L^2}^2}, \quad z \in \mathbb{R}^3, \tag{7.5.7}$$

$$W_2^j(z) = \frac{\left|\left\langle u_s^\infty(\hat{x};\ \Omega^{(m)}),\ A_2^j(\hat{x};\ z)\right\rangle\right|}{\|u_s^\infty(\hat{x},\ \Omega_j^{(e)})\|_{L^2}^2}, \quad z \in \mathbb{R}^3, \tag{7.5.8}$$

$$W_1^j(z) = \frac{\left|\left\langle u^\infty(\hat{x};\ \Omega^{(m)}),\ A_1^j(\hat{x};\ z) + A_2^j(\hat{x};\ z)\right\rangle\right|}{\|u^\infty(\hat{x},\ \Omega_j^{(e)})\|_{L^2}^2}, \quad z \in \mathbb{R}^3, \tag{7.5.9}$$

where for $j = 1,\ 2,\ \ldots,\ l_e$,

$$A_1^j(\hat{x},\ z) := e^{ik_p(d-\hat{x})\cdot z} u_p^\infty(\hat{x};\ d,\ d^\perp, \alpha,\ 0,\ \Omega_j^{(e)}) + e^{i(k_s d - k_p \hat{x})\cdot z} u_p^\infty(\hat{x};\ d,\ d^\perp, 0,\ \beta,\ \Omega_j^{(e)}),$$

and

$$A_2^j(\hat{x}, z) := e^{i(k_p d - k_s \hat{x})\cdot z} u_s^\infty(\hat{x}; d, d^\perp, \alpha, 0, \Omega_j^{(e)}) + e^{ik_s(d-\hat{x})\cdot z} u_s^\infty(\hat{x}; d, d^\perp, 0, \beta, \Omega_j^{(e)}).$$

In what follows, the indicator functions (7.5.7) and (7.5.9) shall be adopted to locate extended scatterers. Since $\Sigma_j \in \mathscr{A}$ is known in advance, W_m^j $(m = 1, 2, 3)$ could be understood as the projection of the scattering measurement data into a space generated by the scattering data from the admissible base scatterers. When the sampling point is at the location of the targets, i.e., $z = z_j$, the term involved z_j would be eliminated in the implementation of the indicators. Thus W_m^j $(m = 1, 2, 3)$ also possesses a local maximum behavior, similar to (7.4.5) and (7.4.7). We formulate the scheme for reconstructing multiple extended scatterers in Algorithm 2.

Algorithm 2 Locating multiple extended scatterers $\Omega^{(m)}$ with W_m, $m = 1, 2, 3$.

Step 1 For the admissible reference scatterer class $\mathscr{A}$ in (7.5.1), formulate the augmented admissible A_h in (7.5.6);

Step 2 Collect in advance the P-part $(m = 1)$, S-part $(m = 2)$ or the full far-field data $(m = 3)$ associated with the admissible reference scatterer class $\mathscr{A}_h$ corresponding to a single incident plane wave of the form (6.1.2). Reorder $\mathscr{A}_h$ if necessary to make it satisfy assumption (ii), and also verify the generic assumption (i);

Step 3 For an unknown scatterer $\Omega^{(e)}$ in (7.5.2), collect the P-part, S-part or the full far-field data by sending the same detecting plane wave;

Step 4 Select a sampling region with a mesh $\mathscr{T}_h$ containing $\Omega^{(e)}$;

Step 5 Set $j = 1$;

Step 6 For each sampling point $z \in \mathscr{T}_h$, calculate $W_m^j(z)$ $(m = 1, 2, 3)$ according to available far-field data for $\Omega^{(e)}$;

Step 7 Locate all those significant local maximum points of $W_m^j(z)$ satisfying $W_m^j(z) \approx 1$ for the scatterer components of the form $z + \widetilde{\Sigma}_j$. Let z_η, $\eta = 1, \ldots, \eta_0$ be the local maximum points found this step;

Step 8 Remove all those $z + \widetilde{\Sigma}_j$ found in Step 6 from the mesh $\mathscr{T}_h$;

Step 9 Update the far-field patterns according to the following formulae

$$u_p^{\infty,new} = u_p^\infty(\hat{x}; d, d^\perp, \alpha, \beta, \Omega^{(e)}) - u_p^\infty(\hat{x}; d, d^\perp, \alpha, 0, \widetilde{\Sigma}_j) \sum_{\eta=1}^{\eta_0} e^{ik_p(d-\hat{x})\cdot z_\eta}$$

$$-u_p^\infty(\hat{x}; d, d^\perp, 0, \beta, \widetilde{\Sigma}_j) \sum_{\eta=1}^{\eta_0} e^{i(k_s d - k_p \hat{x})\cdot z_\eta}, \tag{7.5.10}$$

$$u_s^{\infty,new} = u_s^\infty(\hat{x}; d, d^\perp, \alpha, \beta, \Omega^{(e)}) - u_s^\infty(\hat{x}; d, d^\perp, \alpha, 0, \widetilde{\Sigma}_j) \sum_{\eta=1}^{\eta_0} e^{i(k_p d - k_s \hat{x})\cdot z_\eta}$$

$$-u_p^\infty(\hat{x}; d, d^\perp, 0, \beta, \widetilde{\Sigma}_j) \sum_{\eta=1}^{\eta_0} e^{ik_s(d-\hat{x})\cdot z_\eta}, \tag{7.5.11}$$

$$u^{\infty,new} = u_p^{\infty,new} + u_s^{\infty,new}; \tag{7.5.12}$$

Step 10 If $j = l''$, namely, the maximum number of the reference scatterers reaches, then stop the reconstruction; otherwise set $j = j + 1$, and go to **Step 6**.

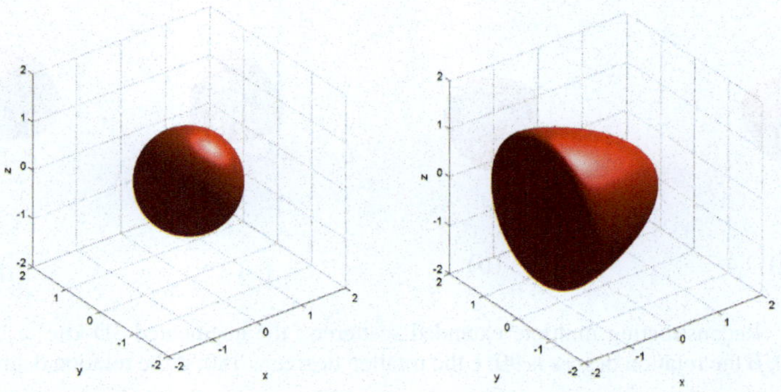

(a) (b)

Fig. 7.11 Reconstructing multiple extended scatterers: base scatterer class

Next, we show the indicating properties of the indicator functions for multiple extended scatterers by a numerical example.

Example 5 Let the base scatterer class $\mathscr{A}$ consist of two base scatterers Σ_1 and Σ_2, where Σ_1 is a traction-free ball of radius 1, and Σ_2 is a rigid 3D-kite, given in Fig. 7.11. The 2D-kite is a typical shape in inverse scattering experiments, whose parameter function is as follows,

$$\begin{cases} x = \cos(t) + 0.65\cos(2t) - 0.65, \\ y = 1.5\sin(t), \end{cases} \quad 0 \le t \le 2\pi.$$

The 3D-kite is generated by revolving the 2D-kite lying in the plane $z = 0$ around x-axis. For simplicity, the augmented data set is obtained by rotating the 3D-kite in the x-z plane every 90°, see, e.g., the four orientations in Fig. 7.12, and by scaling the ball by one half, one and twice. The true scatter in this example is shown in (Fig. 7.13). For the located targets, we assume the multiple extended scatterers consist of two components. One of the components is a ball with radius 1 whose center lies at $(-3,\ 0,\ -3)$; the other one is a 3D-kite with rotation 270 degree lying at $(3,\ 0,\ 3)$. The same as in **Example 2.1**, we set $\mu = 1$ and $\lambda = 2$. Moreover, in this example, we let the incident wave be of the form (6.1.2) with $\omega = 2\pi$, $d = (1,\ 0,\ 0)$, $\alpha = 1$ and $\beta = 0$, and collect the needed far-field data on 590 Lebdev quadrature points for numerical integration.

We adopt W_1 to locate the regular-size scatterer components. By the increasing magnitude of the far field patterns, the Kite reference data is firstly employed for the locating purpose. Figure 7.14a–d show that the first unknown component is a 3D-kite with rotation 270 degree, and the position of the 3D-kite is highlighted. One

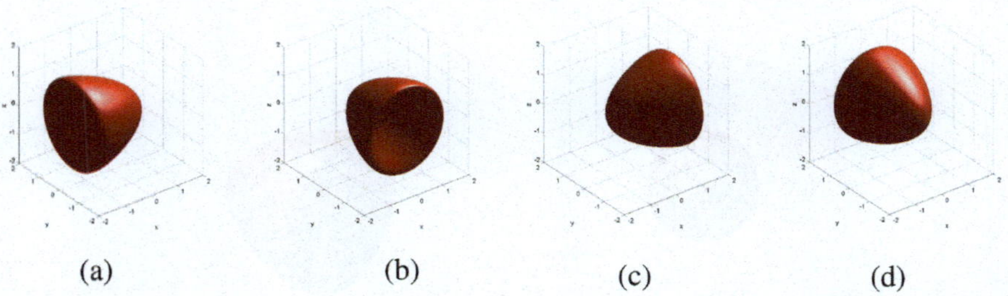

Fig. 7.12 Reconstructing multiple extended scatterers: the augmented 3D-kite. **a** the rotation degree is 0; **b** the rotation degree is 90; **c** the rotation degree is 180; **d** the rotation degree is 270

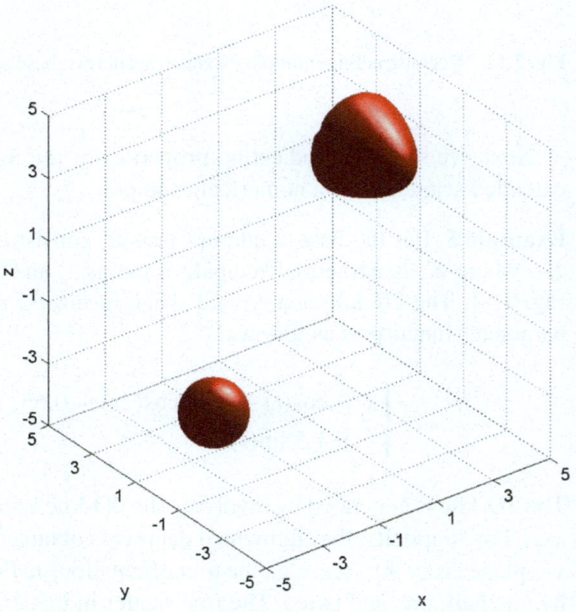

Fig. 7.13 Reconstructing multiple extended scatterers: true scatterer

could find that, some ghosts highlight close to the position of the ball in Fig. 7.14a–d, which is largely due to the similarity between the far-field pattern of the two scatterer components. In the next stage, by subtracting the contribution of the 3D-kite from the total far-field data, we continue our reconstruction work with the far-field associated with the reference 3D-kite and its possible orientations. It is found in Fig. 7.14e–g that the most prominent indicating behavior identifies the size, shape and position of the ball. For comparison, Fig. 7.15 is also displayed to show the indicating behavior of W_2.

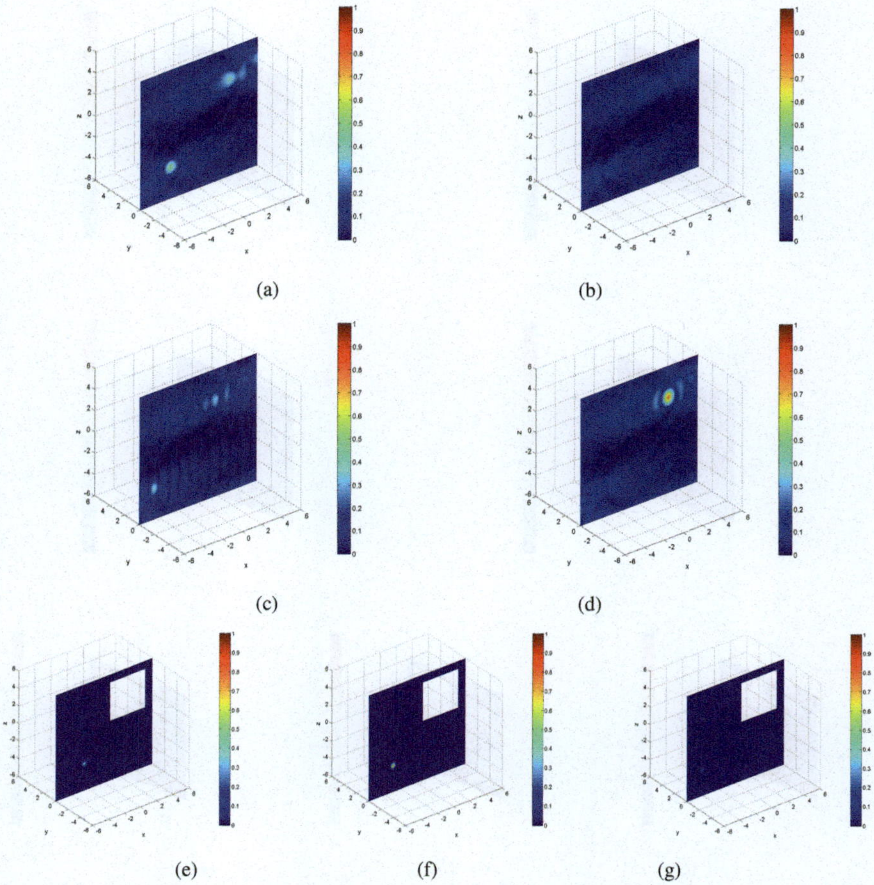

Fig. 7.14 Reconstructing multiple extended scatterers: slice $y = 0$ of W_2

7.6 Reconstructing Multiple Multiscale Scatterers

The elastic scatterer with multiscale components is of the following form

$$\Omega^{(m)} := \Omega^{(s)} \cup \Omega^{(e)}, \tag{7.6.1}$$

where $\Omega^{(s)}$ given in (7.4.3) and $\Omega^{(e)}$ given in (7.5.2) represent, respectively, the collections of small-size and regular-size scatterers. For $\Omega^{(m)}$ introduced above, we assume that $\operatorname{dist}(\Omega^{(s)} \cup \Omega^{(e)}) \gg 1$, which is also a technique consideration on reducing multiple scattering effect. Next, under the a priori knowledge that the base scatterer class $\mathscr{A}$ in (7.5.1) is known in advance, we consider how to recover the multiple multiscale scatterer components of $\Omega^{(m)}$.

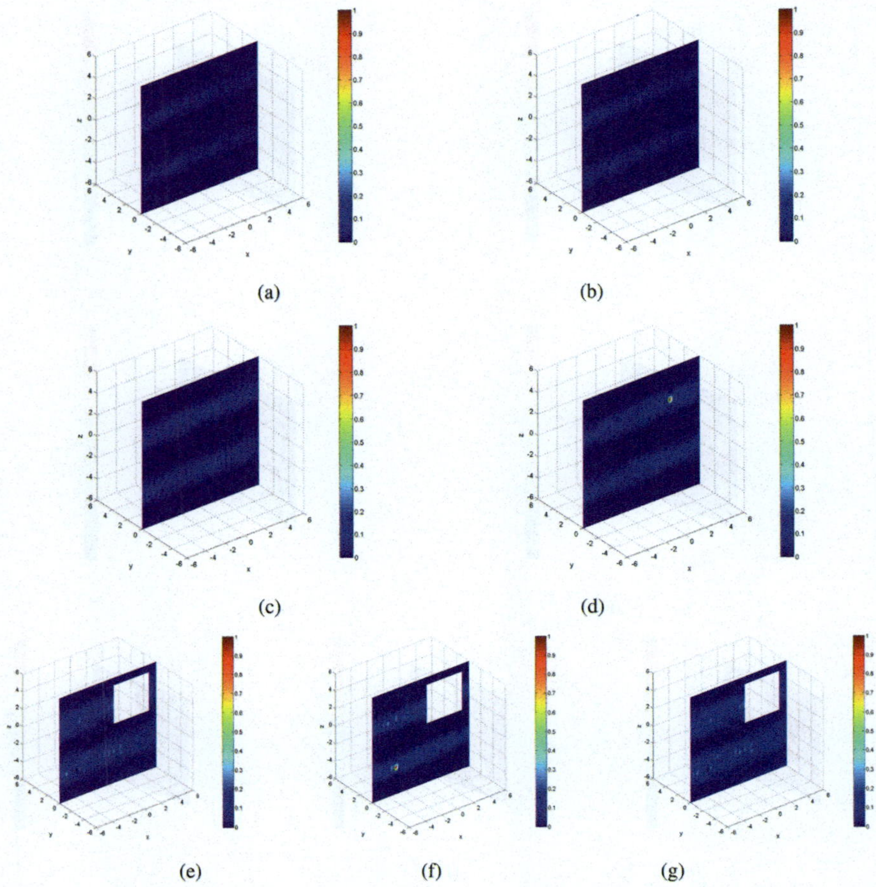

Fig. 7.15 Reconstructing multiple extended scatterers: slice $y = 0$ of W_2

The whole reconstruction procedure is a two-step process. More specifically, we adopt Algorithms 1 and 2 to reconstruct the small and extended scatterer components, respectively. However, reconstructing the small components is really difficult, since the contribution of small scatterer components to the far-field patterns is too small to extract. Therefore, a *local tuning* technique is also incorporated in the second step of the multiscale reconstruction scheme. The exact definition of local tuning could be seen in [29]. In what follows, $\widehat{\Omega^{(e)}}$ is the reconstructed image of the extended scatterer $\Omega^{(e)}$, whereas $\widehat{\widehat{\Omega^{(e)}}}$ is an adjustment of $\widehat{\Omega^{(e)}}$ by locally adjusting the position, orientation and size of each component of $\widehat{\Omega^{(e)}}$. With these notations, we now formulate the scheme of reconstructing multiple multiscale scatterers in Algorithm 3.

In the following, we verify the multiscale scheme by a numerical example.

Algorithm 3 Scheme for locating multiple multiscale scatterers $\Omega^{(m)}$

Step 1 For an unknown scatterer $\Omega^{(m)}$ with multiple components in (7.4.3), collect u_p^∞, u_s^∞ and u^∞ by sending a single detecting plane wave (6.1.2);

Step 2 Select a sampling region with a mesh $\mathscr{T}_h$ containing $\Omega^{(m)}$;

Step 3 Apply Algorithm 2 to reconstruct approximately the extended scatterer $\Omega^{(e)}$, denoted by $\widehat{\Omega^{(e)}}$;

Step 4 For $\widehat{\Omega^{(e)}}$ obtained above, select a local-tuning mesh $\mathscr{L}$;

Step 5 For a tune-up $\widehat{\widehat{\Omega^{(e)}}}$ relative to the local tuning mesh $\mathscr{L}$ calculate

$$\tilde{u}_\tau^\infty(\hat{x}) := u_\tau^\infty(\hat{x};\ \Omega^{(m)}) - u_\tau^\infty(\hat{x};\ ,\ \widehat{\widehat{\Omega^{(e)}}}). \tag{7.6.2}$$

Apply Algorithm 1 with $\tilde{u}_\tau^\infty(\hat{x})$ as the far-field data to locate the significant local maximum points on $\mathscr{T}_h\backslash\mathscr{L}$;

Step 6 Repeat Step 5 by running through all the local tune-ups relative to $\mathscr{L}$. Locate the clustered local maximum points on $\mathscr{T}_h\backslash\mathscr{L}$, which correspond to the small scatterer components;

Step 7 Update $\widehat{\Omega^{(e)}}$ to the local tune-up $\widehat{\widehat{\Omega^{(e)}}}$ which generates the clustered local maximum points in Step 6.

Example 6 Let the true scatterer be composed of a small rigid ball and a regular-size traction-free 3D-kite. The small ball is located at $(-3,\ 0,\ 0)$ whose radius is 0.1, and the 3D-kite is located at $(3,\ 0,\ 0)$ as shown in Fig. 7.16. We assume the augmented base scatterer class is the same with the one in Example 1. Let $\mu = 1$, and $\lambda = 2$, we detect the scatterer by impinging the incident wave with $\alpha = 1$, $\beta = 0$ and $\omega = 2\pi$. In the following, we only present the results by using W_2 since it possesses better decaying properties.

First, we extract the information of the regular-size component using the indicator function $W_2(z)$ by computing the inner product with a priori known far-field data associated with those reference scatterer components with different orientations and sizes. We plot in Fig. 7.17a–d the indicator function values of $W_m(z)$ in one-to-one correspondence with the four orientations of the reference 3D-kite as shown in Fig. 7.12. It can be observed in Fig. 7.17c that the regular-size component in our locating target is the 3D-kite rotated by 180°. Then we build a local tuning mesh to reconstruct the small components in the unknown multiscale scatterers. Figure 7.17g shows that the small ball is located at $(-3,\ 0,\ 0)$; Meanwhile, from the results in Fig. 7.17e–g, we could correct the position of the 3D-kite to be $(3,\ 0,\ 0)$.

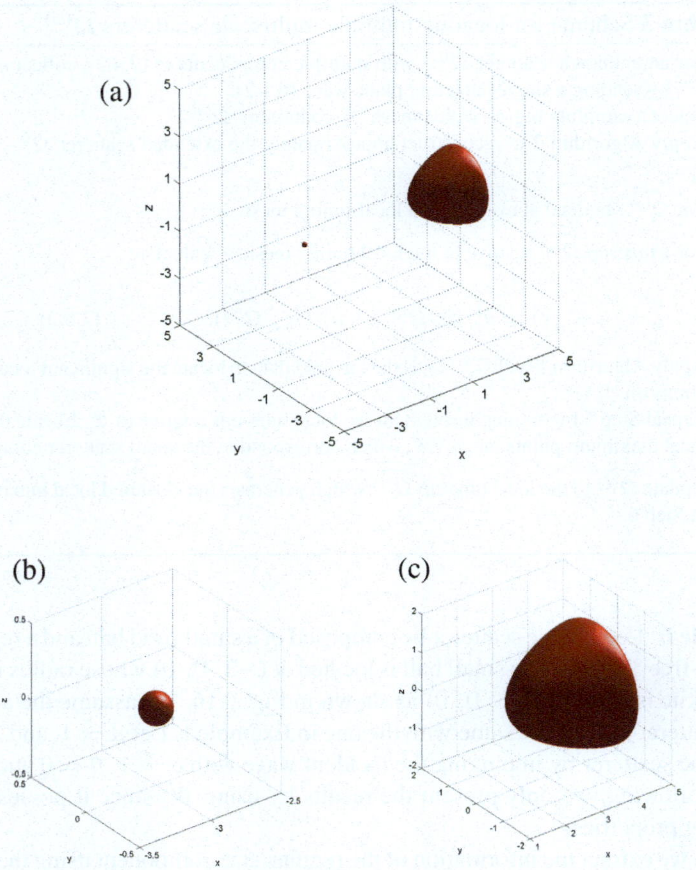

Fig. 7.16 Reconstructing multiple multiscale scatterers: true scatterer

7.7 Two-Stage Fast Imaging of Multiple Multiscale Scatterers

In this section, we consider to speed up the multiple multiscale imaging scheme. As can be observed in the numerical results in last section, one could get several bright spots by evaluating the indicator functions, which characterize the scatterer components. Since the decay of the spots is very fast, the locations we get are accurate. However, to capture the small spots, we have to take a rather fine grid, which brings us a huge computational cost, though most of the cost is spent on sampling in the irrelevant region (The region that is far away from all the scatterer components). Therefore, it is wise to trim most of the irrelevant region before evaluating the indicator values on the fine grid. In the following, we present a two-stage strategy to realize this idea.

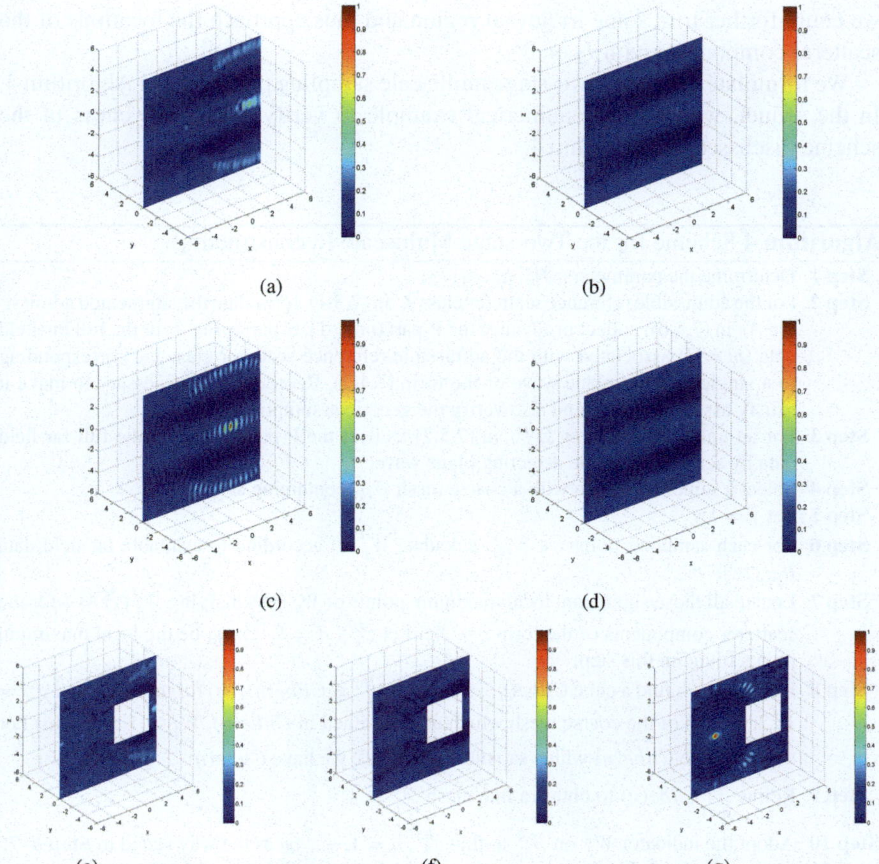

(a) (b) (c) (d) (e) (f) (g)

Fig. 7.17 Reconstructing multiple multiscale scatterers: reconstruction result

The Coarse Stage: Coarse Grid with u_p^∞

It has been analyzed in [41] that, the average decay radius of I_m or W_m around each center of the components is in inverse proportion to the wave number. Supposing $\lambda > 0$ and recalling the fact that $k_s/k_p = \sqrt{(\lambda + 2\mu)/\mu}$, we have $k_s > \sqrt{2}k_p$. Thus we could extract u_p^∞ from u^∞ and detect the unknown inclusions by evaluating I_1 or W_1. In light of the slow decay compared to I_1 or W_1, the corresponding spots posses larger radius, which enable us to take a coarse grid to capture the scatterer components.

The Fine Stage: Fine Grid with u_s^∞

In the fine stage, in order to extract the accurate information of the unknown bodies, we continue the reconstruction by using u_s^∞. That is, we first choose an appropriate threshold value. Comparing the values on the coarse mesh to the threshold value, we could further trim some irrelevant region and thus approach the locations of the scatterer components with I_2 or W_2.

We formulate the above two-stage multiscale sampling procedure in Algorithm 4. In the sequel, we present a numerical example to verify the effectiveness of the scheme discussed in Algorithm 4.

Algorithm 4 Scheme for the Two-stage Multiscale Reconstructing

Step 1 Determine the parameters: H_0, h_0, c_1, c_2;

Step 2 For the admissible reference scatterer class $\mathscr{A}$ in (7.5.1), formulate the augmented admissible A_h in (7.5.6); collect in advance the P-part ($m = 1$), S-part ($m = 2$) or the full far-field data ($m = 3$) associated with the admissible reference scatterer class $\mathscr{A}_h$ corresponding to a single incident plane wave of the form (6.1.2). Reorder $\mathscr{A}_h$ if necessary to make it satisfy assumption (ii), and also verify the generic assumption (i);

Step 3 For an unknown scatterer $\Omega^{(m)}$ in (7.5.2), collect the P-part, S-part or the full far-field data by sending the same detecting plane wave;

Step 4 Select a sampling region with a coarse mesh $\mathscr{T}_{H_0}$ containing $\Omega^{(m)}$;

Step 5 Set $j = 1$;

Step 6 For each sampling point $z \in \mathscr{T}_{H_0}$, calculate $W_1^j(z)$ according to available far-field data for $\Omega^{(e)}$;

Step 7 Locate all those significant local maximum points of $W_1^j(z)$ satisfying $W_1^j(z) \approx 1$ for the scatterer components of the form $z + \widetilde{\Sigma}_j$. Let $z_\eta^{H_0}$, $\eta = 1, \ldots, \eta_0$ be the local maximum points found in this step;

Step 8 For each $z_\eta^{H_0}$, find a cube $Cube(\eta, c_1)$ s.t., (i) $z_\eta^{H_0} \in Cube(\eta)$, (ii) for those nodes $z_j (j = 1, 2, \ldots, l_n)$ of the coarse mesh which are contained in $Cube(\eta)$, $W_1^j(z_j) > c_1$, (iii) For another cube $\widetilde{Cube(\eta)}$ which satisfies (i) and (ii), we have $Cube(\eta) \subset \widetilde{Cube(\eta)}$;

Step 9 Refine $\bigcup_{\eta=1}^{\eta_0} Cube(\eta)$ to obtain a fine mesh $\mathscr{T}_{h_0}^j$;

Step 10 Adopt the indicator W_2^j on $\mathscr{T}_{h_0}^\eta$ to find $z_\eta^{h_0}$, $\eta = 1, \ldots, \eta_0$ as the way stated in **Step 6–7**;

Step 11 Update the far-field patterns according to the formulae (7.5.10)–(7.5.12);

Step 12 If $j = l'$, namely, the maximum number of the reference scatterers reaches, then stop the reconstruction; otherwise set $j = j + 1$, and go to **Step 7**.

Step 13 For $\widehat{\Omega^{(e)}}$ obtained above, select a local-tuning mesh $\mathscr{L}$;

Step 14 For a tune-up $\widehat{\widehat{\Omega^{(e)}}}$ relative to the local tuning mesh $\mathscr{L}$ calculate (7.6.2); Apply Algorithm 1 with $\tilde{u}_p^\infty(\hat{x})$ (Use $I_1(z)$) to locate the significant local maximum points on $\mathscr{T}_{H_0} \backslash \mathscr{L}$;

Step 15 Repeat **Step 14** by running through all the local tune-ups relative to $\mathscr{L}$;

Step 16 Update $\widehat{\Omega^{(e)}}$ to the local tune-up $\widehat{\widehat{\Omega^{(e)}}}$, which generates the clustered local maximum points in **Step 15;** Let $\widetilde{z_\eta^{H_0}} (\eta = 1, \ldots, \eta_1)$ denote the clustered local maximum points, build cube $Cube(\eta, c_2)$ around each $\widetilde{z_\eta^{H_0}} (\eta = 1, \ldots, \eta_1)$;

Step 17 Refine $Cube(\eta, c_2)$ and correct the position of the small scatterers by implementing Algorithm 1 with $\tilde{u}_s^\infty(\hat{x})$ (Use $I_2(z)$) on the fine grid.

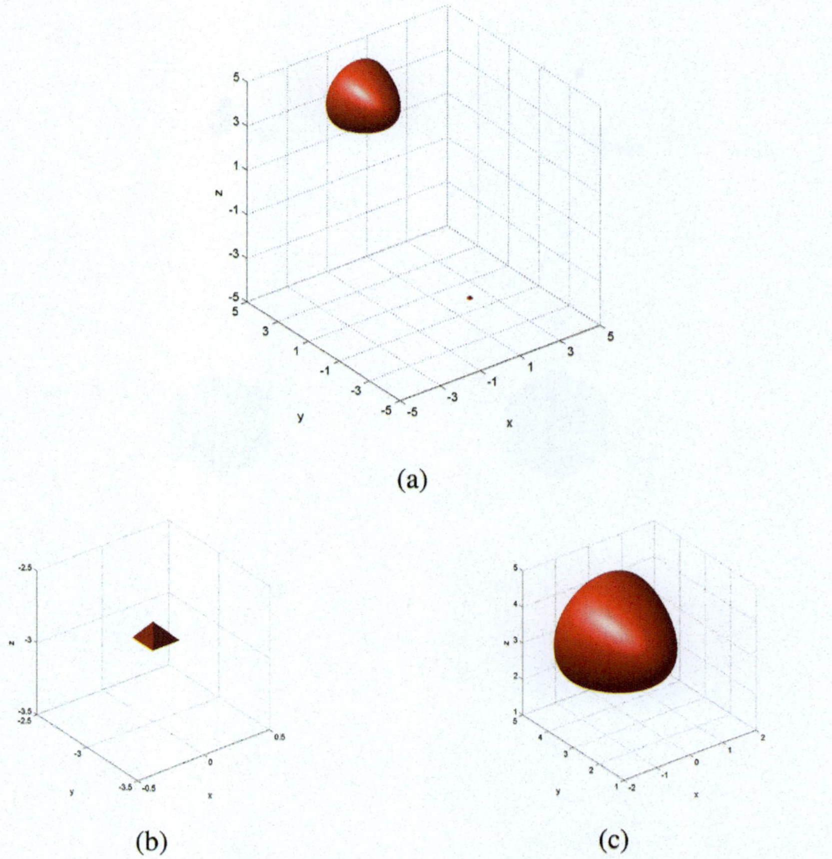

Fig. 7.18 Reconstructing multiscale scatterers with the two-stage sampling: true scatterer

Example 7 The geometries considered in this example are composed of a 3D-kite and a regular tetrahedron, which are shown as in Fig. 7.18. The 3D-kite is located at (0, 3, 3); the tetrahedron is located at (0, −3, −3), and the side length of the tetrahedron is 0.1.

First, we adopt the low-frequency information, that is, we use W_1 to locate the extended component. In Fig. 7.19a–b, we display the isosurface of $W_1 = 0.8$ when the right base far-field pattern (The far-field pattern of a 3D-kite with rotation 180°) is chosen. From the theoretical analysis and inversion algorithms mentioned above, we could know that the center of the 3D-kite is surrounded in the isosurface. Then as stated in Algorithm 4, we properly choose a cube (see in Fig. 7.19c) covering the isosurface. Refine the cube (see in Fig. 7.19d) and continue the sampling process with W_2, we could get the values of the nodes on the fine grid. We display the isosurface $W_2 = 0.8$ in Fig. 7.19e, from which a more accurate position information of the 3D-kite could be obtained. In Fig. 7.19g–h, we also display the slice $x = 0$ of I_1 and I_2

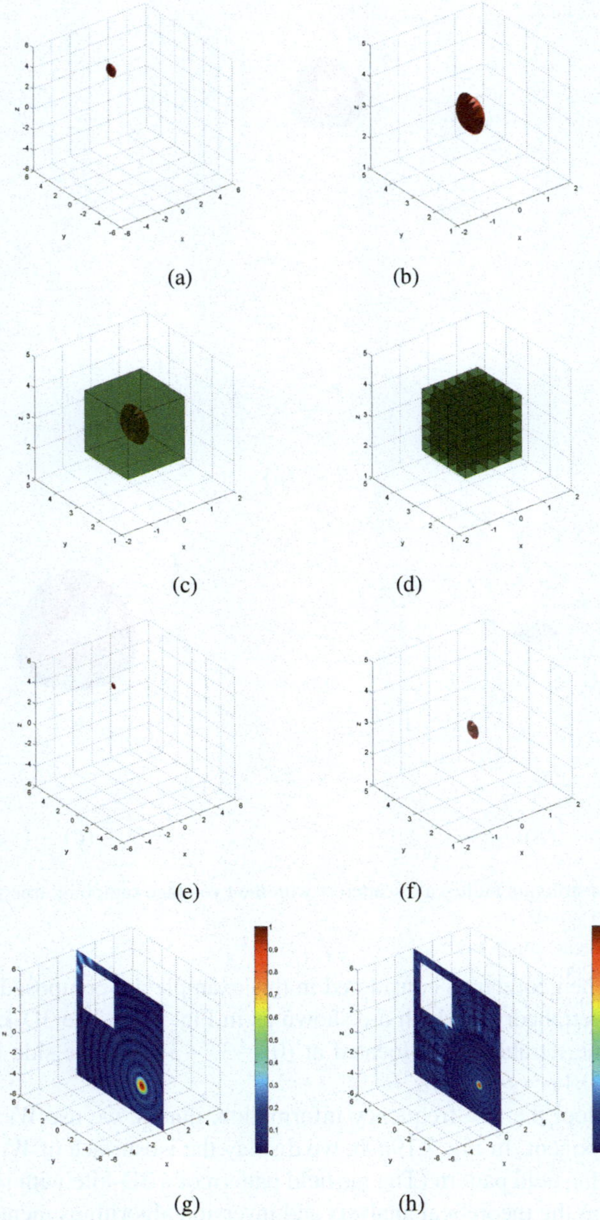
(a)
(b)
(c)
(d)
(e)
(f)
(g)
(h)

◂**Fig. 7.19** Reconstructing multiscale scatterers with the two-stage sampling: reconstruction results. **a** Sampling with W_1 on a coarse grid, where we display the isosurface of W_1 with value $c_1 = 0.8$; **b** Zoom in the result of **a**; **c** Build a cube containing the red body; **d** Refine the cube to obtain a fine sampling grid; **e** Sampling with W_2 on a coarse grid, where we display the isosurface of W_2 with value $c_1 = 0.8$; **f** Zoom in of **e**; **g** Sampling result with I_1 when the extended components are reconstructed correctly; **h** Sampling result with I_2 when the extended components are reconstructed correctly

when the contribution of the 3D-kite is properly subtract from the corresponding far-field pattern. Observing the bright spots in Fig. 7.19g–h, we could find the necessity of two-stage sampling in the local tuning process, since the size of the bright spot is much smaller in Fig. 7.19h.

References

1. I. Abubakar, Scattering of plane elastic waves at rough surface I. Proc. Camb. Philos. Soc. **58**, 136–157 (1962)
2. C. Alves, H. Ammari, Boundary integral formulae for the reconstruction of imperfections of small diameter in an elastic medium. SIAM J. Appl. Math. **62**, 503–531 (2001)
3. C.J. Alves, R. Kress, On the far-field operator in elastic obstacle scattering. IMA J. Appl. Math. **67**, 1–21 (2002)
4. H. Ammari, H. Kang, G. Nakamura, K. Tanuma, Complete asymptotic expansions of solutions of the system of elastostatics in the presence of an inclusion of small diameter and detection of an inclusion. J. Elast. Phys. Sci. Solids **67**, 97–129 (2002)
5. H. Ammari, H. Kang, H. Lee, Asymptotic expansions for eigenvalues of the Lamé system in the presence of small inclusions. Commun. Partial Differ. Equ. **32**, 1715–1736 (2007)
6. H. Ammari, E. Bretin, J. Garnier, W. Jing, H. Kang, A. Wahab, *Localization, stability, and resolution of topological derivative based imaging functionals in elasticity* (SIAM J, Imaging Sci, 2013). ((in press))
7. T. Arens, Linear sampling method for 2D inverse elastic wave scattering. Inverse Prob. **17**, 1445–1464 (2001)
8. Z. Bai, H. Diao, H.Y. Liu, Q. Meng, Stable determination of an elastic medium scatterer by a single far-field measurement and beyond. Calc. Var. Partial. Differ. Equ. **61**(5), 170 (2022)
9. G. Bao, H.Y. Liu, J. Zou, *Nearly cloaking the full Maxwell equations: cloaking active contents with general conducting layers*. Journal de Mathématiques Pures et Appliquées (2013) (In press)
10. R. Benites, K. Aki, K. Yomogida, Multiple Scattering of SH waves in 2D media with many cavities. Pageoph **138**, 353–390 (1992)
11. J.H. Bramble, J.E. Pasciak, D. Trenev, Analysis of a finite PML approximation to the three dimensional elastic wave scattering problem. Math. Comput. **79**, 2079–2101 (2010)
12. D.P. Challa, M. Sini, *The Foldy-Lax approximation of the scattered waves by many small bodies for the Lame system*. arXiv: 1308.3072
13. A. Charalambopoulos, A. Kirsch, K.A. Anagnostopoulos, D. Gintides, K. Kiriaki, The factorization method in inverse elastic scattering from penetrable bodies. Inverse Prob. **23**, 27–51 (2007)
14. D. Colton, R. Kress, *Inverse Acoustic and Electromagnetic Scattering Theory*, 2nd edn. (Springer, New York, 1998)
15. G. Dassios, R. Kleinman, *Low Frequency Scattering* (Clarendon Press, Oxford, 2000)

16. H. Diao, H.Y. Liu, B. Sun, On a local geometric property of the generalized elastic transmission eigenfunctions and application. Inverse Problems **37**(10), 105015 (2021)
17. H. Diao, H.Y. Liu, L. Wang, On generalized Holmgren's principle to the Lamé operator with applications to inverse elastic problems. Calc. Variat. Partial Differ. Equ. **59**(5), 179 (2020)
18. H. Diao, H.Y. Liu, L. Wang, Further results on generalized Holmgren's principle to the Lamé operator and applications. J. Differ. Equ. **309**, 841–882 (2022)
19. J. Elschner, M. Yamamoto, Uniqueness in inverse elastic scattering with finitely many incident waves. Inverse Prob. **26**, 045005 (2010)
20. J.T. Fokkema, P.M. Van den Berg, Elastodynamic diffraction by a periodic rough surface (stress-free boundary). J. Acoust. Soc. Am. **62**, 1095–1101 (1977)
21. J.T. Fokkema, Reflection and transmission of elastic waves by the spatially periodic interface between two solids (theory of the integral-equation method). Wave Motion **2**, 375–393 (1980)
22. D. Gintides, M. Sini, Identification of obstacles using only the scattered P-waves or the Scattered S-waves. Inverse Prob. Imaging **6**, 39–55 (2012)
23. R. Griesmaier, Multi-frequency orthogonality sampling for inverse obstacle scattering problems. Inverse Prob. **27**, 085005 (2011)
24. P. Häner, G.C. Hsiao, Uniqueness theorems in inverse obstacle scattering of elastic waves. Inverse Prob. **9**, 525–534 (1993)
25. G.C. Hsiao, W.L. Wendland, *Boundary Integral Equations*. Applied Mathematical Sciences, vol. 164 (Springer, Berlin, 2008)
26. G. Hu, J. Li, Jingzhi, H.Y. Liu H. Sun, Inverse elastic scattering for multiscale rigid bodies with a single far-field pattern. SIAM J. Imaging Sci. **7**(3), 1799–1825 (2014)
27. G. Hu, J. Li, H.Y. Liu, Q. Wang, A numerical study of complex reconstruction in inverse elastic scattering. Commun. Comput. Phys. **19**(5), 1265–1286 (2016)
28. G. Hu, J. Li, H.Y. Liu, Recovering complex elastic scatterers by a single far-field pattern. J. Differ. Equ. **257**(2), 469–489 (2014)
29. G. Hu, J. Li, H.Y. Liu, H. Sun, Inverse elastic scattering for multiscale rigid bodies with a single far-field pattern. SIAM J. Imag. Sci. **7**(3), 1799–1825 (2014)
30. M. Ikehata, H. Itou, Reconstruction of a linear crack in an isotropic elastic body from a single set of measured data. Inverse Prob. **23**, 589–607 (2007)
31. M. Ikehata, H. Itou, Extracting the support function of a cavity in an isotropic elastic body from a single set of boundary data. Inverse Prob. **25**, 105005 (2009)
32. M. Ikehata, Reconstruction of the shape of the inclusion by boundary measurements. Comm. Partial Differ. Equ. **23**, 1459–1474 (1998)
33. R. Kress, *Linear Integral Equations* (Springer, Berlin, 1989)
34. V.D. Kupradze et al., *Three-dimensional Problems of the Mathematical Theory of Elasticity and Thermoelasticity* (North-Holland, Amsterdam, 1979)
35. J. Lai, H.Y. Liu, J. Xiao, Y. Xu, The decoupling of elastic waves from a weak formulation perspective. East Asian J. Appl. Math. **9**(2), 241–251 (2019)
36. L.D. Landau, E.M. Lifshitz, *Theory of Elasticity* (Pergamon Press, 1986)
37. J. Li, H.Y. Liu, H. Sun, On an inverse elastic wave imaging scheme for nearly incompressible materials. IMA J. Appl. Math. **84**(2), 229–257 (2019)
38. J. Li, H.Y. Liu, Z. Shang, H. Sun, Two single-shot methods for locating multiple electromagnetic scatterers. SIAM J. Appl. Math. **73**, 1721–1746 (2013)
39. J. Li, H.Y. Liu, Q. Wang, Locating multiple multiscale electromagnetic scatterers by a single far-field measurement. SIAM J. Imag. Sci. **6**, 2285–2309 (2013)
40. J. Li, H. Liu, J. Zou, Locating multiple multiscale acoustic scatterers. SIAM Multiscale Model. Simul. **12**, 927–952 (2014)
41. J. Li, J. Zou, A direct sampling method for inverse scattering using far-field data. Inverse Prob. Imaging. arXiv preprint arXiv:1206.0727 (2012)
42. H.Y. Liu, On local and global structures of transmission eigenfunctions and beyond. J. Inverse Ill-Posed Prob. **30**(2), 287–305 (2022)
43. H.Y. Liu, J. Xiao, Decoupling elastic waves and its applications. J. Differ. Equ. **263**(8), 4442–4480 (2017)

44. H. Liu, J. Zou, Uniqueness in an inverse acoustic obstacle scattering problem for both sound-hard and sound-soft polyhedral scatterers. Inverse Prob. **22**, 515–524 (2006)
45. H.Y. Liu, H. Sun, Enhanced near-cloak by FSH lining. Journal de Mathématiques Pures et Appliquées **99**, 17–42 (2013)
46. P.A. Martin, *Multiple scattering, Encyclopedia of Mathematics and its Applications*, vol. 107 (Cambridge University Press, Cambridge, 2006)
47. P.A. Martin, G. Dassios, Karp's theorem in elastodynamic inverse scattering. Inverse Prob. **9**, 97–111 (1993)
48. W. McLean, *Strongly Elliptic Systems and Boundary Integral Equations* (Cambridge University Press, Cambridge, 2000)
49. Q. Meng, Z. Bai, H. Diao, H.Y. Liu, Effective medium theory for embedded obstacles in elasticity with applications to inverse problems. SIAM J. Appl. Math. **82**(2), 720–749 (2022)
50. G. Nakamura, G. Uhlmann, Identification of Lamé parameters by boundary measurements. Am. J. Math. **115**, 1161–1187 (1993)
51. G. Nakamura, G. Uhlmann, Global uniqueness for an inverse boundary problem arising in elasticity. Inventiones Mathematicae **118**, 457–474. Erratum **152**(2003), 205–207 (1994)
52. G. Nakamura, G. Uhlmann, Inverse problems at the boundary for an elastic medium. SIAM J. Math. Anal. **26**, 263–279 (1995)
53. G. Nakamura, T. Kazumi, G. Uhlmann, Layer stripping for a transversely isotropic elastic medium. SIAM J. Appl. Math. **59**, 1879–1891 (1999)
54. L.V. Rachele, Uniqueness of the density in an inverse problem for isotropic elastodynamics. Trans. Am. Math. Soc. **355**, 4781–4806 (2003)
55. J.H. Rose, Elastic wave inverse scattering in nondestructive evaluation. PAGEOPH **131**, 715–739 (1989)
56. V. Sevroglou, The far-field operator for penetrable and absorbing obstacles in 2D inverse elastic scattering. Inverse Prob. **21**, 717–738 (2005)
57. J.W.C. Sherwood, Elastic wave propagation in a semi-infinite solid medium. Proc. Phys. Soc. **71**, 207–219 (1958)
58. T.T. Wu, P.L. Liu, Advancement on the nondestructive evaluation of concrete using transient elastic waves. Ultrasonics **36**, 197–204 (1998)

Chapter 8
Miscellaneous Topics

In this chapter, two enhanced techniques are discussed for the multilevel linear sampling method in Chap. 3 in order to avoid the breakage cells. Inverse acoustic obstacle scattering problem is considered as a model problem for the development of the two enhanced multilevel linear sampling methods. On the one hand, the discussed two methods, expanding MLSM and searching MLSM, are shown to possess the same optimal computational complexity as the original MLSM. On the other hand, the enhanced methods can defeat the breakage cells problem and produce very fine reconstructions. Numerical experiments are conducted to illustrate the effectiveness of our methods. Since the techniques are mainly some computational strategies, they could be straightforwardly extended to MLSM for inverse medium scattering problem and inverse electromagnetic scattering problems. Our discussion follows the treatment in [25, 26].

8.1 Multilevel Linear Sampling Methods

In their original paper [12], Colton and Kirsch developed a 'simple' method for the shape reconstruction in inverse scattering problem which is nowadays known as the *linear sampling method* (LSM). The method has been extensively studied and extended in several directions, and we refer to [28] for a comprehensive review. These works are mainly concerned with an implementation technique with LSM. There are other works which involve the LSM and attempt to either circumvent the cost of sampling or improve the image results, see, e.g., [5]. We take as our model problem the inverse acoustic sound-soft obstacle scattering by time-harmonic plane waves. But like the original linear sampling method, our algorithm can be equally applied to other inverse problems, such as the acoustic sound-hard obstacle scattering or electromagnetic scattering.

J. Li and H. Liu, *Numerical Methods for Inverse Scattering Problems*,
https://doi.org/10.1007/978-981-99-3772-1_8

First, we introduce the far-field operator $F : L^2(\mathbb{S}^{N-1}) \mapsto L^2(\mathbb{S}^{N-1})$ defined by

$$(Fg)(\hat{x}) := \int_{\mathbb{S}^{N-1}} u_\infty(\hat{x}, d)g(d)ds(d), \qquad \hat{x} \in \mathbb{S}^{N-1}. \tag{8.1.1}$$

The linear sampling method is to use g as an indicator and solve the following *far-filed equation*

$$(Fg)(\hat{x}) = \Phi_\infty(\hat{x}, z), \qquad \hat{x} \in \mathbb{S}^{N-1},\ z \in \mathbb{R}^N, \tag{8.1.2}$$

where

$$\Phi_\infty(\hat{x}, z) = \gamma \exp\{-\mathrm{i}k\hat{x} \cdot z\} \tag{8.1.3}$$

with $\gamma = 1/4\pi$ in $\mathbb{R}^3$ and $\gamma = e^{\mathrm{i}\pi/4}/\sqrt{8\pi k}$ in $\mathbb{R}^2$. The following theorem forms the basis of the linear sampling method.

Theorem 8.1.1 *Assume that k^2 is not a Dirichlet eigenvalue for $-\Delta$ in $\mathbf{D}$. Then*

1. *For $z \in \mathbf{D}$ and a fixed $\varepsilon > 0$ there exists a $g_\varepsilon^z \in L^2(\mathbb{S}^{N-1})$ such that*

$$\|Fg_\varepsilon^z - \Phi_\infty(\cdot, z)\|_{L^2(\mathbb{S}^{N-1})} < \varepsilon$$

and

$$\lim_{z \to \partial\mathbf{D}} \|g_\varepsilon^z\|_{L^2(\mathbb{S}^{N-1})} = \infty.$$

2. *For $z \in \mathbb{R}^N \backslash \bar{\mathbf{D}}$ and any given $\varepsilon > 0$, every $g_\varepsilon^z \in L^2(\mathbb{S}^{N-1})$ that satisfies*

$$\|Fg_\varepsilon^z - \Phi_\infty(\cdot, z)\|_{L^2(\mathbb{S}^{N-1})} < \varepsilon$$

ensures

$$\lim_{\varepsilon \to 0} \|g_\varepsilon^z\|_{L^2(\mathbb{S}^{N-1})} = \infty.$$

The LSM turns the reconstruction of the shape of obstacle $\mathbf{D}$ into the process of numerically determining the indicator function g^z in Theorem 8.1.1. The general procedure is stated as follows (see also Chap. 4, [8]):

Algorithm LSM

1. Select a mesh $\mathscr{T}_h$ of sampling points in a region Ω which contains $\mathbf{D}$.
2. Use the Tikhonov regularization and the Morozov discrepancy principle to compute an approximate solution g^z to the far-field equation (8.1.2) for each mesh point z of $\mathscr{T}_h$.
3. Select a cut-off value c; then count $z \in \mathbf{D}$ if $\|g^z\|_{L^2(\mathbb{S}^{N-1})} \leq c$; and $z \notin \mathbf{D}$ if $\|g^z\|_{L^2(\mathbb{S}^{N-1})} > c$.

A mathematical justification was given in [6] for the use of LSM to determines $\mathbf{D}$ through the information of the indicator function g^z. The LSM has been proven to be numerically quite successful and shown to possess several remarkable merits (see

[28] and references therein). However, as one of the most important reconstruction algorithms in inverse scattering theory, it is computationally very expensive, among several other disadvantages. In fact, for a $\mathbb{R}^2$ problem, as one can see, by using an $n \times n$ mesh, we have to solve one linear integral equation (8.1.2) at each mesh point z and this amounts to n^2 totally, while it is n^3 in the $\mathbb{R}^3$ case. In some particular situations, the computational counts will be huge; e.g., to avoid being overambitious, the initial guess region Ω should be chosen to be moderately larger than $\mathbf{D}$; in order to achieve a high-resolution reconstruction of $\mathbf{D}$, we need a very fine mesh over Ω, i.e., n will be moderately large; the scatterer might consist of multiple component obstacles with the components not very close to each other. But in this case, the initial region Ω is still required to contain all these components, which means that Ω has been chosen to be much larger than it is actually needed. It is this computational complexity issue that motivates us to develop a fast numerical procedure of implementing LSM. To our knowledge, this important issue has not been seriously investigated yet.

In the next section, we will address the motivations and implementation details of the algorithm. We will also present some theoretical analysis to show that the discussed method possesses the asymptotically optimal computational complexity. In Sect. 8.1.2, some numerical experiments are performed to illustrate the promising feature of the algorithm in significantly reducing the computational cost of the LSM.

8.1.1 Multilevel Linear Sampling Method

In this section, we will present a multilevel linear sampling method, together with some theoretical analysis. For the sake of simplicity, we will carry out our discussion in $\mathbb{R}^2$, but all the subsequent results can be straightforwardly extended to the three-dimensional case.

Let $\mathbf{D} \subset \mathbb{R}^2$ be a bounded domain as shown in Fig. 8.1 (top-left) and suppose that we are going to use an $n \times n$ mesh for the LSM with $(n-1)^2$ cells of equal size. Clearly, in order to get some satisfactory reconstruction for the profile of $\mathbf{D}$, the mesh must be moderately fine in some sense. However, by performing the LSM on this fine mesh, we have to spend considerable computational cost in finding the indicator functions in those 'remote' cells which are far away from the scatterer $\mathbf{D}$, or in those 'inner' cells which lie deeply inside $\mathbf{D}$; e.g., see those red and blue colored regions in Fig. 8.1 (top-right). So, it would be very advantageous if we could get rid of those remote and inner cells in our computations. This can naturally be realized with a coarser mesh. In fact, this is reasonable since the indicator function g^z has very large norms for z in those remote cells while very small norms for z in those inner cells. Moreover, it is noted that the cut-off value c for the LSM in the fine mesh is still applicable on the coarsened mesh. Here, we would like to remark that, as pointed out in [8], the choice of c is rather heuristic and there is still no standard strategy for it. To be more precise, we first choose a coarse grid covering the sampling region Ω and perform the LSM on this coarse level, then based on the results of the LSM we will label and remove those remote and inner cells. Then, we refine the mesh

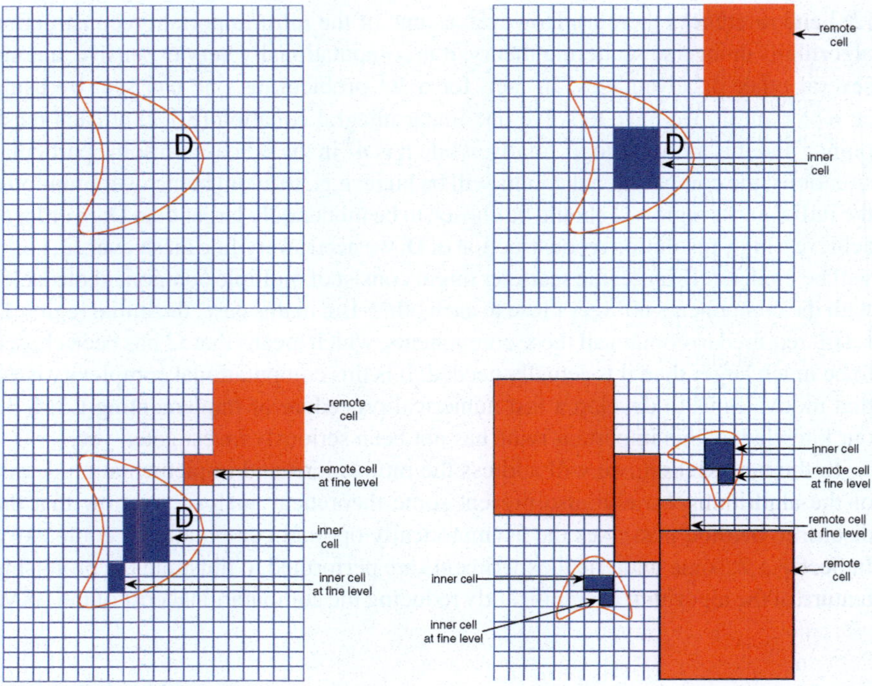

Fig. 8.1 *Label-and-remove* scheme

on the remaining sampling region, and perform the LSM again in this fine level to label and remove those fine remote and inner cells; e.g., see Fig. 8.1 (bottom-left) for those remote and inner cells. By doing this labeling and removing technique in a multilevel way, we can reconstruct the profile of **D** more and more accurately. We would like to remark that in many cases, the trimmed cells could be very large and thus save a lot of computational time; especially, when the scatterer is composed of multiple components which are not very close to each other (see, e.g., Fig. 8.1 (bottom-right)).

Now, we are ready to formulate our algorithm in details. In the following, the sampling region Ω is always chosen to be a square in $\mathbb{R}^2$. Then, let $\{\mathscr{T}_k\}_{k=1}^{L}$ be a nested sequence of meshes on the sampling domain Ω such that $\mathscr{T}_{k+1}$ is a refinement of $\mathscr{T}_k$ for $k = 1, \ldots, L-1$. Throughout, we assume that $\mathscr{T}_{k+1}$ is a $n_{k+1} \times n_{k+1}$ mesh while $\mathscr{T}_k$ is a $n_k \times n_k$ mesh, where $n_{k+1} = 2n_k - 1$ for $k = 1, \ldots, L-1$. That is, we refine the mesh $\mathscr{T}_k$ by equally sub-dividing every subsquare in $\mathscr{T}_k$ into four subsquares of $\mathscr{T}_{k+1}$. Then if the mesh length of $\mathscr{T}_k$ is h_k for $k = 1, \ldots, L-1$, then $h_{k+1} = h_k/2$. Now, the multilevel linear sampling method (MLSM) can be formulated as follows:

Algorithm MLSM

1. Set $k = 0$ and choose an initial mesh for the sampling region Ω;
2. Apply the LSM scheme on the kth-level mesh to investigate those mesh points which have not been examined previously;
3. For a given cut-off value c, independent of the level k, classify and label the kth-level subsquares (cells) into three sets, namely, *'remote cells'*, *'boundary cells'* and *'inner cells'*, based on the cut-off value principle in the LSM: a cell is labeled as 'remote' if the norms of the indicator functions at the vertices of the cell are all larger than c, while a cell is labeled as 'inner' if the norms of the indicator functions at the vertices of the cell are all less than or equal to c, and other remaining cells will be labeled as 'boundary cells'. Then remove the remote and inner cells;
4. Refine the remaining sampling mesh;
5. Set $k = k + 1$ and if $k \le L$, go to Step 2.

It is remarked that in order to exclude the extreme case that the obstacle is trapped into a single subsquare of the sampling mesh, the initial mesh should be chosen to be mildly fine such that both 'remote' cells and 'inner' cells exist.

Next, we will show that the MLSM algorithm is asymptotically optimal in computational complexity. For the purpose, we first present some lemmas. In the following, we denote by Γ a C^2-smooth curve in $\mathbb{R}^2$ which forms the boundary of a bounded domain $\mathbf{G}$. For any $h > 0$, we define two curves parallel to Γ:

$$\Gamma_h^+ := \{x + h\nu(x);\ \ x \in \Gamma \text{ and } \nu(x) \text{ is the unit normal to } \Gamma \text{ at } x \text{ directed to the exterior of } \mathbf{G}\}, \tag{8.1.4}$$

$$\Gamma_h^- := \{x - h\nu(x);\ \ x \in \Gamma \text{ and } \nu(x) \text{ is the unit normal to } \Gamma \text{ at } x \text{ directed to the exterior of } \mathbf{G}\}. \tag{8.1.5}$$

Then we have

Lemma 8.1.1 *There exist constants $h_0^+ > 0$ and $0 < \alpha_0^+ \le 1$ such that*

$$dist(\Gamma, \Gamma_h^+) \ge \alpha_0^+ h \quad whenever\, 0 < h < h_0^+. \tag{8.1.6}$$

Proof Assume contrarily that there are no constants h_0^+ and α_0^+ such that (8.1.6) holds. Then, for $\hat{h}_1 = 1/2$, there must exist a h_1 such that $0 < h_1 < \hat{h}_1$ and dist$(\Gamma, \Gamma_{h_1}^+) < h_1/2$, otherwise Lemma 8.1.1 is true with $h_0^+ = \hat{h}_1$ and $\alpha_0^+ = 1/2$. Next, for $\hat{h}_2 = \min\{h_1, 1/2^2\}$, there must exist a h_2 such that $0 < h_2 < \hat{h}_2$ and dist$(\Gamma, \Gamma_{h_2}^+) < h_2/2^2$, otherwise Lemma 8.1.1 is true with $h_0^+ = \hat{h}_2$ and $\alpha_0^+ = 1/2^2$. Continuing with this procedure, we have by induction that for $\hat{h}_k = \min\{h_{k-1}, 1/2^k\}$ ($k \ge 3$), there exists a h_k such that $0 < h_k < \hat{h}_k$ and dist$(\Gamma, \Gamma_{h_k}^+) < h_k/2^k$. So we obtain a positive sequence $\{h_k\}_{k=1}^{\infty}$ such that

$$\lim_{k\to\infty} h_k = 0 \quad \text{and} \quad \lim_{k\to\infty} \frac{\text{dist}(\Gamma, \Gamma_{h_k}^+)}{h_k} = 0. \tag{8.1.7}$$

Since both Γ and $\Gamma_{h_k}^+$ are compact sets in $\mathbb{R}^2$, there exists $x_k \in \Gamma$ and $y_k^+ \in \Gamma_{h_k}^+$ for any $k \in \mathbb{N}$ such that

$$\operatorname{dist}(\Gamma, \Gamma_{h_k}^+) = |x_k - y_k^+|. \tag{8.1.8}$$

Set

$$y_k = y_k^+ - h_k \nu(y_k^+) \in \Gamma \quad \text{for } k \in \mathbb{N}, \tag{8.1.9}$$

where $\nu(y_k^+)$ is the unit outward normal to $\Gamma_{h_k}^+$ at y_k^+. By extracting subsequences if necessary, we may assume that

$$\lim_{k\to\infty} x_k = x_0 \quad \text{and} \quad \lim_{k\to\infty} y_k = y_0. \tag{8.1.10}$$

By (8.1.9), (8.1.10), we see that $\lim_{k\to\infty} y_k^+ = y_0$, which together with (8.1.7), (8.1.8) implies that

$$x_0 = y_0 = x^* \tag{8.1.11}$$

for some $x^* \in \Gamma$. Noting that $\nu(x)$ is continuous, for an arbitrary $\varepsilon > 0$ there exists $\delta > 0$ such that

$$|\nu(x) - \nu(x^*)| < \varepsilon \quad \forall x \in B_\delta(x^*) \cap \Gamma, \tag{8.1.12}$$

where $B_\delta(x^*) = \{x \in \mathbb{R}^2; |x - x^*| < \delta\}$. By (8.1.10) and (8.1.11), we know that there exists $k_\varepsilon \in \mathbb{N}$ such that

$$x_k \in B_\delta(x^*), \quad y_k \in B_\delta(x^*) \quad \text{for all } k > k_\varepsilon.$$

Furthermore, by (8.1.7), we can assume that k_ε is chosen such that

$$\frac{\operatorname{dist}(\Gamma, \Gamma_{h_k}^+)}{h_k} < \frac{1}{2} \quad \text{for all } k > k_\varepsilon, \tag{8.1.13}$$

namely,

$$|x_k - y_k^+| < \frac{1}{2} h_k \quad \text{for all } k > k_\varepsilon. \tag{8.1.14}$$

It is noted that by (8.1.14) we must have $x_k \neq y_k$ for all $k > k_\varepsilon$, since otherwise we would have $|x_k - y_k^+| = |y_k - y_k^+| = h_k$. Let $\tau(x)$ be the tangential to Γ at x, and we know from (8.1.12) that

$$|\tau(x) - \tau(x^*)| < \varepsilon \quad \forall x \in B_\delta(x^*) \cap \Gamma. \tag{8.1.15}$$

Next, we investigate the angle $\angle(\overrightarrow{x_k y_k}, \tau(y_k))(\in [0, \pi/2])$ between the two vectors $\overrightarrow{x_k y_k}$ and $\tau(y_k)$ for $k > k_\epsilon$. From the geometric interpretation of Lagrange's theorem, we know that there exists $\xi_k \in B_\delta(x^*) \cap \Gamma$ such that $\tau(\xi_k)$ is parallel to $\overrightarrow{x_k y_k}$. By (8.1.15), we know that

$$\lambda_k := <\tau(\xi_k), \tau(y_k)> = <\tau(x^*) - \mathscr{O}(\varepsilon), \tau(x^*) - \mathscr{O}(\varepsilon)> = 1 - \mathscr{O}(\varepsilon) \quad \text{as } \varepsilon \to +0, \tag{8.1.16}$$

where $< \cdot, \cdot >$ is the inner product in $\mathbb{R}^2$. Hence,

$$\theta_k := \angle(\overrightarrow{x_k y_k}, \tau(y_k)) = \arccos \lambda_k = \mathscr{O}(\sqrt{\varepsilon}) \quad \text{as } \varepsilon \to +0. \tag{8.1.17}$$

Now, let $\Delta x_k y_k y_k^+$ denote the triangle with vertices x_k, y_k and y_k^+. It is easily seen that the interior angle of $\Delta x_k y_k y_k^+$ at y_k, namely, $\angle(\overrightarrow{x_k y_k}, \overrightarrow{y_k^+ y_k})$ is either $\pi/2 + \theta_k$ or $\pi/2 - \theta_k$. Then, by (8.1.13) and (8.1.17), we take $\varepsilon_0 > 0$ to be sufficiently small and $k_{\varepsilon_0} \in \mathbb{N}$ be sufficiently large such that for all $k > k_{\varepsilon_0}$,

$$\frac{|x_k - y_k^+|}{h_k} < \frac{1}{2} \quad \text{and} \quad \sin(\frac{\pi}{2} - \theta_k) > \frac{1}{2}. \tag{8.1.18}$$

Then, in the case that $\angle(\overrightarrow{x_k y_k}, \overrightarrow{y_k^+ y_k}) = \pi/2 + \theta_k > \pi/2$,

$$|x_k - y_k^+| > |y_k - y_k^+| = h_k,$$

and in the case that $\angle(\overrightarrow{x_k y_k}, \overrightarrow{y_k^+ y_k}) = \pi/2 - \theta_k$,

$$\frac{|x_k - y_k^+|}{h_k} = \frac{|x_k - y_k^+|}{|y_k - y_k^+|} \geq \sin(\pi/2 - \theta_k) > \frac{1}{2}.$$

In both cases, we have contradiction with the first inequality in (8.1.18). This completes the proof of the Lemma 8.1.1. □

Lemma 8.1.2 *There exist constants $h_0^- > 0$ and $0 < \alpha_0^- \leq 1$ such that*

$$dist(\Gamma, \Gamma_h^-) \geq \alpha_0^- h \quad whenever\, 0 < h < h_0^-. \tag{8.1.19}$$

Proof The lemma can be proved in a completely similar way to that of Lemma 8.1.1. □

Lemma 8.1.3 *There exists constants $h_0 > 0$ and $\alpha_0 > 0$ such that*

$$dist(\Gamma, \Gamma_{\alpha_0 h}^{\pm}) \geq \sqrt{2} h \quad whenever\, 0 < h < h_0.$$

Proof Set

$$\alpha_0 = \frac{\sqrt{2}}{\min(\alpha_0^+, \alpha_0^-)} \quad \text{and} \quad h_0 = \frac{\min(h_0^+, h_0^-)}{\alpha_0},$$

where $\alpha_0^{\pm}$ and $h_0^{\pm}$ are constants given in Lemmas 8.1.1 and 8.1.2. Then, it is easy to verify that when $h < h_0$, namely, $\alpha_0 h < \min(h_0^+, h_0^-)$,

$$\operatorname{dist}(\Gamma, \Gamma^+_{\alpha_0 h}) \geq \alpha_0^+ \alpha_0 h \geq \sqrt{2}h \tag{8.1.20}$$

and

$$\operatorname{dist}(\Gamma, \Gamma^-_{\alpha_0 h}) \geq \alpha_0^- \alpha_0 h \geq \sqrt{2}h. \tag{8.1.21}$$

□

The following theorem is crucial to our subsequent investigation.

Theorem 8.1.2 *Let $\mathscr{T}$ be an $n \times n$ mesh on the sampling region Ω. There exist two constants $\kappa_0 > 0$ and $n_0 \in \mathbb{N}$ such that $\partial\mathbf{D}$ lies on at most $\kappa_0 n$ subsquares of $\mathscr{T}$ for all $n \geq n_0$.*

Proof To ease the discussion, we assume that the scatterer **D** is composed of a single component obstacle. That is, **D** is a bounded domain. But we remark that our subsequent proof can be easily modified to the case that **D** has finitely many connected components.

Let $\Gamma := \partial\mathbf{D}$ in Lemma 8.1.3. Take $n \in \mathbb{N}$ be sufficiently large such that the mesh length h of $\mathscr{T}$ satisfies $h < h_0$. Suppose that $\partial\mathbf{D}$ lies on m subsquares of $\mathscr{T}$. By (8.1.20) and (8.1.21), it is easily seen that these m subsquares must lie in the ring-shaped region formed by $\Gamma^+_{\alpha_0 h}$ and $\Gamma^-_{\alpha_0 h}$. Let s_0 denote the area occupied by this ring-shaped region, $\omega_0 = |\Omega|$ be the area of Ω and $\eta_0 = |\partial\mathbf{D}|$ be the length of the boundary curve $\partial\mathbf{D}$. Then, we have

$$mh^2 \leq s_0 \leq 2\eta_0\alpha_0 h,$$

hence

$$m \leq \frac{2\eta_0\alpha_0}{h}.$$

By noting $n^2h^2 = \omega_0$, we further have

$$m \leq \frac{2\eta_0\alpha_0}{\sqrt{\omega_0}} n.$$

Now, the theorem is seen to be held with

$$\kappa_0 = \left\lceil \frac{2\eta_0\alpha_0}{\sqrt{\omega_0}} \right\rceil,$$

where for a positive number a, $\lceil a \rceil$ denotes the smallest integer not less than a. □

The above theorem shows that for a sufficient fine $n \times n$ mesh, $\partial\mathbf{D}$ lies on at most $\mathscr{O}(n)$ subsquares. We next show that $\partial\mathbf{D}$ also lies on at least $\mathscr{O}(n)$ subsquares.

Theorem 8.1.3 *Let $\mathscr{T}$ be an $n \times n$ mesh on the sampling region Ω. There exist two constants $\beta_0 > 0$ and $m_0 \in \mathbb{N}$ such that $\partial\mathbf{D}$ lies on at least $\beta_0 n$ subsquares of $\mathscr{T}$ for all $n \geq m_0$.*

Fig. 8.2 Illustration of the proof of Theorem 8.1.3

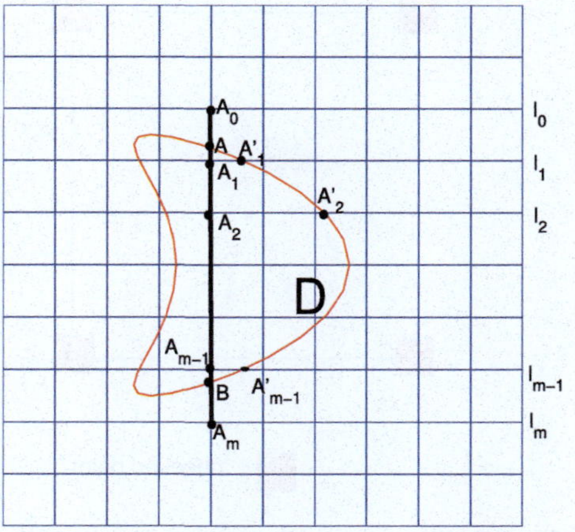

Proof As in Theorem 8.1.2, we need only to consider the simple case that **D** is a connected bounded domain. And the subsequent proof is easily modified to the case that **D** has finitely many connected components.

By our assumption on the sampling mesh, we may choose $\mathscr{T}$ to be fine enough such that there is at least one inner cell. Take one of the edges of this cell and denote its connected extension in **D** by AB with the two endpoints A and B lying on $\partial\mathbf{D}$ (see Fig. 8.2). We suppose that AB lies on m subsquares of $\mathscr{T}$. Let $A_0, A_1, \ldots, A_m$ be the vertices of those subsquares, all lying on the extended line of AB and ordered in the direction from A to B (see Fig. 8.2). By our organization, A is either A_0 or lies between A_0 and A_1, and B is either A_m or lies between A_{m-1} and A_m, whereas A_j, $j = 1, 2, \ldots, m-1$, all lie inside of **D**. We denote by $l_0, l_1, \ldots, l_m$ those line segments of $\mathscr{T}$ in Ω which respectively passes through $A_0, A_1, \ldots, A_m$. Noting that **D** is connected and by the fundamental property of connected set, we know that l_j, $j = 1, 2, \ldots, m-1$ must have intersection with $\partial\mathbf{D}$. We denote by ${A'}_1, {A'}_2, \ldots, {A'}_{m-1}$ those intersection points which lie on one side of AB. It is remarked that A_j for $j = 1, \ldots, m-1$ is not necessarily unique. Now, by the connectedness of $\partial\mathbf{D}$, we know that between A and ${A'}_1$, B and ${A'}_{m-1}$, ${A'}_j$ and ${A'}_{j+1}$ for $j = 1, \ldots, m-2$, there must be a connected part of $\partial\mathbf{D}$ which lies in the stripped region, respectively, formed by l_0 and l_1, l_{m-1} and l_m, and l_j and l_{j+1} for $j = 1, \ldots, m-2$. Therefore, if we suppose that $\partial\mathbf{D}$ lies on m' subsquares of $\mathscr{T}$, then there must be at least one from those subsquares which lies in the stripped region formed by l_j and l_{j+1} for $j = 0, 1, \ldots, m-1$. Hence, we have $m' \le m$. Next, we set $A_0 = A$ and $A_m = B$, and by noting $|A_j A_{j+1}| \le h$ for $j = 0, 1, \ldots, m-1$, we have

$$\sum_{j=0}^{m-1} |A_j A_{j+1}| \le mh, \quad \text{i.e., } |AB| \le mh.$$

Finally, we have by noting $n^2 h^2 = |\Omega|$ that

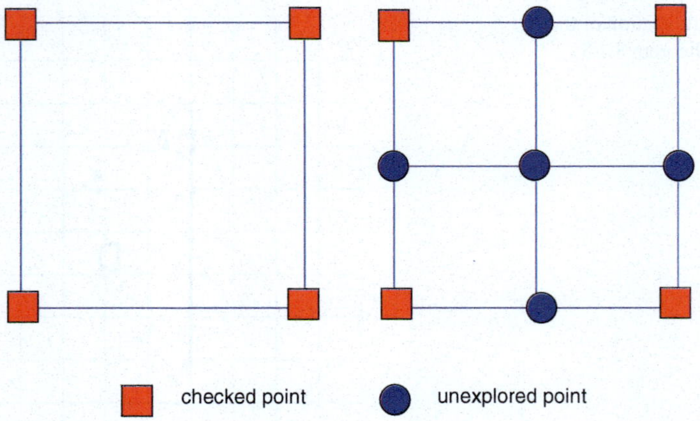

Fig. 8.3 Illustration of the proof of Theorem 8.1.4

$$m' \geq m \geq |AB|\frac{1}{h} \geq \beta_0 n, \tag{8.1.22}$$

with $\beta_0 = |AB|/\sqrt{|\Omega|}$. The theorem is completed. □

Theorems 8.1.2 and 8.1.3 reveal that in order to achieve a good reconstruction of the scatterer **D**, we need at least to solve $\mathscr{O}(n)$ far-field equations (8.1.2) with a fine $n \times n$ sampling mesh. Now, we are ready to present the main result that the algorithm MLSM possesses the asymptotically optimal computational complexity.

Theorem 8.1.4 *Consider an L-level MLSM algorithm with a nested sequence of sampling mesh* $\{\mathscr{T}_k\}_{k=1}^L$. *Suppose for each k,* $\mathscr{T}_k$ *is of size* $n_k \times n_k$ *with mesh length* h_k *such that* $0 < h_1 < h_0$, *where* h_0 *is given in Lemma 8.1.3 corresponding to* $\partial\mathbf{D}$. *Then by using the MLSM to reconstruct* $\partial\mathbf{D}$, *the far-field equation (8.1.2) is solved* $\mathscr{O}(n_L)$ *times in total.*

Proof We denote by $\mathscr{C}_k$, $k = 1, 2, \ldots, L$ the points to be investigated on the k-th level. By Theorem 8.1.1, we know that $\partial\mathbf{D}$ lies on at most $\kappa_0 n_k$ subsquares of $\mathscr{T}_k$. Next, when we turn to the $k+1$th level, by our description of the MLSM, we only need to investigate the mesh points on those subsquares of $\mathscr{T}_{k+1}$ which have not been examined before, which can be easily seen to be at most $5\kappa_0 n_k$ mesh points as shown in Fig. 8.3. Hence, we have

$$\mathscr{C}_k \leq \mathscr{C}_{k-1} + 5\kappa_0 n_{k-1}, \quad k = 2, \ldots, L, \tag{8.1.23}$$

where $n_{k-1} = (n_k + 1)/2$. Recursively, we can obtain

$$\begin{aligned}\mathscr{C}_L &\le \mathscr{C}_{L-1} + 5\kappa_0 n_{L-1},\\ \mathscr{C}_{L-1} &\le \mathscr{C}_{L-2} + 5\kappa_0 n_{L-2},\\ &\cdots\cdots \quad \cdots\cdots\\ \mathscr{C}_2 &\le \mathscr{C}_1 + 5\kappa_0 n_1.\end{aligned}$$

By summing up the above inequalities we get

$$\mathscr{C}_L \le \mathscr{C}_1 + 5\kappa_0[n_{L-1} + n_{L-2} + \cdots + n_1].$$

Since it is easy to deduce that $n_{L-k} = n_L/2^k + \sum_{j=1}^{k} 1/2^j$ for $k = 1, 2, \ldots, L-1$, we see that

$$\mathscr{C}_L \le \mathscr{C}_1 + 5\kappa_0(L + n_L),$$

i.e.,

$$\mathscr{C}_L \le \mathscr{O}(n_L) \quad \text{for sufficiently large } n_L \in \mathbb{N}.$$

This means that the MLSM has the asymptotically optimal computational complexity. The proof is completed. □

Remark 8.1.1 As we have pointed out earlier that all the results in this section can be modified to the $\mathbb{R}^3$ case, where the MLSM algorithm needs to solve far-field equations (8.1.2) $\mathscr{O}(n_L^2)$ times.

8.1.2 Numerical Experiments and Some Discussions

In this section, we perform three tests to illustrate the effectiveness and efficiency of the MLSM algorithm. All the programs in our experiments are written in Matlab and run on a Pentium 3GHz PC.

The scatterer in system will be chosen to be the kite-shaped object which has been widely tested in the inverse scattering problems (see, e.g., [10, 28]). There are totally three tests to be considered, and they are respectively referred to as SK, SKn and DKn. For experiments SK and SKn, the scatterer **D** is composed of a single kite. But in experiment SK, we would not add noise to the synthetic far-field data, and in experiment SKn, we add random noise. For experiment DKn, the scatterer **D** is composed of two kites, and the synthetic far-field data are noise-bound. The other parameters chosen for these experiments are listed in Table 8.1.

It is noted that for experiment DKn, we have taken two cut-off values c_1 and c_2, $c_1 < c_2$, instead of only one cut-off value c. Since in DKn, the scatterer is composed of two kites, it is better to take a range of cut-off values, i.e., $[c_1, c_2]$, which enables us to get a buffer region of locating the boundary of the underlying object. Like in the original LSM, we label as *inner points* those points with the norm of distributed

Table 8.1 Experimental parameters for the tests

	Test 1 (SK)	Test 2 (SKn)	Test 3 (DKn)
Sampling domain Ω	$[-3, 3] \times [-3, 3]$	$[-3, 3] \times [-3, 3]$	$[-4, 8] \times [-4, 8]$
Incident wave number k	1	1	1
Finest level n_L	129	129	129
Upper threshold c_1	0.03	0.032	0.03
Lower threshold c_2	0.03	0.032	0.02
Noise level δ	0	0.10	0.05
No. of incident directions	32		
No. of observation directions	32		

density g less than c_1, and *remote points* those points with the norm of distributed density g greater than c_2.

The synthetic far-field data are generated by solving the layer potential operator equation with the Nyström's method (see Sect. 3.5, Chap. 3 in [13]). We compute the far-field patterns at 32 equidistantly distributed observation points $(\cos t_j, \sin t_j)$, $t_j = 2j\pi/32$, $j = 0, 1, \ldots, 31$, and 32 equidistantly distributed incident directions $(\cos \tau_j, \sin \tau_j)$, $\tau_j = 2j\pi/32$, $j = 0, 1, \ldots, 31$. The far-field patterns we obtain are subjected pointwise to uniform random noise. The uniform random noise is added according to the following formula,

$$u_\infty = u_\infty + \delta r_1 |u_\infty| \exp(i\pi r_2)$$

where r_1 and r_2 are two uniform random numbers, both ranging from -1 to 1, and δ is the noise level. For each mesh point z, the corresponding far-field equation (8.1.2) is solved by using Tikhonov regularization method.

In Tests 1 and 2, the kite-shaped object $\mathbf{D}$ is shown in Fig. 8.4 with the boundary $\partial\mathbf{D}$ given by the following parametric form:

$$x(t) = (\cos t + 0.65 \cos 2t - 0.65, 1.5 \sin t), \quad 0 \le t \le 2\pi. \tag{8.1.24}$$

For Test 3, the two kite-shaped objects are shown in Fig. 8.5 which are derived from the kite in Fig. 8.4 by rigid motions: the bottom-left one is given by the one in Fig. 8.4 after counterclockwise $\pi/2$ rotation, and the top-right one is given by the one in Fig. 8.4 after counterclockwise $\pi/4$ rotation and 5-unit displacement in both longitude and latitude directions.

We now turn to experiment SK. First, we solve the far-field equation (8.1.2) on the finest mesh (129×129) to find g^z with z being a sampling mesh point. In order to have a view of the behavior of this g^z over the sampling mesh, we plot the logarithm of its L^2-norm, namely $\log \|g^z\|_{L^2(\mathbb{S}^1)}$, in a 3D graph (see Fig. 8.6), but such scalings are not needed in our MLSM procedure for those tests. The corresponding contours for $\log \|g^z\|_{L^2(\mathbb{S}^1)}$ is also given in Fig. 8.7 for a 2D view. Then, we can use the cut-off

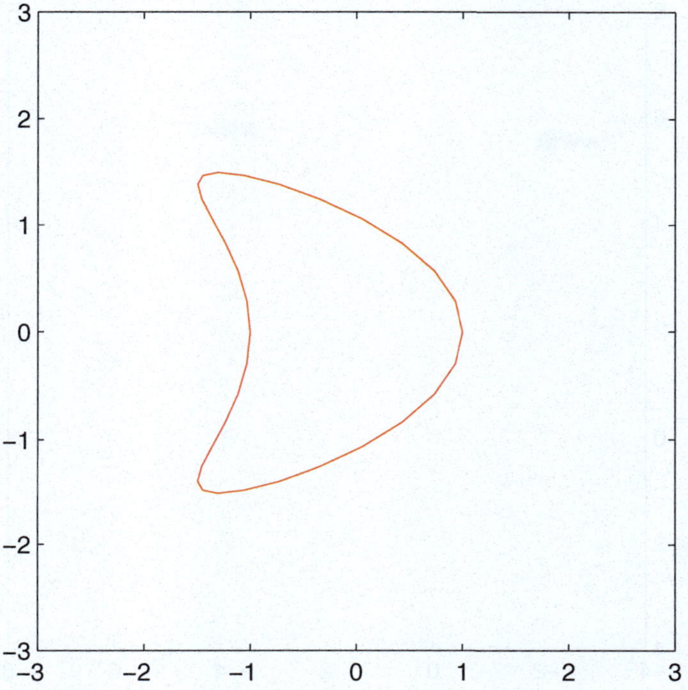

Fig. 8.4 Kite-shaped obstacle

value principle to detect the kite and this gives the original LSM. We remark that the regularization is crucial in the numerical procedure. Even in this noise-free case when $\delta = 0$, regularization is still necessary since the far-field data u_∞ are computed approximately using Nyström's method, and thus there are some approximate errors besides the round-off errors in the various computations involved. We have also plotted the logarithm of the L^2-norm of g^z obtained by solving the far-field equation without regularization, from which it can be seen that the reconstruction would be rather unsatisfactory; see the 3D display and 2D contour curves in Figs. 8.8 and 8.9, respectively.

Next we apply our (6-level) MLSM to this problem with $n_L := n_6 = 129$ and plot the evolution of the detected boundary of the underlying object level-by-level. It must be emphasized in the implementation of MLSM that if some cell labelled as '*remote*' or '*inner*' is determined to be trimmed, we set all sampling points in this cell, including all its boundary points, to be *exploited*. Figure 8.10 demonstrates that the boundary of the kite-shaped object can be approximated in a clearly improving manner as we go from coarse to fine meshes, but the points examined are kept within the order $O(n_L)$.

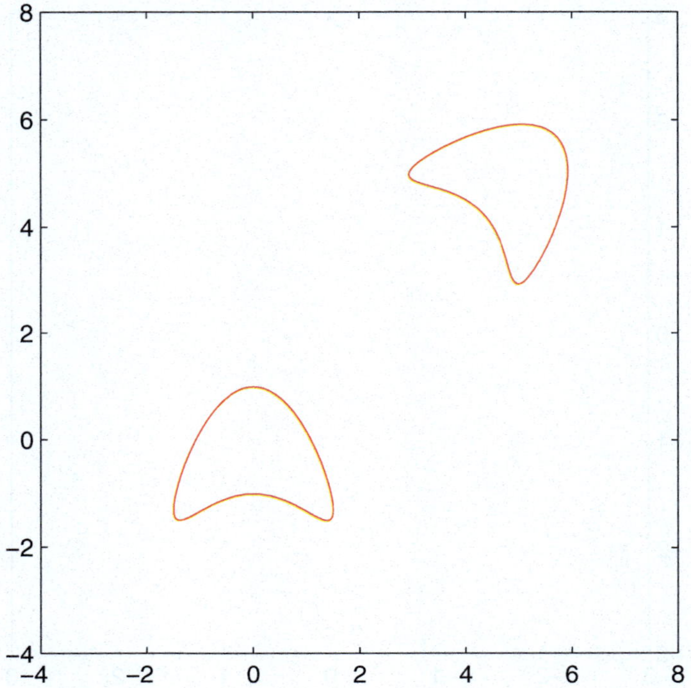

Fig. 8.5 Two kite-shaped objects

This first experiment SK suggests that the MLSM performs as well as the original LSM method but the computational costs have reduced significantly. In fact, we have counted the number of points examined in the MLSM which is listed in Table 8.2; and for Test 1, it is 483 and this is roughly one thirtieth of that for LSM which is 16641 ($= 129 \times 129$). It is worth remarking in the Table 8.2 that for the Test 1, the number 8 in the second level of grid comes from exploiting the eight second-level fine points around the origin as shown in the left subfigure of level 2 in Fig. 8.10, which does not include the other eight second-level fine grid on the boundary who has been set to be exploited at level 1. The same way applies to interpret the numbers at the finer level and the other tests.

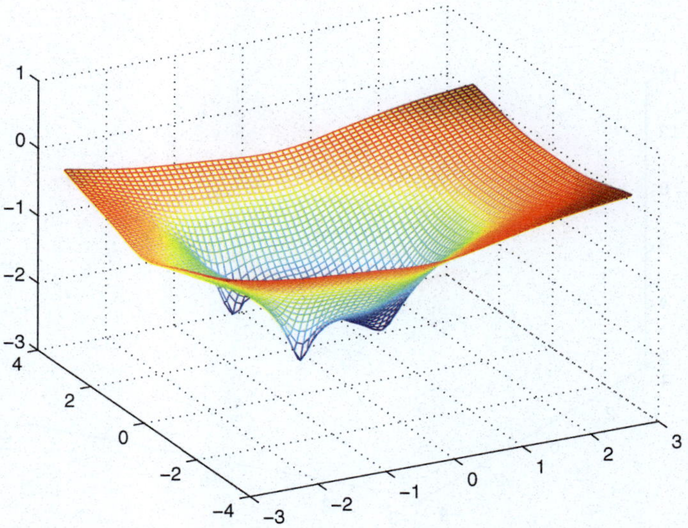

Fig. 8.6 Test 1 (SK): the logarithm of the L^2-norm of g^z plotted in 3D

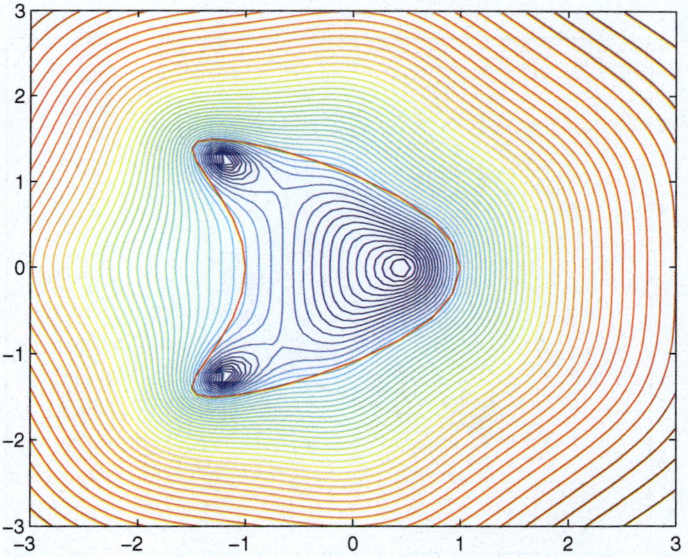

Fig. 8.7 Test 1 (SK): contours of the logarithm of the L^2-norm of g^z

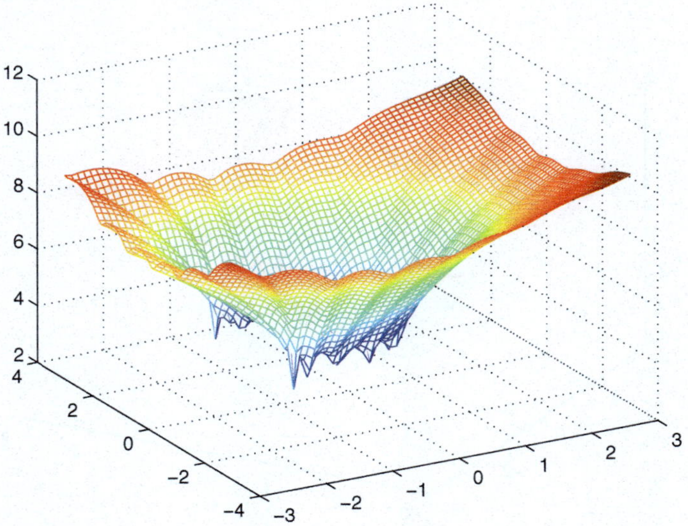

Fig. 8.8 Test 1 (SK): the logarithm of the L^2-norm of g^z plotted in 3D without regularization in deriving g^z

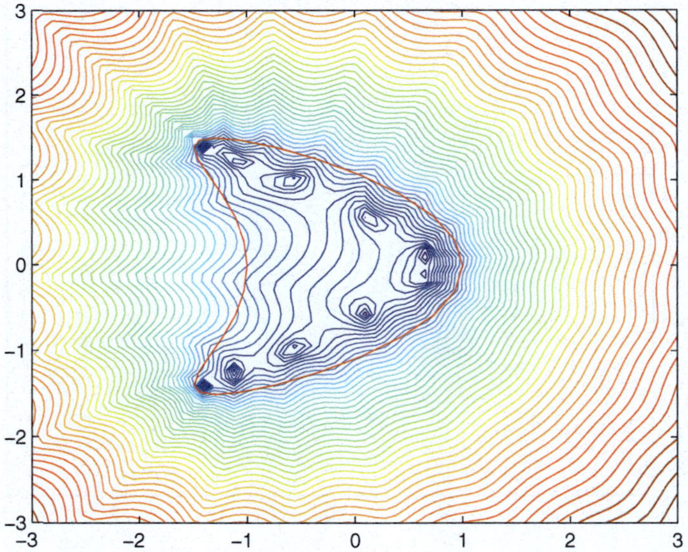

Fig. 8.9 Test 1 (SK): contours of the logarithm of the L^2-norm of g^z without regularization in deriving g^z

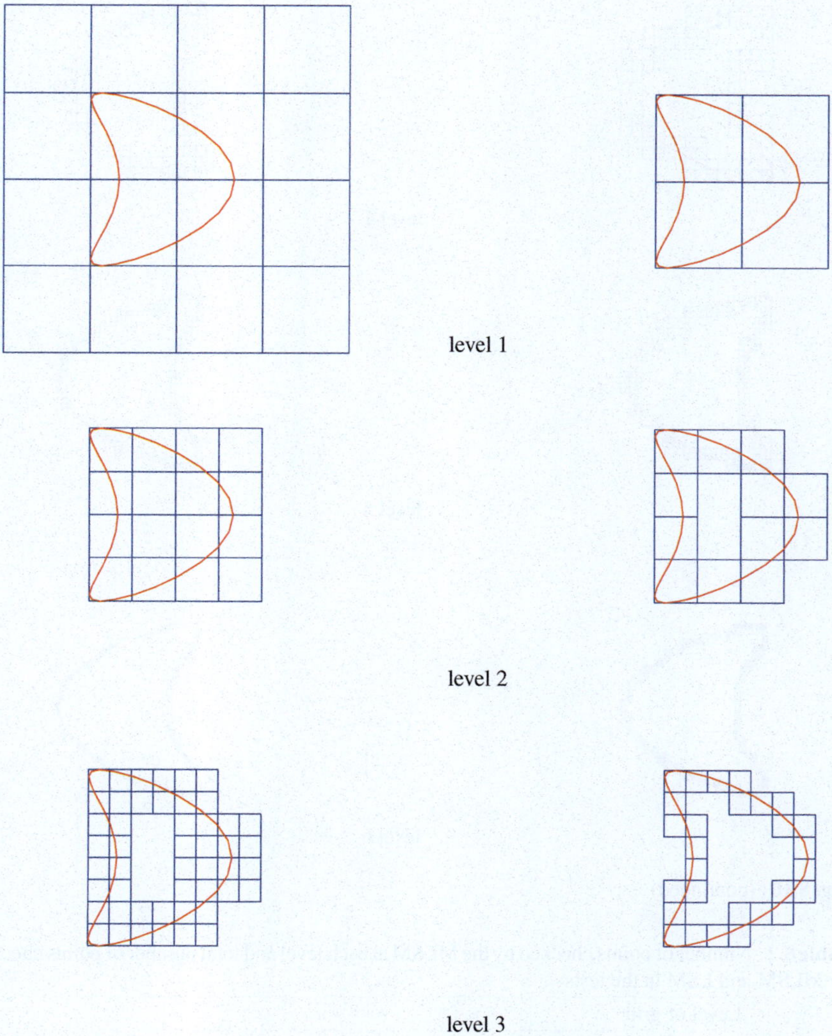

Fig. 8.10 MLSM iteration for Test 1 (SK). Figures on the left: refinement of the previous coarse grid; Figures on the right: the remote and inner cells are removed

level 4

level 5

level 6

Fig. 8.10 (continued)

Table 8.2 Number of points checked by the MLSM at each level and total number of points checked by MLSM and LSM in the tests

	Level of grid							
	1	2	3	4	5	6	MLSM	LSM
Test 1.	25	8	25	56	121	248	483	16641
Test 2.	25	8	34	61	129	260	517	16641
Test 3.	25	24	36	71	163	375	694	16641

Next we add 10% uniform random noise to the far-field data and rerun the MLSM for the SKn case. The evolution of the boundary of the kite is illustrated in Fig. 8.11. We see the total number of the points examined to locate the boundary is 517, almost the same as the previous SK case (see Table 8.2).

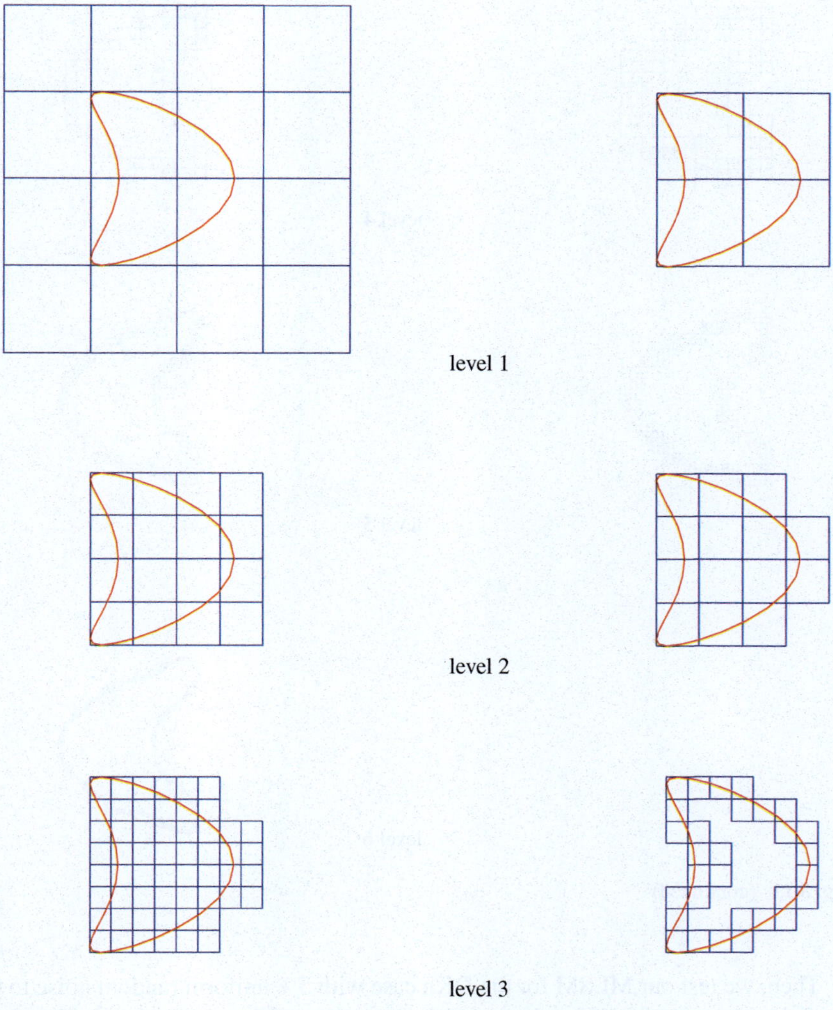

Fig. 8.11 MLSM iteration for Test 2 (SKn). Figures on the left: refinement of the previous coarse grid; Figures on the right: the remote and inner cells are removed

level 4

level 5

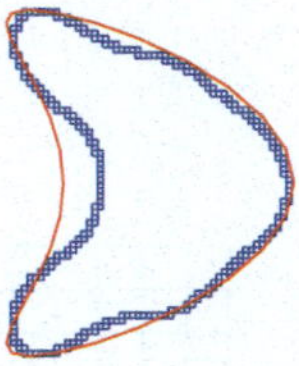

level 6

Fig. 8.11 (continued)

Then, we test our MLSM for the DKn case with 5% uniform random noise to the far-field data and plot the evolution of the boundary of these two kites in Fig. 8.12. Note that there is a slight increase for the number of exploited points which is due to a buffer range of cut-off values used in this test.

Finally, we plot all the subsquares that have been checked in the MLSM procedure in a single figure for all the above experiments; see Fig. 8.13 for Test 1 (SK), Fig. 8.14 for Test 2 (SKn) and Fig. 8.15 for Test 3 (DKn). From those figures, we can have a concrete feeling about how MLSM works to identify the boundary of the underlying object.

For comparison, we list in Table 8.2 the number of points examined at each level in the MLSM procedure and the total numbers points examined by MLSM and LSM for all the three tests. It can be seen from the table that the number of points examined at each level is about $\sigma_0 n_k$ with $\sigma_0 \approx 2$, which is consistent with our theoretical analysis in Sect. 8.1.1.

To consolidate the asymptotically optimal computational complexity, we perform the three tests again with the mesh size n_L in the finest level being 33, 65, 129 and 257, respectively. But for all those experiments, we start with the coarsest mesh given by $n_1 = 5$. Furthermore, we let the cut-off value c be the average of c_1 and c_2 in Table 8.1, and this is to eliminate the possible deterioration due to the additional points checked in the buffer region. The total number of points examined and the time

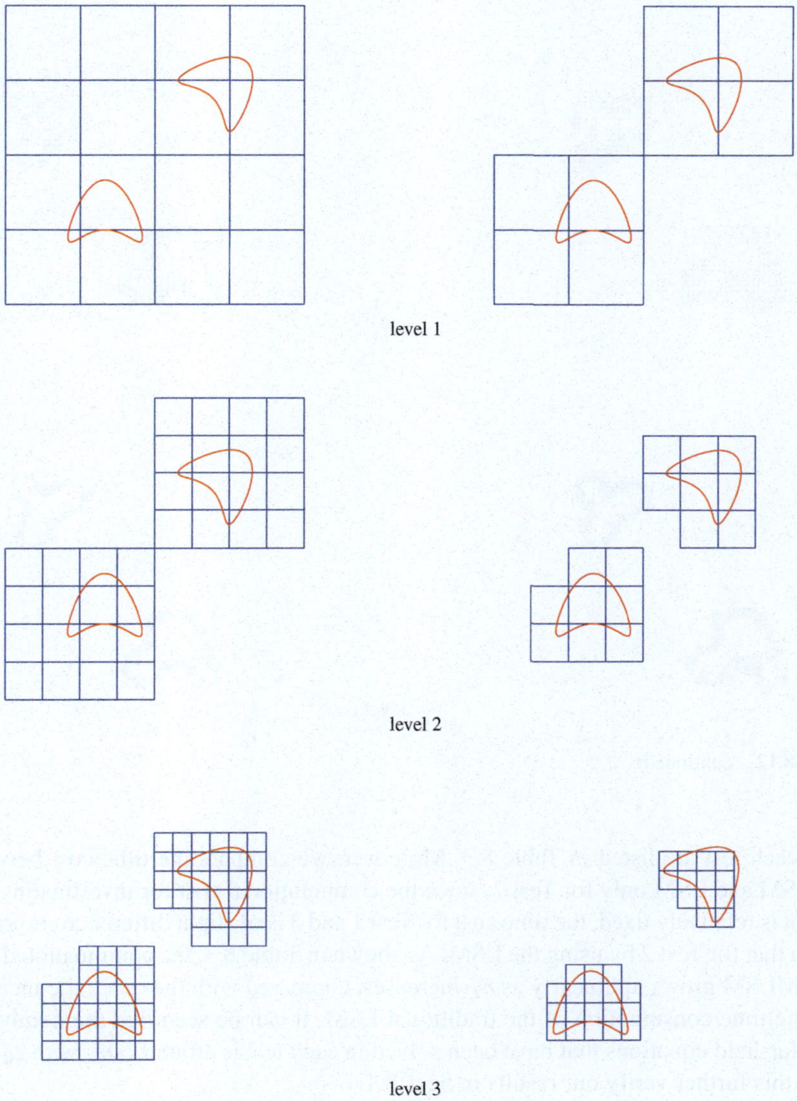

Fig. 8.12 MLSM iteration for Test 3 (DKn)

level 4

level 5

level 6

Fig. 8.12 (continued)

for each test are listed in Table 8.3. Moreover, we compare the time cost between MLSM and LSM only for Test 2, since the computational cost for investigating one point is relatively fixed, the time cost for Test 1 and 3 is of slight difference compared with that for Test 2 by using the LSM. As shown in Table 8.3, the computational cost for MLSM grows up linearly as n_L increases, compared with the quadratic increase of the time consumption of the traditional LSM. It can be seen that the number of the far-field equations that have been solved in each test is around $\zeta_0 n_L$ with $\zeta_0 \approx 4$ and this further verify our results in Sect. 8.1.1.

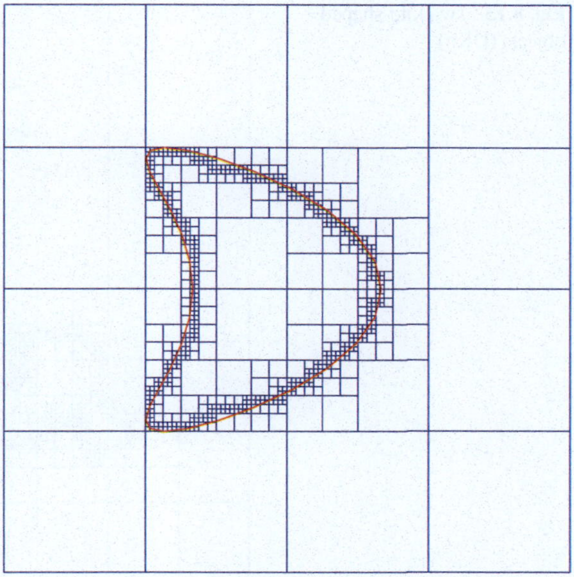

Fig. 8.13 One kite-shaped object (SK)

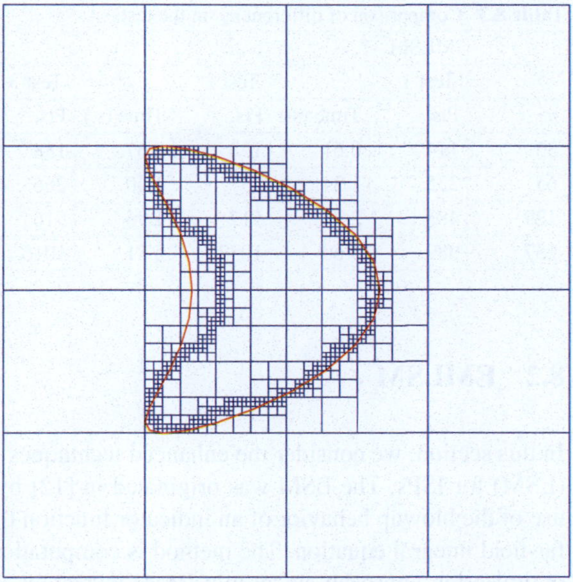

Fig. 8.14 One kite-shaped object (SKn)

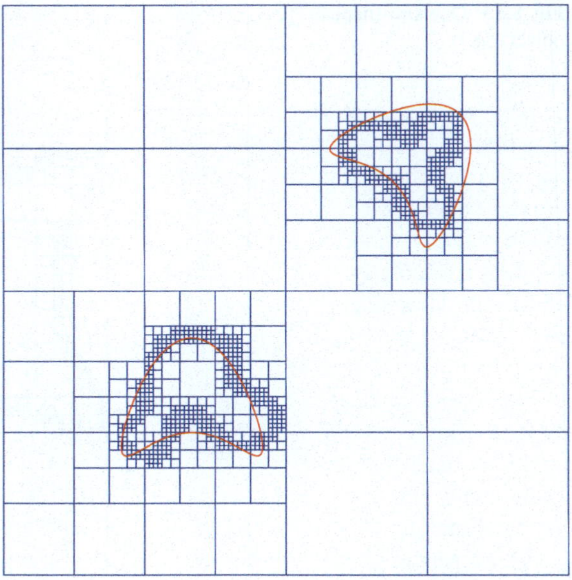

Fig. 8.15 Two kite-shaped objects (DKn)

Table 8.3 Comparison of different n_L in the tests

	MLSM						LSM	
	Test 1		Test 2		Test 3		Test 2	
n_L	Pts.	Time (s.)	Pts.	Time (s.)	Pts.	Time (s.)	Pts.	Time (s.)
33	114	0.61	128	0.67	138	0.72	1089	5.98
65	235	1.29	257	1.40	266	1.43	4225	24.07
129	483	2.77	517	2.96	516	2.95	16641	99.16
257	989	5.60	1037	5.81	1016	5.75	66049	396.90

8.2 EMLSM

In this section, we consider the enhanced techniques of the linear sampling method (LSM) for ISPs. The LSM was originated in [12] by Colton and Kirsch. It makes use of the blowup behavior of an indicator function that can be solved from a linear far-field integral equation. The method is computationally faster than the nonlinear optimization approach since only linear inversions would be involved. Moreover, the LSM requires no a priori knowledge of the physical properties of the underlying target object, which possibly includes, at the same time, sound-soft, sound-hard or impedance-type impenetrable obstacles, or penetrable inhomogeneous mediums. The method has also been shown to work for both inverse acoustic and electromagnetic scattering problems. In [4, 11], strategies on how to choose the critical cut-off values and how to avoid the interior eigenvalues for the LSM have been developed. In [25], a multilevel technique was developed for the LSM, which was shown to possess the optimal computational complexity. Indeed, for an $n \times n$ sampling mesh in $\mathbb{R}^2$ or an $n \times n \times n$ sampling mesh in $\mathbb{R}^3$, the multilevel linear sampling method (MLSM) requires to solve only $\mathscr{O}(n^{N-1})$ far-field equations for an $\mathbb{R}^N$ problem ($N = 2, 3$). This is in sharp contrast to the original LSM which requires to solve n^N far-field equations.

However, the MLSM is shown to suffer some "breakage cells" problem in the present paper. That is, some cells on the sampling mesh which lie on the boundary of the scatterer would be trimmed down undesirably in the MLSM processing. This happens particularly at the boundary of a scatterer where the curvature is very large and at the later stage of the MLSM. The major goal of this section is to discuss two enhanced techniques to avoid the "breakage cells". Specifically, we discuss the "searching" and "expanding" strategies which are shown to effectively defeat the undesirable "breakage cells" problem, even in some extreme situations. On the other hand, the enhanced MLSMs (EMLSMs) are shown to possess the same optimal computational complexity as the original MLSM. Hence, the discussed methods significantly improve the robustness of the MLSM, and meanwhile they bring no extra computational complexity.

We present our EMLSMs based on a model inverse scattering problem of reconstructing the shapes of acoustic obstacles. However, we would like to emphasize that since our discussed techniques are mainly some computational strategies, they can be straightforwardly extended to the LSM for reconstructing the supports of inhomogeneous mediums, and for inverse electromagnetic scattering problems as well.

The rest of the section is organized as follows. In Sect. 8.2.1, we introduce the inverse acoustic obstacle scattering problem that we shall take as the model problem for our subsequent study. Section 8.2.2 is devoted to a brief review of the LSM and MLSM. In Sect. 8.2.3, we present the EMLSMs, together with the computational complexity analysis. Section 8.2.4 contains the numerical results, which illustrate the effectiveness of the discussed EMLSMs.

8.2.1 Inverse Acoustic Obstacle Scattering Problem

In this section, we briefly introduce the time harmonic inverse acoustic obstacle scattering problem that we shall use as a model problem for our subsequent study on the EMLSMs.

Define the operator $\mathscr{F}$, which maps the boundary of the obstacle to the corresponding far-field pattern. Then the inverse problem can be expressed as the following operator equation,

$$\mathscr{F}(\partial \mathrm{D}) = u_\infty(\hat{x}, d), \tag{8.2.1}$$

where $\hat{x},\ d \in \mathbb{S}^{N-1}$. It is easily seen that $\mathscr{F}$ is nonlinear, and moreover, the operator equation (8.2.1) is widely known to be severely ill-posed in the sense of Hadamard (see, e.g., [13]).

8.2.2 *Review of LSM and MLSM*

In this subsection, we will give a brief account of the LSM and MLSM for the inverse obstacle scattering problem (8.2.1).

8.2.2.1 LSM

First, we introduce the far-field operator $F : L^2(\mathbb{S}^{N-1}) \to L^2(\mathbb{S}^{N-1})$ defined by

$$(Fg)\left(\hat{x}\right) := \int_{\mathbb{S}^{N-1}} u_\infty\left(\hat{x},\, d\right) g(d) ds(d) \quad \text{for } \hat{x} \in \mathbb{S}^{N-1},$$

and a function Φ_∞ induced by the fundamental solution,

$$\Phi_\infty\left(\hat{x},\, z\right) = \gamma \exp\left\{-ik\hat{x} \cdot z\right\},$$

with $\gamma = 1/4\pi$ in $\mathbb{R}^3$ and $\gamma = e^{i\pi/4}/\sqrt{8\pi k}$ in $\mathbb{R}^2$. The LSM is to find the solution g to the following far-field equation

$$(Fg)\left(\hat{x}\right) = \Phi_\infty\left(\hat{x},\, z\right), \quad \hat{x} \in \mathbb{S}^{n-1},\ z \in \mathbb{R}^n, \tag{8.2.2}$$

and use the L^2-norm of the solution as an indicator function. As the far-field operator F has a smooth kernel, it is compact in $L^2(\mathbb{S}^{n-1})$, and (8.2.2) does not have a solution in general. But under appropriate assumptions, one can use a density argument to solve the equation approximately to find an approximate function $\tilde{g}$ whose behavior can be used to characterize the boundary of the underlying obstacle. The assumption is that k^2 is not a Laplacian eigenvalue for D, namely, k^2 is not a Dirichlet eigenvalue for $-\Delta$ in $H_0^1(\mathrm{D})$ when D is sound-soft and is not a Neumann eigenvalue for $-\Delta$ in $H^1(\mathrm{D})$ when D is sound-hard. The following theorem forms the basis of the LSM (see [13]).

Theorem 8.2.1 *Assume that k^2 is not a Dirichlet eigenvalue for $-\Delta$ in $H_0^1(\mathrm{D})$ when* D *is sound-soft and is not a Neumann eigenvalue for $-\Delta$ in $H^1(\mathrm{D})$ when* D *is sound-hard. Then the following holds:*

1. For $z \in \mathrm{D}$ and a fixed $\varepsilon > 0$ there exists a $g_\varepsilon^z \in L^2(\mathbb{S}^{N-1})$ such that

$$\|Fg_\varepsilon^z - \Phi_\infty(\cdot,\, z)\|_{L^2(\mathbb{S}^{N-1})} < \varepsilon,$$

and

$$\lim_{z \to \partial \mathrm{D}} \|g_\varepsilon^z\|_{L^2(\mathbb{S}^{N-1})} = \infty.$$

2. For $z \in \mathbb{R}^N \backslash \overline{\mathrm{D}}$ and any given $\varepsilon > 0$, every $g_\varepsilon^z \in L^2(\mathbb{S}^{N-1})$ satisfying

$$\|Fg_\varepsilon^z - \Phi_\infty(\cdot, z)\|_{L^2(\mathbb{S}^{N-1})} < \varepsilon,$$

ensures

$$\lim_{\varepsilon \to 0} \|g_\varepsilon^z\|_{L^2(\mathbb{S}^{N-1})} = \infty.$$

Here, we would like to remark again that the blow-up behavior of g_ε^z in Theorem 8.2.1 would occur disregarding the physical properties of the underlying obstacle. The LSM turns elegantly the reconstruction of the shape of the obstacle D into the process of numerically determining the indicator function g_ε^z in Theorem 8.2.1. The general procedure can be stated as Algorithm 8.1.

Algorithm 8.1 LSM

1. Select a mesh $\mathcal{T}_h$ of sampling points in a region Ω which contains D.
2. Use the Tikhonov regularization and the Morozov discrepancy principle to compute an approximate solution g^z to the far-field equation (8.2.2) for each mesh point z in $\mathcal{T}_h$.
3. Select a cut-off value c; then count $z \in$ D if $\log \|g^z\|_{L^2(\mathbb{S}^{N-1})} \leq c$ and $z \notin$ D if $\log \|g^z\|_{L^2(\mathbb{S}^{N-1})} > c$.

The LSM works to reconstruct the shape of the obstacle, provided one could solve (8.2.2) approximately for a "valid" indicator function g^z; we refer to [12, 25], for numerous numerical illustrations. It is emphasized that the LSM remains valid for noisy far-field data. In view of practical applications, the far-field data is measured and thus the noise is inevitable. Hence it is of crucial importance that the LSM should be robust with respect to noise, and this is guaranteed by intensive theoretical and numerical analysis in existing work mentioned earlier.

8.2.2.2 MLSM

In this subsection, the MLSM will be presented together with some theoretical results, and we refer interested readers to [25] for more details. Following [25], we would carry out our discussion in $\mathbb{R}^2$, but all of the subsequent results could be straightforwardly extended to the three-dimensional case.

First, we recall the intuition of the MLSM. Let $D \subset \mathbb{R}^2$ be a bounded domain as shown in Fig. 8.16a and suppose that we are going to use an $n \times n$ mesh for the LSM with $(n-1)^2$ cells of equal size. In order to get some satisfactory reconstruction for the profile of D, the mesh must be moderately fine in some sense. Hence, when implementing LSM, considerable computational costs would be spent in finding the indicator functions in those "remote" cells which are far away from the scatterer D, or in those "inner" cells which lie deeply inside D; e.g., see the red and blue colored regions in Fig. 8.16b. Therefore, it would be advantageous if one could get rid of the remote and inner cells in our computations. It was proposed in [25] that this can be realized by doing the labeling and removing technique in a multilevel manner (See Fig. 8.16c).

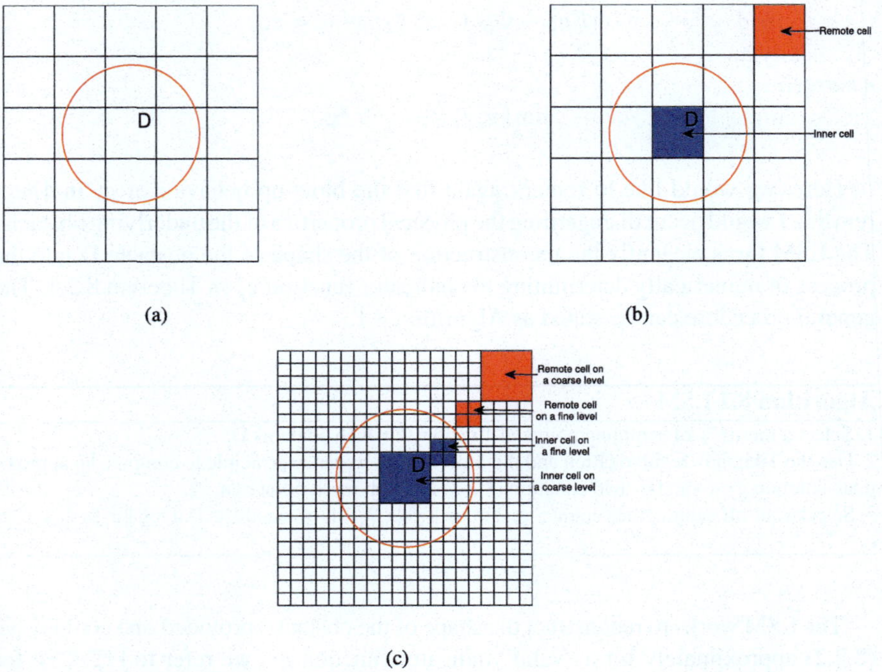

Fig. 8.16 Remote cells and inner cells

We are in a position to present the MLSM. First, we assume Ω is a square in $\mathbb{R}^2$. Let $\{\mathscr{T}_k\}_{k=1}^{L}$ be a nested sequence of meshes on the sampling domain Ω such that $\mathscr{T}_{k+1}$ is a refinement of $\mathscr{T}_k$ for $k = 1, \ldots, L-1$. Throughout, we assume that $\mathscr{T}_{k+1}$ is an $n_{k+1} \times n_{k+1}$ mesh while $\mathscr{T}_k$ is an $n_k \times n_k$ mesh, where $n_{k+1} = 2n_k - 1$ for $k = 1, \ldots, L-1$. That is, we refine the mesh $\mathscr{T}_k$ by equally subdividing every subsquare in $\mathscr{T}_k$ into four subsquares of $\mathscr{T}_{k+1}$. Then if the mesh length of $\mathscr{T}_k$ is h_k for $k = 1, \ldots, L-1$, then $h_{k+1} = h_k/2$. The MLSM is formulated as Algorithm 8.2.

Algorithm 8.2 MLSM

1. Set $k = 0$ and choose an initial mesh for the sampling region Ω.
2. Apply the LSM scheme on the kth-level mesh to investigate those mesh points which have not been examined previously.
3. For a given cut-off value c, independent of the level k, classify and label the kth-level subsquares (cells) into three sets, namely, *remote cells*, *boundary cells*, and *inner cells* (based on the cut-off value principle in the LSM). A cell is labeled as *remote* if the norms of the indicator functions at the vertices of the cell are larger than c, while a cell is labeled as *inner* if the norms of the indicator functions at the vertices of the cell are all less than or equal to c, and other remaining cells will be labeled as *boundary cells*. Then remove the remote and inner cells.
4. Refine the remaining sampling mesh.
5. Set $k = k + 1$, and if $k \leq L$, go to Step 2.

At the end of this subsection, we present the related results on the computational complexity analysis of MLSM. To that end, we denote by Γ a C^2-smooth curve in $\mathbb{R}^2$ which forms the boundary of a bounded domain G. For any $h > 0$, we define two curves parallel to Γ as follows

$$\begin{aligned}\Gamma_h^+ :&= \{x + h\nu(x),\ x \in \Gamma,\ \text{and } \nu(x) \text{ is the unit normal to } \Gamma \text{ at } x\\ &\quad \text{directed to the exterior of G}\},\\ \Gamma_h^- :&= \{x - h\nu(x),\ x \in \Gamma,\ \text{and } \nu(x) \text{ is the unit normal to } \Gamma \text{ at } x\\ &\quad \text{directed to the exterior of G}\}.\end{aligned}$$

Then we have (cf. [25]),

Lemma 8.2.1 *There exist constants $h_0 > 0$ and $\alpha_0 > 0$ such that* dist $(\Gamma, \Gamma^{\pm}_{\alpha_0 h}) \geq \sqrt{2}h$ *whenever* $0 < h < h_0$.

With Lemma 8.2.1, the following Theorem could be readily proved by following the proof of Theorem 2.6 in [25],

Theorem 8.2.2 *Consider an L-level MLSM algorithm with a nested sequence of sampling mesh $\{\mathcal{T}_k\}_{k=1}^L$. Suppose that for each k, $\mathcal{T}_k$ is of size $n_k \times n_k$ with mesh length h_k such that $0 < h_1 < h_0$, where h_0 is given in Lemma 8.2.1 corresponding to ∂D. Then, by using the MLSM to reconstruct ∂D, the far-field equation is solved $\mathcal{O}(n_L)$ times in total.*

8.2.3 Enhanced Multilevel-LSMs

The MLSM is remarkable in that it reduces the computational complexity from solving $\mathcal{O}(n^N)$ far-field equations to only solving $\mathcal{O}(n^{N-1})$ far-field equations for an N-dimensional inverse scattering problem. This is certainly of great significance in the implementation of the LSM. However, it is occasionally observed that the MLSM might fail in some practical situations. In this section, we first give a detailed description about when the MLSM fails and then present the EMLSMs to improve the performance.

First, we introduce the definition of *breakage cells*.

Definition 8.2.1 *Breakage cells* are the cells that one removes during the process of MLSM, but they are actually on the boundary of the obstacles.

The *breakage cells* often occur when the curvature of the obstacles are large or the background grid is not fine enough, which are related to the intrinsic nature of MLSM. As is known that the LSM and MLSM are "point sampling", and this means that we approximate the boundary of scatterers by values of indicator function only on nodes of sampling mesh. If the value of the indicator function on boundary nodes of a cell

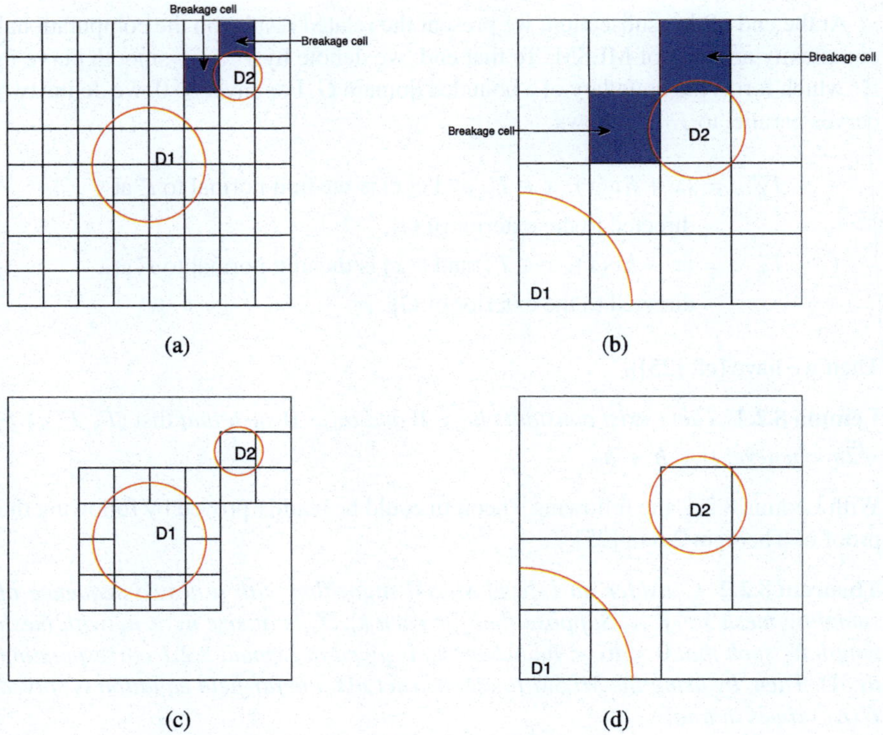

Fig. 8.17 Illustration of a failed case by the MLSM

could not correctly indicate the properties of the whole cell, then the MLSM process would make wrong decisions on the labeling process, and the reconstructed image breaks up. It is numerically observed that such breakage cell phenomenon happens in particular on the fine mesh levels at the later stage of the MSLM algorithm.

For illustration of the breakage problem, we give an example in the sequel. Assume the scatterer are composed of two disk-shaped obstacles and we call the larger disk D_1 and the smaller disk D_2. The computational region has been divided into a relatively fine grid (see Fig. 8.17a). Then by a careful inspection of the top-right circle in our example, we note that there are two "disputed" cells on the boundary of D_2 (see blue marked cell in Fig. 8.17b). On the one hand, these cells are definitely on the boundary of the scatterer at a first glance, so we should keep these cells for finer level reconstruction in MLSM. On the other hand, the four boundary nodes of both the two square cells are all outside the scatterer, and hence the values of the indicator function on these nodes are all larger than the cut-off value, so the two cells would be removed undesirably during the MLSM process and they are *breakage cells* (see Fig. 8.17 c and d).

To avoid *breakage cells*, we discuss two solutions. The first idea is to avoid the emergence of *breakage cells*. This could be realized by adding "sampling

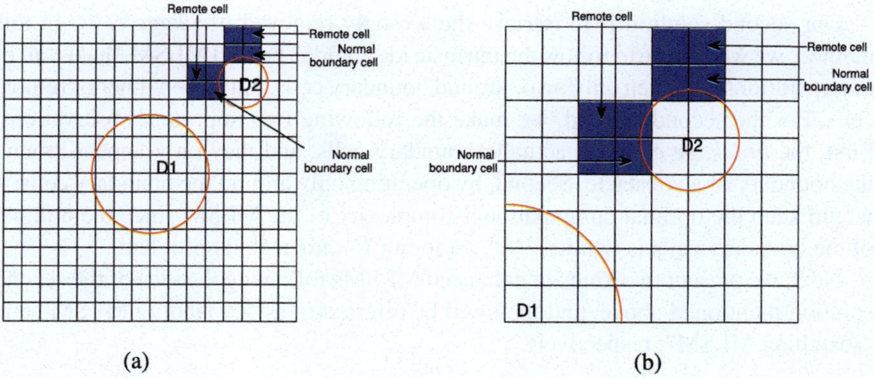

Fig. 8.18 A finer initial grid

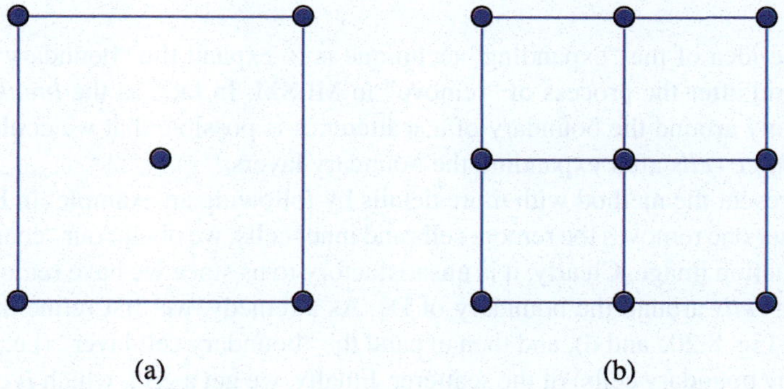

Fig. 8.19 The use of more sampling points in a cell

points". For example, we could set the initial grid relatively finer. As one can see in Fig. 8.18 a and b, the *breakage cells* in our example would change into four finer cells, all of which are not *breakage cells* anymore. Alternatively, we could also increase sampling points in a particular cell. For example, we could compute the values of the indicator function on five points or nine points in a cell (see Fig. 8.19a and b). With the information from more points, we could naturally approximate the boundary more precisely.

However, we have to emphasize that, "adding sampling points" might not be an optimal solution in curing the *breakage cells*. Since the MLSM is discussed for computing values of indicator function on fewer nodes, this kind of "adding sampling points" violates its "spirit". In fact, when adding sampling points, we increase the computational complexity on the whole sampling region. At the same time, it is also a problem on where one should place the "added points". Since the shapes of the scatterers differ sharply in different cases, without sufficient *a prior* information, we may introduce useless points. A simple example for this is that the "five sampling solution" in Fig. 8.19, which apparently doesn't work in our failed case in Fig. 8.17.

Our second solution is to retrieve the wrongly removed *breakage cells*. In this method, we would like to follow the intrinsic idea hidden behind MLSM, that is, all of our operations are taken only on or around boundary cells, but not on inner or remote cells. For the second method, we make the following two important observations. First, the *breakage cells* are actually boundary cells, and they only appear around the boundary of an obstacle. Second, by operating only around the boundary cells, it would keep the optimal computational complexity of the MLSM since the number of the boundary cells is within $\mathscr{O}(n^{N-1})$ for an N-dimensional problem.

Next, we present two kinds of enhanced MLSMs following the idea of the second solution mentioned above, and they will be referred to as "expanding MLSM" and "searching MLSM", respectively.

8.2.3.1 Expanding MLSM

The basic idea of the "expanding" technique is to expand the "boundary cells" on a finer grid after the process of "remove" in MLSM. In fact, as the *breakage cells* appear only around the boundary of a scatterer, it is possible that we could retrieve the *breakage cells* after expanding the boundary layers.

We present the method with more details by following an example. In Fig. 8.20a and b, after one removes the remote cells and inner cells, we obtain our "coarse-level" reconstruction image. Clearly, it is unsatisfactory to us since we have removed some *breakage cells* around the boundary of D_2. As a remedy, we first refine the current grid (see Fig. 8.20c and d), and then expand the "boundary cell layer" (i.e., the layer formed by boundary cells) of the scatterer. Finally, we get a grid, which is composed of two kinds of cells, namely the cells coming from the refining process, and the other cells coming from the expanding process. The resultant images are contained in Fig. 8.20e and f, where we have marked the "expanded layer", which is composed of "inner expanded layer" and "outer expanded layer", in blue color. It is easily seen that, the *breakage cells* on D_2 are retrieved with the "outer expanded layer" on a finer grid.

We formulate the expanding MLSM as Algorithm 8.3. The notations in our algorithm are kept consistent with Algorithm 8.2.

Algorithm 8.3 Expanding MLSM

1. Set $k = 0$ and choose an initial mesh for the sampling region Ω.
2. Apply the LSM scheme on the kth-level mesh to investigate those mesh points which have not been examined previously.
3. For a given cut-off value c, independent of the level k, classify and label the kth-level subsquares (cells) into three sets, namely, remote cells, boundary cells, and inner cells (based on the cut-off value principle in the LSM). Then remove the remote and inner cells.
4. Refine the remaining sampling mesh.
5. Implement the "expanding process". Expand the sampling mesh in both interior and outer directions along the boundary layer.
6. Set $k = k + 1$, and if $k \leq L$, go to Step 4.

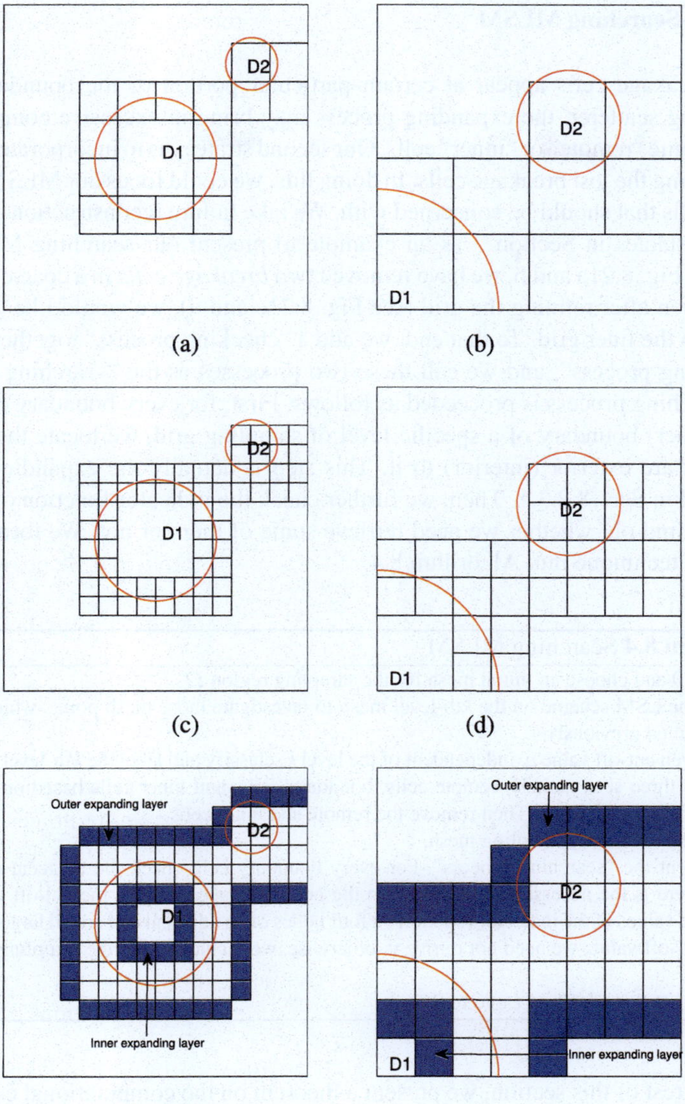

Fig. 8.20 Illustration of the expanding MLSM algorithm

8.2.3.2 Searching MLSM

Since breakage cells appear at certain particular portion of the boundary of the underlying scatterer, the expanding process may bring in excessive computational cost on some "remote" or "inner" cells. Our second strategy is to incorporate a process of retrieving the lost breakage cells. In doing this, we could focus our MLSM process on the cells that should be concerned with. We take failure reconstruction of the two circle obstacles in Section 3 as an example to present our searching MLSM. As shown in Fig. 8.21a and b, we have removed two *breakage cells* in a coarse sampling mesh. Then after refining the grid (see Fig. 8.21c and d), we would like to retrieve them into the finer grid. To that end, we add a "checking process" together with the "expanding process", and we call these two processes as the "searching process". The searching process is proceeded as follows. First, for every boundary cell on the outer (inner) boundary of a specific level of sampling grid, we locate the adjacent cells that are exterior (interior) to it. This step is actually the expanding process discussed in Sect. 8.2.3.1. Then, we further check the indicator function on the four nodes to find out whether we need retrieve some of them or not. We formulate the proposed technique into Algorithm 8.4.

Algorithm 8.4 Searching MLSM

1. Set $k = 0$ and choose an initial mesh for the sampling region Ω.
2. Apply the LSM scheme on the kth-level mesh to investigate those mesh points which have not been examined previously.
3. For a given cut-off value c, independent of the level k, classify and label the kth-level subsquares (cells) into three sets, namely, remote cells, boundary cells, and inner cells based on the cut-off value principle in the LSM. Then remove the remote and inner cells.
4. Refine the remaining sampling mesh.
5. Implement the "searching process". For every boundary cell, check the adjacent cells to see whether there is the necessity to retrieve into the next level according to the cut-off principle of LSM: if the values of the indicator function on four nodes of an adjacent cell are all larger or smaller than the cut-off value, we need not retrive it, otherwise, we include it into the sampling grid of the next level.
5. Set $k = k + 1$, and if $k \le L$, go to Step 4.

In the rest of this section, we present a theorem on the computational complexity of the discussed EMLSMs. As we have mentioned earlier, the enhanced operations in EMLSMs are performed only on the cells around the boundary of the underlying scatterer, and therefore one can readily see that the computational complexity of EMLSMs is the same as that of the LSM, namely $\mathscr{O}(n^{N-1})$ for an N-dimensional problem. In conclusion, we have

Theorem 8.2.3 *Consider an L-level EMLSM algorithm with a nested sequence of sampling mesh* $\{\mathscr{T}_k\}_{k=1}^{L}$. *Suppose that for each k,* $\mathscr{T}_k$ *is of size* $n_k \times n_k$ *with mesh length* h_k *such that* $0 < h_1 < h_0$, *where* h_0 *is given in Lemma 8.2.1 corresponding to* ∂D. *Then, the EMLSM is of the same computational complexity as the MLSM; that*

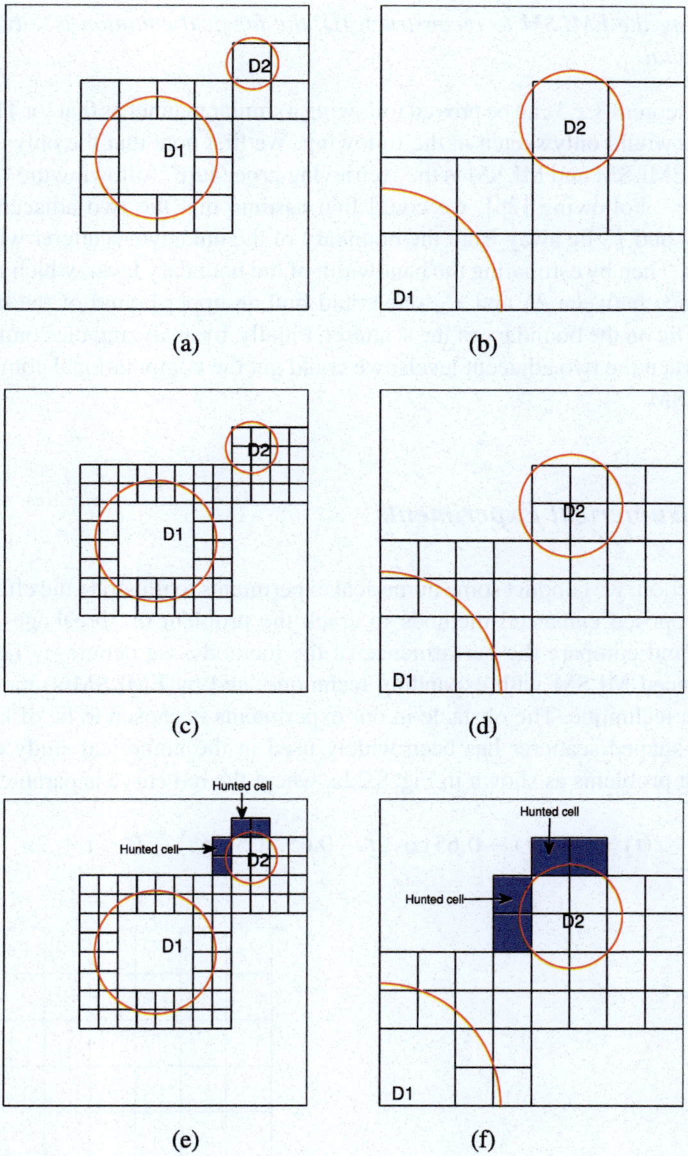

Fig. 8.21 Illustration of the searching MLSM algorithm

is, by using the EMLSM to reconstruct ∂D, *the far-field equation is solved* $\mathscr{O}(n_L)$ *times in total.*

Proof Theorem 8.2.3 can be proved following a similar manner as that for Theorem 3, which we would only sketch in the following. We first note that the only difference between EMLSM and MLSM is the "retrieving procedure" following the "removing procedure". Following [25], we could first assume that the two adjacent parallel curves Γ_1 and Γ_2 lie away from the boundary of the unknown scatterer with certain distances. Then by estimating the bandwidth of the boundary layer, which is actually the distance between Γ_1 and Γ_2, one could find an upper bound of the number of cells that lie on the boundary of the scatterer. Finally, by analyzing the computational cost between the two adjacent levels, we could get the computational complexity of the EMLSM. □

8.2.4 Numerical Experiments

In this section, we conduct some numerical experiments to illustrate the effectiveness of the proposed enhanced methods to crack the problem of "breakage cells". To evaluate and compare the performance of the methods, we denote by EMLSM(e) the enhanced MLSM with expanding technique, and by EMLSM(s) the one with searching technique. The obstacle in our experiments is chosen to be of kite shape. The kite-shaped scatterer has been widely used in the numerical study of inverse scattering problems as shown in Fig. 8.22a, where the red curve is parametrized by

$$x(t) = (\cos(t) + 0.65\cos 2t - 0.65,\ 1.5\sin t),\quad 0 \le t \le 2\pi.$$

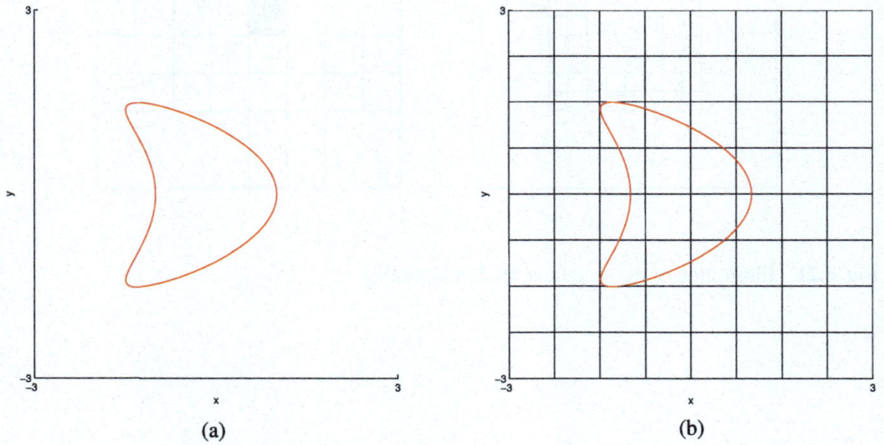

Fig. 8.22 One kite-shaped scatterer **a** and the initial mesh **b** in Example SK

Two groups of numerical tests will be given and they will be referred to as Single Kite (SK) and Double Kites (DK), respectively. The exact far-field data of the direct problems are synthesized by solving the combined-layer potential operator equation with Nyström method (see [13]). One may refer to an adaptive finite element method using the DtN map [18] for the truncation of the unbounded domain. We compute the far-field patterns at 64 equidistantly distributed observation points $\big(\cos t_j,\ \sin t_j\big)$, $t_j = 2j\pi/64$, $j = 0, 1, \ldots, 63$, corresponding to 64 equidistantly distributed incident directions $\big(\cos \tau_j,\ \sin \tau_j\big)$, $\tau_j = 2j\pi/64$, $j = 0, 1, \ldots, 63$, around the unit circle. Then the data corrupted by a uniform random noise of 5% noise level are used as our measurement data. For each mesh point z, the corresponding far-field equation (8.2.2) is solved by using the Tikhonov regularization method (cf. [13]), with the regularization parameter determined by the L-curve method.

Example SK

In this test, the scatterer consists of a single kite-shaped component as shown in Fig. 8.22a. The initial mesh is chosen to be fixed for all schemes and coarse enough, but it also ensures that there are some grid points inside the scatterer, see Fig. 8.22b.

The far-field equation (8.2.2) is solved with resort to the Tikhonov regularization approach and applying the MLSM, EMLSM(e) and EMLSM(s), respectively. Without the multilevel idea, the finest mesh (129×129 grid points) will lead to over ten thousands solutions of the integral equation, which is formidable compared with a few hundreds of solutions by multilevel LSM of all three versions.

First, we apply a 5-level MLSM to obtain a reconstructed image. One could see from Fig. 8.23 that after a removing process in the third level, breakage cells indeed occur there. Hence, the final reconstructed image in the bottom right plot of Fig. 8.23 is degenerated into two broken chains of small cells.

Next, we implement the EMLSM(e) and EMLSM(s) to the same example. Once the expanding or searching procedure is triggered, some additional cells are appended and marked in blue on each level of the sampling mesh. We can see that both EMLSMs successfully avoid the breakage cells and produce very satisfactory reconstructions.

It can be pointed out that from Fig. 8.24 that two layers of refined cells are appended at each level to guarantee no breakage cells. Moreover, it is emphasized from Fig. 8.25 that only four additional cells are appended on the two rear wings of large curvature after the first refinement step, there are no more cells are introduced to keep the continuity of the boundary. This interesting observation suggests that the number of appended cells by the EMLSM(s) is much less than that by the EMLSM(e), which holds true for multiple scatterers.

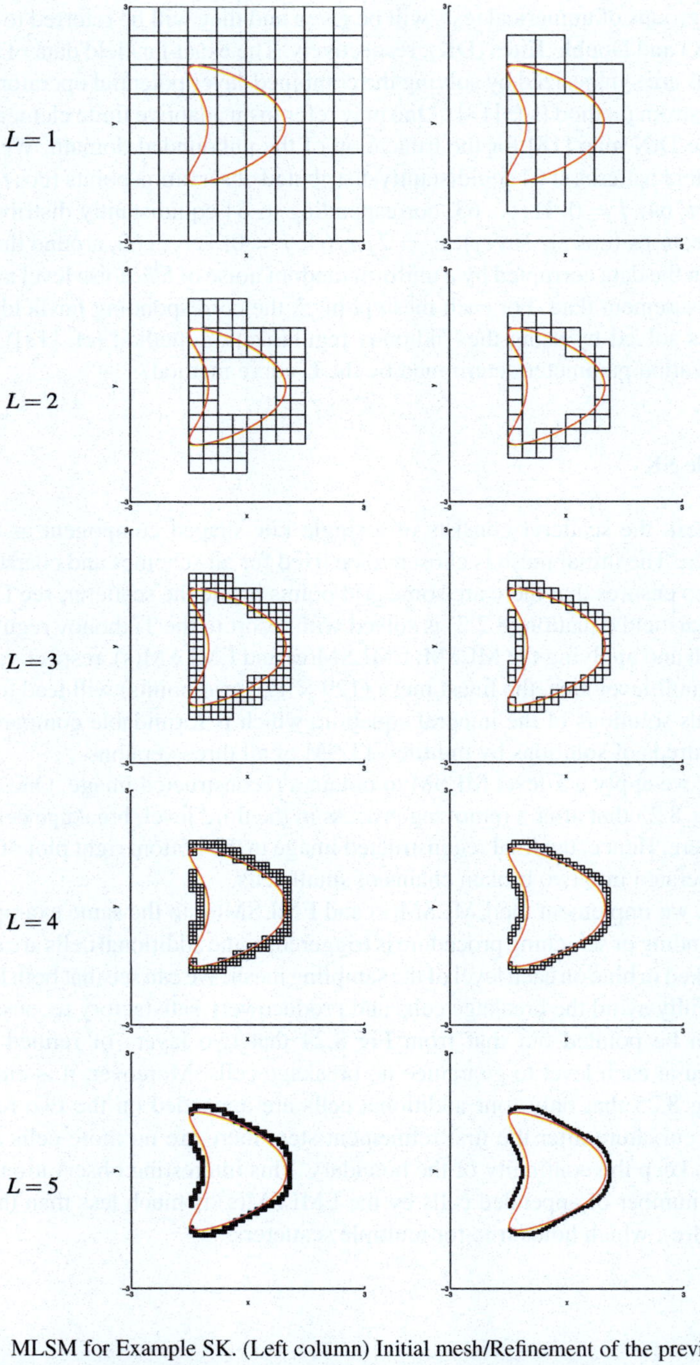

Fig. 8.23 MLSM for Example SK. (Left column) Initial mesh/Refinement of the previous level. (Right column) Reconstructed image after removing process on the current level

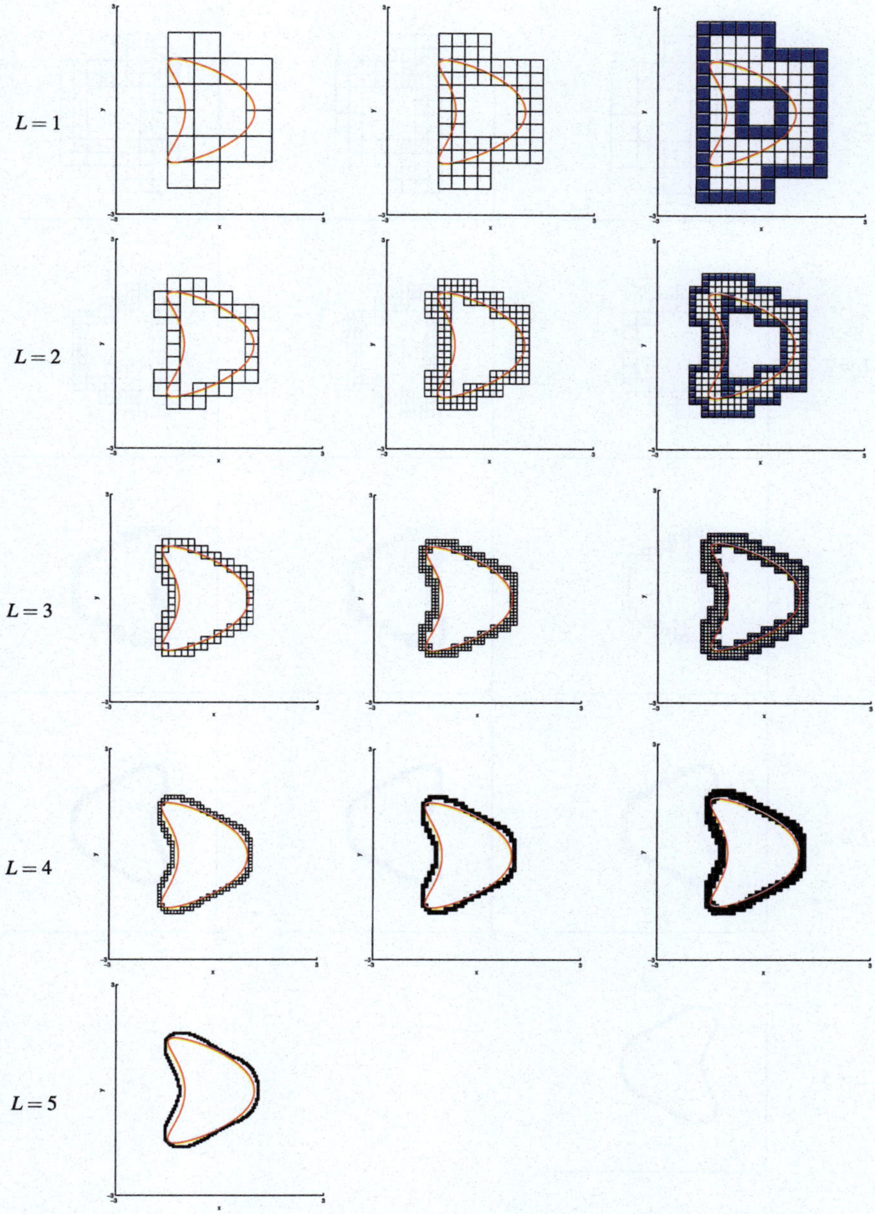

Fig. 8.24 EMLSM(e) for Example SK. (Left column) Reconstructed image after removing process of the initial/previous level. (Middle column) Refinement process. (Right column) Appended cells (in blue) after "expanding process"

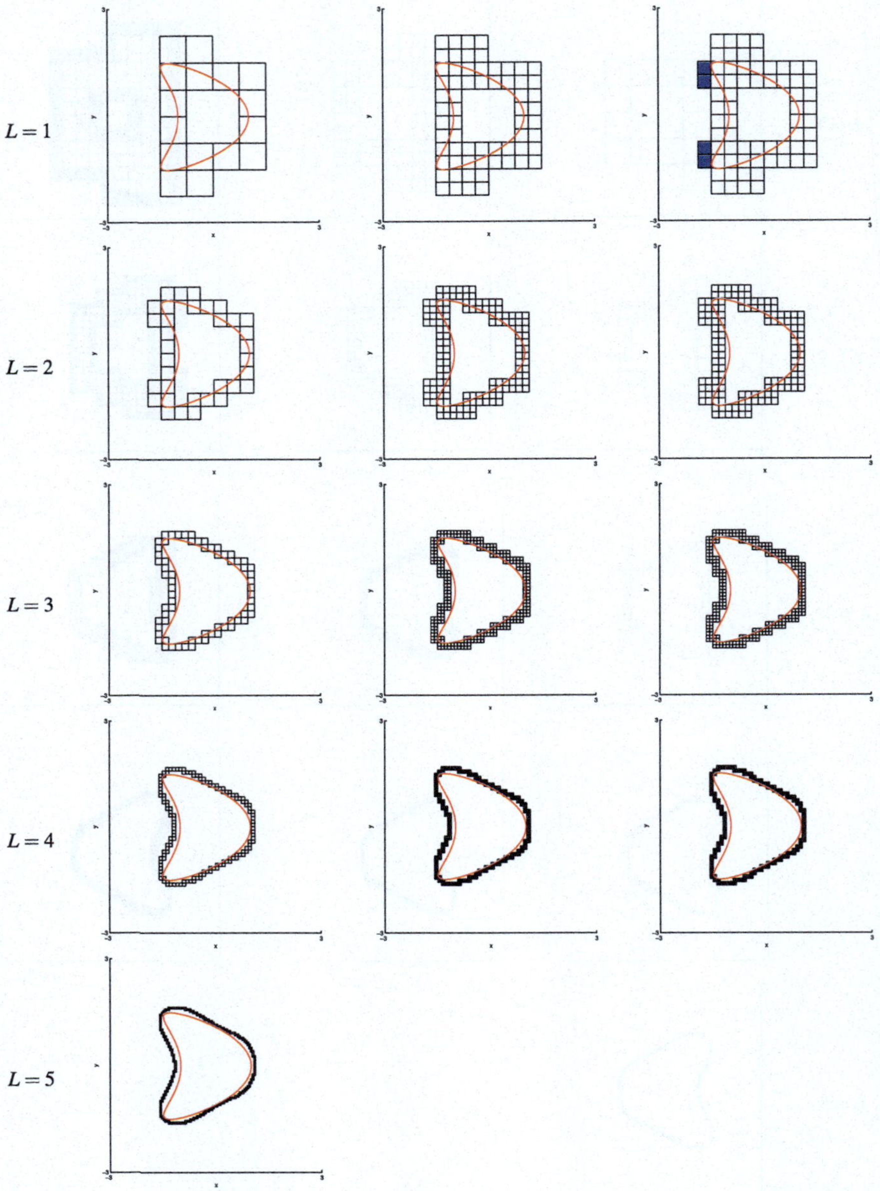

Fig. 8.25 EMLSM(s) for Example SK. (Left column) Reconstructed image after removing process of the initial/previous level. (Middle column) Refinement process. (Right column) Appended cells (in blue) after "expanding process"

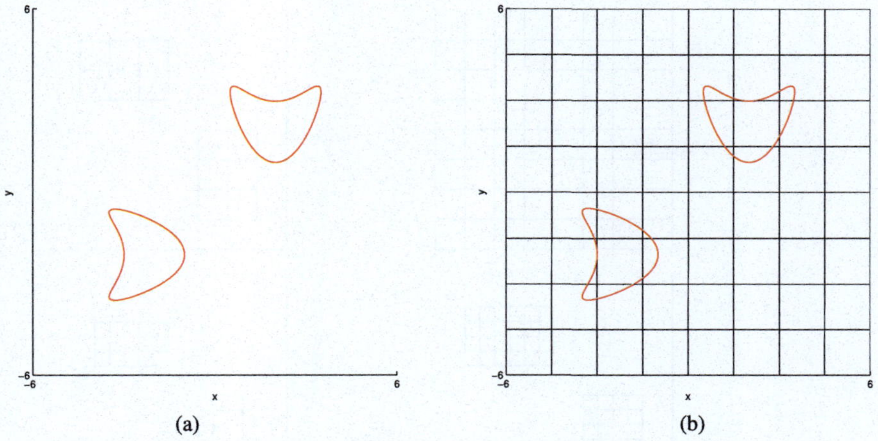

Fig. 8.26 Double kite-shaped scatterers **a** and the initial mesh **b** in Example DK

Example DK

In this test, we try a scatterer with multiple component of kite-shaped obstacles; One of them is the original kite displaced at $(-2,\ -2)$, and the other is displaced at $(2,\ 2)$ after a clockwise rotation by $\pi/2$ radian as shown in Fig. 8.26a. The sampling domain is $(-6,\ 6) \times (-6,\ 6)$ with an initial grid of 8×8 square cells, see Fig. 8.26b.

We perform the comparison between MLSM, EMLSM(e) and EMLSM(s), respectively, as those for the Example SK. As seen from Fig. 8.27, the "breakage cell" phenomenon occurs first on the third level of the removing step, and then happens again on the four level, which ends up with four detached chains of cells on the final level. The final reconstructed results of the EMLSM(e) and EMLSM(s) for Example DK are shown in the bottom left plots of Figs. 8.28 and 8.29, respectively. Both EMLSM(e) and EMLSM(s) keep the continuity of the boundary chain of cells and yield pretty good approximation of the exact boundary. The significant reduction of the appended cells in blue by the EMLSM(s) in Fig. 8.29 is observed again, compared with those appended ones in Fig. 8.28.

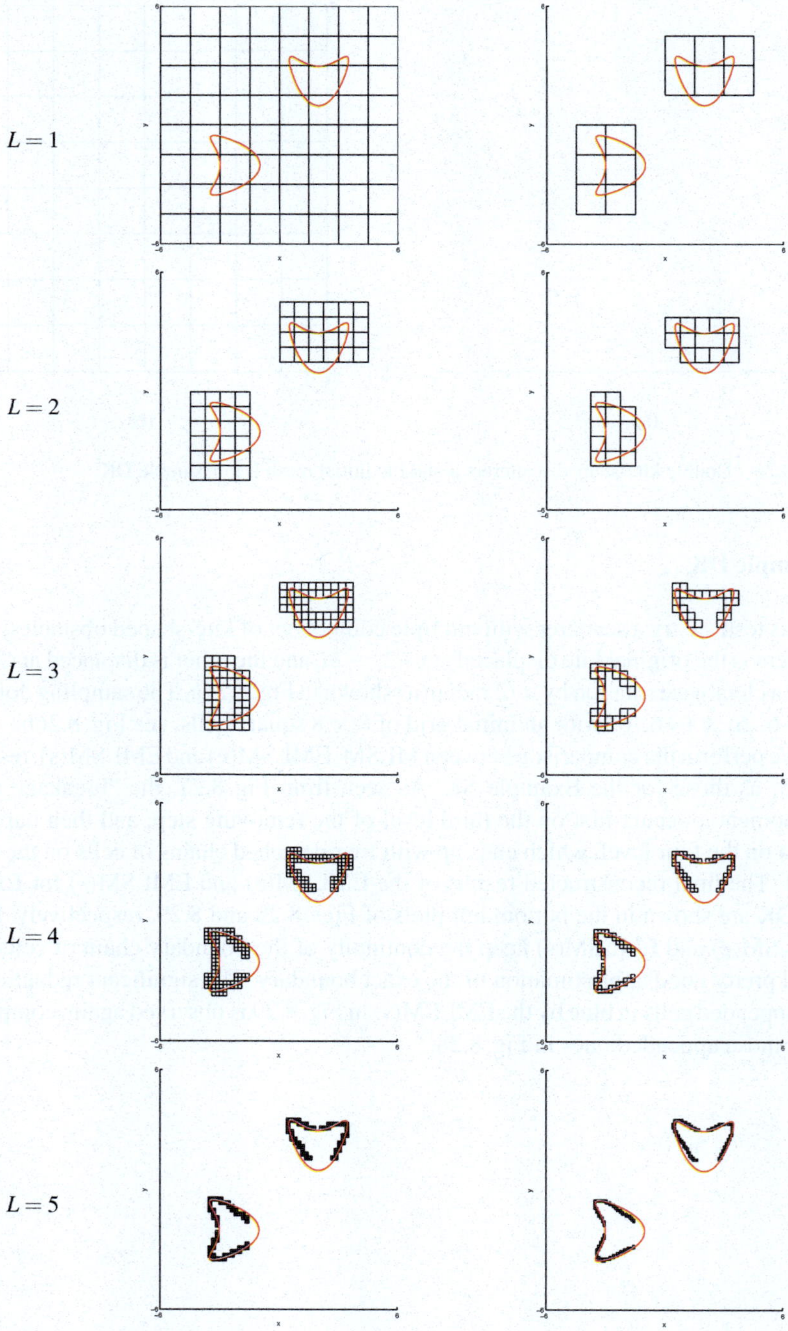

Fig. 8.27 MLSM for Example DK. (Left column) Initial mesh/Refinement of the previous level. (Right column) Reconstructed image after removing process on the current level

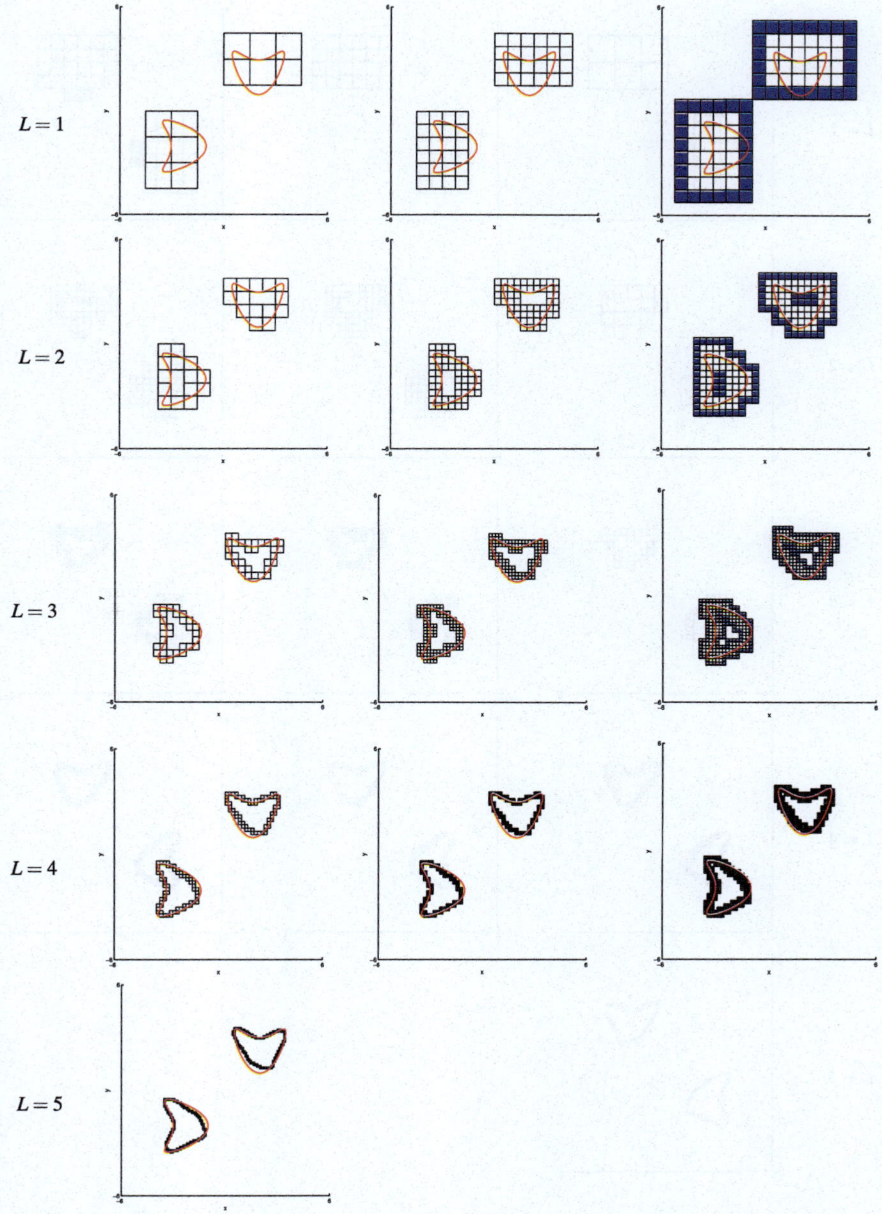

Fig. 8.28 EMLSM(e) for Example DK. (Left column) Reconstructed image after removing process of the initial/previous level. (Middle column) Refinement process. (Right column) Appended cells (in blue) after "expanding process"

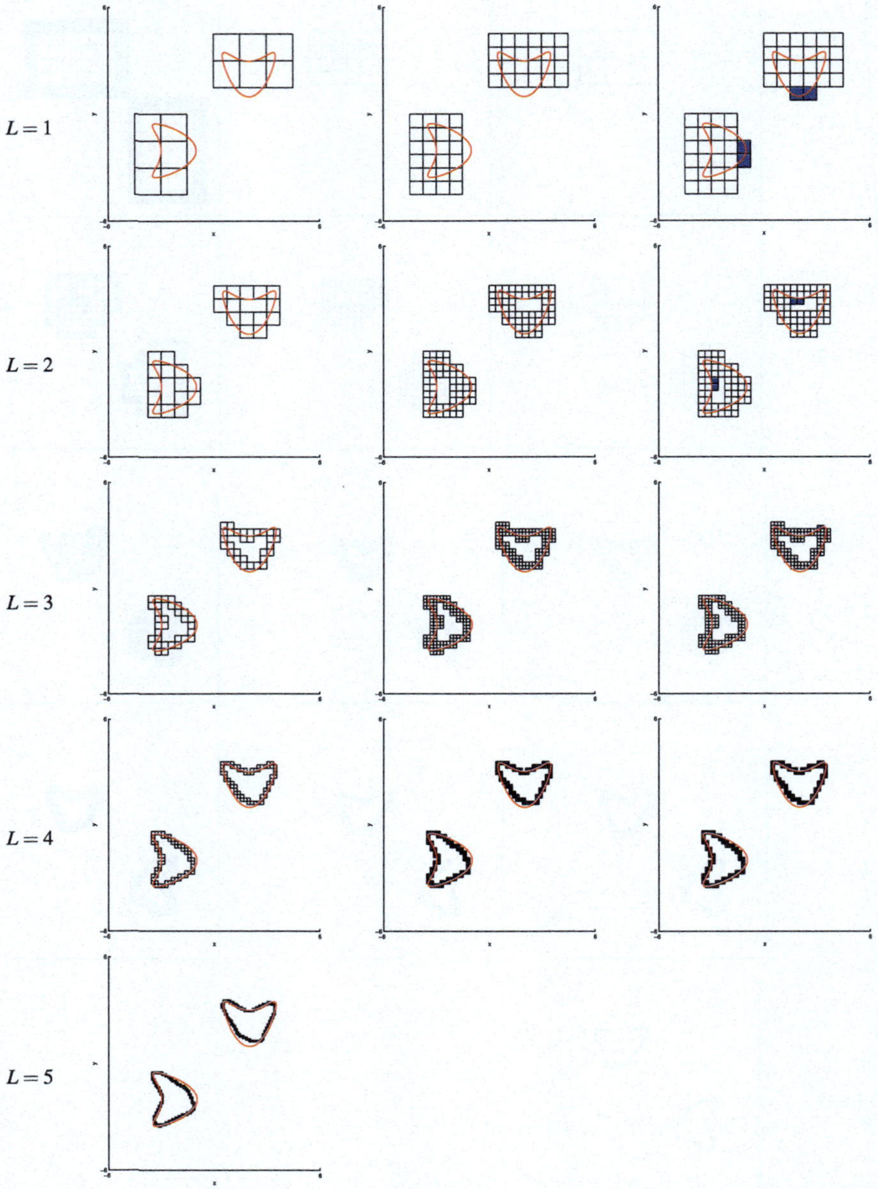

Fig. 8.29 EMLSM(s) for Example DK. (Left column) Reconstructed image after removing process of the initial/previous level. (Middle column) Refinement process. (Right column) Appended cells (in blue) after "expanding process"

References

1. A. Abubakar, P.M. van den Berg, Iterative forward and inverse algorithms based on domain integral equations for three-dimensional electric and magnetic objects. J. Comput. Phys. **195**, 236–262 (2004)
2. H. Ammari, *An Introduction to Mathematics of Emerging Biomedical Imaging* (Springer, Berlin, 2008)
3. H. Ammari, E. Iakovleva, D. Lesselier, A MUSIC algorithm for locating small inclusions buried in a half-space from the scattering amplitude at a fixed frequency. SIAM Multiscale Model. Simul. **3**, 597–628 (2005)
4. R. Aramini, M. Brignone, J. Coyle, M. Piana, Postprocessing of the linear sampling method by means of deformable models. SIAM J. Sci. Comput. **30**, 2613–2634 (2008)
5. R. Aramini, M. Brignone, M. Piana, The linear sampling method without sampling. Inverse Prob. **22**(6), 2237–2254 (2006)
6. T. Arens, Why linear sampling works. Inverse Prob. **20**, 163–173 (2004)
7. H. Barucq, C. Bekkey, R. Djellouli, Full aperture reconstruction of the acoustic far-field pattern from few measurements. Commun. Comput. Phys. **11**, 647–659 (2012)
8. F. Cakoni, D. Colton, *Qualitative Methods in Inverse Scattering Theory* (Springer, Berlin, 2006)
9. M. Cheney, The linear sampling method and the MUSIC algorithm. Inverse Prob. **17**, 591–595 (2001)
10. D. Colton, J. Coyle, P. Monk, Recent developments in inverse acoustic scattering theory. SIAM Rev. **42**, 369–414 (2000)
11. D. Colton, K. Giebermann, P. Monk, A regularized sampling method for solving three-dimensional inverse scattering problems. SIAM J. Sci. Comput. **21**, 2316–2330 (2006)
12. D. Colton, A. Kirsch, A simple method for solving inverse scattering problems in the resonance region. Inverse Prob. **12**, 383–393 (1996)
13. D. Colton, R. Kress, *Inverse Acoustic and Electromagnetic Scattering Theory* (Cambridge University Press, Cambridge, 1998)
14. T. Hohage, Fast numerical solution of the electromagnetic medium scattering problem and applications to the inverse problem. J. Comput. Phys. **214**, 224–238 (2006)
15. M. Ikehata, Enclosing a polygonal cavity in a two-dimensional bounded domain from cauchy data. Inverse Prob. **15**, 1231–1241 (1999)
16. M. Ikehata, Reconstruction of obstacles from boundary measurements. Wave Motion **3**, 205–223 (1999)
17. V. Isakov, *Inverse Problems for Partial Differential Equations*, Applied Mathematical Sciences, vol. 127, 2nd edn. (Springer, New York, 2006)
18. X. Jiang, P. Li, W. Zheng, Numerical solution of acoustic scattering by an adaptive DtN finite element method. Commun. Comput. Phys. **13**(259), 1227–1244 (2013)
19. A. Kirsch, The MUSIC algorithm and the factorization method in inverse scattering theory for inhomogeneous media. Inverse Prob. **18**, 1025–1040 (2002)
20. A. Kirsch, N. Grinberg, *The Factorization Method for Inverse Problems* (Oxford University Press, Oxford, 2008)
21. R.E. Kleinman, P.M. van den Berg, A modified gradient method for two-dimensional problems in tomography. J. Comput. Appl. Math. **42**, 17–35 (1992)
22. J. Li, J. Zou, A direct sampling method for inverse scattering using far-field data. Inverse Prob. Imaging to appear (2013)
23. J. Li, H.Y. Liu, H. Sun, J. Zou, Imaging acoustic obstacles by singular and hypersingular point sources. Inverse Prob. Imaging **7**, 545–563 (2013)
24. J. Li, H.Y. Liu, J. Zou, Strengthened linear sampling method with a reference ball. SIAM J. Sci. Comput. **31**, 4013–4040 (2009)
25. J. Li, H.Y. Liu, J. Zou, Multilevel linear sampling method for inverse scattering problems. SIAM J. Sci. Comput. **30**, 1228–1250 (2008)

26. J. Li, H.Y. Liu, J. Zou, Multilevel linear sampling method for inverse scattering problems. SIAM J. Sci. Comput. **30**(3), 1228–1250 (2008)
27. Y. Olshansky, E. Turkel, Simultaneous scatterer shape estimation and partial aperture far-field pattern denoising. Commun. Comput. Phys. **11**, 271–284 (2012)
28. R. Potthast, A survey on sampling and probe methods for inverse problems. Inverse Prob. **22**(2), R1–R47 (2006)
29. W. Tabbara, B. Duchene, C. Pichot, D. Lesselier, L. Chommeloux, N. Joachimowicz, Diffraction tomography: contribution to the analysis of some applications in microwaves and ultrasonics. Inverse Prob. **4**, 305–331 (1988)
30. G. Uhlmann, (ed.), *Inside Out: Inverse Problems and Applications*. Math. Sci. Res. Inst. Publ., vol. 47 (Cambridge University Press, Cambridge, 2003), pp. 193–218

Chapter 9
Others

We consider detecting objects on a flat ground by using the electromagnetic (EM) measurement made from a height. We conduct the study in a very general and practical setting. The number of the target scatterers is not required to be known in advance, and each scatterer could be either an inhomogeneous medium or an impenetrable perfectly conducting (PEC) obstacle. Moreover, there might be multiscale components of small-size and extended-size (compared to the detecting wavelength) presented simultaneously. Some a priori information is required on scatterers of extended-size. The inverse problem is nonlinear and ill-conditioned. We propose a "direct" locating method by using a single EM far-field measurement. Our discussion in this chapter follows the treatment in [30, 32].

9.1 Ground Detection by a Single Electromagnetic Far-Field Measurement

This section concerns locating objects on a ground by using the electromagnetic (EM) scattering measurement made from a height. In Fig. 9.1, we give a schematic illustration of our study, where one wants to detect the multiple objects on the ground $\mathscr{G}$. To that end, one sends certain detecting wave fields and then measures the scattered wave fields from a height, from which to infer knowledge about the target objects. A practical scenario for our study is the scoutplane detection in the battlefield.

In what follows, we present the mathematical formulation for the current study. The detecting waves are chosen to be the time-harmonic electromagnetic plane waves of the following form

$$E^i(x) = pe^{\mathrm{i}\omega x\cdot d}, \quad H^i(x) = \frac{1}{\mathrm{i}\omega}\nabla \wedge E^i(x), \quad x \in \mathbb{R}^3 \tag{9.1}$$

J. Li and H. Liu, *Numerical Methods for Inverse Scattering Problems*,
https://doi.org/10.1007/978-981-99-3772-1_9

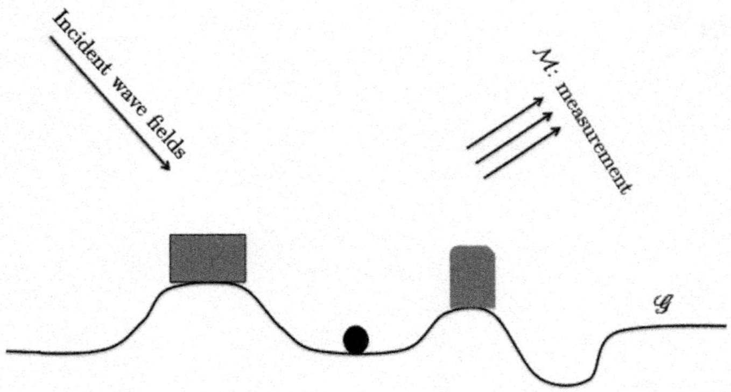

Fig. 9.1 Schematic illustration of locating multiscale ground objects

where $\mathrm{i} = \sqrt{-1}$, $\omega \in \mathbb{R}_+$ denotes the frequency, $d \in \mathbb{S}^2 := \{x \in \mathbb{R}^3; |x| = 1\}$ denotes the impinging direction, and $p \in \mathbb{R}^3$ denotes the polarization with $p \cdot d = 0$. E^i and H^i are entire solutions to the Maxwell equations in the free space

$$\nabla \wedge E^i - \mathrm{i}\omega H^i = 0, \quad \nabla \wedge H^i + \mathrm{i}\omega E^i = 0.$$

The ground $\mathscr{G}$ is assumed to be perfectly electric conducting (PEC). The EM waves cannot penetrate inside the ground and propagate only in the space above the ground. If there is no object presented on the ground, one would have a reflected wave field $E^i_{\mathscr{G}}$ such that the total wave field $E = E^i - E^i_{\mathscr{G}}$ satisfies the following PEC boundary condition

$$\nu \wedge E = \nu \wedge (E^i - E^i_{\mathscr{G}}) = 0 \quad \text{on } \mathscr{G}, \tag{9.2}$$

where ν is the unit upward normal vector to $\mathscr{G}$. If $\mathscr{G}$ is flat, the reflected wave field $E^i_{\mathscr{G}}$ is well-understood through the work [38, 39], and if $\mathscr{G}$ is non-flat/rough, the reflection would be much more complex. Throughout the present section, we assume that $\mathscr{G}$ is flat. Furthermore, without loss of generality, we assume that $\mathscr{G} := \{x := (\mathrm{x}_1, \mathrm{x}_2, \mathrm{x}_3) \in \mathbb{R}^3; x' := (\mathrm{x}_1, \mathrm{x}_2) \in \mathbb{R}^2,\ \mathrm{x}_3 = 0\}$. Denote $\mathbb{R}^3_\pm := \{x := (\mathrm{x}_1, \mathrm{x}_2, \mathrm{x}_3) \in \mathbb{R}^3; x' := (\mathrm{x}_1, \mathrm{x}_2) \in \mathbb{R}^2,\ \mathrm{x}_3 \lessgtr 0\}$ and $\mathbb{S}^2_\pm := \mathbb{R}^3_\pm \cap \mathbb{S}^2$. Moreover, we let Π denote the usual reflection with respect to $\mathscr{G}$, i.e., $\Pi v = (v_1, v_2, -v_3)$ for a generic 3-vector $v = (v_1, v_2, v_3)$. Then, we have that (cf. [39, 40])

$$E^i_{\mathscr{G}} = \Pi \circ E^i \circ \Pi. \tag{9.3}$$

Next, we consider that there are EM objects presented on the ground. Let $\psi(x')$, $x' \in \mathbb{R}^2$, be a non-negative Lipschitz continuous function such that $\psi(x') = 0$ for $|x'| > R$, where R is a large enough positive constant. Let $\Sigma' := \{x' \in \mathbb{R}^2; \psi(x') > 0\}' := \cup_{j=1}^{l} \Sigma'_j$, where Σ'_j, $j = 1, 2, \ldots, l$ denote the simply connected components of Σ'. Define

$$\Sigma_j^+ := \{(x', x_3) \in \mathbb{R}_+^3;\, x' \in \Sigma_j',\, 0 < x_3 < \psi(x')\},\ 1 \le j \le l; \quad \Sigma^+ := \bigcup_{j=1}^{l} \Sigma_j^+. \tag{9.4}$$

Each Σ_j^+, $1 \le j \le l$, represents an EM object on the ground, and will be referred to as a *scatterer* in the sequel. Let ε_j, μ_j and σ_j be the EM parameters for the object supported in Σ_j, respectively, representing the electric permittivity, magnetic permeability and electric conductivity. It is assumed that ε_j, μ_j and σ_j are all constants, satisfying $0 < \varepsilon_j < +\infty$, $0 < \mu_j < +\infty$ and $0 \le \sigma_j \le +\infty$. Furthermore, it is assumed that $|\varepsilon_j - 1| + |\mu_j - 1| + |\sigma_j| > 0$ for $j = 1, 2, \ldots, l$. If $\sigma_j = +\infty$, then Σ_j^+ is taken to be a PEC obstacle, disregarding ε_j and μ_j. In the free space, $\varepsilon = \mu = 1$ and $\sigma = 0$. We set

$$(\varepsilon(x), \mu(x), \sigma(x)) := \begin{cases} (\varepsilon_j, \mu_j, \sigma_j) & \text{when } x \in \Sigma_j^+,\ j = 1, 2, \ldots, l; \\ (1, 1, 0) & \text{when } x \in \mathbb{R}_+^3 \backslash \Sigma^+. \end{cases} \tag{9.5}$$

The presence of the scatterer $(\Sigma; \varepsilon, \mu, \sigma)$ on the ground would further perturb the propagation of the EM field $E^i - E^i_{\mathscr{G}}$, inducing the so-called *scattered* wave field E^s in $\mathbb{R}_+^3$. The scattered wave field is radiating in nature, characterized by the Silver-Müller radiation condition

$$\lim_{|x| \to +\infty} |x| \left| (\nabla \wedge E^s)(x) \wedge \frac{x}{|x|} - \mathrm{i}\omega E^s(x) \right| = 0, \tag{9.6}$$

which holds uniformly for all directions $\hat{x} := x/|x| \in \mathbb{S}_+^2$. The total electric wave field $E := E^i - E^i_{\mathscr{G}} + E^s$, together with the corresponding magnetic wave field H, is governed by the following Maxwell system

$$\nabla \wedge E - \mathrm{i}\omega\mu H = 0, \quad \nabla \wedge H + \mathrm{i}\omega \left(\varepsilon + \mathrm{i}\frac{\sigma}{\omega} \right) E = 0 \ \text{ in } \mathbb{R}_+^3, \tag{9.7}$$

where ε, μ and σ are given in (9.5). Similar to (9.2), we have that

$$\nu \wedge E = \nu \wedge (E^i - E^i_{\mathscr{G}} + E^s) = 0 \ \text{ on } \mathscr{G}. \tag{9.8}$$

We seek a pair of solutions $(E, H) \in H_{loc}(\mathrm{curl}, \mathbb{R}_+^3) \wedge H_{loc}(\mathrm{curl}, \mathbb{R}_+^3)$ to the scattering system (9.6)–(9.8). Particularly, the radiating wave field $E^s(x)$ has the following asymptotic expansion as $|x| \to +\infty$,

$$E^s(x) = \frac{e^{i\omega|x|}}{|x|} A\left(\frac{x}{|x|}; d, p, \omega \right) + \mathscr{O}\left(\frac{1}{|x|^2} \right). \tag{9.9}$$

$A(\hat{x}; d, p, \omega)$ with $\hat{x} := x/|x| \in \mathbb{S}_+^2$ is known as the electric far-field pattern, which encodes the scattering measurement illustrated in Fig. 9.1.

The ground detection problem can be abstractly formulated as

$$\mathscr{F}((\Sigma^{+}; \varepsilon, \mu, \sigma)) = A(\hat{x}; d, p, \omega), \tag{9.10}$$

where $\mathscr{F}$ is the operator sending the scatterer to the corresponding far-field pattern, defined by the Maxwell system (9.6)–(9.8). It is easily verified that $\mathscr{F}$ is nonlinear and moreover it is ill-conditioned since $\mathscr{F}$ is completely continuous (cf. [19]). In what follows, $A(\hat{x}; d, p, \omega)$ is always assumed to be given with all $\hat{x} \in \mathbb{S}^2_+$. Furthermore, if $d \in \mathbb{S}^2_+$, $p \in \mathbb{R}^3$ and $\omega \in \mathbb{R}_+$ are all fixed, then $A(\hat{x}; d, p, \omega)$ is called a *single EM measurement*; otherwise, it is called *multiple EM measurements*. In practice, a single EM measurement can be obtained by sending a single incident plane wave, and then collecting the scattered electric wave in all the observation angles. Throughout the present study, we shall take a single EM measurement for the ground detection. Moreover, our study shall be conducted in a very general and practical setting. The number of the target scatterers is not required to be known in advance, and each scatterer could be either an inhomogeneous medium or an impenetrable perfectly conducting (PEC) obstacle. Furthermore, there might be multiscale components of small-size and extended-size (compared to the detecting wavelength) presented simultaneously. Some realistic a priori information is required on scatterers of extended-size. We discuss a "direct" locating method without any inversion involved. To our best knowledge, both the direct scattering model and the inverse scattering schemes are new to the literature. The results extend those obtained in [31, 33] for locating multiscale EM scatterers located in a homogeneous space. The present study is closely related to the inverse electromagnetic scattering problems from rough surfaces; see, e.g., [4–6, 12, 13, 29]. We also refer to [6–9, 19, 21, 23, 34–36, 45–47] for the recent progress on the inverse scattering theory and numerical study. It remarked that for highly conductive objects with large σ (9.7) reduces to the eddy current model, which is justified in [1]. A similar asymptotic formalism for detecting and characterizing small conductive inclusions from electromagnetic induction data is developed in [3]. The mechanisms discussed in this section could be migrated to the eddy current model to detect complicated settings, e.g., multiple regular-size or even multiscale conductive inclusions.

The rest of the section is organized as follows. In Sect. 9.1.1, we present some results concerning the direct scattering problem for our subsequent use. Section 9.1.2 is devoted to the inverse scattering scheme. Numerical results and discussion are presented in Sect. 9.1.3.

9.1.1 Scattering from Multiscale Ground Objects

In this section, we consider the scattering from multiscale ground objects. In order to ease the exposition, throughout the rest of the section, we assume that $\omega \sim 1$, and hence the size of an EM object can be interpreted in terms of its Euclidean diameter.

Let Σ^+ and $\cup_{j=1}^{l}(\Sigma_j^+; \varepsilon_j, \mu_j, \sigma_j)$ be as introduced in (9.4) and (9.5). We further assume that there exists $l_s, l_r \in \mathbb{N} \cup \{0\}$ such that $l_s + l_r = l$. Let

$$(\Sigma^+; \varepsilon, \mu, \sigma) = (\Sigma_s^+; \varepsilon, \mu, \sigma) \cup (\Sigma_r^+; \varepsilon, \mu, \sigma), \tag{9.11}$$

where by reordering if necessary, we assume that

$$(\Sigma_r^+; \varepsilon, \mu, \sigma) = \bigcup_{j=1}^{l_r}(\Sigma_j^+; \varepsilon_j, \mu_j, \sigma_j) \quad \text{and} \quad (\Sigma_s^+; \varepsilon, \mu, \sigma) = \bigcup_{j=l_r+1}^{l}(\Sigma_j^+; \varepsilon_j, \mu_j, \sigma_j) \tag{9.12}$$

Furthermore, we assume that $\mathrm{diam}(\Sigma_j^+) \sim 1$ for $1 \le j \le l_r$, whereas $\Sigma_j^+ \sim \rho \ll 1$ for $l_r + 1 \le j \le l$. Σ_r^+ contains all the extended-size scatterer components, whereas Σ_s^+ contains all the small-size scatterer components. For the subsequent discussion, we also classify the target scatterer $(\Sigma^+; \varepsilon, \mu, \sigma)$ in terms of physical properties as follows

$$(\Sigma^+; \varepsilon, \mu, \sigma) = (\Sigma_M^+; \varepsilon, \mu, \sigma) \cup (\Sigma_P^+; \varepsilon, \mu, \sigma), \tag{9.13}$$

where $\sigma(x) < +\infty$ when $x \in \Sigma_M^+$, and $\sigma(x) = +\infty$ when $x \in \Sigma_P^+$. That is, Σ_M^+ contains all the inhomogeneous medium components, whereas Σ_P^+ contains all the PEC obstacle components of the target scatterer. We remark that it may happen that $\Sigma_M^+ = \emptyset$ or $\Sigma_P^+ = \emptyset$. The EM scattering corresponding to the target scatterer described in (9.13) is governed by the following Maxwell system

$$\begin{cases} \nabla \wedge E - \mathrm{i}\omega\mu H = 0, \quad \nabla \wedge H + \mathrm{i}\omega\left(\varepsilon + \mathrm{i}\dfrac{\sigma}{\omega}\right)E = 0 \text{ in } \mathbb{R}_+^3 \backslash \overline{\Sigma_P^+}, \\ E(x) = E^i(x) - E_{\mathscr{G}}^i(x) + E^s(x) \qquad x \in \mathbb{R}_+^3 \backslash \overline{\Sigma_P^+}, \\ \nu \wedge E = 0 \qquad \text{on } \partial(\mathbb{R}_+^3 \backslash \overline{\Sigma_P^+}), \\ \lim\limits_{|x| \to +\infty} |x| \left| (\nabla \wedge E^s)(x) \wedge \dfrac{x}{|x|} - \mathrm{i}\omega E^s(x) \right| = 0. \end{cases} \tag{9.14}$$

Next, we introduce an auxiliary scattering system defined over the whole space $\mathbb{R}^3$. Starting from now on, for an EM scatterer $(O^+; \varepsilon_0, \mu_0, \sigma_0)$, where $O^+ \subset \mathbb{R}_+^3$, ε_0, μ_0 and σ_0 are all constants with σ_0 possibly being $+\infty$, we define $(O^-; \varepsilon_0, \mu_0, \sigma_0) = (\Pi O^+; \varepsilon_0, \mu_0, \sigma_0)$ and $(O; \varepsilon_0, \mu_0, \sigma_0) = (O^+; \varepsilon_0, \mu_0, \sigma_0) \cup (O^-; \varepsilon_0, \mu_0, \sigma_0)$. Here, ΠO^+ denotes the reflected domain of O^+ with respect to $\mathscr{G}$. For the ground scatterer $(\Sigma^+; \varepsilon, \mu, \sigma)$ in (9.13), we consider the following scattering system in the whole space $\mathbb{R}^3$ corresponding to the scatterer $(\Sigma; \varepsilon, \mu, \sigma) = (\Sigma_M; \varepsilon, \mu, \sigma) \cup (\Sigma_P; \varepsilon, \mu, \sigma)$,

$$\begin{cases} \nabla \wedge \mathscr{E} - \mathrm{i}\omega\mu\mathscr{H} = 0, \quad \nabla \wedge \mathscr{H} + \mathrm{i}\omega\left(\varepsilon + \mathrm{i}\dfrac{\sigma}{\omega}\right)\mathscr{E} = 0 & \text{in } \mathbb{R}^3\backslash\overline{\Sigma_P}, \\ \mathscr{E}(x) = E^i(x) - E^i_{\mathscr{G}} + \mathscr{E}^s(x) & x \in \mathbb{R}^3\backslash\overline{\Sigma_P}, \\ \nu \wedge \mathscr{E} = 0 & \text{on } \partial\Sigma_P, \\ \lim\limits_{|x|\to+\infty} |x| \left| (\nabla \wedge \mathscr{E}^s)(x) \wedge \dfrac{x}{|x|} - \mathrm{i}\omega\mathscr{E}^s(x) \right| = 0. \end{cases} \tag{9.15}$$

We refer to [44] for the well-posedness of the Maxwell system (9.15).

We are in a position to establish an important relationship between the two scattering systems (9.14) and (9.15).

Theorem 9.1.1 *Let* $(E, H) \in H_{loc}(\mathrm{curl}; \mathbb{R}^3_+\backslash\overline{\Sigma_P^+}) \wedge H_{loc}(\mathrm{curl}; \mathbb{R}^3_+\backslash\overline{\Sigma_P^+})$, *and* $(\mathscr{E}, \mathscr{H}) \in H(\mathrm{curl}; \mathbb{R}^3\backslash\overline{\Sigma_P}) \wedge H(\mathrm{curl}; \mathbb{R}^3\backslash\overline{\Sigma_P})$ *be, respectively, solutions to the scattering systems* (9.14) *and* (9.15). *Then we have that*

$$\mathscr{E} = -\Pi \circ \mathscr{E} \circ \Pi, \quad \mathscr{H} = \Pi \circ \mathscr{H} \circ \Pi, \tag{9.16}$$

and

$$E = \mathscr{E}|_{\mathbb{R}^3_+}, \quad H = \mathscr{H}|_{\mathbb{R}^3_+}. \tag{9.17}$$

Proof Set

$$\widetilde{\mathscr{E}} = -\Pi \circ \mathscr{E} \circ \Pi, \quad \widetilde{\mathscr{H}} = \Pi \circ \mathscr{H} \circ \Pi,$$

Since $(\mathscr{E}, \mathscr{H}) \in H(\mathrm{curl}; \mathbb{R}^3\backslash\overline{\Sigma_P}) \wedge H(\mathrm{curl}; \mathbb{R}^3\backslash\overline{\Sigma_P})$, it is easily seen that $(\widetilde{\mathscr{E}}, \widetilde{\mathscr{H}}) \in H(\mathrm{curl}; \mathbb{R}^3\backslash\overline{\Sigma_P}) \wedge H(\mathrm{curl}; \mathbb{R}^3\backslash\overline{\Sigma_P})$. Moreover, since $\mathscr{E}$ and $\mathscr{H}$ satisfy the Maxwell equations,

$$\nabla \wedge \mathscr{E} - \mathrm{i}\omega\mu\mathscr{H} = 0, \quad \nabla \wedge \mathscr{H} + \mathrm{i}\omega\left(\varepsilon + \mathrm{i}\frac{\sigma}{\omega}\right)\mathscr{E} = 0 \quad \text{in } \mathbb{R}^3\backslash\overline{\Sigma_P}, \tag{9.18}$$

one can verify by straightforward computations that

$$\nabla \wedge \widetilde{\mathscr{E}} - \mathrm{i}\omega\mu\widetilde{\mathscr{H}} = 0, \quad \nabla \wedge \widetilde{\mathscr{H}} + \mathrm{i}\omega\left(\varepsilon + \mathrm{i}\frac{\sigma}{\omega}\right)\widetilde{\mathscr{E}} = 0 \quad \text{in } \mathbb{R}^3\backslash\overline{\Sigma_P}. \tag{9.19}$$

Next, by the geometric interpretation of vector product, we note the following fact for any two vectors **a** and **b**,

$$\mathbf{a} \wedge \mathbf{b} = \Pi\mathbf{a} \wedge \Pi\mathbf{b}. \tag{9.20}$$

Hence, for any $x \in \partial\Sigma_P$ and the corresponding normal vector $\nu(x)$, by using (9.20), we have

$$\nu(x) \wedge (\Pi \circ \mathscr{E} \circ \Pi)(x) = (\Pi^{-1}\nu)(x) \wedge (\mathscr{E} \circ \Pi)(x). \tag{9.21}$$

On the other hand, by noting $\Pi\Sigma_p = \Sigma_p$, we see that $\Pi^{-1}\nu(x) = (\nu \circ \Pi)(x)$, which together with (9.21) further implies that

$$(\nu \wedge \widetilde{\mathscr{E}})(x) = -\nu(x) \wedge (\Pi \circ \mathscr{E} \circ \Pi)(x) = -(\nu \wedge \mathscr{E})(\Pi x) = 0, \quad x \in \partial\Sigma_p. \tag{9.22}$$

Now, we set

$$\mathbf{E} := \mathscr{E} + \Pi \circ \mathscr{E} \circ \Pi \quad \text{and} \quad \mathbf{H} := \mathscr{H} - \Pi \circ \mathscr{H} \circ \Pi. \tag{9.23}$$

By (9.19)–(9.22), we readily see that

$$\nabla \wedge \mathbf{E} - \mathrm{i}\omega\mu\mathbf{H} = 0, \quad \nabla \wedge \mathbf{H} + \mathrm{i}\omega\left(\varepsilon + \mathrm{i}\frac{\sigma}{\omega}\right)\mathbf{E} = 0 \text{ in } \mathbb{R}^3\backslash\overline{\Sigma_P}, \quad \nu \wedge \mathbf{E} = 0 \text{ on } \partial\Sigma_p. \tag{9.24}$$

On the other hand, by (9.14),

$$\mathscr{E}(x) = E^i(x) - E^i_{\mathscr{G}} + \mathscr{E}^s(x), \qquad x \in \mathbb{R}^3\backslash\overline{\Sigma_P}. \tag{9.25}$$

It is directly verified that

$$\mathbf{E} = \mathscr{E} + \Pi \circ \mathscr{E} \circ \Pi = \mathscr{E}^s + \Pi \circ \mathscr{E}^s \circ \Pi. \tag{9.26}$$

Since $\mathscr{E}^s$ satisfies the Silver-Müller radiation condition, (9.26) clearly implies that $\mathbf{E}$ satisfies the Silver-Müller radiation condition, namely,

$$\lim_{|x|\to+\infty} |x| \left|(\nabla \wedge \mathbf{E})(x) \wedge \frac{x}{|x|} - \mathrm{i}\omega\mathbf{E}(x)\right| = 0. \tag{9.27}$$

By the well-posedness of the Maxwell system (9.26)–(9.27), we immediately have that

$$\mathbf{E} = \mathbf{H} = 0, \tag{9.28}$$

which in turn implies (9.16).

Finally, let $\nu_0 = (0, 0, 1)$ be the unit upward normal vector to $\mathscr{G}\backslash\overline{\Sigma_p}$. By using (9.16) and straightforward calculation, one can show that

$$\nu_0 \wedge \mathscr{E} = 0 \quad \text{on } \mathscr{G}\backslash\overline{\Sigma_p}, \tag{9.29}$$

which together with the fact that $\nu \wedge \mathscr{G} = 0$ on $\partial\Sigma_p$ in (9.15) that

$$\nu \wedge \mathscr{E} = 0 \quad \text{on } \partial(\mathbb{R}^3_+\backslash\overline{\Sigma^+_p}). \tag{9.30}$$

By using (9.30), and comparing the Maxwell systems (9.14) and (9.15), one easily seen that E and H in (9.14) are actually the restrictions of $\mathscr{E}$ and $\mathscr{H}$ in (9.15) in the upper half space.

The proof is completed.

By Theorem 9.1.1, one readily has

Corollary 9.1.1 *Let $A(\hat{x}; d, p, \omega)$, $\hat{x} \in \mathbb{S}^2_+$, and $\mathscr{A}(\hat{x}; d, p, \omega)$, $\hat{x} \in \mathbb{S}^2$, be the scattering amplitudes corresponding to the Maxwell systems* (9.14) *and* (9.15), *respectively. Then we have*

$$\mathscr{A} = -\Pi \circ \mathscr{A} \circ \Pi \quad \text{and} \quad A = \mathscr{A}|_{\mathbb{S}^2_+}. \tag{9.31}$$

Next, we consider the scattering from multiple small and extended scatterers, respectively. We first consider the scattering from multiple small scatterers of the form $(\Sigma_s^+; \varepsilon, \mu, \sigma)$ in (9.12) by taking $l_r = 0$. Let the small scatterer components be given as

$$\Sigma_j^+ = z_j + \rho D_j^+, \quad j = 1, 2, \ldots, l_s, \tag{9.32}$$

where $z_j \in \mathscr{G}$, and $\overline{D_j^+}$, $j = 1, 2, \ldots, l_s$, are simply-connected C^2 domains in $\mathbb{R}^3_+$ that contain the origin and

$$L_s := \min_{1 \le j, j' \le l_s, j \ne j'} \operatorname{dist}(z_j, z_{j'}) \gg 1. \tag{9.33}$$

We have

Lemma 9.1.1 *Let $(\Sigma_s^+; \varepsilon, \mu, \sigma)$ be described as above and $A(\hat{x}) := A(\hat{x}; d, p, \omega, (\Sigma_s^+; \varepsilon, \mu, \sigma))$, $\hat{x} \in \mathbb{S}^2_+$. Set*

$$\mathbf{A}(\hat{x}) = \begin{cases} A(\hat{x}) & \text{when } \hat{x} \in \mathbb{S}^2_+, \\ -\Pi A(\Pi \hat{x}) & \text{when } \hat{x} \in \mathbb{S}^2_-. \end{cases} \tag{9.34}$$

Then we have

$$\mathbf{A}(\hat{x}) = \sum_{j=1}^{l_s} e^{i\omega(d-\hat{x})\cdot z_j} \left[(\omega\rho)^3 \left(\sum_{m=-1,0,1} a_{1,m}^j U_1^m(\hat{x}) + b_{1,m}^j V_1^m(\hat{x}) \right) + \mathscr{O}((\omega\rho)^4) \right] + \mathscr{O}(L_s^{-1}), \tag{9.35}$$

where U_1^m and V_1^m are the vectorial spherical harmonics (cf. [19])

$$\begin{cases} U_n^m(\hat{x}) := \dfrac{1}{\sqrt{n(n+1)}} \operatorname{Grad} Y_n^m(\hat{x}) \\ V_n^m(\hat{x}) := \theta \wedge U_n^m(\hat{x}), \end{cases} \quad n \in \mathbb{N}, \; m = -n, \cdots, n, \tag{9.36}$$

and $a_{1,m}^j$ and $b_{1,m}^j$ $(m = -1, 0, 1,\ j = 1, 2, \ldots, l_s)$, are constants independent of $\omega\rho$, z_j and L_s.

Proof By Corrollary 9.1.1, we clearly have that

$$\mathbf{A}(\hat{x}) = \mathscr{A}(\hat{x}; d, p, \omega, (\Sigma_s; \varepsilon, \mu, \sigma)), \quad \hat{x} \in \mathbb{S}^2.$$

Hence, Lemma 9.1.1 follows directly from Lemmas 3.1 and 3.3 in [31].

In the rest of this section, we consider the scattering from multiple extended scatterers of the form $(\Sigma_r^+; \varepsilon, \mu, \sigma)$ in (9.12) by taking $l_s = 0$. Let Γ^+ be a simply-connected domain in $\mathbb{R}_+^3$ that touches $\mathscr{G}$ at the origin. Henceforth, we denote by $\mathscr{R}^+ := \mathscr{R}(\theta) \in SO(3)$ the 3D rotation matrix around the x_3-axis. Here, $\theta \in [0, 2\pi]$ denotes the corresponding Euler angle. Moreover, we define a dilation/scaling operator as follows

$$\Lambda_r \Gamma^+ := \{rx; x \in \Gamma^+\}, \quad r \in \mathbb{R}_+.$$

Next, we introduce

$$\mathscr{D} := \{\Gamma_j^+\}_{j=1}^{l'}, \quad l' \in \mathbb{N} \tag{9.37}$$

where $\Gamma_j^+ \subset \mathbb{R}_+^3$ and $\overline{\Gamma_j^+}$ is a bounded simply-connected Lipschitz domain that contains the origin. $\mathscr{D}$ is called a *base scatterer class*, and each base scatterer Γ_j^+, $1 \le j \le l'$, could be an inhomogeneous medium or a PEC obstacle. Next, we introduce the multiple extended scatterers for our study via the base class $\mathscr{D}$ in (9.37). Set $r_j \in \mathbb{R}_+$ such that

$$r_j \in [R_0, R_1], \quad 0 < R_0 < R_1 < +\infty, \quad R_0, R_1 \sim \mathscr{O}(1),$$

and moreover, let $\theta_j \in [0, 2\pi]$, $j = 1, 2 \ldots, l_r$, be l_r Euler angles. For $z_j \in \mathbb{R}^3$, we let

$$\Sigma_r^+ = \bigcup_{j=1}^{l_r} \Sigma_j^+, \quad \Sigma_j^+ := z_j + \mathscr{R}_j^+ \Lambda_{r_j} \Gamma_j^+, \quad \Gamma_j^+ \in \mathscr{D}, \quad \mathscr{R}_j^+ := \mathscr{R}(\theta_j). \tag{9.38}$$

The EM parameters of Σ_j^+ is inherited from those of the base scatterer Γ_j^+. For technical purpose, we impose the following sparsity assumption on the extended scatterer Σ_r^+ introduced in (9.38),

$$L_e = \min_{j \neq j', 1 \le j, j' \le l_r} \operatorname{dist}(\Sigma_j^+, \Sigma_{j'}^+) \gg 1. \tag{9.39}$$

Lemma 9.1.2 *Let $(\Sigma_r^+; \varepsilon, \mu, \sigma)$ be described as above. Then we have*

$$\begin{aligned} &A(\hat{x}; d, p, \omega, \bigcup_{j=1}^{l_r} (\Sigma_j^+; \varepsilon_j, \mu_j, \sigma_j)) \\ &= \sum_{j=1}^{l_r} e^{\mathrm{i}\omega(d-\hat{x})\cdot z_j} r_j A((\mathscr{R}_j^+)^T \hat{x}; (\mathscr{R}_j^+)^T d, (\mathscr{R}_j^+)^T p, r_j\omega, \bigcup_{j=1}^{l_r} (\Gamma_j^+; \varepsilon_j, \mu_j, \sigma_j)). \end{aligned} \tag{9.40}$$

Proof The proof follows from a similar argument to that for Proposition 3.1 in [33] by using change of variables for the corresponding Maxwell equations. The only point one should note with attention is that for $z \in \mathscr{G}$, one clearly has $z = \Pi z$.

9.1.2 Locating Multiscale Ground Objects

With the preparations made earlier, we are ready to present the inverse scattering scheme of detecting the multiscale ground objects introduced in (9.11). Our result extends those developed in [31, 33] for locating multiscale space objects to this interesting case of ground detection.

We first consider locating multiple small ground objects of $(\Sigma_s^+; \varepsilon, \mu, \sigma)$ as described in (9.32)–(9.33). Let $A(\hat{x}) := A(\hat{x}; d, p, \omega, (\Sigma_s^+; \varepsilon, \mu, \sigma))$, $\hat{x} \in \mathbb{S}_+^2$. The next theorem underlies the foundation for our first locating scheme.

Theorem 9.1.2 *For $z \in \mathscr{G}$, we define*

$$I_s(z) := \frac{1}{\|\mathbf{A}(\hat{x})\|^2_{L^2(\mathbb{S}^2)^3}} \sum_{T=U,V} \sum_{m=-1,0,1} \left(\left| \left\langle A(\hat{x}), e^{\mathrm{i}\omega(d-\hat{x})\cdot z} T_1^m(\hat{x}) \right\rangle_{L^2(\mathbb{S}^2_+)^3} \right|^2 + \left| \left\langle \Pi A(\Pi\hat{x}), e^{\mathrm{i}\omega(d-\hat{x})\cdot z} T_1^m(\hat{x}) \right\rangle_{L^2(\mathbb{S}^2_-)^3} \right|^2 \right), \tag{9.41}$$

where $\mathbf{A}(\hat{x})$ is given in (9.34). *Then, z_j, $j = 1, \ldots, l_s$, are local maximizers for $I_s(z)$.*

Proof By using Lemma 9.1.1, the proof follows from a similar argument to that of Theorem 2.1 in [31].

Based on Theorem 9.1.2 we can formulate Scheme S to locate the multiple small scatterer components of $(\Sigma_s^+; \varepsilon, \mu, \sigma)$ in (9.32)–(9.33).

Scheme S	Locating small scatterers of $(\Sigma_s^+; \varepsilon, \mu, \sigma)$ in (9.32)–(9.33).
Step 1	For an unknown scatterer $(\Sigma_s^+; \varepsilon, \mu, \sigma)$, collect the far-field pattern $A(\hat{x})$ by sending a single pair of detecting plane waves (9.1).
Step 2	Select a sampling region with a mesh $\mathscr{T}_h$ containing $\mathscr{G} \cap \overline{\Sigma_s^+}$.
Step 3	For each sampling point $z \in \mathscr{T}_h$, calculate $I_s(z)$ in (9.41).
Step 4	Locate all the local maximizers of $I_s(z)$ on $\mathscr{T}_h$, which represent the locations of the scatterer components.

Next, we consider the locating of multiple extended scatterers of $(\Sigma_r^+; \varepsilon, \mu, \sigma)$ in (9.38)–(9.39). To that end, we need assume that the base scatterer class $\mathscr{D}$ in (9.37) is known in advance. Let $\mathscr{D}'$ denote the set consisting of the scatterers of the form $\mathscr{R}^+\Lambda_r\Sigma_j^+$ with all possible $\mathscr{R}^+ \in SO(3)$ around the x_3-axis, $r \in [R_0, R_1]$ and $\Sigma_j^+ \in \mathscr{D}$. Then, we introduce $\widetilde{\mathscr{D}}$ with

$$\widetilde{\mathscr{D}} := \{\widetilde{\Gamma}_j^+\}_{j=1}^{l''}, \quad l'' \in \mathbb{N} \tag{9.42}$$

such that $\widetilde{\mathscr{D}}$ is an ε-net of $\mathscr{D}'$, where $\varepsilon \in \mathbb{R}_+$ is sufficiently small. Moreover, it is assumed that $\widetilde{\mathscr{D}}$ is admissible in the following sense:

1. $\|A(\hat{x}; d, p, \omega, \widetilde{\Gamma}_j^+)\|_{L^2(\mathbb{S}_+^2)^3} \geq \|A(\hat{x}; d, p, \omega, \widetilde{\Gamma}_{j+1}^+)\|_{L^2(\mathbb{S}_+^2)^3}, \qquad 1 \leq j \leq l'' - 1;$
2. $A(\hat{x}; d, p, \omega, \widetilde{\Gamma}_j^+) \neq A(\hat{x}; d, p, \omega, \widetilde{\Gamma}_{j'}^+), \qquad 1 \leq j, j' \leq l''.$

The first condition above can be fulfilled by reordering if necessary, whereas the second condition is the related to the uniqueness in the corresponding inverse problem. However, the uniqueness result is not available in the literature. Nevertheless, even if two base scatterers $\widetilde{\Gamma}_j^+$ and $\widetilde{\Gamma}_{j'}^+$ produce the same the far-field pattern, the locating scheme developed in the sequel would still work, and the only problem one would encounter is that if both of them are presented as the target ground objects, then the locating scheme cannot distinguish them.

Theorem 9.1.3 *For $z \in \mathscr{G}$, we define*

$$I_e^j(z) := \frac{1}{\|A(\hat{x}; \widetilde{\Gamma}_j^+)\|^2_{L^2(\mathbb{S}_+^2)^3}} \left| \left\langle A(\hat{x}; \Sigma_r^+), e^{\mathrm{i}\omega(d-\hat{x})\cdot z} A(\hat{x}; \widetilde{\Gamma}_j^+) \right\rangle_{L^2(\mathbb{S}_+^2)^3} \right|, \quad j = 1, 2, \ldots, l''. \tag{9.43}$$

Consider the indicator function I_e^1 introduced above. If for some scatterer component, say $\Sigma_{j_0}^+$, $1 \leq j_0 \leq l_r$, of $(\Sigma_r^+; \varepsilon, \mu, \sigma)$, one has that

$$d_H(\Sigma_{j_0}^+, \widetilde{\Gamma}_1^+) \leq \varepsilon, \tag{9.44}$$

where d_H denotes the Hausdorff distance. Then

$$I_e^1(z_{j_0}) = 1 + \mathscr{O}(\varepsilon + L_e^{-1}), \tag{9.45}$$

and moreover, z_{j_0} is a local maximizer for $I_e^1(z)$.

Proof By using Lemma 9.1.2, the proof follows from a similar argument to that for Theorem 3.1 in [33].

Based on Theorem 9.1.3 we can formulate Scheme R to locate the multiple extended scatterer components of $(\Sigma_r^+; \varepsilon, \mu, \sigma)$ in (9.38)–(9.39).

Scheme R	Locating extended scatterers of $(\Sigma_r^+; \varepsilon, \mu, \sigma)$ in (9.38)–(9.39).
Step 1	For the base scatterer class $\mathscr{D}$ in (9.37), augment it into $\widetilde{\mathscr{D}}$ and then collect the far-field pattern $A(\hat{x}; \widetilde{\Gamma}_j^+)$ for every scatterer $\widetilde{\Gamma}_j^+$ in $\widetilde{\mathscr{D}}$ in advance.
Step 2	For an unknown scatterer $(\Sigma_r^+; \varepsilon, \mu, \sigma)$, collect the far-field pattern $A(\hat{x}; \Sigma_r^+)$ by sending a single pair of detecting plane waves (9.1).
Step 3	Set $j = 1$.
Step 4	Select a sampling region with a mesh $\mathscr{T}_h$ containing $\mathscr{G} \cap \overline{\Sigma_r^+}$.
Step 5	For each sampling point $z \in \mathscr{T}_h$, calculate $I_e^j(z)$ in (9.43).
Step 6	Locate all the local maximizers of $I_e^j(z)$ on $\mathscr{T}_h$ with values around 1, which represent the locations of the scatterer components.
Step 7	End if $j = l''$, otherwise $j = j + 1$ and go to Step 4.

Finally, we consider the locating of the general multiscale scatterers of $(\Sigma^+; \varepsilon, \mu, \sigma)$ in (9.11), by concatenating Schemes S and R through the *local tuning technique* introduced in [33], which we briefly describe in the following. Let $A(\hat{x}; \Sigma^+)$ be the far-field pattern of $(\Sigma^+; \varepsilon, \mu, \sigma)$ collected corresponding to a single pair of incident plane waves (9.1). Then, we apply $A(\hat{x}; \Sigma^+)$ as the far-field pattern to Scheme R to approximately locate the positions of the extended scatterer components of $(\Sigma_r^+; \varepsilon, \mu, \sigma)$. Suppose that $\widetilde{z}_j + \widetilde{\Gamma}_j^+$, $j = 1, 2, \ldots, l_r$, are the approximate scatterers found above. We next refine the mesh $\mathscr{T}_h$ around each $\widetilde{z}_j$, and the ε-net $\widetilde{\mathscr{D}}$ around each $\widetilde{\Gamma}_j^+$. After that, for each set of sampling points from the refined mesh made above, say $\widehat{z}_j$, $j = 1, 2, \ldots, l_r$, and each set of base scatterers from the refined ε-net, say $\widehat{\Gamma}_j^+$, $j = 1, 2, \ldots, l_r$, one calculates

$$\widehat{A}(\hat{x}) := A(\hat{x}; \Sigma_r^+) - \sum_{j=1}^{l_r} e^{i\omega(d-\hat{x})\cdot\widehat{z}_j} A(\hat{x}; \widehat{\Gamma}_j^+). \tag{9.46}$$

By using $\widehat{A}(\hat{x})$ as the far-field pattern for Scheme S, and by running through all the possible sampling points and base scatterers from the refined sampling mesh and ε-net, one can locate the clustered local maximizers, which represent the locations of the small scatterer components of $(\Sigma_s^+; \varepsilon, \mu, \sigma)$. We refer to the locating procedure sketched above as **Scheme M**.

9.1.3 *Numerical Experiments and Discussions*

We present extensive numerical experiments in this section to illustrate the salient features of the locating schemes (**Scheme S, R** and **M**) for the inverse EM scattering problem with locally perturbed ground objects in three dimensions. An oblique EM plane wave with the incident direction of polar angle $\pi/6$ radian and azimuthal angle $\pi/3$ radian is employed as the detecting wave field incident on the ground objects and it yields a perturbed EM wave scattering off from ground objects to the infinity. In the examples below, we set $\varepsilon_0 = \mu_0 = 1$ and $\sigma_0 = 0$ outside the scatterer and above the ground, and hence the wavelength is unitary (namely $\lambda = 1$) in the homogeneous background. Unless otherwise specified, all the ground objects are either PEC conductors or inhomogeneous media with all other parameters the same as those in the homogeneous background except $\varepsilon = 4$.

In the sequel, the exact EM far field data are synthesized in the following way. The scattered field is obtained by solving the Maxwell system (9.14) by a forward solver using H(curl)-conforming quadratic edge finite elements of Nedéléc's first kind on a truncated domain. The computational domain is an upper semi-sphere centered at the origin, and enclosed by a semi-spherical PML layer to damp the reflection and a PEC boundary on the ground, namely the $x - y$ plane. Local adaptive refinement scheme within the inhomogeneous scatterer or around the PEC components is adopted to enhance the accuracy of the scattered wave. The far-field data are approximated by the Stratton-Chu integral equation representation using the spherical Lebedev quadrature (cf. [28, 45]). The computation is carried out on a sequence of successively refined meshes till the relative maximum error of successive groups of far-field data is below 0.1%. The synthetic far field data on the finest mesh are taken as the exact one.

The electric far-field patterns $A(\hat{x}, \Omega)$, $\Omega = \Omega^{(r)}$ or $\Omega^{(m)}$, are observed at Lebedev quadrature points distributed on the unit upper half-sphere $\mathbb{S}^2_+$ with sufficient order of accuracy (cf. [28] and references therein). The exact far-field data $A(\hat{x}, \Omega)$ are corrupted point-wise by the formula

$$A_\delta(\hat{x}, \Omega) = A(\hat{x}, \Omega) + \delta\zeta_1 \max_{\hat{x}} |A(\hat{x}, \Omega)| \exp(i2\pi\zeta_2)\,, \tag{9.47}$$

where δ refers to the relative noise level, and both ζ_1 and ζ_2 follow the uniform distribution ranging from -1 to 1. The values of the indicator functions have been normalized between 0 and 1 to highlight the positions identified.

In the following, we demonstrate the effectiveness and efficiency of Schemes **S**, **R** and **M** by three groups of experiments. The first group of experiments is on locating small-size, or *partially-small* ground objects in various scenarios by Scheme **S**, and the second group of experiments on testing Scheme **R** for locating extended-size ground objects. In the third group of experiments, we shall test the performance of Scheme **M** on locating multi-scale multiple ground objects.

9.1.3.1 Scheme S

Example 9.1.1 (*A mini-rocket on the ground*) In this example, we consider a mini-rocket of height 0.1λ and located at the origin, with the EM parameters given by $\varepsilon = 4$, $\mu = 1$ and $\sigma = 0$, see Fig. 9.2a by zooming-in around the origin. The orthogonal slices of the contours of the indicator function $I(z)$ for Scheme **S** are given in Fig. 9.2b. It can be seen that the position of the mini-rocket is highlighted as predicted. Scheme **S** can locate the small scatterer in a very accurate and stable manner even if 20% random noise is attached to the measurement data.

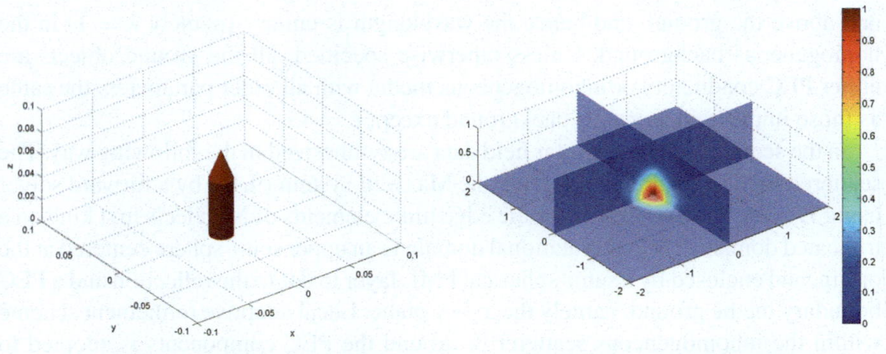

(a) Mini-rocket at the origin (zoomed in). (b) Slice plots of indicator function.

Fig. 9.2 Example 1

Example 9.1.2 (*A mini-rocket and a mini-tank on the ground*) In this example, we consider a scatterer of multiple components: a mini-rocket of height 0.1λ and located at $(-2\lambda,\ -2\lambda,\ 0)$ with the EM parameters given by $\varepsilon = 4$, $\mu = 1$ and $\sigma = 0$, and a PEC mini-tank of height 0.075λ located at $(2\lambda,\ 2\lambda,\ 0)$, see Fig. 9.3a–b by zooming in around the respective ground object. For this example, the orthogonal slices of the contours of the indicator function $I(z)$ for Scheme **S** are shown in Fig. 9.3c. Scheme **S** yields a very accurate identification of the location of both scatterers even if the measurement data is significantly perturbed to a high-level noise of 20%. This example also demonstrates that Scheme **S** can locate the multiple scatterer components without knowing the physical property of each component in advance.

Example 9.1.3 (*Two close small half-balls on the ground*) The scatterer consists of two small PEC half-balls of radius 0.1, located at $(-0.35,\ 0,\ 0)$ and $(0.35,\ 0,\ 0)$, respectively, as shown in Fig. 9.4a. We shall investigate the lower resolution limit between the ground objects for Scheme S. The results are shown in Fig. 9.4b. It can be seen in this case, namely the distance between the two components is of a half wavelength, Scheme S can locate both scatterer components and separate them very well. If we further reduce the distance between the two components (less than a half wavelength), Scheme S can no longer separate the two scatterer components, although it can still roughly locate their average position.

In the next three examples, we test some *partially-small* ground objects, which are only small in certain but not all dimensions.

Example 9.1.4 (*A half-cylinder crack bar*) The scatterer consists of a half-cylinder crack bar over the $x - y$ plane with axial direction along the x-axis, located at origin with height 2λ and bottom radius 0.1λ, respectively, see Fig. 9.5a. A single impinging EM wave is enough to provides us all important information of the scatterer, e.g., the rough position and approximate length as shown in Fig. 9.5b.

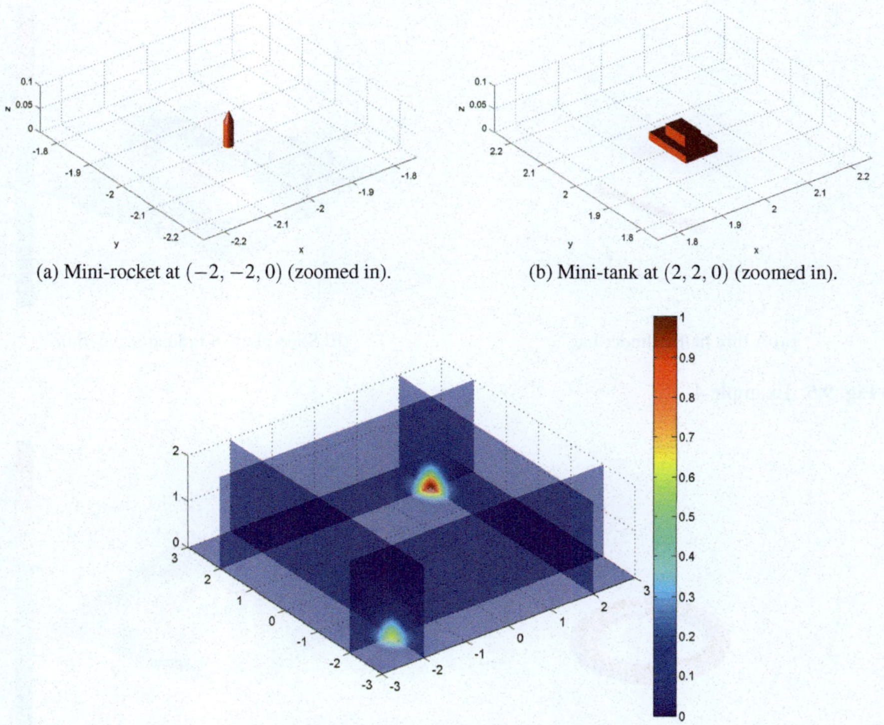

(a) Mini-rocket at $(-2, -2, 0)$ (zoomed in).

(b) Mini-tank at $(2, 2, 0)$ (zoomed in).

(c) Slice plots of the indicator function of two ground components.

Fig. 9.3 Example 2

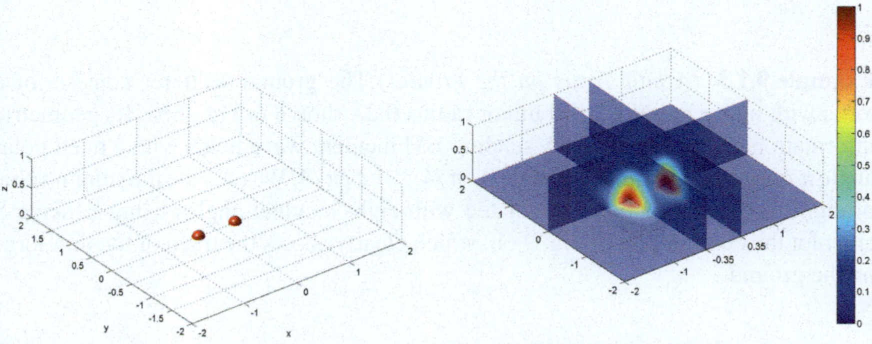

(a) Two half-balls.

(b) Slice plots of indicator function.

Fig. 9.4 Example 3

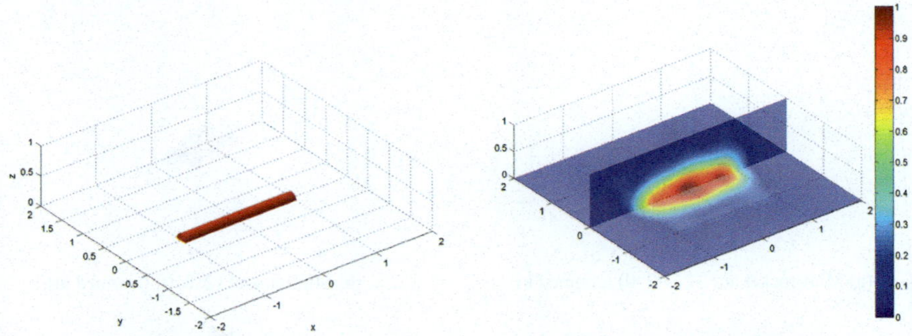

(a) A thin half-cylinder bar. (b) Slice plots of indicator function

Fig. 9.5 Example 4

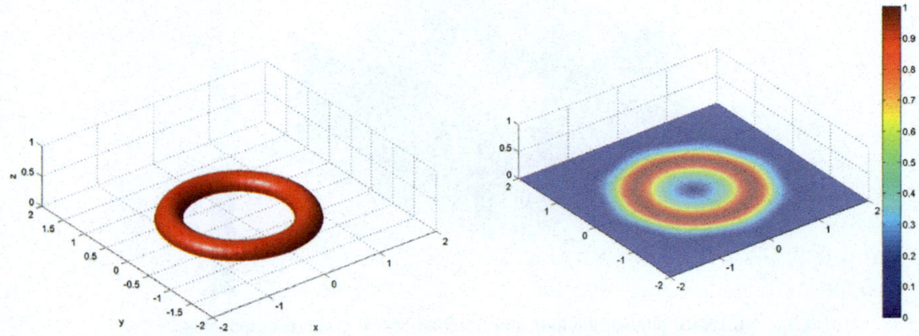

(a) A thin half torus. (b) Slice plots of indicator function.

Fig. 9.6 Example 5

Example 9.1.5 (*A half Torus on the ground*) The ground scatterer consists of a torus with major radius 1λ and minor radius 0.2λ shown in Fig. 9.6a. Its geometric information can be extracted by sending EM incident wave fields with a fixed polar angle $\pi/6$ and eight azimuthal angle $0,\ \pi/4, \cdots, 7\pi/4$. We take the maximum value of eight indicator functions associated with eight incident angles using Scheme S and plot the contour plot in Fig. 9.6b, which clearly shows the trace of the thin torus on the ground.

Example 9.1.6 (*A thin-aircraft*) In this example we consider a PEC thin aircraft of height 0.1, see Fig. 9.7a. Following the same multi-static data of up to eight incident angles, the position and its geometric information can be extracted from the indicating behavior in Fig. 9.7b.

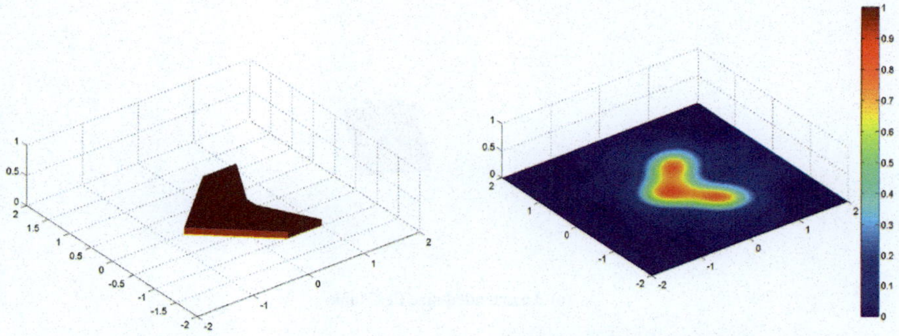

(a) A thin aircraft.

(b) Slice plots of indicator function.

Fig. 9.7 Example 6

9.1.3.2 Scheme R

In this subsection, we consider two examples to demonstrate the capability of Scheme R for locating extended-size ground objects. The synthesized far field data are corrupted by a noise level of 5%.

Example 9.1.7 (*A extended-size tank on the ground*) As shown in Fig. 9.8a, we consider an extended-size PEC tank of length 1.5, width 1, height 0.5 displaced at (, 0, 0). The far field data are collected in advance associated with the augmented reference set, namely four different orientations with azimuthal angles 0, $\pi/2$, π and $3\pi/2$ (see Fig. 9.8b–e). Reconstruction results using Scheme R are illustrated in Fig. 9.8(f)-(i). It is found that only Fig. 9.8(i) gives a significantly highlighted region (achieving the maximum magnitude of unit) around the exact position (, 0, 0), it further tells us what orientation the tank takes with the information carried in the augmented shape set.

Example 9.1.8 (*Extended-size ellipsoid and peanut*) In this example, we consider a ground scatterer consists of two PEC components as shown in Fig. 9.9: (1) an extended semi-ellipsoid with the ground center chosen at (−2.5, −2.5, 0) with semi-axis radius (2, ,) along x-, $y-$ and $z-$axes, which is further rotated by $\pi/2$; and (2) an extended peanut parametrized by $x(t) = \sqrt{3\cos^2(t)+1}\cos(t)$, $y(t) = \sqrt{3\cos^2(t)+1}\sin(t)$ in the $x - y$ plane and revolved around x-axis, which is then displaced at $(2.5\lambda,\ 2.5\lambda,\ 0)$. The reference shape set is further augmented by four different orientations as shown in Fig. 9.10. Using the far field data associated with those a priori augmented admissible shapes, Scheme R is then implemented for imaging purpose.

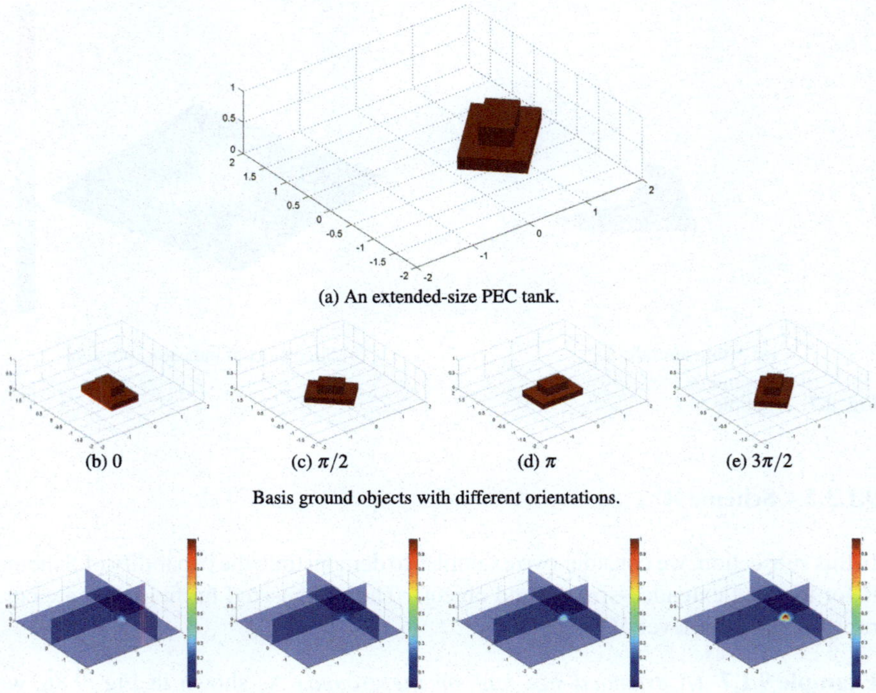

(a) An extended-size PEC tank.

(b) 0 (c) $\pi/2$ (d) π (e) $3\pi/2$

Basis ground objects with different orientations.

(f) 0 (g) $\pi/2$ (h) π (i) $3\pi/2$

Slice results associated with different orientations.

Fig. 9.8 Example 7

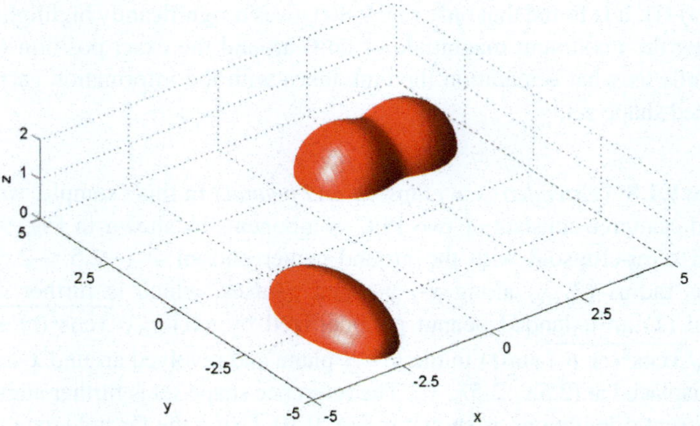

Fig. 9.9 Example 8. A two-component ground scatterer

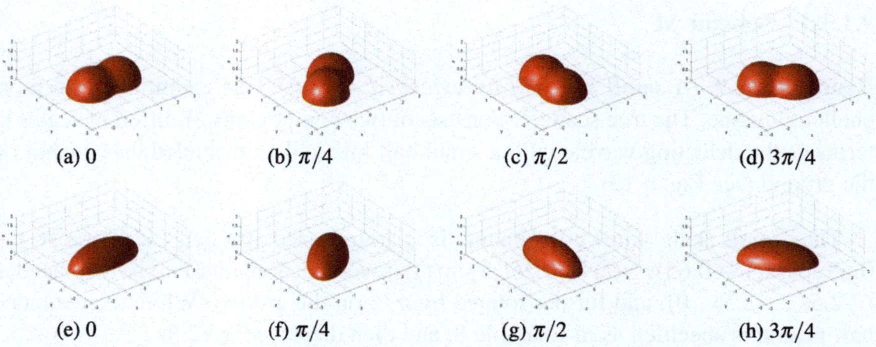

Fig. 9.10 Augmented reference shape set in Example 8

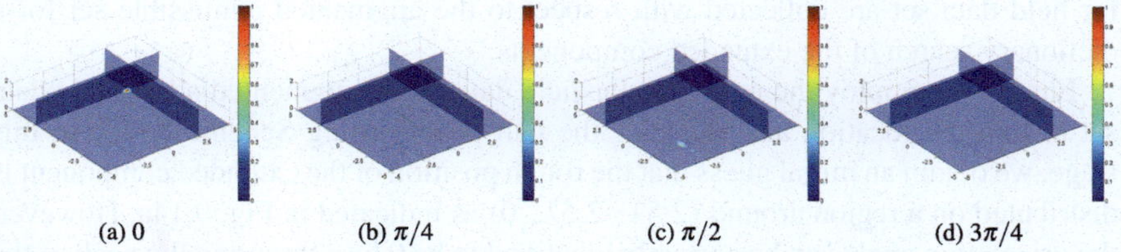

Fig. 9.11 Example 8: Locating the first component

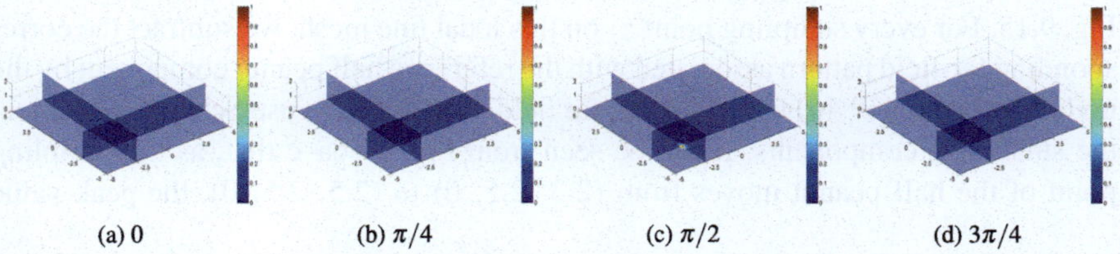

Fig. 9.12 Example 8: Locating the second component

By the magnitude of the far field pattern, those associated with the half peanut are first examined. The slice plots of the indicator function using Scheme R are shown in Fig. 9.11a–d. For noisy far-field data, Scheme R can successfully determine the location of the peanut with the right configuration by the dark red part as shown in Fig. 9.11a. After the peanut is determined and trimmed from the sampling domain, if one continues Scheme R by a dictionary match of the augmented far field data associated with the reference half ellipsoid component, the significant peak value in Fig. 9.12c indicates the position and orientation of the second component.

9.1.3.3 Scheme M

Example 9.1.9 (*A small kite and an extended peanut*) This example is the most challenging one. The true scatterer consists of two components of different scales in terms of the detecting wavelength: a small half kite and an extended half peanut on the ground (see Fig. 9.13).

The small half kite component is parametrized by $x(t) = 0.2(\cos(t) + 0.65\cos(2t) - 0.65)$, $y(t) = 0.2(1.5\sin(t))$, revolved along the x-axis, displaced at $(-2.5\lambda, -2.5\lambda, 0)$, and further rotated by $\pi/2$ on the ground. While the extended half peanut is specified as in Example 8, and then displaced at $(2.5\lambda, 2.5\lambda, 0)$.

The reference shape of the extended scatterer (half-peanut) in the admissible set is augmented by four different orientations as shown in Fig. 9.10a–d, and then the far field data set are collected with respect to the augmented admissible set for a dictionary search of the extended components.

Firstly, we employ the reference far-field data in the augmented admissible data set to find the location and shape of the half-peanut using Scheme R. Up to this stage, we obtain an initial guess that the rough position of the extended component is distributed on a region around $(2.5\lambda, 2.5\lambda, 0)$ as indicated in Fig. 9.14a. However, the orientation angle can be correctly identified to be 0 by comparing the peak of the indicator function in Fig. 9.14 with the reference set in Fig. 9.10a–d.

Next, we define a locally finer sampling mesh around (2.5, 2.5, 0) shown in Fig. 9.15. For every sampling point z_j on this local fine mesh, we subtract the corresponding far-field pattern associated with the reference half peanut component by the updating formula (9.46) from the total far field data, and then use Scheme S to detect the small-size components. It can be seen from Fig. 9.16a–c that, as the sampling point of the half peanut moves from (2.3, 2.5, 0) to (2.5, 2.5, 0), the peak value

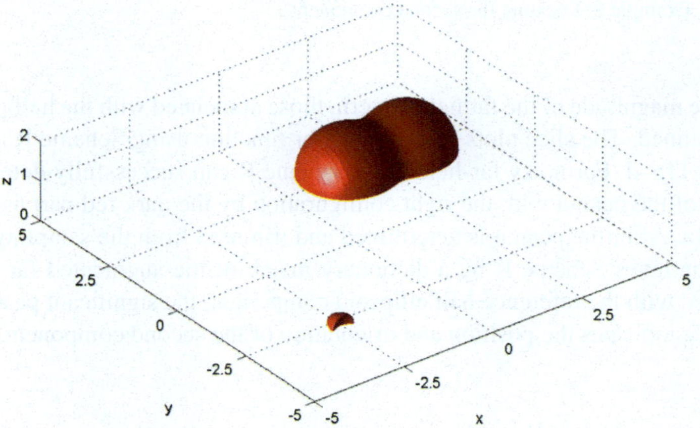

Fig. 9.13 Example 9. A multiscale scatterer

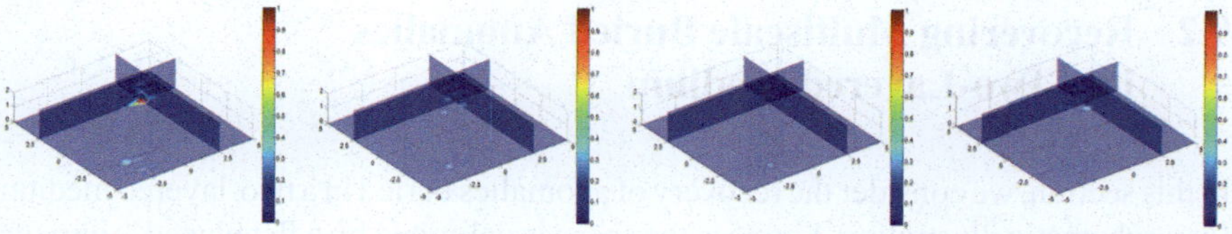

(a) 0 (b) $\pi/4$ (c) $\pi/2$ (d) $3\pi/4$

Fig. 9.14 Example 9: (Step 1) Locating the extended component

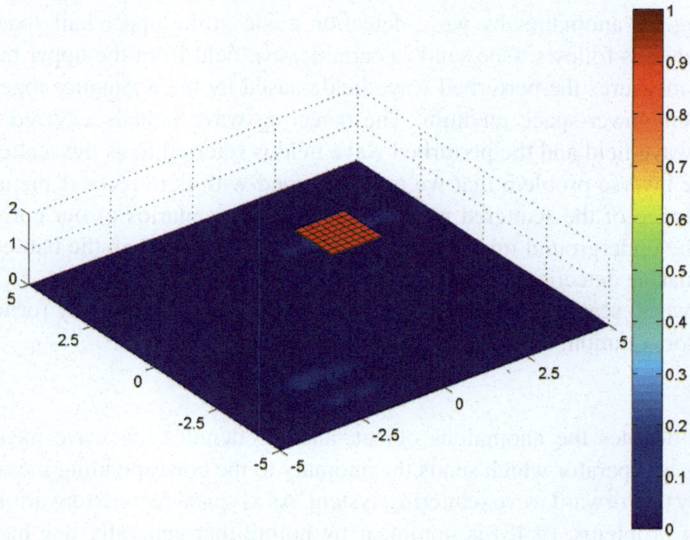

Fig. 9.15 Example 9. A local fine mesh around the rough position of the half peanut

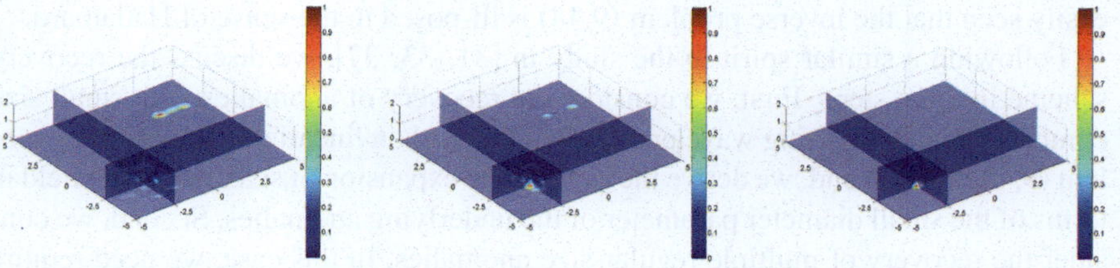

(a) (2.3, 2.5, 0) (b) (2.4, 2.5, 0) (c) (2.5, 2.5, 0)

Fig. 9.16 Example 9: (Step 2) Local fine tuning and locating the small-size component

and focused highlighted region are centered at $(-2.5,\ -2.5,\ 0)$, which indicates the existence of a small-size component lying at the position $(-2.5,\ -2.5,\ 0)$.

9.2 Recovering Multiscale Buried Anomalies in a Two-Layered Medium

In this section, we consider the recovery of anomalies buried in a two-layered medium for a schematic illustration. Suppose the space is delimited by a flat plane Γ_0 into two half-spaces: the upper one and the lower one. The two half-spaces are occupied by two different (homogeneous) mediums. It is further supposed that some inhomogeneous anomalies are buried or immersed in the lower half-space. We are interested in recovering the anomalies by wave detection made in the upper half-space, which is proceeded as follows. One sends a certain wave field from the upper half-space, and then measures the perturbed wave field caused by the anomalies together with the ambient lower-space medium. The detecting wave field is referred to as the incident wave field and the perturbed wave field is referred to as the scattered wave filed. The inverse problem that we are concerned with is to recover the anomalies by knowledge of the scattered wave field. Practical scenarios of our current study include the underground mineral prospection, mines locating in the battlefield, and anti-submarine detection.

The inverse scattering problem described above can be abstractly formulated as an operator equation,

$$F(\mathrm{O}) = \mathscr{M}, \tag{9.48}$$

where O denotes the anomalous object, and $\mathscr{M}$ denotes the wave measurement data. F is an operator which sends the anomaly to the corresponding measurement, defined by the forward wave scattering system. As a typical feature for various inverse scattering problems, (9.48) is nonlinear by noting that generally one has $F(\mathrm{O}_1 \cup \mathrm{O}_1) \neq F(\mathrm{O}_1) + F(\mathrm{O}_2)$, where O_1 and O_2 are two different anomalies. This is mainly due to the multiple wave scattering interaction between O_1 and O_2. Moreover, it is easily seen that the inverse problem (9.48) is ill-posed in the sense of Hadamard.

Following a similar spirit to the study in [31, 33, 37], we discuss the recovery scheme in three steps. First, we consider the recovery of anomalies with small size compared to the detecting wavelength. This is based on linearizing the inverse problem (9.48). To that end, we derive the asymptotic expansion of scattered wave field in terms of the small diameter parameter of the underlying anomalies. Second, we consider the recovery of multiple regular-size anomalies. In this case, we need require that the anomalies are from an admissible class, which is known in advance. The recovery is based on projecting the measured far-field pattern into a space of far-field patterns generated by the admissible scatterers. Finally, by concatenating the above two procedures via a local tuning technique, one can recover multiple multiscale buried anomalies. We would like to mention in passing that similar inverse problems of recovering buried objects were also considered in [9, 18] with different methods.

9.2.1 Mathematical Formulation

For $x = (x^1, x^2, \cdots, x^n) \in \mathbb{R}^n$, $(n = 2, 3)$, we let

$$\mathbb{R}^n_+ := \{x \in \mathbb{R}^n : x^n > 0\} \quad \text{and} \quad \mathbb{R}^n_- := \{x \in \mathbb{R}^n : x^n < 0\},$$

be, respectively, the upper and lower half-spaces. The interface between the two layers, namely $\{x \in \mathbb{R}^n; x^n = 0\}$, is denoted by $\mathbb{R}^n_0$. Let k_+ and k_- be the wavenumbers in $\mathbb{R}^n_+$ and $\mathbb{R}^n_-$, respectively. Denote by Ω an impenetrable obstacle that is completely buried in the lower half-space. It is assumed that Ω is a bounded Lipschitz domain with connected complement. In what follows, we let ν denote the unit outward normal vector to $\partial\Omega$, as well as the unit upward normal vector to $\mathbb{R}^n_0$, which should be clear from the context.

Let u^i be a time-harmonic incident plane wave given by

$$u^i(x) = e^{ik_+ x\cdot d^i},$$

where $d^i \in \mathbb{S}^{n-1} := \{x \in \mathbb{R}^n; |x| = 1\}$ denotes the incident direction. In what follows, we set $\mathbb{S}^{n-1}_\pm := \mathbb{S}^{n-1} \cap \mathbb{R}^n_\pm$ and, let (r, θ) and (r, θ, φ) denote the standard polar coordinates in $\mathbb{R}^2$ and $\mathbb{R}^3$, respectively. Denote by $\theta_c \in (-\pi, 0]$ the critical incident angle which is defined by $\cos\theta_c = k_-/k_+$ if $k_- < k_+$ and $\theta_c = 0$ if $k_- \geq k_+$. We take the incident direction

$$d^i := \begin{cases} (\cos\theta_0 \cos\varphi_0,\ \cos\theta_0 \sin\varphi_0,\ \sin\theta_0), & n = 3; \\ (\cos\theta_0, \sin\theta_0), & n = 2, \end{cases} \tag{9.49}$$

where $\theta_0 \in (-\pi - \theta_c, \theta_c)$ such that $d^i \in \mathbb{S}^{n-1}_-$, and $\varphi_0 \in (0, 2\pi)$. The interface $\mathbb{R}^n_0$ generates the reflected and transmitted wave fields, which shall be denoted as u^r and u^t, respectively. By the Fresnel formula, we have

$$u^r(x) = R(\theta_0) e^{ik_+ x\cdot d^r}, \quad x \in \mathbb{R}^n_+ \quad u^t(x) = T(\theta_0) e^{ik_- x\cdot d^t}, \quad x \in \mathbb{R}^n_-, \tag{9.50}$$

where d^r is the reflected direction and d^t is the transmitted direction, while $R(\theta_0)$ and $T(\theta_0)$ are the reflection and transmission coefficients, respectively. The corresponding reflected direction is given by

$$d^r = \begin{cases} (\cos\theta_0 \cos\varphi_0,\ \cos\theta_0 \sin\varphi_0,\ -\sin\theta_0), & n = 3 \\ (\cos\theta_0, -\sin\theta_0), & n = 2, \end{cases} \tag{9.51}$$

and the transmitted direction is given by

$$d^t = \begin{cases} (\cos\chi_0 \cos\varphi_0,\ \cos\chi_0 \sin\varphi_0,\ \sin\chi_0), & n = 3 \\ (\cos\chi_0, \sin\chi_0), & n = 2. \end{cases} \tag{9.52}$$

In (9.52), $\chi_0 \in (-\pi, 0)$ stands for the transmitted angle into $\mathbb{R}^n_-$, and is implicitly given by the Snell relation

$$k_+ \cos\theta_0 = k_- \cos\chi_0. \tag{9.53}$$

Set

$$u_0(x) = \begin{cases} u^i(x) + u^r(x), & x^n > 0 \\ u^t(x), & x^n < 0. \end{cases}$$

It is required that both u_0 and $\partial u_0/\partial \nu$ are continuous across the interface $\mathbb{R}_0$. That is, $u^i + u^r - u^t = 0$ and $\partial(u^i + u^r)/\partial\nu - \partial(u^t)/\partial\nu = 0$ on $\mathbb{R}^n_0$. By using such continuities, one can deduce that the coefficients $R(\theta_0)$ and $T(\theta_0)$ are given according to the following formulas:

$$R(\theta) = \frac{k_+ \sin\theta - k_- \sin\chi}{k_+ \sin\theta + k_- \sin\chi} \quad \text{and} \quad T(\theta) = \frac{2k_+ \sin\theta}{k_+ \sin\theta + k_- \sin\chi}. \tag{9.54}$$

In particular, we note that if $k_+ = k_-$, then $R(\theta_0) = 0$, $T(\theta_0) = 1$ and $u_0 = u^i$.

With above preparations, the forward problem of the scattering due to the buried impenetrable anomalies in a two-layered medium can be described as finding the scattered wave field $u^s \in H^1_{loc}(\mathbb{R}^n\backslash\overline{\Omega})$ such that

$$\begin{aligned} &\Delta u^s + (k_\pm)^2 u^s = 0 \quad \text{in } \mathbb{R}^n_\pm\backslash\overline{\Omega}, \\ &[u^s] = 0, \quad \left[\frac{\partial u^s}{\partial \nu}\right] = 0 \quad \text{on } \mathbb{R}_0, \\ &\mathscr{B}(u^s) = -\mathscr{B}(u_0) \quad \text{on } \partial\Omega, \\ &\lim_{r\to\infty} \int_{\mathbb{S}_{r,\pm}} \left|\frac{\partial u^s}{\partial r} - ik_\pm u^s\right|^2 ds = 0, \end{aligned} \tag{9.55}$$

where $[\cdot]$ denotes the jump in its argument across the interface $\mathbb{R}^n_0$, $\mathbb{S}_{r,\pm} = \{x \in \mathbb{R}^n_\pm;\ |x| = r\}$ is the half sphere/circle of radius r centered at the origin in $\mathbb{R}^n_\pm$ and $\mathscr{B}$ denotes one of the following three boundary conditions:

$$\mathscr{B}(u^s) = u^s \text{ on } \partial\Omega; \quad \mathscr{B}(u^s) = \frac{\partial u^s}{\partial \nu} \text{ on } \partial\Omega; \quad \mathscr{B}(u^s) = \frac{\partial u^s}{\partial \nu} + i\lambda u^s \text{ on } \partial\Omega \tag{9.56}$$

corresponding, respectively, to the case when the anomaly Ω is sound-soft, sound-hard, and of impedance type. In (9.56), $\lambda \in C(\partial\Omega)$, $(\lambda \geq 0)$, is the surface impedance. By a variational approach and following essentially a similar argument as in [17] or [43, Sect. 12.4], one can establish the well-posedness of the scattering problem (9.49)–(9.56). For $x \in \mathbb{R}^n_+$, $u^s(x)$ admits the following asymptotic expansion

$$u^s(x) = \gamma_n \frac{e^{ik_+ r}}{r^{\frac{n-1}{2}}} \left\{ u^\infty(\hat{x}) + \mathscr{O}\left(\frac{1}{r}\right) \right\} \quad \text{as } r = |x| \to \infty \tag{9.57}$$

with

$$\gamma_n = \frac{1}{4\pi}, \quad n = 3; \qquad \frac{e^{i\frac{\pi}{4}}}{\sqrt{8k_+\pi}}, \quad n = 2,$$

uniformly for all directions $\hat{x} := x/|x|$. In (9.57), $u^\infty(\hat{x})$ defined on the upper half unit sphere/circle $\mathbb{S}_+^{n-1}$ is known as the scattering amplitude or far-field pattern with $\hat{x} \in \mathbb{S}_+^{n-1}$ denoting the observation direction.

The inverse scattering problem that we are concerned with is to recover Ω by knowledge of $u_\infty(\hat{x})$. In terms of the abstract operator equation (9.48), Ω is the unknown O, u_∞ is the measurement data set $\mathscr{M}$, and F is defined by the direct scattering system as described in (9.49)–(9.57). Throughout the current study, we shall take d^i fixed. That is, the measurement $u^\infty(\hat{x})$ is obtained by sending a single incident plane wave, and we call it a single far-field measurement.

9.2.2 *Results on Direct Scattering Problem*

9.2.2.1 Green's Function and Its Asymptotic Behaviour at Infinity

For the subsequent use, we briefly present the Green function $G(x, y)$ of the two-layered scattering problem, i.e., the fundamental solution of the unperturbed problem (9.49)–(9.55) with $\Omega = \emptyset$, and discuss its asymptotic behaviour at infinity.

For an observation point $x = (x^1, x^2, \cdots, x^n) \in \mathbb{R}^n_\pm$ and a source point $y = (y^1, \cdots, y^n) \in \mathbb{R}^n_-$, define $x' := (x^1, x^2, \cdots, x^{n-1})$ and $y' := (y^1, y^2, \cdots, y^{n-1})$. Let $\xi = (\xi^1, \cdots, \xi^{n-1}) \in \mathbb{R}^{n-1}$. Define $\eta_\pm := \sqrt{k_\pm^2 - |\xi|^2}$ with $\Im(\eta_\pm) \geq 0$. By using the transmission conditions across $\mathbb{R}^n_0$ and the Fourier transformation technique, one can derive that the Green function for $y^n < 0$ is given by (see e.g., [10])

$$G(x, y) = \begin{cases} G^t(x, y), & x \in \mathbb{R}^n_+ \\ G^i(x, y) + G^r(x, y), & x \in \mathbb{R}^n_-,\ x \neq y \end{cases} \tag{9.58}$$

with

$$G^i(x, y) := \begin{cases} \frac{e^{ik_-|x-y|}}{4\pi|x-y|}, & n = 3; \\ \frac{i}{4} H_0^{(1)}(k_-|x-y|), & n = 2, \end{cases} \tag{9.59}$$

$$G^t(x, y) := \frac{i}{2\pi} \int_{\mathbb{R}^{n-1}} \frac{e^{i(\eta_+ x^n - \eta_- y^n)}}{\eta_+ + \eta_-} e^{i\xi \cdot (x' - y')} d\xi \tag{9.60}$$

and

$$G^r(x,y) := \frac{i}{4\pi}\int_{\mathbb{R}^{n-1}} \frac{\eta_- - \eta_+}{\eta_-(\eta_+ + \eta_-)} e^{-i\eta_-(x^n+y^n)} e^{i\xi\cdot(x'-y')} d\xi, \tag{9.61}$$

where $H_0^{(1)}$ is the Hankel function of the first kind and order zero.

Denote by $\hat{\theta}_c \in [0,\pi)$ the critical observation angle which is defined by $\cos\hat{\theta}_c = k_-/k_+$ if $k_- < k_+$ and $\hat{\theta}_c = 0$ if $k_- \geq k_+$. Since the observation for the inverse problem shall be made in the upper half-space $\mathbb{R}^n_+$, we are mainly interested in the asymptotic formula of $G^t(x,y)$, which is given by (cf. [10])

$$G^t(x,y) = \gamma_n \frac{e^{ik_+ r}}{r^{\frac{n-1}{2}}} \left\{ T(\theta_{\hat{x}}) e^{-ik_- \widehat{x^t}\cdot y} + O\left(\frac{1}{r}\right)\right\}, \quad x = r\widehat{x} \tag{9.62}$$

with

$$\widehat{x} := \begin{cases} (\cos\theta_{\hat{x}}\cos\phi_{\hat{x}},\ \cos\theta_{\hat{x}}\sin\phi_{\hat{x}},\ \sin\theta_{\hat{x}}), & n=3; \\ (\cos\theta_{\hat{x}}, \sin\theta_{\hat{x}}), & n=2, \end{cases} \tag{9.63}$$

where $\theta_{\hat{x}} \in (\hat{\theta}_c, \pi - \hat{\theta}_c)$, $\phi_{\hat{x}} \in (0, 2\pi)$ and

$$\widehat{x^t} := \begin{cases} (\cos\chi_{\hat{x}}\cos\phi_{\hat{x}},\ \cos\chi_{\hat{x}}\sin\phi_{\hat{x}},\ \sin\chi_{\hat{x}}), & n=3; \\ (\cos\chi_{\hat{x}}, \sin\chi_{\hat{x}}), & n=2, \end{cases} \tag{9.64}$$

with $\chi_{\hat{x}} \in (0,\pi)$ uniquely determined by the relation

$$k_+\cos\theta_{\hat{x}} = k_-\cos\chi_{\hat{x}}. \tag{9.65}$$

9.2.2.2 Translation Relation

We first fix some notations that shall be used throughout the rest of the subsection. Let D be a bounded simply connected Lipschitz domain in $\mathbb{R}^n_-$. For any $z \in \mathbb{R}^n_-$ and $\rho \in \mathbb{R}_+$, we define $z + D := \{z + x; x \in D\}$ and $\rho D := \{\rho x; x \in D\}$. Moreover, for a unitary rotation matrix $U \in SO(n)$, we define $UD := \{Ux; x \in D\}$. Let $\Omega = z + \rho U D$. We write the quaternion $(D; z, \rho, U)$ to represent the scatterer Ω. D is referred to as a *base*, and z, ρ and U are respectively referred to as the *location, size* and *orientation* of the scatterer Ω with respect to the base scatterer D. Throughout, we assume that $\Omega \subset \mathbb{R}^n_-$. Furthermore, it is assumed that the physical property of the scatterer Ω is inherited from the base scatterer D. That is, if D is sound-soft (resp. sound-hard or of impedance type), then Ω is also sound-soft (resp. sound-hard or of impedance type). We write D^s, D^h and D^i to indicate that the scatterer is sound-soft, sound-hard and of impedance type, respectively. In the case that D is of impedance type with the surface impedance parameter $\lambda(x)$ for $x \in \partial D$, then the surface impedance parameter for $\Omega = (D; z, \rho, U)$ is given by $\lambda(\frac{1}{\rho}U^T(x-z))$. Without the superscript indication, D could be a scatter of any of the three types.

In this section, we consider the scattering due to a translated obstacle $\Omega = (D; z) := (D; z, 1, I)$. In the sequel, we write the scattered wave as $u^s(\cdot; \Omega)$ in order to indicate its dependence on the underlying scatterer Ω. We need make use of the following Green formula,

$$u^s(x; \Omega) = \int_{\partial\Omega} \left\{ u^s(y; \Omega) \frac{\partial G(x, y)}{\partial \nu(y)} - G(x, y) \frac{\partial u^s(y; \Omega)}{\partial \nu(y)} \right\} ds(y), \quad x \in \mathbb{R}^n \backslash \overline{\mathbb{R}^n_0 \cup \Omega}, \tag{9.66}$$

whose proof follows from a similar argument to that of Theorem 2.5 in [16]. Applying the asymptotic formula (9.62) of $G^t(x, y)$ to the Green formula (9.66), one can show by straightforward calculations that the scattering amplitude of $u^s(\cdot; \Omega)$ on $\mathbb{S}^{n-1}_+$ is given by

$$u^\infty(\widehat{x}; \Omega) = T(\theta_{\widehat{x}}) \int_{\partial\Omega} \left\{ u^s(y; \Omega) \frac{\partial e^{-ik_-\widehat{x^t}\cdot y}}{\partial \nu(y)} - e^{-ik_-\widehat{x^t}\cdot y} \frac{\partial u^s(y; \Omega)}{\partial \nu(y)} \right\} ds(y), \quad \widehat{x} \in \mathbb{S}^{n-1}_+. \tag{9.67}$$

Next, we present a relation of the scattering amplitude due to the translation of the underlying scatterer.

Lemma 9.2.1 *Let $\Omega = (D; z) \subset \mathbb{R}^n_-$. Then we have*

$$u^\infty(\widehat{x}; \Omega) = e^{ik_-(d^t - \widehat{x^t})\cdot z} u^\infty(\widehat{x}; D), \tag{9.68}$$

where d^t, $\widehat{x}$ and $\widehat{x^t}$ are given by (9.52), (9.63) and (9.64), respectively, satisfying the relations (9.53) and (9.65).

Proof Since $e^{ik_- d^t\cdot z}$ is a constant, the boundary condition (9.56) implies that, for $y \in \partial\Omega$,

$$\mathscr{B}u^s(y; \Omega) = -\mathscr{B}u^t(y; \Omega) = -e^{ik_- d^t\cdot z}\mathscr{B}u^t(x; D) = e^{ik_- d^t\cdot z}\mathscr{B}u^s(x; D),$$

where $x = y - z \in \partial D$. By the uniqueness of the direct scattering problem one has

$$u^s(y; \Omega) = e^{ik_- d^t\cdot z} u^s(x; D), \quad \forall\, y \in \mathbb{R}^n \backslash \overline{\Omega}.$$

Therefore, we have for $y \in \partial\Omega$

$$u^s(y; \Omega) = e^{ik_- d^t\cdot z} u^s(x; D) \quad \text{and} \quad \frac{\partial u^s(y; \Omega)}{\partial \nu(y)} = e^{ik_- d^t\cdot z} \frac{\partial u^s(x; D)}{\partial \nu(x)},$$

which implies that

$$
\begin{aligned}
& u^{\infty}(\widehat{x}; \Omega) \\
&= T(\theta_{\hat{x}}) \int_{\partial\Omega} \left\{ u^s(y;\Omega) \frac{\partial e^{-ik_-\widehat{x^i}\cdot y}}{\partial\nu(y)} - e^{-ik_-\widehat{x^i}\cdot y} \frac{\partial u^s(y;\Omega)}{\partial\nu(y)} \right\} ds(y) \\
&= e^{ik_-(d^t-\widehat{x^i})\cdot z} T(\theta_{\hat{x}}) \int_{\partial D} \left\{ u^s(y;D) \frac{\partial e^{-ik_-\widehat{x^i}\cdot y}}{\partial\nu(y)} - e^{-ik_-\widehat{x^i}\cdot y} \frac{\partial u^s(y;D)}{\partial\nu(y)} \right\} ds(y) \\
&= e^{ik_-(d^t-\widehat{x^i})\cdot z} u^{\infty}(\widehat{x}; D).
\end{aligned}
$$

The proof is complete.

9.2.2.3 Scattering From Sparse Scatterers

In this section, we consider the scattering from sparse scatterers. Let Ω_1 and Ω_2 be two scatterers contained in $\mathbb{R}^n_-$ with

$$
L := \operatorname{dist}(\Omega_1, \Omega_2) \gg 1. \tag{9.69}
$$

In order to ease the exposition, we assume that both Ω_1 and Ω_2 are bounded C^2 domains with connected complements. For any $\mathbf{a} \in C(\partial\Omega)$, we introduce the single- and double-layer operators $S_\Omega : C(\partial\Omega) \to C(\partial\Omega)$ and $K_\Omega : C(\partial\Omega) \to C(\partial\Omega)$, defined respectively by

$$
(S_\Omega \mathbf{a})(x) = \int_{\partial\Omega} G(x,y)\mathbf{a}(y)\, ds(y), \ (K_\Omega \mathbf{a})(x) = \int_{\partial\Omega} \frac{\partial G(x,y)}{\partial\nu(y)} \mathbf{a}(y)\, ds(y), \ x \in \partial\Omega. \tag{9.70}
$$

Moreover, by changing the integration domain in (9.70) to $\partial\Omega_1$, we denote the resulting operators by S_1 and K_1, respectively; and by changing the integration domain to $\partial\Omega_2$, we denote the resulting operators by S_2 and K_2, respectively. We refer to [16] for related mapping properties of these operators.

Lemma 9.2.2 *Let Ω_1 and Ω_2 be two scatterers buried in $\mathbb{R}^n_-$ as described above. Then we have*

$$
u^{\infty}(\widehat{x}; \Omega_1 \cup \Omega_2) = u^{\infty}(\widehat{x}; \Omega_1) + u^{\infty}(\widehat{x}; \Omega_2) + \mathscr{O}(L^{\frac{1-n}{2}}). \tag{9.71}
$$

Proof We first consider the case that both Ω_1 and Ω_2 are sound-soft. The scattered field $u^s(x; \Omega_1 \cup \Omega_2)$ can be represented in the form

$$
\begin{aligned}
u^s(x; \Omega_1 \cup \Omega_2) = & \int_{\partial\Omega_1} \left\{ \frac{\partial G(x,y)}{\partial\nu(y)} - iG(x,y) \right\} \mathbf{a}_1(y)\, ds(y) \\
& + \int_{\partial\Omega_2} \left\{ \frac{\partial G(x,y)}{\partial\nu(y)} - iG(x,y) \right\} \mathbf{a}_2(y)\, ds(y), \quad x \in \mathbb{R}^n \backslash \overline{\Omega_1 \cup \Omega_2},
\end{aligned}
$$

with two densities $\mathbf{a}_1 \in C(\partial\Omega_1)$ and $\mathbf{a}_2 \in C(\partial\Omega_2)$. By using the homogeneous Dirichlet boundary condition, the densities $\mathbf{a}_1$ and $\mathbf{a}_2$ satisfy the following system of integral equations

$$\frac{1}{2}\mathbf{a}_1 + K_1\mathbf{a}_1 - iS_1\mathbf{a}_1 + K_2\mathbf{a}_2 - iS_2\mathbf{a}_2 = f_1 \quad \text{on} \quad \partial\Omega_1, \tag{9.72}$$

$$\frac{1}{2}\mathbf{a}_2 + K_2\mathbf{a}_2 - iS_2\mathbf{a}_2 + K_1\mathbf{a}_1 - iS_1\mathbf{a}_1 = f_2 \quad \text{on} \quad \partial\Omega_2, \tag{9.73}$$

where $f_1 := -u_0$ on $\partial\Omega_1$ and $f_2 := -u_0$ on $\partial\Omega_2$. Since the distance $L \gg 1$, by straightforward calculations, it is easily verified that (see e.g., [22])

$$|G(x,y)| = \mathscr{O}(L^{\frac{1-n}{2}}) \quad \text{and} \quad \left|\frac{G(x,y)}{\nu(y)}\right| = \mathscr{O}(L^{\frac{1-n}{2}}).$$

and therefore

$$\|K_2 - iS_2\|_{\mathscr{L}(C(\partial\Omega_2),C(\partial\Omega_1))} = \mathscr{O}(L^{\frac{1-n}{2}}). \tag{9.74}$$

It is noted that the operator $\frac{1}{2}I + K_1 - iS_1 : C(\partial\Omega_1) \to C(\partial\Omega_1)$ is bijective and the inverse $(\frac{1}{2}I + K_1 - iS_1)^{-1}$ is bounded (see e.g.,[16]). Then from (9.72), using the behaviour (9.74), one has

$$\mathbf{a}_1 = \left(\frac{1}{2}I + K_1 - iS_1\right)^{-1}[f_1 - (K_2 - iS_2)\mathbf{a}_2] = \widetilde{\mathbf{a}}_1 + \mathscr{O}(L^{\frac{1-n}{2}}), \tag{9.75}$$

with $\widetilde{\mathbf{a}}_1 = (\frac{1}{2}I + K_1 - i\eta S_1)^{-1} f_1$. Similarly, one can show that

$$\mathbf{a}_2 = \widetilde{\mathbf{a}}_2 + \mathscr{O}(L^{\frac{1-n}{2}}), \tag{9.76}$$

with $\widetilde{\mathbf{a}}_2 = (\frac{1}{2}I + K_2 - i\eta S_2)^{-1} f_2$. Therefore, we have

$$\begin{aligned} u^\infty(\widehat{x};\Omega_1 \cup \Omega_2) =& T(\theta_{\hat{x}}) \int_{\partial\Omega_1} \left\{ \frac{\partial e^{-ik_-\widehat{x^t}\cdot y}}{\partial\nu(y)} - e^{-ik_-\widehat{x^t}\cdot y} \right\} \mathbf{a}_1(y)\, ds(y) \\ &+ T(\theta_{\hat{x}}) \int_{\partial\Omega_2} \left\{ \frac{\partial e^{-ik_-\widehat{x^t}\cdot y}}{\partial\nu(y)} - e^{-ik_-\widehat{x^t}\cdot y} \right\} \mathbf{a}_2(y)\, ds(y) \\ =& u^\infty(\widehat{x};\Omega_1) + u^\infty(\widehat{x};\Omega_2) + \mathscr{O}(L^{\frac{1-n}{2}}), \end{aligned}$$

where we have used the estimates (9.75)-(9.76) and the fact that

$$u^\infty(\widehat{x};\Omega_j) = T(\theta_{\hat{x}}) \int_{\partial\Omega_1} \left\{ \frac{\partial e^{-ik_-\widehat{x^t}\cdot y}}{\partial\nu(y)} - e^{-ik_-\widehat{x^t}\cdot y} \right\} \widetilde{\mathbf{a}}_j(y)\, ds(y), \quad j = 1, 2.$$

For the other cases that if Ω_1 and Ω_2 are obstacles of other types, we seek the solution in the form

$$u^s(x;\Omega_1\cup\Omega_2)=\int_{\partial\Omega_1}G(x,y)\mathbf{a}_1(y)+\int_{\partial\Omega_2}G(x,y)\mathbf{a}_2(y)\,ds(y),\quad x\in\mathbb{R}^n\backslash\overline{\Omega_1\cup\Omega_2},$$

if we assume that k_-^2 is not a Dirichlet eigenvalue of $-\Delta$ in $\Omega_1\cup\Omega_1$. To remove the additional assumption, one may consider some more complex form, see e.g., (3.29) in [16]. Then (9.71) follows from similar the arguments as in the sound-soft case.

The proof is complete. □

Finally, we would like to remark that by using the mapping properties of the single- and double-layer boundary integral operators in [42], one can show that similar results to Lemma 9.2.2 hold when Ω_1 and Ω_2 are Lipschitz domains.

9.2.2.4 Scattering From Multiple Small Scatterers

Starting from now and throughout the rest of the subsection, we assume that the wave number $k=\mathcal{O}(1)$. Hence, the size of a scatterer Ω can be characterized by its Euclidean diameter. Let $D_j\subset\mathbb{R}^n_-$, $j=1,2,\ldots,l$ be a family of base scatterers. For technical reasons, we assume that ∂D_j, $j=1,2,\ldots,l$ are C^2 continuous in this section. Let $\rho\in\mathbb{R}_+$ with $\rho\ll 1$, and

$$\Omega_j^p=(D_j^p;z_j,\rho):=z_j+\rho D_j^p,\quad j=1,2,\ldots,l,\tag{9.77}$$

where $z_j\in\mathbb{R}^n_-$ and $p=s,h$ or i, representing the type of the scatterer. It is assumed that $\Omega_j^p\subset\mathbb{R}^n_-$. Set

$$\Omega^p=\bigcup_{j=1}^{l}\Omega_j^p,\tag{9.78}$$

which represents the multiple small scatterers for our inverse scattering reconstruction. For the scatterer Ω^p introduced in (9.78), we further assume that

$$L:=\min_{1\le j,j'\le l,j\ne j'}\operatorname{dist}(z_j,z_{j'})\gg 1.\tag{9.79}$$

This means, the obstacle components of Ω^p in (9.78) are sparsely distributed.

Let $Y_\alpha^\beta(\cdot)$ for $\alpha\in\mathbb{N}\cup\{0\}$ and $\beta=-\alpha,\ldots,\alpha$ be the spherical harmonics which form a complete orthonormal system in $L^2(\mathbb{S}^{n-1})$ (cf. [16]). In particular, we recall the spherical harmonics $Y_\alpha^\beta(\hat{x})$ of order $\alpha=0,1$, for $\hat{x}=(\hat{x}^l)_{l=1}^n\in\mathbb{S}^{n-1}$. In the three-dimensional case,

$$Y_0^0(\hat{x})=\sqrt{\frac{1}{4\pi}},\ Y_1^{-1}(\hat{x})=\sqrt{\frac{3}{8\pi}}(\hat{x}^1-i\hat{x}^2),\ Y_1^0(\hat{x})=\sqrt{\frac{3}{4\pi}}\hat{x}^3,\ Y_1^1(\hat{x})=\sqrt{\frac{3}{8\pi}}(\hat{x}^1+i\hat{x}^2).$$

In the two-dimensional case, Y_1^0 does not exist and

$$Y_0^0(\hat{x}) = \sqrt{\frac{1}{2\pi}}, \quad Y_1^{-1}(\hat{x}) = \sqrt{\frac{1}{2\pi}}(\hat{x}^1 - i\hat{x}^2), \quad Y_1^1(\hat{x}) = \sqrt{\frac{1}{2\pi}}(\hat{x}^1 + i\hat{x}^2).$$

We are in a position to present the main result of this section on the scattering from multiple small scatterers.

Theorem 9.2.1 *Let Ω^p be the multiple small scatterers as described in (9.77)–(9.79). Let $u^\infty(\widehat{x}; \Omega^p) \in L^2(\mathbb{S}_+^{n-1})$ be the scattering amplitude corresponding to a single incident plane wave $u^i(x) = e^{ik_+ x\cdot d^i}$. Then, for sufficiently large L, as $\rho \to +0$, the scattering amplitude corresponding to the sound-soft case satisfies*

$$\begin{aligned} u^\infty(\widehat{x}; \Omega^s) =& \rho^{n-2}(\ln\rho)^{n-3} T(\theta_{\hat{x}}) Y_0^0(\widehat{x^t}) \sum_{j=1}^{l} c_j^s e^{ik_-(d^t - \widehat{x^t})\cdot z_j} \\ &+ \mathscr{O}\left(\rho^{2n-4}(\ln\rho)^{2n-6} + L^{\frac{1-n}{2}}\right), \end{aligned} \tag{9.80}$$

where c_j^s are constants depending on D_j, k_- and d^i, but independent of ρ. In the case when $p = i$, we have

$$\begin{aligned} u^\infty(\widehat{x}; \Omega^i) =& \rho^{n-1} T(\theta_{\hat{x}}) Y_0^0(\widehat{x^t}) \sum_{j=1}^{l} c_j^e \overline{\lambda}_j e^{ik_-(d^t - \widehat{x^t})\cdot z_j} \\ &+ \mathscr{O}\left(\rho^n (\ln\rho)^{3-n} + L^{\frac{1-n}{2}}\right), \end{aligned} \tag{9.81}$$

where $\widehat{\lambda}_j := \int_{\partial\Omega_j} \lambda_j ds / |\partial\Omega_j|$ and c_j^e are constants depending on D_j, k_- and d^i, but independent of ρ. In the case when $p = h$, the scattering amplitude satisfies

$$\begin{aligned} u^\infty(\widehat{x}; \Omega^h) =& \rho^n T(\theta_{\hat{x}}) \sum_{j=1}^{l} \sum_{\alpha=0}^{1} \sum_{\beta=-\alpha}^{\alpha} c_{\alpha,\beta,j}^h e^{ik_-(d^t - \widehat{x^t})\cdot z_j} Y_\alpha^\beta(\widehat{x^t}) \\ &+ \mathscr{O}\left(\rho^{n+1}(\ln\rho)^{3-n} + L^{\frac{1-n}{2}}\right), \end{aligned} \tag{9.82}$$

where $c_{\alpha,\beta,j}^h$ are constants depending on D_j, k_- and d^i, but independent of ρ, and Y_1^0 should be removed from the summation in (9.82) for the two-dimensional case.

Proof By Lemma 9.2.2, it is sufficient to prove the case with a single scatterer of the form $\Omega = z + \rho D$. We present the proof mainly for the sound-soft case, namely (9.80), and remark the major modifications required for the sound-hard and impedance cases, namely (9.81) and (9.82), after the proof of Lemma 9.2.3.

We first derive the asymptotic expansion of the Green function $G(x, y)$ introduced in Section 9.2.2.1 as $|x-y| \to +0$. Using the series expansions of the Hankel functions and the exponential function we find that for $|x-y| \to +0$,

$$G(x,y) = G_0(x,y) + G^r(x,y) + \begin{cases} \frac{ik_-}{4\pi} + \mathcal{O}(|x-y|), & n=3; \\ \frac{i}{4} - \frac{1}{2\pi}\ln\frac{k_-}{2} - \frac{C_E}{2} + \mathcal{O}(|x-y|^2 \ln|x-y|), & n=2, \end{cases} \tag{9.83}$$

and

$$\nabla_y G(x,y) = \nabla_y G_0(x,y) + \nabla_y G^r(x,y) + \begin{cases} \mathcal{O}(|x-y|^2), & n=3; \\ \frac{-k_-^2}{4\pi}(x-y)\ln(|x-y|) + \mathcal{O}(|x-y|), & n=2, \end{cases} \tag{9.84}$$

where $C_E := \lim_{p\to\infty}\left\{\Sigma_{m=1}^{p}\frac{1}{m} - \ln p\right\} \approx 0.57721566$ denotes Euler's constant and

$$G_0(x,y) := \frac{1}{4\pi|x-y|}, \quad n=3; \quad \frac{1}{2\pi}\ln\frac{1}{|x-y|}, \quad n=2,$$

represents the fundamental solution of the Laplace equation. In the above derivation, we have used the fact that $G^r(\cdot, y)$ is an analytic function in $\mathbb{R}^n_\pm$.

Next, for any $\mathbf{a} \in C(\partial B)$, $(B = D, \Omega)$, we introduce the boundary integral operators $S_B : C(\partial B) \to C(\partial B)$, $K_B : C(\partial B) \to C(\partial B)$ and $K'_B : C(\partial B) \to C(\partial B)$ defined by

$$\begin{aligned} (S_B\mathbf{a})(x) &= \int_{\partial B} \mathbf{a}(y) G(x,y)\, ds(y), \quad x \in \partial B, \\ (K_B\mathbf{a})(x) &= \int_{\partial B} \mathbf{a}(y) \frac{\partial G(x,y)}{\partial \nu(y)}\, ds(y), \quad x \in \partial B, \\ (K'_B\mathbf{a})(x) &= \int_{\partial B} \mathbf{a}(y) \frac{\partial G(x,y)}{\partial \nu(x)}\, ds(y), \quad x \in \partial B, \end{aligned} \tag{9.85}$$

respectively. Similarly, we let S_B^0, K_B^0 and $K_B^{0,\prime}$ be the corresponding operators introduced in (9.85) when the integral kernel $G(x, y)$ is replaced by $G_0(x, y)$. Finally, we define $M_B^0 : C(\partial B) \to C(\partial B)$ by

$$(M_B^0\mathbf{a})(x) = \int_{\partial B} \mathbf{a}(y)\, ds(y), \quad x \in \partial B.$$

In the following, for any $x \in \partial\Omega$ we define the one-to-one corresponding point $\xi_x \in \partial D$ by $\xi_x := (x - z)/\rho$. Accordingly, for any $\mathbf{a}_D \in C(\partial D)$, we define $\mathbf{a}_\Omega \in C(\partial\Omega)$ as

$$\mathbf{a}_\Omega(x) := \mathbf{a}_D(\xi_x), \quad x \in \partial\Omega. \tag{9.86}$$

With the above preparations, we let $u^s(x;\Omega^s)$ denote the scattered wave field corresponding to Ω^s, and make use of the following ansatz

$$u^s(x;\Omega^s)=\int_{\partial\Omega}\left\{\frac{\partial G(x,y)}{\partial\nu(y)}-i\varrho G(x,y)\right\}\mathbf{a}_\Omega(y)\,ds(y),\quad x\in\mathbb{R}^n\backslash\overline{\Omega},\tag{9.87}$$

where the coupling parameter ϱ is chosen to be

$$\varrho:=\rho^{-1}(\ln\rho)^{n-3}.\tag{9.88}$$

By using the homogeneous Dirichlet boundary condition and the jumping properties of the integral operators, we see that $\mathbf{a}_\Omega\in C(\partial\Omega)$ satisfies

$$\left(\frac{1}{2}I+K_\Omega-i\varrho S_\Omega\right)\mathbf{a}_\Omega=-u^t\quad\text{on }\partial\Omega,\tag{9.89}$$

where I denotes the identity operator. Note that the operator $\frac{1}{2}I+K_\Omega-i\varrho S_\Omega:C(\partial\Omega)\to C(\partial\Omega)$ is bijective and hence we have for $x\in\mathbb{R}^n\backslash\overline{\Omega}$

$$u^s(x;\Omega^s)=-\int_{\partial\Omega}\left(\frac{\partial G(x,y)}{\partial\nu(y)}-i\varrho G(x,y)\right)\left[\left(\frac{1}{2}I+K_\Omega-i\varrho S_\Omega\right)^{-1}u^t\right](y)\,ds(y),$$

which further implies by using (9.62) that the corresponding scattering amplitude $u_\rho^\infty(\widehat{x},d^i)$ for $\widehat{x}\in\mathbb{S}_+^{n-1}$ is given by

$$\begin{aligned}&u^\infty(\hat{x};\Omega^s)\\&=-T(\theta_{\hat{x}})\int_{\partial\Omega}\left(\frac{\partial e^{-ik_-\widehat{x^t}\cdot y}}{\partial\nu(y)}-i\varrho e^{-ik_-\widehat{x^t}\cdot y}\right)\left[\left(\frac{1}{2}I+K_\Omega-i\varrho S_\Omega\right)^{-1}u^t(\cdot,d^i)\right](y)\,ds(y)\\&=iT(\theta_{\hat{x}})\int_{\partial\Omega}(k_-\widehat{x^t}\cdot\nu(y)+\varrho)e^{-ik_-\widehat{x^t}\cdot y}\left[\left(\frac{1}{2}I+K_\Omega-i\varrho S_\Omega\right)^{-1}u^t(\cdot,d^i)\right](y)\,ds(y).\end{aligned}\tag{9.90}$$

Next, we introduce an operator $A_D:C(\partial D)\to C(\partial D)$ by

$$A_D:=\frac{1}{2}I+K_D^0-iS_D^0,\ n=3;\quad\frac{1}{2}I+K_D^0+iM_D^0,\ n=2.\tag{9.91}$$

Then by applying Lemma 9.2.3 in the following and Theorem 10.1 in [27] to (9.90), we have by direct calculations that

$$
\begin{aligned}
&u^\infty(\widehat{x};\Omega^s)\\
&= iT(\theta_{\hat{x}})\int_{\partial\Omega}(k\_\widehat{x^t}\cdot\nu(y)+\varrho)e^{-ik\_\widehat{x^t}\cdot y}\left[\left(\frac{1}{2}I+K_\Omega-i\varrho S_\Omega\right)^{-1}u^t(\cdot,d^i)\right](y)\,ds(y)\\
&= iT(\theta_{\hat{x}})\int_{\partial D}(k\_\widehat{x^t}\cdot\nu(\xi_y)+\varrho)[e^{-ik\_\widehat{x^t}\cdot z}+\mathscr{O}(\rho)]\\
&\qquad\left\{A_D^{-1}[u^t(z,d^i)+\mathscr{O}(\rho)]+\mathscr{O}(\rho^{n-2}(\ln\rho)^{n-3})\right\}\rho^{n-1}\,ds(\xi_y),
\end{aligned}
$$

which readily implies (9.80).

The proof is complete. □

The following lemma was required in the proof of Theorem 9.2.1.

Lemma 9.2.3 *Let $\Omega = z + \rho D$ and let A_D be defined in* (9.91). *For $\mathbf{a}_D \in C(\partial D)$, we let $\mathbf{a}_\Omega \in C(\partial\Omega)$ be defined as in (9.86). Let ϱ be given in (9.88). Then there holds*

$$
\left(\frac{1}{2}I+K_\Omega-i\varrho S_\Omega\right)\mathbf{a}_\Omega = A_D\mathbf{a}_D+\mathscr{O}(\rho^{n-2}(\ln\rho)^{n-3}) \tag{9.92}
$$

uniformly on $\partial\Omega$.

Proof Using change of variables in the integrals, we have by direct computations that

$$
\begin{aligned}
&(S_\Omega^0\mathbf{a}_\Omega)(x) = \int_{\partial\Omega}G_0(x,y)\mathbf{a}_\Omega(y)ds(y)\\
&=\frac{1}{4\pi}\int_{\partial\Omega}\frac{1}{|x-y|}\mathbf{a}_\Omega(y)\,ds(y),\ n=3;\ -\frac{1}{2\pi}\int_{\partial\Omega}\ln(|x-y|)\mathbf{a}_\Omega(y)\,ds(y),\ n=2\\
&=\frac{1}{4\pi}\int_{\partial D}\frac{1}{\rho|\xi_x-\xi_y|}\mathbf{a}_D(\xi_y)\rho^2\,ds(\xi_y),\ n=3;\ -\frac{1}{2\pi}\int_{\partial D}\ln(\rho|\xi_x-\xi_y|)\mathbf{a}_\Omega(y)\rho\,ds(\xi_y), n=2\\
&=\rho(S_D^0\mathbf{a}_D)(\xi_x),\ n=3;\ -\rho\ln\rho(M_D^0\mathbf{a}_D)(\xi_x)+O(\rho),\ n=2.
\end{aligned} \tag{9.93}
$$

and

$$
\begin{aligned}
&(K_\Omega^0\mathbf{a}_\Omega)(x) = \int_{\partial\Omega}\frac{\partial G_0(x,y)}{\partial\nu(y)}\mathbf{a}_\Omega(y)\,ds(y) = \frac{1}{2^{n-1}\pi}\int_{\partial\Omega}\frac{\nu(y)\cdot(x-y)}{|x-y|^n}\mathbf{a}_\Omega(y)\,ds(y)\\
&=\frac{1}{2^{n-1}\pi}\int_{\partial D}\frac{\nu(\xi_y)\cdot(\xi_x-\xi_y)}{\rho^{n-1}|\xi_x-\xi_y|^n}\phi_\Omega(\xi_y)\rho^{n-1}\,ds(\xi_y) = (K_D^0\mathbf{a}_D)(\xi_x).
\end{aligned} \tag{9.94}
$$

Since $G^r(\cdot,y)$ is an analytic function in $\mathbb{R}^n_-$, we see

$$
G^r(x,y)=\mathscr{O}(1)\quad\text{and}\quad|\nabla_y G^r(x,y)|=\mathscr{O}(1)\qquad\text{as }|x-y|\to+0
$$

and hence by (9.83) and (9.84) we can show that

$$G(x,y) - G_0(x,y) = \mathscr{O}(1) \quad \text{and} \quad |\nabla_y[G(x,y) - G_0(x,y)]| = \mathscr{O}(1) \qquad \text{as } |x-y| \to 0.$$

This further implies that as $\rho \to +0$

$$\begin{aligned} \Big((S_\Omega - S_\Omega^0)\mathbf{a}_\Omega\Big)(x) &= \int_{\partial\Omega} [G(x,y) - G_0(x,y)]\mathbf{a}_\Omega(y)ds(y) \\ &= \int_{\partial D} \mathscr{O}(1)\mathbf{a}_D(\xi_y)\rho^{n-1}\, ds(\xi_y) = \mathscr{O}(\rho^{n-1}), \end{aligned} \tag{9.95}$$

and

$$\begin{aligned} \Big((K_\Omega - K_\Omega^0)\mathbf{a}_\Omega\Big)(x) &= \int_{\partial\Omega} \left(\frac{\partial[G(x,y) - G_0(x,y)]}{\nu(y)}\right)\mathbf{a}_\Omega(y)\, ds(y) \\ &= \int_{\partial D} \mathscr{O}(1)\mathbf{a}_D(\xi_y)\rho^{n-1}\, ds(\xi_y) = \mathscr{O}(\rho^{n-1}). \end{aligned} \tag{9.96}$$

Finally, by combining (9.93)–(9.96), we readily have (9.92).

The proof is complete. □

In the rest of this section, we give the necessary modifications required for proving Theorem 9.2.1 in the sound-hard and impedance cases, namely (9.81) and (9.82). In doing so, we take the Neumann boundary condition in (9.56) a special impedance boundary condition with $\lambda \equiv 0$ on $\partial\Omega$. We make use of the following asantz for the scattered wave field

$$u^s(x;\Omega^i) = \int_{\partial\Omega} G(x,y)\mathbf{a}_\Omega(y)ds(y), \quad x \in \mathbb{R}^n\backslash\overline{\Omega}, \tag{9.97}$$

where $\mathbf{a}_\Omega \in C(\partial\Omega)$ can be shown to satisfy the following integral equation

$$\left(-\frac{1}{2}I + K_\Omega' + i\lambda S_\Omega\right)\mathbf{a}_\Omega = -f_\Omega := -\frac{\partial u^t}{\partial\nu} - i\lambda u^t \quad \text{on } \partial\Omega. \tag{9.98}$$

Similar to the sound-soft case, we define $f_D \in C(\partial D)$ by $f_D(\xi_x) := f_\Omega(x)$ for all $\xi_x \in \partial D$. The operator $-\frac{1}{2}I + K_\Omega' + i\lambda S_\Omega : C(\partial\Omega) \to C(\partial\Omega)$ is bijective and the inverse $(-\frac{1}{2}I + K_\Omega' + i\lambda S_\Omega)^{-1} : C(\partial\Omega) \to C(\partial\Omega)$ is bounded provided that k_-^2 is not a Dirichlet eigenvalue of $-\Delta$ in Ω. Such an assumption on k_-^2 can always be fulfilled by noting that we are only interested in the asymptotic behaviour of u_ρ^s as $\rho \to +0$, and hence we may choose that $\rho < \pi/(2k_-)$ which ensures that k_-^2 is not a Dirichlet eigenvalue of $-\Delta$ in Ω (see Sect. 5.1 in [16]). The next lemma is an important ingredient and is a counterpart to Lemma 9.2.3, whose proof follows from a similar argument to that of Lemma 9.2.3.

Lemma 9.2.4 *For $\mathbf{a}_D \in C(\partial D)$, we let $\mathbf{a}_\Omega \in C(\partial\Omega)$ be defined as in (9.86). Then there holds*

$$\begin{aligned}&\left(-\frac{1}{2}I + K'_\Omega + i\lambda S_\Omega\right)\mathbf{a}_\Omega\\ &= \left(-\frac{1}{2}I + K_D^{0,'}\right)\mathbf{a}_D + \mathscr{O}(\rho^{2-\tau_\lambda}(\ln\rho)^{3-n})\end{aligned} \tag{9.99}$$

uniformly on $\partial\Omega$. In (9.99), $\tau_\lambda := 0$ if $\lambda \equiv 0$, otherwise $\tau_\lambda := 1$.

The operator $-\frac{1}{2}I + K_D^{0,'} : C(\partial D) \to C(\partial D)$ is bijective and its inverse is bounded (see Theorem 6.26 in [27]). Using Theorem 10.1 in [27], Lemma 9.2.4 implies that for any $\mathbf{a}_D \in C(\partial D)$,

$$\begin{aligned}&\left(-\frac{1}{2}I + K'_\Omega + i\lambda S_\Omega\right)^{-1}\mathbf{a}_\Omega\\ &= \left(-\frac{1}{2}I + K_D^{0,'}\right)^{-1}\mathbf{a}_D + \mathscr{O}(\rho^{2-\tau_\lambda}(\ln\rho)^{3-n}).\end{aligned} \tag{9.100}$$

By (9.97)–(9.100), the scattering amplitude of the scattered field u_ρ^s on $\mathbb{S}_+^{n-1}$ is given by

$$\begin{aligned}u^\infty(\widehat{x};\Omega^i) =& -T(\theta_{\widehat{x}})\int_{\partial\Omega} e^{-ik_-\widehat{x^t}\cdot y}\left[\left(-\frac{1}{2}I + K'_\Omega + i\lambda S_\Omega\right)^{-1} f_\Omega\right](y)\,ds(y)\\ =& -T(\theta_{\widehat{x}})\int_{\partial D} e^{-ik_-\widehat{x^t}\cdot y}\left[\left(-\frac{1}{2}I + K_D^{0,'}\right)^{-1} f_D\right.\\ &\left. + \mathscr{O}(\rho^{2-\tau_\lambda}(\ln\rho)^{3-n})\right](\xi_y)\rho^{n-1}\,ds(\xi_y).\end{aligned} \tag{9.101}$$

Using the series expansion, one has

$$\begin{aligned}e^{-ik_-\widehat{x^t}\cdot y} =& e^{-ik_-\widehat{x^t}\cdot(z+\rho\xi_y)} = e^{-ik_-\widehat{x^t}\cdot z}\left[1 - \rho ik_-\widehat{x^t}\cdot\xi_y + \mathscr{O}(\rho^2)\right],\\ u^t(y) =& T(\theta)e^{ik_-(z+\rho\xi_y)\cdot d^t} = T(\theta)e^{ik_-z\cdot d^t}\left[1 + \mathscr{O}(\rho)\right],\end{aligned}$$

and

$$\frac{\partial u^t}{\partial\nu}(y) = T(\theta)ik_-(\nu(\xi_y)\cdot d^t)e^{ik_-z\cdot d^t}\left[1 + \rho ik_-(\xi_y\cdot d^t) + \mathscr{O}(\rho^2)\right]$$

as $\rho \to +0$. Inserting these asymptotic estimates into (9.101), we finally have by straightforward calculations that

$$
\begin{aligned}
u_\rho^\infty(\widehat{x};\Omega^i) = & \rho^{n-1} T(\theta_{\hat{x}}) e^{ik_-z\cdot(d^t-\widehat{x^t})}\widehat{\lambda} c_1 + \rho^n T(\theta_{\hat{x}}) e^{ik_-z\cdot(d^t-\widehat{x^t})}\left(c_2 + \widehat{x^t}\cdot C_3\right) \\
& + \mathscr{O}\left(\rho^{n+1-\tau_\lambda}(\ln\rho)^{3-n}\right),
\end{aligned}
\tag{9.102}
$$

where $\widehat{\lambda} := \int_{\partial\Omega}\lambda(y)ds(y)/|\partial\Omega|$, c_1, c_2 are constants and C_3 is a constant vector depending on D, k_- and d^i, but independent of ρ. (9.102) readily implies (9.81) and (9.82).

9.2.3 Recovery Scheme

The present section is devoted to our study on recovering multiple multiscale anomalies buried in a two layered medium. The general structure is described as follows. First, we present Scheme S to recover multiple small scatterers buried in a two layered medium. Then, we discuss Scheme R to recover multiple buried anomalies of regular size. Finally, using a local tuning technique to combine Schemes S and R, we obtain the final Scheme M to recover multiple multiscale buried anomalies. With the preparations in the previous section, the general strategy follows from those developed in [31, 33, 37] for recovering scatterers located in a homogeneous space. Hence, we shall focus on the necessary modifications and be sketchy at certain points by referring to [31, 33, 37] for more details.

Let Ω^p be the multiple small scatterers as described in (9.77)–(9.79). Let $u^\infty(\widehat{x};\Omega^p) \in L^2(\mathbb{S}_+^{n-1})$ be the scattering amplitude corresponding to a single incident plane wave $u^i(x) = e^{ik_+x\cdot d^i}$. We next show Scheme S to recover Ω^p by knowledge of $u^\infty(\widehat{x};\Omega^p)$. To that end, we introduce the following imaging functional

$$
J_S^p(z) := \frac{1}{\|u^\infty(\widehat{x};\Omega^p)\|^2_{L^2(\mathbb{S}_+^{n-1})}} \sum_{\alpha=0}^{\kappa(p)}\sum_{\beta=-\alpha}^{\alpha}\left|\left\langle u^\infty(\widehat{x};\Omega^p), T(\theta_{\hat{x}})e^{ik_-(d^t-\widehat{x^t})\cdot z}Y_\alpha^\beta(\widehat{x^t})\right\rangle_{L^2(\mathbb{S}_+^{n-1})}\right|^2,
\tag{9.103}
$$

where $\langle u, v\rangle_{L^2(\mathbb{S}_+^{n-1})} = \displaystyle\int_{\mathbb{S}_+^{n-1}} u\cdot\overline{v}\,ds(\widehat{x})$, and $\kappa(p) = 0$ if $p = s, i$, and $\kappa(p) = 1$ if $p = h$. It is emphasized that there is no harmonic function $Y_1^0(\widehat{x})$ in the two-dimensional case, so it should be removed from the summation in (9.103) in defining $J_S^p(z)$. Clearly, $J_S^p(z)$ is a nonnegative function for $z \in \mathbb{R}_-^n$.

The next theorem about the local maximum behaviour of $J_S^p(z)$, $p = s, i$, is the core of our Scheme S.

Theorem 9.2.2 *Let Ω^p and $J_S^p(z)$ be described as in (9.78) and (9.103), respectively. Set*

$$
\Theta_j^p := \frac{\|u^\infty(\widehat{x};\Omega_j^p)\|^2_{L^2(\mathbb{S}_+^{n-1})}}{\|u^\infty(\widehat{x};\Omega^p)\|^2_{L^2(\mathbb{S}_+^{n-1})}}, \quad j = 1,\ldots,l.
$$

Then we have for $p = s, i$,

$$\Theta_j^p = \Theta_j^{0,p} + \mathcal{O}\left(\frac{1}{L} + \rho^{n-2}(\ln\rho)^{n-3}\right), \quad j = 1, \dots, l, \tag{9.104}$$

where $\Theta_j^{0,p}$ is a positive number independent of L and ρ. Moreover, there exists an open neighborhood of z_j, neigh(z_j), $1 \le j \le l$, such that

$$J_S^p(z) \le \Theta_j^{0,p} + \mathcal{O}\left(\frac{1}{L} + \rho^{n-2}(\ln\rho)^{n-3}\right) \quad \textit{for } z \in \textit{neigh}(z_j), \tag{9.105}$$

where the equality holds only at $z = z_j$. That is, z_j is a local maximizer of $J_S^p(z)$ in $neigh(z_j)$.

Proof Without loss of generality, we only consider the local maximum behaviour of $J_S^s(z)$ in $B_\rho(z_1)$, a ball of radius ρ centered at z_1. Clearly, one has

$$|z_j - z| \ge L \gg 1 \quad \text{for } z \in B_\rho(z_1) \text{ and } j = 2, 3, \dots, l. \tag{9.106}$$

Hence, by (9.106) and (9.80), and using the Riemann-Lebesgue Lemma about oscillatory integrals, we have by direct calculations that

$$J_S^s(z) = \Theta_1^p + \mathcal{O}\left(\frac{1}{L}\right) \quad \text{for } z \in B_\rho(z_1). \tag{9.107}$$

Next, by (9.80), we clearly have that

$$\|u^\infty(\widehat{x}; \Omega^s)\|^2_{L^2(\mathbb{S}_+^{n-1})} = \rho^{2n-4}(\ln\rho)^{2n-6}\Upsilon_0 + \mathcal{O}\left(L^{1-n} + \rho^{4n-8}(\ln\rho)^{4n-12}\right), \tag{9.108}$$

where Υ_0 is a positive constant independent of L and ρ. Using (9.80) again, we see that for $z \in B_\rho(z_1)$

$$\begin{aligned}
&\left|\left\langle u^\infty(\widehat{x}; \Omega^s), T(\theta_{\hat{x}})e^{ik_-(d^t - \widehat{x^t})\cdot z} Y_0^0(\widehat{x^t})\right\rangle_{L^2(\mathbb{S}_+^{n-1})}\right| \\
=&\rho^{n-2}(\ln\rho)^{n-3}\left|\left\langle c_1^e T(\theta_{\hat{x}})e^{ik_-(d^t - \widehat{x^t})\cdot z_1} Y_0^0(\widehat{x^t}), T(\theta_{\hat{x}})e^{ik_-(d^t - \widehat{x^t})\cdot z} Y_0^0(\widehat{x^t})\right\rangle_{L^2(\mathbb{S}_+^{n-1})}\right. \\
&\left. + \mathcal{O}\left(\frac{1}{L} + \rho^{n-2}(\ln\rho)^{n-3}\right)\right| \\
\le&\rho^{n-2}(\ln\rho)^{n-3}\left(|c_1^e| \int_{\mathbb{S}_+^{n-1}} [T(\theta_{\hat{x}})Y_0^0(\widehat{x^t})]^2 \, ds(\widehat{x}) + \mathcal{O}\left(\frac{1}{L} + \rho^{n-2}(\ln\rho)^{n-3}\right)\right),
\end{aligned} \tag{9.109}$$

where the last relation was obtained by the fact that

$$|e^{ik_-(d^t - \widehat{x^t})\cdot(z_1 - z)}| \le 1 \quad \text{for all } z \in B_\rho(z_1).$$

Furthermore, the equality in the last estimate in (9.109) holds only at $z = z_1$. Finally, by combining (9.106)–(9.109), the proof can be completed by taking

$$\Theta_1^{0,s} = \frac{1}{\Upsilon_0}|c_1^e| \int_{\mathbb{S}_+^{n-1}} [T(\theta_{\hat{x}})Y_0^0(\widehat{x^t})]^2 \, ds(\widehat{x}).$$

Remark 9.2.1 By Theorem 9.2.1 and (9.103), we readily see that only the leading order term of the scattered wave field $u^\infty(\widehat{x}; \Omega^s)$ is used for the imaging functional $J_S^p(z)$. This means, we have linearized the nonlinear inverse scattering problem (9.48) in this case with small scatterers. In Theorem 9.2.2, we only justified the indicating behaviours of $J_S^p(z)$ for $p = s, i$. The indicating behaviour of $J_S^h(z)$ for the sound-hard case is not so evident as the sound-soft and impedance case. Indeed, by following a completely similar argument as that for the proof of Theorem 9.2.2, due to the integral terms involving the coupling of Y_α^β and $Y_{\alpha'}^{\beta'}$ with $\alpha \neq \alpha'$ and/or $\beta \neq \beta'$, one does not have the local maximum behaviour in general. This is also evidenced by our numerical examples in the following (see Fig. 9.19b), where the reconstructions of sound-hard scatterers are in general not as good as those for the sound-soft and impedance-type scatterers. This is mainly caused by the inhomogeneous background and in sharp difference from those considered in [31, 37] with a homogeneous background space. The problematic issue can be remedied by making use of multiple far-field measurements.

Using Theorem 9.2.2, we are now ready to formulate our first imaging scheme of locating multiple small scatterer components.

Scheme S

---

1. For an unknown scatterer Ω^p in (9.78), collect the far-field data by sending a single incident plane wave $u^i(x) = e^{ikx\cdot d^i}$ with fixed k and d^i.
2. Select a sampling region in $\mathbb{R}_-^n$ with a mesh $\mathscr{T}_h$ containing Ω^p.
3. For each sampling point $z \in \mathscr{T}_h$, compute the index value $J_S(z)$.
4. Locate all the significant local maxima of $J_S(z)$ on $\mathscr{T}_h$, which represent the locations of the scatterer components.

We proceed to discuss Scheme R of recovering multiple regular-size anomalies. The multiple regular-size anomalies buried in the lower half-space are first introduced as follows. Let Λ_j, $j = 1, 2, \ldots, m'$ be bounded simply connected C^2 domains in $\mathbb{R}_-^n$ that form the class of base scatterers. Set

$$\mathscr{A} := \{\Lambda_j\}_{j=1}^{m'}. \tag{9.110}$$

In the sequel, $\mathscr{A}$ is referred to as the admissible class, and it is noted that since we do not specify the type of each base scatterer Λ_j with the superscription, it could be either sound-soft, sound-hard of of impedance type. Let R_0 and R_1 be two fixed positive constants with $R_0 < R_1$, and $r_j \in [R_0, R_1]$, $j = 1, 2, \ldots, l$. Let $\Gamma_j \in \mathscr{A}$, $y_j \in \mathbb{R}^n_-$ and $U_j \in SO(n)$, $j = 1, 2, \ldots, m$ and set

$$\Sigma_j := (\Gamma_j; y_j, r_j, U_j), \quad j = 1, 2, \ldots, m. \tag{9.111}$$

It is assumed that $\Sigma_j \subset \mathbb{R}^n_-$. We define

$$\Sigma := \bigcup_{j=1}^{m} \Sigma_j, \tag{9.112}$$

which represents the multiple regular-size anomalies for our study. The sparsity condition is also imposed as that

$$L = \min_{1 \le j, j' \le m, j \ne j'} \operatorname{dist}(y_j, y_{j'}) \gg 1. \tag{9.113}$$

We emphasize again that since the type of each component obstacle of Σ is not specified with the superscription, it could be either sound-soft, sound-hard, of impedance type, or consisting of mixed-type scatterers. Scheme R discussed in the following shall work to recover the scatterers in Σ. In doing so, we need further require that the admissible class $\mathscr{A}$ is known in advance. Loosely speaking, the discussed Scheme R could only be used to (approximately) recover the location, scale and orientation of each scatterer component Σ_j. It is pointed out that, in the recovery process, one needs not know exactly the base scatterer Γ_j as long as it is from the admissible class $\mathscr{A}$, and its exact information will also be recovered. Moreover, we note that it is not necessary for $m' = m$. This means that certain admissible base scatterers might appear more than once or do not appear in Σ. The scenario described above covers some important applications from practice. For example, in the anti-submarine detection, the possible models of the hostile submarines are known in advance which provide the admissible class $\mathscr{A}$.

We are in a position to present Scheme R. It begins with augmenting the admissible class $\mathscr{A}$ as follows. Let $\tau \in \mathbb{R}_+$ and $\tau \ll 1$. Let $\mathscr{I}_1$ be a suitably chosen finite index set, such that $\{U_j\}_{j \in \mathscr{I}_1}$ is a τ-net of $SO(n)$. That is, for any rotation matrix $U \in SO(n)$, there exists $j \in \mathscr{I}_1$ such that $\|U_j - U\| \le \tau$. In a similar manner, let $\mathscr{I}_2$ be a finite index set such that $\{r_j\}_{j \in \mathscr{I}_2}$ is a τ-net of $[R_0, R_1]$. We define

$$\widetilde{\mathscr{A}} = \bigcup_{j=1}^{m'} \bigcup_{j_1 \in \mathscr{I}_1} \bigcup_{j_2 \in \mathscr{I}_2} (\Lambda_j; r_{j_2}, U_{j_1}) := \{\widetilde{\Lambda}_j\}_{j=1}^{m''}. \tag{9.114}$$

The following two assumptions shall be imposed on the augmented admissible class $\widetilde{\mathscr{A}}$,

1. $u^\infty(\widehat{x}; \widetilde{\Lambda}_j) \ne u^\infty(\widehat{x}; \widetilde{\Lambda}_{j'})$ for $j \ne j'$ and $1 \le j, j' \le m''$, $\widehat{x} \in \mathbb{S}^{n-1}_+$;
2. $\|u^\infty(\widehat{x}; \widetilde{\Lambda}_j)\|_{L^2(\mathbb{S}^{n-1}_+)} \ge \|u^\infty(\widehat{x}; \widetilde{\Lambda}_{j'})\|_{L^2(\mathbb{S}^{n-1}_+)}$ for $j < j'$ and $1 \le j, j' \le m''$.

Assumption (ii) can be fulfilled by reordering if necessary. For assumption (i), we recall the following conjecture in the theory of the inverse acoustic scattering problem: Assumption (ii) can be fulfilled by reordering if necessary. For assumption (i), we recall the following conjecture in the theory of the inverse acoustic scattering problem:

$$u^{\infty}(\widehat{x};\Lambda) \neq u^{\infty}(\widehat{x};\widetilde{\Lambda}) \quad \text{if and only if} \quad \Lambda = \widetilde{\Lambda}, \tag{9.115}$$

where Λ and $\widetilde{\Lambda}$ are two obstacles. The formula (9.115) states that a single far-field measurement can uniquely determine an acoustic obstacle. There is a widespread belief that (9.115) holds true, but there is no progress in the literature. A closely related uniqueness result can be found in [41].

For Scheme R, we introduce the following m'' imaging functionals

$$J_R^j(z) = \frac{\left|\left\langle u^{\infty}(\widehat{x};\Sigma), e^{ik_-(d^t-\widehat{x}^t)\cdot z} u^{\infty}(\widehat{x};\widetilde{\Lambda}_j)\right\rangle_{L^2(\mathbb{S}_+^{n-1})}\right|}{\|u^{\infty}(\widehat{x};\widetilde{\Lambda}_j)\|^2_{L^2(\mathbb{S}_+^{n-1})}}, \quad \widetilde{\Lambda}_j \in \widetilde{\mathscr{A}}, \ j = 1, 2, \ldots, m''. \tag{9.116}$$

The following theorem contains the indicating behaviours of the imaging functionals introduced above.

Theorem 9.2.3 *Suppose that $\widetilde{\Lambda}_1 \in \widetilde{\mathscr{A}}$ is of the following form*

$$\widetilde{\Lambda}_1 = (\Lambda_{j_0}; r_{\alpha_0}, U_{\beta_0}) \quad \Lambda_{j_0} \in \mathscr{A}, \beta_0 \in \mathscr{I}_1, \alpha_0 \in \mathscr{I}_2.$$

Suppose that in Σ, there exists $I_0 \subset \{1, 2, \ldots, m\}$ such that for $j \in I_0$, the component $\Sigma_j = (\Gamma_j; y_j, r_j, U_j)$ satisfies

$$(i)\ \Gamma_j = \Lambda_{j_0}; \quad (ii)\ \|U_j - U_{\beta_0}\| \leq \tau; \quad (iii)\ \|r_j - r_{\alpha_0}\| \leq \tau; \tag{9.117}$$

whereas for $j \in \{1, 2, \ldots, m\}\backslash I_0$, at least one of the conditions in (9.117) *is not fulfilled by the scatterer component Σ_j. Then for each y_j, $j = 1, 2, \ldots, m$, there exists an open neighborhood of y_j, $neigh(y_j)$, such that*

(i) if $j \in I_0$, then

$$J_R^1(z) \leq 1 + \mathscr{O}\left(\frac{1}{L} + \tau\right) \quad \forall z \in neigh(y_j). \tag{9.118}$$

Moreover, the equality holds in the above relation only when $z = y_j$. That is, y_j is a local maximum point for $J_R^1(z)$.

(ii) if $j \in \{1, 2, \ldots, m\}\backslash I_0$, then there exists $\epsilon_0 > 0$ such that

$$J_R^1(z) \leq 1 - \epsilon_0 + \mathscr{O}\left(\frac{1}{L} + \tau\right) \quad \forall z \in neigh(y_j). \tag{9.119}$$

Proof By using the translation relation provided in Lemma 9.2.1, the proof follows from a similar argument to that for Theorem 3.1 in [37]

Based on Theorem 9.2.3, Scheme R for successively recovering the multiple regular-size anomalous components in Σ is formulated as follows.

Scheme R

---

1. For the admissible scatterer class $\mathscr{A}$ in (9.110), formulate the augmented admissible class $\widetilde{\mathscr{A}}$ as that given in (9.114).
2. Collect in advance the scattering amplitudes associated with the admissible scatterer class $\widetilde{\mathscr{A}}$ corresponding to a single incident plane wave $e^{ik_+ x\cdot d^i}$ with a fixed d^i, and reorder $\widetilde{\mathscr{A}}$ if necessary so that assumptions (i) and (ii) are satisfied.
3. For an unknown scatterer Σ in (9.112), collect the scattering amplitude corresponding to the single incident plane wave as specified in 2).
4. Select a sampling region with a mesh $\mathscr{T}_h$ in $\mathbb{R}^n_-$ containing Σ.
5. Set $j = 1$ and $u_1^\infty(\widehat{x}) = u^\infty(\widehat{x};\, \Sigma)$
6. If $j \neq 1$, then update $u^\infty(\widehat{x};\, \Sigma)$ to be $u_j^\infty(\widehat{x})$.
7. For each sampling point $z \in \mathscr{T}_h$, compute the value of the imaging functional $J_R^j(z)$.
8. Locate all those significant local maxima of $J_R^j(z)$ satisfying $J_R^j(z) \approx 1$ for the scatterer components of the form $y + \widetilde{\Lambda}_j$.
9. Remove all the sampling points inside those identified components $y + \widetilde{\Lambda}_j$, say $j = 1, 2, \ldots, N_j$, found in 7) from $\mathscr{T}_h$. Subtract the individual scattering amplitudes associated with those already recovered components in 7) and their respective identified locations y_j from the total scattering amplitude according to the following formula,

$$u_j^\infty(\widehat{x}) := u_j^\infty(\widehat{x}) - \sum_{j=1}^{N_j} e^{ik_+(d^i - \widehat{x}^i)\cdot y_j} u^\infty(\widehat{x};\, \widetilde{\Lambda}_j) \tag{9.120}$$

10. If $\mathscr{T}_h = \emptyset$ or $j = m''$, then stop the reconstruction; otherwise, set $j := j + 1$, and go to 6).

Finally, we consider the recovery of multiple multiscale anomalies of the form

$$\mathbf{M} = \Omega^p \cup \Sigma, \tag{9.121}$$

where Ω^p is the scatterer in (9.78) and Σ is the one in (9.112). Additionally, we assume that

$$L = \mathrm{dist}(\Omega^p, \Sigma) \gg 1. \tag{9.122}$$

The corresponding recovery scheme is based on combining Schemes S and R by using a local tuning technique and shall be referred to as Scheme M. With Schemes

S and R discussed above, the combination can be down by following a completely similar manner as that in [33, 37] and we skip it here.

9.2.4 Numerical Experiments

In this section, some numerical tests are presented to demonstrate and verify the applicability of the three schemes (S, R and M) discussed for locating anomalies in a two-layered medium in both two and three dimensions. In all the tests, the exact far-field data are obtained by solving the Helmholtz system within a two-layered medium (9.49)–(9.57) using the quadratic finite elements on a truncated circular (2D) or spherical (3D) domain enclosed by a PML layer. The forward equation is solved on a sequence of successively refined meshes till the relative error of two successive finite element solutions between the two adjacent meshes is below 0.1%. Then the scattered data are transformed into the far-field data on $\mathbb{S}_+^{n-1}$ by employing the integral representation formula using (9.67) on a closed circle (2D) or surface (3D) enclosing the scatterer. For scatterers of small and regular size, we always add to the exact far-field data a uniform noise of 5% and 1%, respectively, and use them as the measurement data in our numerical tests.

It is pointed out that the ratio between k_- and k_+ is implicitly fixed due to that of the squared refractive indices in physics and has nothing to do with the inherent frequency of the detecting wave. In the sequel, we always set $k_- = 2k_+$. When there exist no anomalies, our forward solver shows the periodic structure of the transmitted wave in the lower half-space and superimposed patterns of the incident and reflected waves in the upper half-space in two and three dimensions in Fig. 9.17. In this case, the scattered wave is solely due to the two-layered medium.

Example 1: Scheme S for Locating Multiple Small Anomalies
The anomalies buried underground are multiple small sound-soft obstacles lying in the lower half-space. In the first 2D test, three little ball anomalies with radius 0.1 (in red) are laid at $(-2,\ -2)$, $(0,\ -1)$ and $(2,\ -1.5)$, respectively, as shown in Fig. 9.18a.

We set the detecting wave number $k_+ = \pi$ and choose the incident direction downward with $\pi/4$ radian below the horizon $y = 0$. The detecting wave length is significantly larger than the sizes of all the components. Figure 9.18b shows the indicating behaviour using the indicator function (9.103) of Scheme S. The three components of the unknown scatterer are reasonably located using a single detecting plane wave field. Since refraction and reflection take place at the same time on the interface, the crucial issue is that we could only collect far-field pattern in some limited aperture, which leads to the incomplete resolution of the depth information of the scatterer in the vertical direction, which explains the slender unfocused images of the positions of those anomalies in Fig. 9.18b. But this side effect could be overcome by increasing the frequency of the detecting wave, which will yield sharper resolution

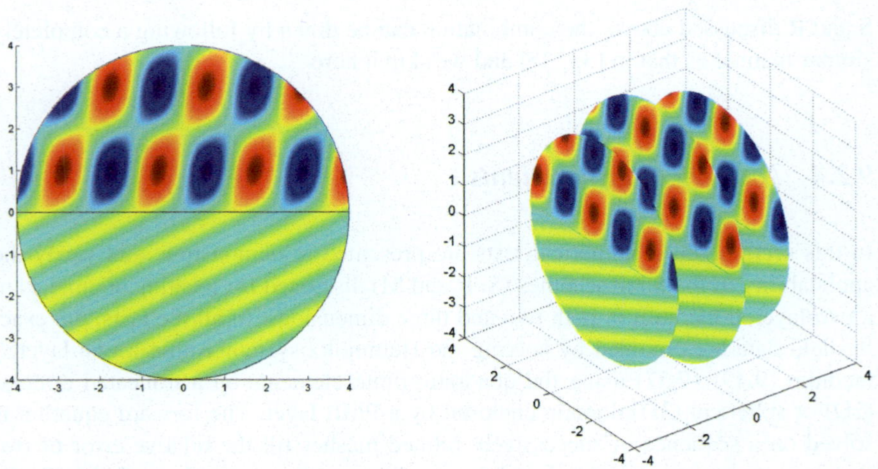

Fig. 9.17 Plane wave scattering with no anomalies by a two-layered medium with $k_+ = \pi$ and $k_- = 2\pi$. Total wave plots in 2D (left) and in 3D (right)

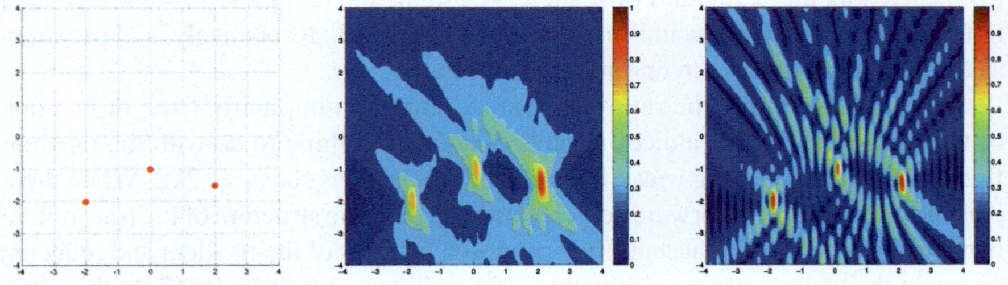

(a) (b) (c)

Fig. 9.18 plane wave scattering by interface: **a** True anomalies in 2D; **b** Imaging of anomalies with ($k_+ = \pi$); **c** Imaging of anomalies with $k_+ = 5\pi$

of the positions. This observation holds true for the remaining tests. By further increasing the wave number $k_+ = 5\pi$, we find that each component of the scatterer is now well captured as a local maximum highlighted as red dots as shown in Fig. 9.18c.

In the second 3D test, The underground anomalies consist of a impedance-type sphere with impedance coefficient $\lambda = 5$ with radius 0.1 located at $(-2,\ 0,\ -1.5)$, and a sound-hard square with sidelength 0.2 located at $(2,\ 0,\ -1)$. The incident wave is pointing downward with $\pi/4$ radian within the $x - z$ plane below the horizontal ground $z = 0$.

The resulting indicator function value distribution is plotted on a pair of orthogonal slice planes $x = \pm 2$ and $y = 0$ in Fig. 9.19b. Clearly, the positions of the respective detected components match reasonably well with the ones of the exact components. As one can see, the spherical anomaly is well located, while the position of the sound-

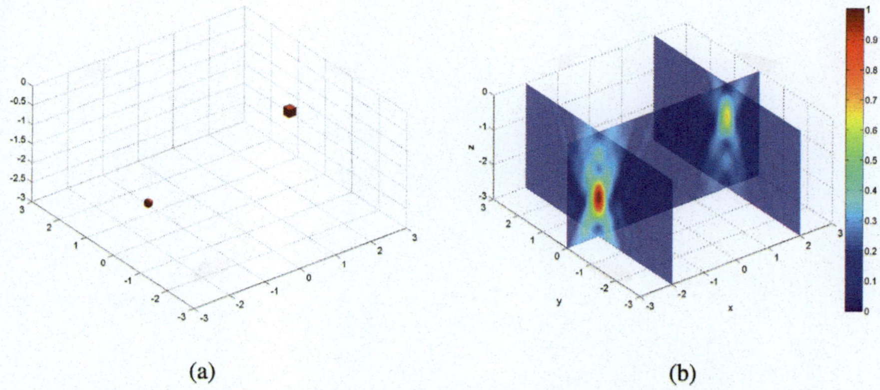

Fig. 9.19 **a** True small anomalies in 3D; **b** Imaging of anomalies by Scheme S with $(k^+ = \pi)$

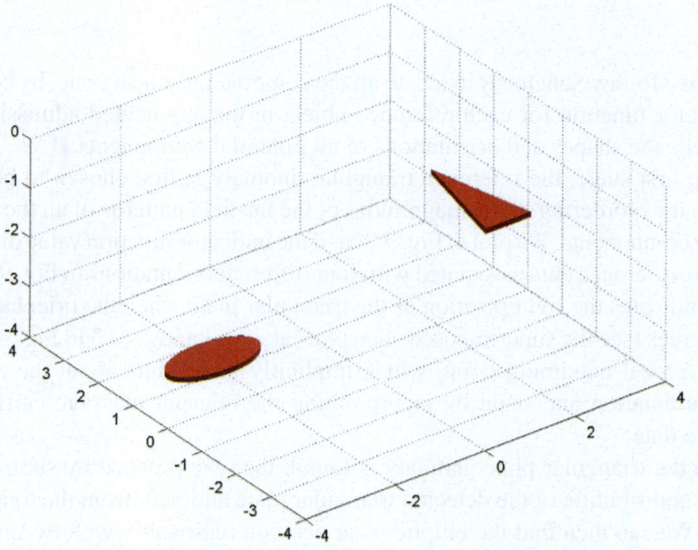

Fig. 9.20 True regular-size anomalies: one triangular component lying at (3, 0, −2), and an elliptic component at (−3, 0, −3)

hard square is much dimer compared with its counterpart, though both are visualized in the highlighted part (local maxima). This verifies our theoretical observations made in Remark 9.2.1.

Example 2: Scheme R for Locating Multiple Regular-Size Anomalies
We adopt two regular-size anomalies as shown in Fig. 9.20, one is a triangular plate and the other is an elliptic one, both are of thickness 0.2. These two reference anomalies have six orientations as shown in Fig. 9.21.

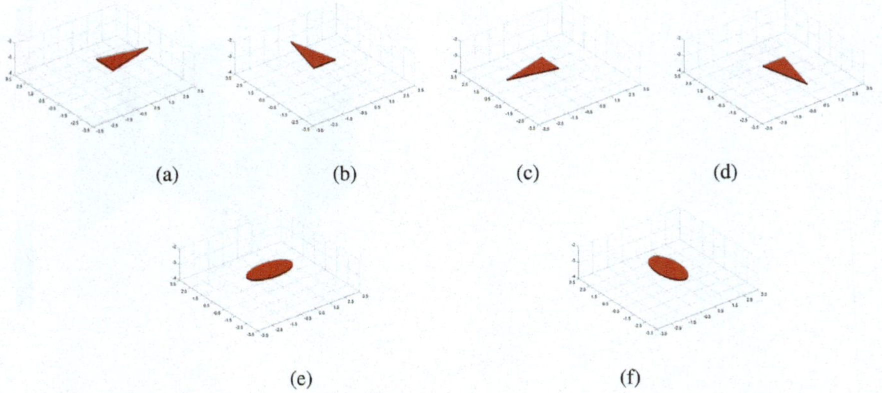

Fig. 9.21 Reference set: **a–d** Triangular plate with rotation 0, 90, 180 and 270 degrees; (e)-(f) Elliptic screen with rotation 0 and 90 degrees

Now we follow Scheme R to locate all the components, one by one, by computing an indicator function for each reference object in the augmented admissible class, which tells the shapes and orientations of all potential components.

In the first stage, the reference triangular anomaly is first chosen to be located, based on the reordering of the magnitudes of the far-field patterns of all the reference scatterer components. We plot in Fig. 9.22a–d the indicator function value distribution by testing reference data associated with four different orientations in Fig. 9.21a–d. It clearly indicates the right position of the triangular plate when the orientation angle is 90 degrees (see the superimposed slice plots at $x = 3$ and $y = 0$ in Fig. 9.22b) and there is a local maximum point, which implicitly gives hints about the anomaly's shape, orientation and scale by incorporating the relevant message carried in the reference data.

Once the triangular plate anomaly is found, then we proceed by subtracting the far-field contribution of the detected triangular plate anomaly from the total far-field pattern. We can then find the elliptic plate position reasonably well by showing the superimposed slice plots at $x = -3$ and $y = 0$; see Fig. 9.21(e)-(f) and 9.22(e)-(f), respectively. We see that only the configuration with 0 degree maximizes the indicator function to achieve the maximum and indicates the position of the detected elliptic plate anomaly very clearly.

Example 3: Scheme M for Locating Multiple Multiscale Anomalies

We now consider a scenario of multiple 3D multiscale buried anomalies consisting of two components: a small spherical anomaly with radius 0.1 and a large triangular plate; see Fig. 9.23.

In the first stage, we extract the information of the regular-size component using the indicator function of Scheme R by computing the inner product between the collected far-field data and a priori known far-field patterns associated with those reference scatterer components with different orientations and sizes. We can find the

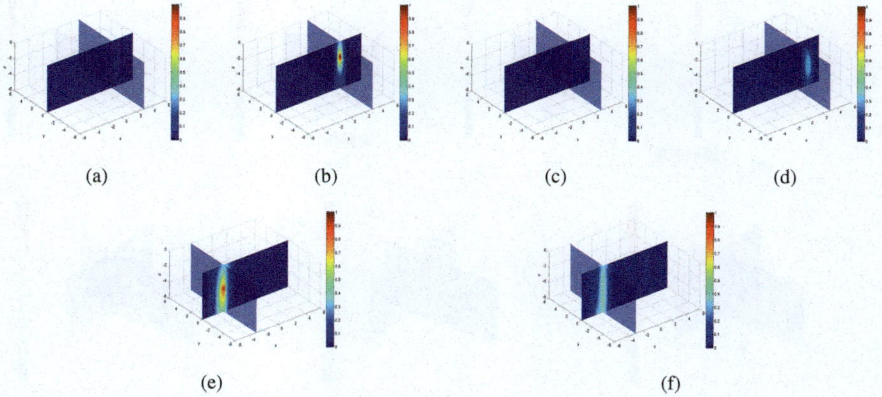

Fig. 9.22 Results of locating two regular-sized scatterer

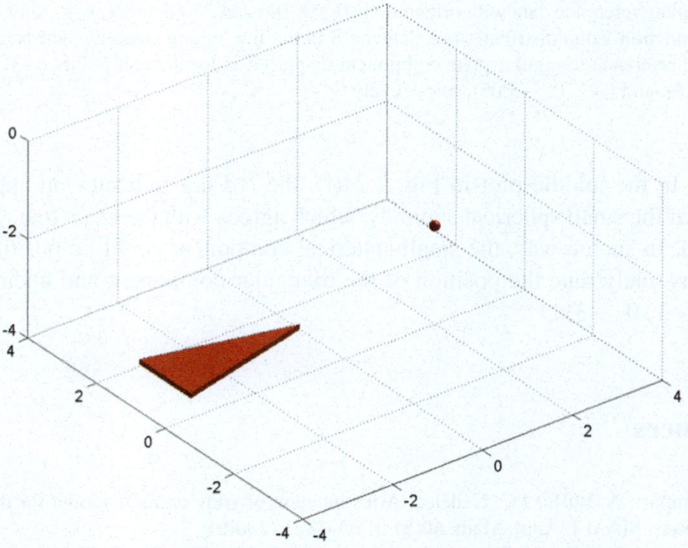

Fig. 9.23 Underground scenario of multiple multiscale anomalies: one triangular plate anomaly lying at $(-3,\ 0,\ -3)$, and a small spherical anomaly lying at $(3,\ 0,\ -2)$

approximate position of the larger triangular component of regular size when the reference scatterer is its orientation of 0 degree as shown in Fig. 9.24a.

In the next stage, the location of the small spherical component can be obtained by performing a local tuning technique via searching grid points in some local cubic mesh around covering the local maximum in Fig. 9.24a.

In Fig. 9.24e–g, as the searching grid-points approach gradually from $(-3\,,0\,,-2.95)$ to $(-3\,,0\,,-3.05)$ (from left to right), the value distribution of the indicator function in Scheme S displays an interesting change of the highlighted

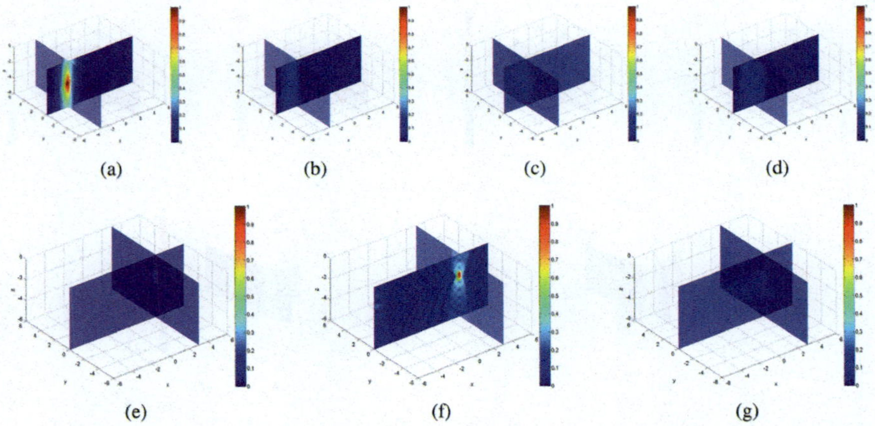

Fig. 9.24 **a–d**: Slice plots of the indicator function value distribution of Scheme R associated with triangular plate reference data with orientations 0, 90, 180 and 270 degrees. **e–g**: Slice plots of the indicator function value distribution of Scheme S in the fine tuning stage by subtracting far-field data of the reference triangular plate component displaced at local mesh points $(-3,\ 0,\ -2.95)$, $(-3,\ 0,\ -3)$, and $(-3,\ 0,\ -3.05)$, respectively

position. In the middle plot in Fig. 9.24(f), the red dot indicates an approximate position of the small spherical anomaly, which agrees with the exact one $(3\,,0\,,-3)$ very well. In such a way, the small spherical component could be positioned, and it helps us finely tune the position of the triangular component and update it to be around $(-3\,,0\,,-3)$.

References

1. H. Ammari, A. Buffa, J.C. Nedelec, A justification of eddy currents model for the Maxwell equations. SIAM J. Appl. Math. **60**(5), 1805–1823 (2000)
2. H. Ammari, P. Calmon, E. Iakovleva, Direct elastic imaging of a small inclusion. SIAM J. Imag. Sci. **1**, 169–187 (2008)
3. H. Ammari, J. Chen, Z. Chen, J. Garnier, D. Volkov, Target detection and characterization from electromagnetic induction data. Journal de mathématiques pures et appliquées **101**(1), 54–75 (2014)
4. H. Ammari, S. Moskow, M.S. Vogelius, Boundary integral formulae for the reconstruction of electric and electromagnetic inhomogeneities of small volume, ESAIM: Control. Optim. Calcul. Variat. **9**, 49–66 (2003)
5. H. Ammari, M. Vogelius, D. Volkov, Asymptotic formulas for per-turbations in the electromagnetic fields due to the presence of inhomogeneities of small diameter. II. The full Maxwell equations. Journal de Mathematiques Pures et Appliques **80**(9), 769–814 (2001)
6. H. Ammari, E. Iakovleva, D. Lesselier, G. Perrusson, MUSIC-type electromagnetic imaging of a collection of small three-dimensional inclusions. SIAM J. Sci. Comput. **29**, 674–709 (2007)
7. H. Ammari, H. Kang, *Reconstruction of Small Inhomogeneities from Boundary Measurements*. Lecture Notes in Mathematics, vol. 1846. (Springer, Berlin, 2004)

8. H. Ammari, H. Kang, *Polarization and Moment Tensors with Applications to Inverse Problems and Effective Medium Theory*. Applied Mathematical Sciences (Springer, New York, 2007)
9. H. Ammari, E. Iakovleva, D. Lesselier, A MUSIC algorithm for locating small inclusions buried in a half-space from the scattering amplitude at a fixed freuqency. Multiscale Model. Simul. **3**, 597–628 (2005)
10. H. Ammari, E. Iakovleva, D. Lesselier, A MUSIC algorithm for locating small inclusions buried in a half-space from the scattering amplitude at a fixed frequency. SIAM Multiscale Model. Simul. **3**, 597–628 (2005)
11. C. Bellis, M. Bonnet, F. Cakoni, Acoustic inverse scattering using topological derivative of far-field measurements-based L^2 cost functionals. Inverse Prob. **29**, 075012 (2013)
12. M. Bonnet, Inverse acoustic scattering by small-obstacle expansion of a misfit function. Inverse Prob. **24**, 035022 (2008)
13. M. Bonnet, B.B. Guzina, Sounding of finite solid bodies by way of topological derivative. Int. J. Numer. Meth. Eng. **61**, 2344–2373 (2004)
14. M. Cheney, The linear sampling method and the MUSIC algorithm. Inverse Prob. **17**, 591–595 (2001)
15. D. Colton, J. Coyle, P. Monk, Recent developments in inverse acoustic scattering theory. SIAM Rev. **42**, 369–414 (2000)
16. D. Colton, R. Kress, *Inverse Acoustic and Electromagnetic Scattering Theory*, 3rd edn. (Springer, Berlin, 2013)
17. P.M. Cutzach, C. Hazard, Existence, uniqueness and analyticity properties for electromagnetic scattering in a two-layered medium. Math. Methods Appl. Sci. **21**, 433–461 (1998)
18. F. Delbary, K. Erhard, R. Kress, R. Potthast, J. Schulz, Inverse electromagnetic scattering in a two-layered medium with an application to mine detection. Inverse Prob. **24**, 015002 (2008)
19. D. Colton, R. Kress, *Inverse Acoustic and Electromagnetic Scattering Theory*, 2nd edn. (Springer, New York, 1998)
20. A.J. Devaney, *Super-resolution Processing of Multi-static Data Using Time Reversal and MUSIC*, preprint (Northeastern University, 2000)
21. T. Hohage, Fast numerical solution of the electromagnetic medium scattering problem and applications to the inverse problem. J. Comput. Phys. **214**, 224–238 (2006)
22. E. Iakovliva, Inverse scattering from small inhomogeneities, PhD Thesis, Ecole Polytechnique (2004)
23. V. Isakov, *Inverse Problems for Partial Differential Equations*, Applied Mathematical Sciences, vol. 127, 2nd edn. (Springer, New York, 2006)
24. A. Kirsch, The MUSIC algorithm and the factorization method in inverse scattering theory for inhomogeneous media. Inverse Prob. **18**, 1025–1040 (2002)
25. A. Kirsch, N. Grinberg, *The Factorization Method for Inverse Problems* (Oxford University Press, Oxford, 2008)
26. A. Kirsch, X. Liu, The factorization method for inverse acoustic scattering by a penetrable anisotropic obstacle. Math. Methods Appl. Sci. **37**, 1159–1170 (2014)
27. R. Kress, *Linear Integral Equations* 3rd edn. (Springer, 2014)
28. V.I. Lebedev, D.N. Laikov, A quadrature formula for the sphere of the 131st algebraic order of accuracy. Dokl. Math. **59**, 477–481 (1999)
29. P. Li, An inverse cavity problem for Maxwell's equations. J. Differ. Equ. **252**, 3209–3225 (2012)
30. J. Li, P. Li, H.Y. Liu, X. Liu, Recovering multiscale buried anomalies in a two-layered medium. Inverse Prob. **31**(10), 105006 (2015)
31. J. Li, H.Y. Liu, Z. Shang, H. Sun, Two single-shot methods for locating multiple electromagnetic scatterers. SIAM J. Appl. Math. **73**, 1721–1746 (2013)
32. J. Li, H.Y. Liu, Y. Sun, Q. Wang, Ground detection by a single electromagnetic far-field measurement. J. Comput. Phys. **273**, 472–487 (2014)
33. J. Li, H.Y. Liu, Q. Wang, Locating multiple multiscale electromagnetic scatterers by a single far-field measurement. SIAM J. Imag. Sci. **6**, 2285–2309 (2013)

34. J. Li, H.Y. Liu, Q. Wang, Enhanced multilevel linear sampling methods for inverse scattering problems. J. Comput. Phys. **257**, 554–571 (2014)
35. J. Li, H.Y. Liu, J. Zou, Multilevel linear sampling method for inverse scattering problems. SIAM J. Sci. Comput. **30**, 1228–1250 (2008)
36. J. Li, H. Liu, J. Zou, Strengthened linear sampling method with a reference ball. SIAM J. Sci. Comput. **31**, 4013–4040 (2009)
37. J. Li, H. Liu, J. Zou, .Locating multiple multiscale acoustic scatterers. SIAM Multiscale Model. Simul. **12**, 927–952 (2014)
38. H.Y. Liu, M. Yamamoto, J. Zou, Reflection principle for the Maxwell equations and its application to inverse electromagnetic scattering. Inverse Prob. **23**, 2357–2366 (2007)
39. H.Y. Liu, M. Yamamoto, J. Zou, New reflection principles for Maxwell equations and their applications. Numer. Math. Theory Methods Appl **2**, 1–17 (2009)
40. H.Y. Liu, M. Yamamoto, J. Zou, Reflection principle for Maxwell's equations and its application to inverse electromagnetic scattering. Inverse Prob. **23**, 2357–2366 (2007)
41. X. Liu, B. Zhang, A uniqueness result for inverse electromagnetic scattering problem in a two-layered medium. Inverse Prob. **26**, 105007 (2010)
42. W. McLean, *Strongly Elliptic Systems and Boundary Integral Equations* (Cambridge University Press, Cambridge, 2000)
43. P. Monk, *Finite Element Methods for Maxwell's Equations* Numerical Mathematics and Scientific Computation. (Oxford University Press, New York, 2003)
44. J.-C. Nédélec, *Acoustic and Electromagnetic Equations: Integral Representations for Harmonic Problems* (Springer, New York, 2001)
45. R. Pike, P. Sabatier (eds.), *Scattering : Scattering and Inverse Scattering in Pure and Applied Science* (Academic, 2002)
46. G. Uhlmann, (ed.), *Inside Out: Inverse Problems and Applications*, MSRI Publications, vol. 47 (Cambridge University Press, 2003)
47. M. Zhao, A fast high order iterative solver for the electromagnetic scattering by open cavities filled with the inhomogeneous media. Adv. Appl. Math. Mech. **5**, 235–257 (2013)

The manufacturer's authorised representative in the EU is Springer Nature Customer Service Centre GmbH, Europaplatz 3, 69115 Heidelberg, Germany. If you have any concerns regarding our products, please contact ProductSafety@springernature.com

Printed and bound by CPI Group (UK) Ltd, Croydon, CR0 4YY
01/10/2026
02225870-0013